Surveys of Modern Mathematics
Volume VI

Open Problems and Surveys of Contemporary Mathematics

edited by

Lizhen Ji
Yat-Sun Poon
Shing-Tung Yau

International Press
www.intlpress.com

高等教育出版社
HIGHER EDUCATION PRESS

Surveys of Modern Mathematics, Volume VI
Open Problems and Surveys of Contemporary Mathematics

Editors:
Lizhen Ji, University of Michigan
Yat-Sun Poon, University of California at Riverside
Shing-Tung Yau, Harvard University

ISBN: 978-1-57146-278-7

Printed in the United States of America.

17 16 15 14 13   1 2 3 4 5 6 7 8 9

SURVEYS OF MODERN MATHEMATICS

Contents

Preface

This book consists mainly of lecture notes of some talks and courses at Mathematical Sciences Center (MSC) of Tsinghua University together with several other papers.

Since it was founded in December 2009, one of the missions of MSC has been to teach both undergraduate and graduate students important ideas, theories and results of contemporary mathematics. These lecture notes reflect this philosophy. They are expository and accessible to both students and nonexperts. On the other hand, they also contain novel ideas or presentations of important topics in mathematics. Therefore, this book is also useful to experts. Especially we would like to point out that the last paper in this book *Open Problems in Differential Geometry* by the third editor of this book is only the first three of many lectures given by him in both Beijing and Taipei, which can be considered as a reviewing and updating of the very influential open problem lists by him.

Besides these lecture notes from MSC, this book also contains four other papers. The first is a paper by James Milne based on his talk at the seminar "What is ..." at University of Michigan. The concept of motives is important and difficult, and the talk and this paper are attempts by an expert to explain it in concrete terms. The second is a master thesis in 2002 by Joris van Hoboken who gives a coherent and accessible exposition of the ubiquity of the important ADE classification in mathematics, which originally occurred in the classification of simple complex Lie algebras. Joris van Hoboken switched to study law right after obtaining his Master degree and is now a senior researcher at a law school. The ADE classification occurs at many different situations, and it is still a mystery whether there are some deep, intrinsic connections between them. This master thesis was never published and has been highly cited and circulated on the web. We are grateful that Dr. van Hoboken has given us permission to include it in the current book. We hope that this will make the ADE classification better known to the reader and also give a permanent record of this beautiful master thesis. The other two are reprints of papers of the third editor. The short paper *A note on the distribution of critical points of eigenfunctions* considered a novel question. As it is well-known, the location and distribution of the zero sets (i.e., nodal sets) of eigenfunctions of Riemannian manifolds have been extensively and intensively studied. Critical points of eigenfunctions are also special and deserve to be understood better. Analysis on nonsmooth spaces has been becoming quite important and applied to several subjects in mathematics. The paper is one of the early papers in this subject.[1] Due to inaccessibility and no review of it in MathSciNet, this paper has been largely unknown. We hope that its inclusion in this book will be valuable to the reader as well.

It has been a lot of work for the speakers at MSC to write up their lecture notes. We would like to thank them, especially the four note-takers and co-authors (Hui Ma, Chun-Jun Tsai, Mu-Tao Wang, En-Tao Zhao) of the last paper in this

[1] For a recent survey, see J. Heinonen, *Nonsmooth calculus*. Bull. Amer. Math. Soc. (N.S.) 44 (2007), no. 2, 163–232.

book, for their efforts and contributions. We would also like to thank reviewers of the papers in this book for their help.

This book marks the beginning of publication from MSC and we hope and expect that future volumes will appear regularly.

Editors: Lizhen Ji, Yat-Sun Poon, Shing-Tung Yau

May 30, 2013

Open Problems and Surveys of Contemporary Mathematics
SMM 6, pp. 1–44

Lectures on Mirror Symmetry and Topological String Theory

Murad Alim*

Abstract

These are notes of a series of lectures on mirror symmetry and topological string theory given at the Mathematical Sciences Center of Tsinghua University. The $\mathcal{N} = 2$ superconformal algebra, its deformations and its chiral ring are reviewed. A topological field theory can be constructed whose observables are only the elements of the chiral ring. When coupled to gravity, this leads to topological string theory. The identification of the topological string A- and B-models by mirror symmetry leads to surprising connections in mathematics and provides tools for exact computations as well as new insights in physics. A recursive construction of the higher genus amplitudes of topological string theory expressed as polynomials is reviewed.

2000 Mathematics Subject Classification: 14J33, 81T30, 81T45

Keywords and Phrases: Mirror symmetry, String and superstring theories, Topological field theories

Contents

*Jefferson Physical Laboratory, Harvard University, Cambridge, MA 02138, USA. Email: alim@physics.harvard.edu

1 Introduction

The study of supersymmetric gauge theories and string theory has been a very fruitful arena of interaction between mathematics and physics which has pushed the boundaries of knowledge and understanding on both sides. A key lesson learned on the physics side is that it is very useful to think of physical theories in terms of families, where members of a family of theories are related by deformations. This amounts to studying theories together with their space of couplings. Physical theories are usually formulated using a Lagrangian for the light degrees of freedom and their interactions. When the interactions are governed by a small coupling, perturbative techniques can be used to evaluate the correlation functions of the physical observables. Perturbation theory does however not take into account non-perturbative effects which are nontrivial field configurations with finite action which have to be considered as various backgrounds for perturbation theory and are in general hard to study. As the coupling of the theory is varied, perturbation theory breaks down as the non-perturbative effects become more and more important signaling that the original choice of light degrees of freedom is no longer valid. Hence, in different regions in the coupling space, perturbative and non-perturbative degrees of freedom may interchange.

A great leap forward in the understanding of the interplay of perturbative and non-perturbative degrees of freedom and the physical manifestation of their interchange in terms of dualities followed the pioneering work of Seiberg and Witten (SW) [1]. In this work the exact low energy effective theory was determined including all non-perturbative effects using holomorphicity and the expected behavior at singularities in the space of the effective couplings due to physical particles becoming massless. The key insight was that the problem of determining the exact couplings could be mapped to an equivalent mathematical problem of determining periods of a curve. This auxiliary geometry could be given a physical meaning, identifying it with part of the Calabi-Yau (CY) threefold geometry of a type IIB string compactification [2], embedding this understanding into the larger framework of string dualities (See [3] for a review).

One particular physical duality which had emerged earlier is mirror symmetry. It was observed that an exchange of signs of some generators of the $\mathcal{N} = (2, 2)$ superconformal algebra underlying some string compactifications leads to an isomorphism [4]. This isomorphism identifies furthermore a ring of operators in the superconformal algebra, the chiral ring [5]. The deformation problem and the isomorphism defined abstractly using the superconformal algebra can be given a mathematical meaning when this algebra is realized by nonlinear sigma models defined on different target spaces Z and Z^*, see [6] and references therein.[1] Truncating the states of a representation of the superconformal algebra to those created by operators in the chiral ring and summing over all $2d$ topologies defines topological string theory, the two isomorphic versions are called the A- or the B-model respectively. These probe Kähler and complex structure deformations of mirror Calabi-Yau threefolds Z and Z^*. The B-model is more accessible to computations since its deformations are the complex structure deformations of Z^* which are captured by the variation of Hodge structure. Mirror symmetry is established by providing the mirror maps which are a distinguished set of local coordinates in a given patch of the deformation space. These provide the map to the A-model, since they are naturally associated with the chiral ring.

At special loci in the moduli space, the A-model data provides enumerative information of the CY Z. Using methods [11] to construct the mirror manifold of the quintic, it was thus possible [12] to compute the data associated to the variation of Hodge structure in the B-model and make predictions of the Gromov-Witten (GW) invariants. The higher genus GW invariants can be resumed to give integer multiplicities of BPS states in a five-dimensional theory obtained from an M-theory compactification on Z [13, 14]. Moreover, the special geometry governing the deformation spaces allows one to compute the prepotential $F_0(t)$ which governs the exact effective action of the four dimensional theories obtained from compactifying type IIA(IIB) string theory on $Z(Z^*)$, respectively. Various $4d$ gauge theories can be geometrically engineered [15, 16] and mirror symmetry can be used to compute their exact effective actions. (See also [17] and references therein.)

The prepotential is the genus zero free energy of topological string theory,

[1] By now there is a vast literature on the subject of mirror symmetry, see in particular [7, 8, 9, 10] and references therein for a more complete account of the developments leading to mirror symmetry.

which is defined perturbatively in a coupling constant governing the higher genus expansion. The partition function of topological string theory indicating its dependence on local coordinates in the deformation space has the form:

$$\mathcal{Z}(t,\bar{t}) = \exp\left(\sum_g \lambda^{2g-2}\mathcal{F}^{(g)}(t,\bar{t})\right). \tag{1.1}$$

In [18, 19], Bershadsky, Cecotti, Ooguri and Vafa (BCOV) developed the theory and properties of the higher genus topological string free energies putting forward recursive equations, the holomorphic anomaly equations along with a method to solve these in terms of Feynman diagrams. For the full partition function these equations take the form of a heat equation [19, 20] and can be interpreted [20] as describing the background independence of the partition function when the latter is interpreted as a wave function associated to the geometric quantization of $H^3(Z^*)$.

Using the special geometry of the deformation space a polynomial structure of the higher genus amplitudes in a finite number of generators was proven for the quintic and related one parameter deformation families [21] and generalized to arbitrary target CY manifolds [22], see also [23, 24, 25]. The polynomial structure supplemented by appropriate boundary conditions enhances the computability of higher genus amplitudes. Moreover the polynomial generators are expected to bridge the gap towards constructing the appropriate modular forms for a given target space duality group which is reflected by the special geometry of the CY manifold.

The aim of these notes is to provide an accessible concise introduction into some of the ideas outlined above. The presentation is far from being self-contained and the topics which are covered represent a small sample of a vast amount of developments in this field, reference will be given to the original literature and to many of the excellent reviews on the subject. By the nature of it, the material presented here is very close to own contributions and in fact it relies heavily on [22, 26, 27, 28].

The outline of these notes is as follows. In Sect. 2 the $\mathcal{N} = 2$ superconformal algebra and its chiral ring as well as its deformations are reviewed. It is discussed how to construct a bundle of states out of the subring of the chiral ring spanned by the deformation operators. Furthermore, the geometric realization of the $\mathcal{N} = (2,2)$ superconformal algebra as a nonlinear sigma model is described. In Sect. 3 it is outlined how topological string theory can be obtained by first restricting the physical observables to the chiral ring and then coupling to gravity. The two inequivalent ways of doing so, the A- and B-model are discussed as well as the geometric interpretation of the chiral ring on both sides. Special geometry and its B-model realization in the variation of Hodge structure are reviewed. The quintic and its mirror are discussed as an example of constructing mirror geometries and finding the mirror map. In Sect. 4, the BCOV anomaly equation and its recursive solution are reviewed. Using special geometry, a polynomial structure of the higher genus amplitudes can be proven and used for easier computations when the physically expected boundary conditions are implemented. An example of applying this structure to the quintic is given.

2 The superconformal algebra and the chiral ring

Mirror symmetry originates from representations of the $\mathcal{N}=(2,2)$ superconformal algebra, which refers to two copies, left- and right-moving of the $\mathcal{N}=2$ superconformal algebra which is discussed in the following. The topological string A- and B-models which are identified by mirror symmetry refer to a truncation of the states of a representation of $\mathcal{N}=(2,2)$ superconformal algebra to the chiral ring [5]. The following is based on [27, 29, 6, 30, 31, 10, 32].

2.1 $\mathcal{N}=2$ superconformal algebra

The $\mathcal{N}=2$ superconformal algebra is generated by the energy momentum tensor $T(z)$, which has conformal weight $h=2$, by two anti-commuting currents $G^{\pm}(z)$ of conformal weight $3/2$ and a $U(1)$ current $J(z)$ under which the $G^{\pm}(z)$ carry charge $\pm$ and z is a local coordinate on the $2d$ world-sheet. These currents satisfy the following operator product expansions (OPE):

$$G^{\pm}(z)G^{\mp}(w)=\frac{\frac{2}{3}c}{(z-w)^3}\pm\frac{2J(w)}{(z-w)^2}+\frac{2T(w)\pm\partial_w J(w)}{(z-w)}+\cdots\,, \tag{2.1}$$

$$J(z)\,G^{\pm}(w)=\pm\frac{G^{\pm}(w)}{(z-w)}+\cdots\,, \tag{2.2}$$

$$J(z)\,J(w)=\frac{\frac{1}{3}c}{(z-w)^2}+\cdots\,, \tag{2.3}$$

$$T(z)\,J(w)=\frac{J(w)}{(z-w)^2}+\frac{\partial_w J(w)}{(z-w)}+\cdots\,, \tag{2.4}$$

$$T(z)\,G^{\pm}(w)=\frac{\frac{3}{2}G^{\pm}(w)}{(z-w)^2}+\frac{\partial_w G^{\pm}(w)}{(z-w)}+\cdots\,, \tag{2.5}$$

$$T(z)\,T(w)=\frac{\frac{1}{2}c}{(z-w)^4}+\frac{2T(w)}{(z-w)^2}+\frac{\partial_w T(w)}{(z-w)}+\cdots\,, \tag{2.6}$$

where the dots refer to the addition of regular terms in the limit $z\to w$, c is the central charge. The boundary conditions which must be imposed for the currents $G^{\pm}(z)$ can be summarized as follows

$$G^{\pm}(e^{2\pi i}z)=-e^{\mp 2\pi i a}G^{\pm}(z)\,, \tag{2.7}$$

with a continuous parameter a which lies in the range $0\leq a<1$. For integral and half integral a one recovers anti-periodic and periodic boundary conditions corresponding to the Ramond and Neveu-Schwarz sectors, respectively. The currents can be expanded in Fourier modes

$$T(z) = \sum_n \frac{L_n}{z^{n+2}}, \quad G^{\pm}(z) = \sum_n \frac{G^{\pm}_{n\pm a}}{z^{n\pm a+\frac{3}{2}}}, \quad J(z) = \sum_n \frac{J_n}{z^{n+1}}. \tag{2.8}$$

The $\mathcal{N} = 2$ superconformal algebra can be expressed in terms of the operator product expansion of the currents or by the commutation relations of their modes. The latter reads:

$$\begin{aligned}
[L_m, L_n] &= (m-n)L_{m+n} + \frac{c}{12}m(m^2-1)\delta_{m+n,0}\,,\\
[J_m, J_n] &= \frac{c}{3}m\delta_{m+n,0}\,,\\
[L_n, J_m] &= -mJ_{m+n}\,,\\
\left[L_n, G^{\pm}_{m\pm a}\right] &= \left(\frac{n}{2} - (m\pm a)\right) G^{\pm}_{m+n\pm a}\,,\\
\left[J_n, G^{\pm}_{m\pm a}\right] &= \pm G^{\pm}_{n+m\pm a}\,,\\
\left\{G^{+}_{n+a}, G^{-}_{m-a}\right\} &= 2L_{m+n} + (n-m+2a)J_{n+m}\\
&\quad + \frac{c}{3}\left((n+a)^2 - \frac{1}{4}\right)\delta_{m+n,0}\,.
\end{aligned} \tag{2.9}$$

The algebras obtained for every value of the continuous parameter a are isomorphic. This isomorphism induces an operation on the states which is called *spectral flow*. This operation shows the equivalence of the Ramond (R) and Neveu-Schwarz (NS) sectors as the states in each sector are continuously related by the flow. Moreover, as NS and R sectors give rise to space-time bosons and space-time fermions respectively, this isomorphism of the algebra induces space-time supersymmetry.

2.2 Chiral ring

The representation theory of the $\mathcal{N} = 2$ superconformal algebra is equipped with an interesting additional structure, namely a finite sub-sector of the operators creating the highest weight states carries an additional ring structure which is discussed in the following.

The unitary irreducible representations of the algebra can be built up from highest weight states by acting on these with creation operators which are identified with the negative modes in (2.8). Similarly all the modes with positive indices can be thought of as annihilation operators which lower the L_0 eigenvalue of a state. A highest weight state is thus one which satisfies,

$$L_n|\phi\rangle = 0\,, \quad G^{\pm}_r|\phi\rangle = 0\,, \quad J_m|\phi\rangle = 0\,, \quad n, r, m > 0\,. \tag{2.10}$$

The zero index modes L_0 and J_0 modes can be used to label the states by their eigenvalues

$$L_0|\phi\rangle = h_\phi|\phi\rangle\,, \quad J_0|\phi\rangle = q_\phi|\phi\rangle\,. \tag{2.11}$$

In the Ramond sector there are furthermore the modes $G_0^{\pm}$. If a state is annihilated by these then it is called a Ramond ground state. A highest weight state is created by a primary field ϕ

$$\phi|0\rangle = |\phi\rangle\,. \tag{2.12}$$

The subset of primary fields which will be of interest is constituted of the *chiral primary* fields. States which are created by those satisfy furthermore

$$G^+_{-1/2}|\phi\rangle = 0\,. \tag{2.13}$$

The name anti-chiral primary will be used for the primary fields annihilated by $G^-_{-1/2}$. In combination with the representations of the anti-holomorphic currents $\overline{G}^{\pm}$ this leads to the notions of $(c,c), (a,c), (a,a)$, and (c,a) primary fields, where c and a stand for chiral and anti-chiral and the pair denotes the conditions in the holomorphic and anti-holomorphic sectors. Considering

$$\langle\phi|\{G^-_{1/2}, G^+_{-1/2}\}|\phi\rangle = ||G^+_{-1/2}|\phi\rangle||^2 = \langle\phi|2L_0 - J_0|\phi\rangle \geq 0\,, \tag{2.14}$$

implies

$$h_\phi \geq \frac{q_\phi}{2}\,, \tag{2.15}$$

with equality holding for chiral states. This property of chiral primary states is an analog of the BPS bound for physical states. Now looking at the operator product expansion of two chiral primary fields ϕ and χ

$$\phi(z)\chi(w) = \sum_i (z-w)^{h_{\psi_i} - h_\phi - h_\chi}\psi_i\,, \tag{2.16}$$

the $U(1)$ charges add $q_{\psi_i} = q_\phi + q_\chi$ and hence $h_{\psi_i} \geq h_\phi + h_\chi$. The operator product expansion has thus no singular terms and the only terms which survive in the expansion when $z \to w$ are the ones for which ψ_i is itself chiral primary. It is thus shown that the chiral primary fields give a closed non-singular ring under operator product expansion. Furthermore, one can show the finiteness of this ring by considering

$$\langle\phi|\{G^-_{3/2}, G^+_{-3/2}\}|\phi\rangle = \langle\phi|2L_0 - 3J_0 + \frac{2}{3}c|\phi\rangle \geq 0\,, \tag{2.17}$$

to see that the conformal weight of a chiral primary is bounded by $c/6$.

For the world-sheet description of certain string compactifications two copies of the $\mathcal{N}=2$ superconformal algebra are required, these are usually referred to as left and right moving or holomorphic and anti-holomorphic, the currents of the latter will be denoted by $\overline{T}, \overline{G}^{\pm}, \overline{J}$. The algebra is then referred to as the $\mathcal{N}=(2,2)$ superconformal algebra.

In the $\mathcal{N}=(2,2)$ superconformal algebra there are now four finite rings, depending on the different combinations of chiral and antichiral in the right and left moving algebras, this gives the (c,c), (a,c), (a,a) and (c,a) rings, one sees that the latter two are charge conjugates of the first two. The relation between charge and conformal weight for an anti-chiral primary becomes $h_\psi = -\frac{q_\psi}{2}$. Denoting the set of chiral primary fields by ϕ_i where the index i runs over all chiral primaries, the ring structure can be formulated as follows[2]

$$\phi_i \phi_j = C_{ij}^k \phi_k \,. \tag{2.18}$$

2.3 Deformation families

Mirror symmetry is a symmetry relating the deformation family of the (a,c) chiral ring with the deformation family of the (c,c) chiral ring. To make this statement more precise the deformation family of the (c,c) chiral ring will be discussed in the following. Deformations of a conformal field theory are achieved by adding marginal operators to the original action, these are operators having conformal weight $h + \overline{h} = 2$. In the following spinless operators will be studied which have $h = 1$, $\overline{h} = 1$. The operators which maintain their $h = \overline{h} = 1$ conformal weights after perturbation of the theory are called truly marginal operators. Such operators can be constructed from the chiral primary operators in two steps. For instance in the (c,c) ring, starting from an operator of charge $q = \overline{q} = 1$, $h = \overline{h} = 1/2$ one can first construct

$$\phi^{(1)}(w,\overline{w}) = \left[G^-(z), \phi(w,\overline{w})\right] = \oint dz\, G^-(z)\phi(w,\overline{w})\,, \tag{2.19}$$

which now has $h = 1, q = 0$. In the next step

$$\phi^{(2)}(w,\overline{w}) = \left\{\overline{G}^-(z), \phi^{(1)}(w,\overline{w})\right\} = \oint d\overline{z}\, \overline{G}^-(\overline{z})\phi^{(1)}(w,\overline{w})\,, \tag{2.20}$$

which has $h = \overline{h} = 1$ and zero charge and is hence a truly marginal operator and can be used to perturb the action of the theory

$$\delta S = t^i \int \phi_i^{(2)} + \overline{t}^{\overline{i}} \int \phi_{\overline{i}}^{(2)}\,, \quad i = 1, \dots, n\,, \tag{2.21}$$

where a priori also the deformations coming from the (a,a) operators are included and $n = \dim \mathcal{H}^{(1,1)}$ denotes the dimension of the subspace of the Hilbert space of the theory spanned by the states which are created by the charge $(1,1)$ operators. A similar construction can be done for the (a,c) chiral ring. The superscript notation is borrowed from topological field theories where an analogous construction gives the two form descendants which can be used to perturb the topological theory, see for example [33]. The deformations constructed in this fashion span a deformation space $\mathcal{M}$, the moduli space of the SCFT.

---

[2] Formulas for products of operators are understood to hold within correlation functions.

2.4 The deformation subring

For the study of the geometric realization of the equivalence of the SCFTs a holomorphic vector bundle $V_{\mathbb{C}}$ over the deformation space $\mathcal{M}$ is introduced. Its fiber corresponds to a subset of the ground-states of the theory. Its importance lies in the fact that the ground-states of the theory do not change over the space of deformations [9]. However, there is a certain way of splitting the bundle which varies smoothly over the moduli space. This bundle collects the states of the theory which are created by the sub-ring of the chiral ring spanned by the charge $(1,1)$ operators. A basis for this sub-ring is denoted by $(\phi_0, \phi_i, (\phi_D)_i, (\phi_D)_0)$, $i = 1, \ldots, n$. ϕ_0 is the identity operator of charge $(0,0)$ and $(\phi_D)_i$ are the charge $(2,2)$ fields and $(\phi_D)_0$ is an element of charge $(3,3)$.

The threepoint function on the sphere is the correlator of 3 chiral fields of charge $(1,1)$:[3]

$$C_{ijk} = \langle \phi_i \phi_j \phi_k \rangle \,, \quad i,j,k = 1, \ldots, \dim \mathcal{H}^{1,1} \,. \tag{2.22}$$

Using the chiral ring relation (2.18) this allows the following manipulation:

$$C_{ijk} = \langle \phi_i \phi_j \phi_k \rangle = \langle \phi_i C_{jk}^m (\phi_D)_m \rangle = C_{jk}^m \langle \phi_i (\phi_D)_m \rangle = C_{jk}^m \, \eta_{im} \,, \tag{2.23}$$

defining the topological metric η_{im} which gives a pairing between the charge $(1,1)$ and charge $(2,2)$ operators. This metric can be used to raise and lower indices and identify the structure constants of the chiral ring with the threepoint function.[4]

Using $\phi^i = \eta^{ij} (\phi_D)_j$ and $\langle (\phi_D)_0 \rangle = 1 \,, \eta_{00} = 1$, the ring structure can now be put in matrix form

$$\phi_i \begin{pmatrix} \phi_0 \\ \phi_j \\ \phi^j \\ \phi^0 \end{pmatrix} = \begin{pmatrix} 0 & \delta_i^k & 0 & 0 \\ 0 & 0 & C_{ijk} & 0 \\ 0 & 0 & 0 & \delta_i^j \\ 0 & 0 & 0 & 0 \end{pmatrix} \begin{pmatrix} \phi_0 \\ \phi_k \\ \phi^k \\ \phi^0 \end{pmatrix} . \tag{2.24}$$

The states created by this sub-ring are organized in a vector bundle

$$\mathcal{V} = \mathcal{H}^{0,0} \oplus \mathcal{H}^{1,1} \oplus \mathcal{H}^{2,2} \oplus \mathcal{H}^{3,3} \,, \tag{2.25}$$

where $\mathcal{H}^{(i,i)}$ denotes the subspace of the Hilbert space of states created by charge (i,i) operators. The splitting of the bundle is given by the charge grading. The variation of this grading over the moduli space $\mathcal{M}$ and its geometric realization is going to be a central theme in the study of mirror symmetry. The operator-state correspondence can be used to obtain a basis of the vector bundle from the basis of the chiral ring operators. Denoting by $|e_0\rangle \in \mathcal{H}^{0,0}$ the unique ground state of charge $(0,0)$ up to scale, a basis for $\mathcal{V}$ can be obtained as follows

$$|e_i\rangle = \phi_i |e_0\rangle \,, \quad |e^i\rangle = \phi^i |e_0\rangle \,, \quad |e^0\rangle = \phi^0 |e_0\rangle \,. \tag{2.26}$$

[3] The charges of all operators in a correlation function in the topological theory have to add up to $(3,3)$ due to an anomaly in the $U(1)$ current (See for example [30, 10]).

[4] See [33] for a review with more details on the topological metric.

The metric on this bundle is given by the topological metric

$$\langle e_a | e^b \rangle = \delta^b_a \quad a, b = 0, \dots, n\,. \tag{2.27}$$

The representation of the chiral ring in this basis reads

$$\phi_i \begin{pmatrix} |e_0\rangle \\ |e_j\rangle \\ |e^j\rangle \\ |e^0\rangle \end{pmatrix} = \underbrace{\begin{pmatrix} 0 & \delta_i^k & 0 & 0 \\ 0 & 0 & C_{ijk} & 0 \\ 0 & 0 & 0 & \delta_i^j \\ 0 & 0 & 0 & 0 \end{pmatrix}}_{:=C_i} \begin{pmatrix} |e_0\rangle \\ |e_k\rangle \\ |e^k\rangle \\ |e^0\rangle \end{pmatrix}\,. \tag{2.28}$$

This structure will turn out to be crucial for the understanding of mirror symmetry. In terms of coordinates on the moduli space $\mathcal{M}$ an insertion of the chiral field ϕ_i is equivalent to an infinitesimal displacement in moduli space and hence can be obtained by a derivative $\frac{\partial}{\partial t^i}$. Denoting the matrix on the right hand side by C_i, the whole equation can be read as a connection on the bundle $\nabla = \partial_i - C_i$ which is flat.[5]

$$[\nabla_i, \nabla_j] = 0\,. \tag{2.29}$$

This connection is called the Gauss-Manin connection since the geometric realization of it on the B-model side coincides with the Gauss-Manin connection defined in the variation of Hodge structure.

2.5 Geometric realization of SCFT

A geometric realization of the $\mathcal{N} = (2,2)$ SCFT is provided by the nonlinear sigma model [35, 6], see also the reviews in [9, 31, 10]. The focus here is on the geometric realization of the (a,c) and (c,c) rings. The nonlinear sigma model that will be considered is a field theory of bosons ϕ and fermions ψ, χ living on a $2d$ Riemann surface Σ, being related by supersymmetry where the bosonic fields are considered as coordinates of some target space X, i.e., $\phi : \Sigma \to X$. In order for the theory to have $\mathcal{N} = (2,2)$ supersymmetry the target space spanned by the bosonic fields has to be a Kähler manifold, which in turn allows one to split its tangent bundle $TX = T^{1,0}X \oplus T^{0,1}X$. The action of the theory is (as given in [6, 10])[6]

$$\begin{aligned} S = \frac{1}{4\pi} \int_\Sigma d^2 z \big(& g_{i\bar{j}} \left(\partial_z \phi^i \partial_{\bar{z}} \phi^{\bar{j}} + \partial_{\bar{z}} \phi^i \partial_z \phi^{\bar{j}} \right) + B_{i\bar{j}} \left(\partial_z \phi^i \partial_{\bar{z}} \phi^{\bar{j}} - \partial_{\bar{z}} \phi^i \partial_z \phi^{\bar{j}} \right) \\ & + i g_{i\bar{j}} \psi^i D_z \psi^{\bar{j}} + i g_{i\bar{j}} \chi^i D_{\bar{z}} \chi^{\bar{j}} + R_{i\bar{k}j\bar{l}} \psi^i \psi^{\bar{k}} \chi^j \chi^{\bar{l}} \big)\,. \end{aligned} \tag{2.30}$$

[5] See [19, 9, 34] for more details.

[6] The notation is chosen differently from [6, 10], in order to be more suitable for the discussion of higher genus topological strings, where the notation of [19] is adopted. The superscript $\pm$ refers to the $U(1)$ charge of the superconformal algebra currents whereas the bar on top of the expressions refers to the right-moving sector. For the fermions, the left moving fermions are denoted by ψ whereas the right moving fermions are denoted by χ.

Denoting by K the canonical bundle on Σ the fermions are sections of

$$\begin{aligned} \psi^i &\in \Gamma(K^{1/2} \otimes \phi^* T^{1,0}X)\,, & \psi^{\bar\imath} &\in \Gamma(K^{1/2} \otimes \phi^* T^{0,1}X)\,, \\ \chi^i &\in \Gamma(\overline{K}^{1/2} \otimes \phi^* T^{1,0}X)\,, & \chi^{\bar\imath} &\in \Gamma(\overline{K}^{1/2} \otimes \phi^* T^{0,1}X)\,. \end{aligned} \tag{2.31}$$

The covariant derivative in the action is with respect to these bundles. The fermions ψ and χ correspond to the left and right moving sectors. The action is conformally invariant if the Ricci tensor is vanishing, together with the Kähler condition this yields a Calabi-Yau target space. The supersymmetries are given by the following transformations of the fields:

$$\delta\,\phi^i = i\varepsilon^-\,\psi^i + i\bar\varepsilon^-\,\chi^i\,, \tag{2.32}$$

$$\delta\,\phi^{\bar\imath} = i\varepsilon^+\,\psi^{\bar\imath} + i\bar\varepsilon^+\,\chi^{\bar\imath}\,, \tag{2.33}$$

$$\delta\psi^i = -\varepsilon^+\partial\phi^i - i\bar\varepsilon^-\chi^j\,\Gamma^i_{jk}\,\psi^k\,, \tag{2.34}$$

$$\delta\psi^{\bar\imath} = -\varepsilon^-\,\partial\phi^i - i\bar\varepsilon^+\chi^{\bar\jmath}\,\Gamma^{\bar\imath}_{\bar\jmath\bar k}\,\psi^{\bar k}\,, \tag{2.35}$$

$$\delta\chi^i = -\bar\varepsilon^+\bar\partial\phi^i - i\varepsilon^-\psi^j\,\Gamma^i_{jk}\,\chi^k\,, \tag{2.36}$$

$$\delta\chi^{\bar\imath} = -\bar\varepsilon^-\,\bar\partial\phi^{\bar\imath} - i\varepsilon^+\psi^{\bar\jmath}\,\Gamma^{\bar\imath}_{\bar\jmath\bar k}\,\chi^{\bar k}\,, \tag{2.37}$$

where $\varepsilon^-, \bar\varepsilon^-, \varepsilon^+, \bar\varepsilon^+$ are sections of $K^{-1/2}, \overline{K}^{-1/2}, K^{-1/2}, \overline{K}^{-1/2}$, respectively. The geometric realization of the currents (2.1) is given by:

$$T = -g_{i\bar\jmath}\partial_z\phi^i\,\partial_z\phi^{\bar\jmath} + \frac{1}{2}g_{i\bar\jmath}\psi^i\partial_z\psi^{\bar\jmath} + \frac{1}{2}g_{i\bar\jmath}\psi^{\bar\jmath}\partial_z\psi^i\,, \tag{2.38}$$

$$G^+ = \frac{1}{2}g_{i\bar\jmath}\psi^i\partial_z\phi^{\bar\jmath}\,, \tag{2.39}$$

$$G^- = \frac{1}{2}g_{i\bar\jmath}\psi^{\bar\jmath}\partial_z\phi^i\,, \tag{2.40}$$

$$J = \frac{1}{4}g_{i\bar\jmath}\psi^i\psi^{\bar\jmath}\,, \tag{2.41}$$

and similarly for $\overline{T}, \overline{G}^+, \overline{G}^-, \overline{J}$. The supersymmetry variation in terms of the charges of the supersymmetry currents reads:[7]

$$\delta = \varepsilon^+\,Q^+ + \varepsilon^-\,Q^- + \bar\varepsilon^+\,\overline{Q}^+ + \bar\varepsilon^-\,\overline{Q}^-\,. \tag{2.42}$$

3 Topological string theory and mirror symmetry

The geometric realization of the isomorphic (a,c) and (c,c) chiral rings of the superconformal algebra are studied using topological string theory which is a $2d$ topological field theory coupled to gravity.

[7] $Q^\pm = \frac{1}{2\pi i}\oint G^\pm dz\,, \quad \overline{Q}^\pm = \frac{1}{2\pi i}\oint \overline{G}^\pm d\bar z.$

3.1 Topological field theory

In the following some features of topological sigma models of cohomological type [35] will be outlined. These are sigma models, i.e., theories of maps $\phi : \Sigma_g \to X$ into some target space X, the case of interest is where Σ_g denotes a Riemann surface of genus g. The maps are realized physically as bosons and fermions which are related by supersymmetry. The additional structure of cohomological topological field theories is the existence of a Grassmann, scalar symmetry operator $\mathcal{Q}$, which is sometimes called the BRST operator, with the following properties:

- $\mathcal{Q}$ is nilpotent $\mathcal{Q}^2 = 0$.
- The energy momentum tensor is $\mathcal{Q}$ exact,

$$T_{\mu\nu} = \{\mathcal{Q}, G_{\mu\nu}\}.$$

A direct consequence of the last property is that the correlation functions do not depend on the two dimensional metric. To see this one notices that a variation with respect to the metric is equivalent to inserting the energy momentum tensor in the correlators, this one being exact and $\mathcal{Q}$ being a symmetry of the theory shows the metric independence. Furthermore, physical states of the theory correspond to cohomology classes of the operator $\mathcal{Q}$ and the semi-classical approximation for evaluating the path integral is exact. Further background as well as derivations of these properties can be found in [36, 9].

3.2 Twisting and topological string theory

A topological field theory can be constructed from the $\mathcal{N} = 2$ superconformal field theories whose states are those created by operators in the chiral ring. This is achieved by treating the annihilators of the states in question as BRST operators and considering only those states as physical states which are annihilated by the BRST operator. From the definition of a chiral state in Eq. (2.13) and its anti-chiral counterpart it is clear that if one wants to restrict to the $(a,c)/(c,c)$ chiral rings then the operators in question should be:

$$\underset{(a,c)}{Q_A = G^-_{-1/2} + \overline{G}^+_{-1/2}}, \quad \underset{(c,c)}{Q_B = \quad G^+_{-1/2} + \overline{G}^+_{-1/2}}. \tag{3.1}$$

These operators square to zero in the algebra $(Q_{A/B})^2 = 0$ fulfilling the first requirement of (3.1). The energy momentum tensor using either operator is however not $Q-$exact and furthermore the operators are fermionic.

Topological string theory is obtained by coupling the resulting theory to $2d$ gravity which involves an integration over all $2d$ geometries and metrics. The two dimensional geometries are the Riemann surfaces which are organized by their genus g. One has in particular to be able to define the BRST operator globally on every such Riemann surface. The charges corresponding to the SCFT $G^{\pm}$

currents are however fermions. In order to have a global well defined charge on every Riemann surface the *topological twist* is introduced. Since the spin of a current is determined by its conformal weight, which appears in the operator product expansion with the energy momentum tensor, the idea of the twist is to modify the energy momentum tensor. The new energy momentum tensor is a combination of the old one and the $U(1)$ current J. Depending on which ring one wants to restrict to, this is achieved by

A	B
$T \to T + \frac{1}{2}\partial J$	$T \to T - \frac{1}{2}\partial J$
$\overline{T} \to \overline{T} - \frac{1}{2}\overline{\partial J}$	$\overline{T} \to \overline{T} - \frac{1}{2}\overline{\partial J}$
(a, c)	(c, c)

(3.2)

yielding Grassmann valued scalars Q_A and Q_B. Restricting to the chiral ring for example one now has:

$$T_{\text{top}} = T - \frac{1}{2}\partial J\,, \tag{3.3}$$

from the OPE (2.1) it is possible to see that G^+ has now conformal weight 1. And furthermore one has:

$$T_{\text{top}} = \{Q^+, G^-\}\,, \tag{3.4}$$

provided that the $U(1)$ charge in correlators is $c/3$, where c is the central charge. For the geometric realizations of the superconformal algebra as nonlinear sigma models into a CY threefold target space $c = 9$. The cohomology of these operators gives the desired restriction to the finite rings. As the notation suggests restricting to the (a, c) ring gives the *A-model* and restricting to the (c, c) ring gives the *B-model*. The coupling to gravity is achieved by defining a correlator for every genus of a Riemann surface and integrating the correlator over the moduli space of Riemann surfaces. The moduli space of a Riemann surface at genus g is described by $3g - 3$ Beltrami differentials $\mu_{\bar{z}}^{z}$, these are anti-holomorphic forms taking values in the holomorphic tangent bundle $\mu_a \in H^{0,1}(\Sigma_g, T\Sigma_g)$ and their complex conjugates $\overline{\mu}_{z}^{\bar{z}}$. These differentials are contracted with the two currents which get their conformal weight augmented to $h = 2\,(\bar{h} = 2)$. This construction is motivated from the bosonic string where the anti-ghosts are contracted with the Beltrami differentials in a similar manner. The genus g amplitudes or free energies are now defined by (here for the B-model and $g > 1$)[8]

$$\mathcal{F}^g = \int_{\mathcal{M}_g} [dm\, d\bar{m}] \langle \prod_{a=1}^{3g-3} (\int_\Sigma \mu_a G^-)(\int_\Sigma \mu_{\bar{a}} \overline{G}^-) \rangle_{\Sigma_g}\,, \tag{3.5}$$

[8]The amplitudes at $g = 0, 1$ need separate treatment, in particular $\mathcal{F}^0$ denotes the prepotential and can be calculated from the geometry of the deformation bundle.

where $\langle \dots \rangle_{\Sigma_g}$ denotes the CFT correlator and $dmd\bar{m}$ are dual to the Beltrami differentials.[9]

3.3 Mirror symmetry

Mirror symmetry is the identification of the topological field (string-) theories which have as their physical states the (a, c) and the (c, c) chiral rings. Its origin in the superconformal algebra is the sign flip of the left-moving $U(1)$ current $J \leftrightarrow -J$. The power of mirror symmetry comes from the fact that the geometrical interpretation of the deformations and the chiral rings in the non-linear sigma model realizations are very different. The deformation of the A-model seeing the (a, c) chiral ring are deformations of the complexified Kähler form of a CY threefold Z. The physical coordinate on the moduli space of theories is[10]

$$t^a = \int_{C^a} B + i\mathfrak{J}\,, \quad a = 1, \dots, h^{1,1}(Z)\,, \tag{3.6}$$

where $C^a \in H_2(Z), B \in H^2(Z, \mathbb{C})/H^2(Z, \mathbb{Z})$ and $\mathfrak{J}$ is the Kähler form.

The deformations on the B-side are complex structure deformations of a CY Z^*. The good physical coordinates are certain periods of the unique holomorphic $(3, 0)$ form $\tilde{\Omega}$, with a certain normalization

$$t^a = \int_{\alpha^a} \tilde{\Omega}\,, \quad a = 1, \dots, h^{2,1}(Z^*)\,, \tag{3.7}$$

with $\alpha^a \in H_3(Z^*)$. The appropriate choice of coordinates on the moduli space of complex structures as well as the normalization of $\tilde{\Omega}$ is determined by the geometrical realization of the chiral ring and will be discussed later in these lectures. Mirror symmetry is about identifying mirror families of CY spaces Z_t and $Z^*_{t^*}$ such that for every member of the family $h^{1,1}(Z) = h^{2,1}(Z^*)$ and the deformations match. The identification of the right physical coordinates on the moduli spaces of the two sides is achieved by realizing the geometrical counterpart of the deformation subring of the chiral ring on both sides.

3.4 A-model

3.4.1 A-twist

The effect of the twist (3.2) on the non-linear sigma model with target space Z is changing the bundles of which the fermions are sections (See [6, 9, 10])

$$\begin{aligned} \psi^i &\in \Gamma(\phi^* T^{1,0} Z)\,, & \psi^{\bar{\imath}}_z &\in \Gamma(K \otimes \phi^* T^{0,1} Z)\,, \\ \chi^i_{\bar{z}} &\in \Gamma(\overline{K} \otimes \phi^* T^{1,0} Z)\,, & \chi^{\bar{\imath}} &\in \Gamma(\phi^* T^{0,1} Z)\,. \end{aligned} \tag{3.8}$$

[9] More details on topological strings can be found in [9, 37, 38, 39, 40, 36, 41].
[10] It is assumed that $h^{2,0}(Z) = 0$.

This has the effect that $\psi^i, \chi^{\bar{\imath}}$ become scalars on the $2d$ base manifold of the non-linear sigma model. To analyze the cohomology of

$$Q_A = Q^- + \overline{Q}^+ ,$$

one sets $\varepsilon^- = \overline{\varepsilon}^+ = \varepsilon$ and $\overline{\varepsilon}^- = \varepsilon^+ = 0$ in (2.32)–(2.37) such that the supersymmetry variation δ is given by

$$\delta = \varepsilon\, Q_A .$$

The variations of the scalar quantities become (2.32), (2.34) and (2.33), (2.37)

$$\delta\phi^i = i\varepsilon\,\psi^i , \quad \delta\psi^i = 0 , \tag{3.9}$$

$$\delta\phi^{\bar{\imath}} = i\varepsilon\,\chi^{\bar{\imath}} , \quad \delta\chi^{\bar{\imath}} = 0 , \tag{3.10}$$

which suggests the identification:

$$\psi^i \leftrightarrow du^i , \quad \chi^{\bar{\imath}} \leftrightarrow d\bar{u}^{\bar{\imath}} , \quad Q_A \leftrightarrow d = \partial + \overline{\partial} , \tag{3.11}$$

where $u^i, \bar{u}^{\bar{\imath}}$ are local coordinates on Z and d is the de Rham operator. Operators of the topological field theory are associated to points on the manifold and can hence only be made out of $2d$ scalars, these would have the form

$$\omega(\phi)_{i_1,\ldots,i_p,\bar{\imath}_1,\ldots,\bar{\imath}_q}\, \psi^{i_1} \ldots \psi^{i_p} \chi^{\bar{\imath}_1} \ldots \chi^{\bar{\imath}_q} , \tag{3.12}$$

which can then be identified with

$$\omega(u)_{i_1,\ldots,i_p,\bar{\imath}_1,\ldots,\bar{\imath}_q}\, du^{i_1} \cdots du^{i_p} d\bar{u}^{\bar{\imath}_1} \cdots d\bar{u}^{\bar{\imath}_q} . \tag{3.13}$$

The operators of the topological theory are in one to one correspondence with the de Rham cohomology of Z. Furthermore the action can now be written as:

$$S = i \int_\Sigma \{Q_A, \mathcal{D}\} - 2\pi i \int_\Sigma \phi^*(B + i\mathcal{J}) , \tag{3.14}$$

with the Kähler form $\mathcal{J} = i g_{i\bar{\jmath}}\, du^i \wedge d\bar{u}^{\bar{\jmath}}$ and

$$\mathcal{D} = 2\pi g_{i\bar{\jmath}} \left(\psi_z^{\bar{\jmath}}\, \bar{\partial}\phi^i + \partial\phi^{\bar{\jmath}}\, \chi_{\bar{z}}^i \right) . \tag{3.15}$$

The second term depends only on the class $\beta = \phi_*(\Sigma) \in H_2(Z, \mathbb{Z})$. The partition and correlation functions can hence be written as:

$$Z = \int [\mathrm{d}\phi\, \mathrm{d}\psi\, \mathrm{d}\chi\,]\, e^{-S} = \sum_{\beta \in H_2} e^{-2\pi i t_\beta} \int_{\text{fixed } \beta} [\mathrm{d}\phi\, \mathrm{d}\psi\, \mathrm{d}\chi\,]\, e^{-i \int \{Q_A, \mathcal{D}\}} . \tag{3.16}$$

By scaling the term in the exponential in the integrand to infinity it is argued that the dominant contributions come from maps such that $\mathcal{D} = 0$. These are holomorphic maps

$$\bar{\partial}\phi = 0 ,$$

and it can be shown that these contributions are exact, which is the idea of localization (See [9, 10] and references therein for more details). In these cases the path integral can be defined as integrals over moduli spaces of holomorphic maps and gives mathematical invariants which characterize the target space Z, the Gromov-Witten invariants. Physically the sum over holomorphic maps wrapping various 2-cycles in the target space corresponds to summing over different world-sheet instanton sectors.

3.4.2 Quantum geometry

It was shown that the operators of the topological field theory obtained from the A-twist are in correspondence with the de Rham cohomology and moreover the $U(1)$ charges are identified with the degrees of the forms. The geometric interpretation of the chiral ring is given by the wedge product of forms or equivalently by the intersection ring of Z. However, due to the summation over the all the instanton sectors as in Eq. (3.16), this ring gets modified, giving the quantum cohomology ring, which can be thought of as a quantum mechanical correction to the intersection ring taking into account a counting of holomorphic maps. The introduction of notions from quantum mechanics into geometry is referred to as quantum geometry. Another deviation from the classical understanding of geometry is the notion of the complexified Kähler moduli space. Classically the deformations of the Kähler form would define the Kähler cone. However, the sigma model suggests that the correct quantity to be considered is the complexified Kähler class and the enlarged Kähler moduli space, which can be understood in terms of different regions, called phases which may relate different CY geometries [42, 43, 44].

In the following a short exposition of the deformation subring of the chiral ring will be given, more mathematical details and background can be found in [45, 8]. This exposition follows [46]. Quantum cohomology is developed at large volume which refers to the region where the instanton sum in (3.16) has a sensible meaning. Mathematical notions of mirror symmetry in other regions in moduli space are still being developed, see for example [47] and references therein for mirror symmetry involving the orbifold region in moduli space. The deformation subring corresponds to the even cohomology of the CY Z. The complexified Kähler classes are elements in $H^2(Z,\mathbb{C})$. The pairing is given by the wedge product of forms. To recover the chiral ring structure, an interpretation should be given to following diagram

$$H^0(Z,\mathbb{C}) \xrightarrow{\nabla_A} H^2(Z,\mathbb{C}) \xrightarrow{\nabla_A} H^4(Z,\mathbb{C}) \xrightarrow{\nabla_A} H^6(Z,\mathbb{C})\,. \tag{3.17}$$

To define the connection let η_0 be the generator of $H^0(Z,\mathbb{Z})$, η_a , $a=1,\ldots,n=h^{1,1}(Z)$ a basis of $H^2(Z,\mathbb{Z})$, χ_b , $b=1,\ldots,n$ a basis of $H^4(Z,\mathbb{Z})$ which is dual with respect to the symplectic pairing $\langle\eta_a,\chi_b\rangle=\delta_{ab}$ and finally let $\chi_0\in H^6(Z,\mathbb{Z})$

be the dual to η_0. The connection ∇_A is now defined by

$$\begin{aligned}
\nabla_A \eta_0 &= \sum_{a=1}^{n} \eta_a \otimes \frac{\mathrm{d}q_a}{q_a}\,, \\
\nabla_A \eta_c &= \sum_{a,b=1}^{n} C_{abc}\, \chi_b \otimes \frac{\mathrm{d}q_a}{q_a}\,, \\
\nabla_A \chi_b &= \chi_0 \frac{\mathrm{d}q_b}{q_b}\,, \\
\nabla_A \chi_0 &= 0\,.
\end{aligned} \tag{3.18}$$

The coefficients C_{abc} are power series in $q_1, \ldots, q_n$ defined by rational curves, whose class is given by $\beta \in H_2(Z, \mathbb{Z})$

$$q_\beta := \exp\left(2\pi i \int_\beta B + i\mathcal{J}\right)\,,$$

$$C_{abc} = \int_Z \eta_a \wedge \eta_b \wedge \eta_c + \sum_{\beta \in H_2} n_{[\beta]} \langle \beta, \eta_a \rangle \langle \beta, \eta_b \rangle \langle \beta, \eta_c \rangle \frac{q^{[\beta]}}{1 - q^{[\beta]}}\,, \tag{3.19}$$

where $q^{[\beta]} = q^{c_1} \ldots q_n^{c_m}$, $c_a = \langle \beta, \eta_a \rangle$. And where

$$N_{[\beta]}(\eta_a, \eta_b, \eta_c) = n_{[\beta]} \langle \beta, \eta_a \rangle \langle \beta, \eta_b \rangle \langle \beta, \eta_c \rangle\,, \tag{3.20}$$

are rational numbers called the Gromov-Witten invariants. One sees from (3.19), that the geometric interpretation of the threepoint function involves a classical part given by the triple intersection as well as a quantum part given by the sum over instantons.

3.5 B-model

3.5.1 B-twist

Returning to the B-twist (3.2), the modification of the bundles of which the fermions are sections becomes

$$\begin{aligned}
\psi^i_z &\in \Gamma(K \otimes \phi^* T^{1,0} Z)\,, & \psi^{\bar\imath} &\in \Gamma(\phi^* T^{0,1} Z)\,, \\
\chi^i_{\bar z} &\in \Gamma(\overline{K} \otimes \phi^* T^{1,0} Z)\,, & \chi^{\bar\imath} &\in \Gamma(\phi^* T^{0,1} Z)\,.
\end{aligned} \tag{3.21}$$

The world-sheet scalars are grouped in the following form:

$$\eta^{\bar\imath} = \psi^{\bar\imath} + \chi^{\bar\imath}\,, \quad \theta_j = g_{\bar\imath j}\left(\psi^{\bar\imath} - \chi^{\bar\imath}\right)\,, \tag{3.22}$$

and ψ^i_z, χ^i_z are considered to be the $(1,0)$ and $(0,1)$ components of a one-form ρ^i. To analyze the cohomology of

$$Q_B = Q^+ + \overline{Q}^+\,,$$

one sets $\varepsilon^+ = \overline{\varepsilon}^+ = \varepsilon$ and $\overline{\varepsilon}^+ = \varepsilon^+ = 0$ in (2.32) – (2.37) such that the supersymmetry variation δ is given by

$$\delta = \varepsilon \, Q_B \, .$$

The supersymmetry variations suggest the identification [6]

$$\eta^{\bar{\imath}} \leftrightarrow d\bar{u}^{\bar{\imath}} \, , \quad \theta_i \leftrightarrow \frac{\partial}{\partial u^i} \, , \quad Q_B \leftrightarrow \bar{\partial} \, , \tag{3.23}$$

Operators of the topological field theory made out of $2d$ scalars have the form

$$\omega(\phi)^{i_1,\dots,i_p}_{\bar{\imath}_1,\dots,\bar{\imath}_q} \, \eta^{\bar{\imath}_1} \dots \eta^{\bar{\imath}_q} \theta_{i_1} \dots \theta_{i_p} \, , \tag{3.24}$$

which can then be identified with

$$d\bar{u}^{\bar{\imath}_1} \cdots d\bar{u}^{\bar{\imath}_q} \, \omega(u)^{i_1,\dots,i_p}_{\bar{\imath}_1,\dots,\bar{\imath}_q} \, \frac{\partial}{\partial u^{i_1}} \cdots \frac{\partial}{\partial u^{i_p}} \, . \tag{3.25}$$

The operators of the topological theory are in one to one correspondence with $(0,q)$ forms on Z^* valued in $\wedge^p TZ^*$. Furthermore, the action can now be written as:

$$S = i \int_\Sigma \{Q_B, \mathcal{D}\} + U \, , \tag{3.26}$$

with

$$\mathcal{D} = g_{j\bar{k}} \left(\rho^j_z \, \bar{\partial}\phi^{\bar{k}} + \rho^j_{\bar{z}} \, \partial\phi^{\bar{k}} \right) \, , \tag{3.27}$$

$$U = \int_\Sigma \left(-\theta_j D\rho^j - \frac{i}{2} R_{j\bar{\jmath}k\bar{k}} \rho^j \wedge \rho^k \, \eta^{\bar{\jmath}} \theta_l \, g^{l\bar{k}} \right) \, . \tag{3.28}$$

Setting $\mathcal{D} = 0$ the maps ϕ become trivial since $\partial\phi^{\bar{k}} = \bar{\partial}\phi^{\bar{k}} = 0$, these maps map the $2d$ world-sheet to a point in Z^*. The B-model receives thus no instanton corrections. The correlation functions correspond to classical integrals. Furthermore the unique up to scale holomorphic $(3,0)$ form Ω of the CY threefold Z^* can be used to transform the $(0,q)$ forms valued in $\wedge^p TZ^*$ into $(3-p,q)$ forms. The deformation operators having charge $(1,1)$ get hence identified with elements in $H^{2,1}(Z^*)$, which parameterize the complex structure deformations.

3.5.2 Variation of Hodge structure

The moduli space on the B-side corresponds to the moduli space of complex structures of the target space Z^*. For a CY threefold, the bundle corresponding to the deformation subring is the middle dimensional cohomology $H^3(Z^*, \mathbb{C})$. This space has a natural splitting once a given complex structure is chosen, i.e., at a specific point in the complex structure moduli space. The split is

$$H^3(Z^*, \mathbb{C}) \simeq \bigoplus_{p+q=3} H^{p,q}(Z^*) \, . \tag{3.29}$$

This split identifies the unique up to scale holomorphic $(3,0)$ form Ω. It permits furthermore a natural notion of complex conjugation, namely $\overline{H^{p,q}(Y)} = H^{q,p}(Y)$. The term Hodge structure refers to $H^3(Z^*,\mathbb{C})$, together with the split (3.29) and with a lattice given by $H^3(Z^*,\mathbb{Z})$ which generates $H^3(Z^*,\mathbb{C})$ upon tensoring with $\mathbb{C}$ (See [48, 49] and references therein for more details). The split (3.29) does however not vary holomorphically when the complex structure moduli are varied. The non-holomorphicity of this split is at the origin of the holomorphic anomaly equations which will be discussed later. There is however a different split of the bundle which varies holomorphically over the moduli space of complex structures. This split is given by the Hodge filtration $F^\bullet(Z^*) = \{F^p(Z^*)\}_{p=0}^3$, where the spaces in brackets are defined by

$$H^3 = F^0 \supset F^1 \supset F^2 \supset F^3 \supset F^4 = 0\,, \quad F^p(Z^*) = \bigoplus_{a\geq p} H^{a,3-a}(Z^*) \subset H^3\,. \tag{3.30}$$

To recover the splitting (3.29) one can intersect with the anti-holomorphic filtration

$$H^{p,q}(Z^*) = F^p(Z^*) \cap \overline{F^q(Z^*)}\,. \tag{3.31}$$

Instead of a fixed target space, one can consider a variation family by varying the complex structure of Z^*. In this case the filtration is equipped with a flat connection ∇ which is called the Gauss-Manin connection with the property $\nabla F^p \subset F^{p-1}$, which is called Griffiths transversality.

This property permits an identification of the derivatives of [11] $\Omega(x) \in F^3$ with elements in the lower filtration spaces. The whole filtration can be spanned by taking multiderivatives of the holomorphic $(3,0)$ form. Fourth order derivatives can then again be expressed by the elements of the basis, which is reflected by the fact that periods of $\Omega(x)$ are annihilated by a system of differential equations of fourth order called the Picard-Fuchs (PF) equations. The PF equations capture the variation of Hodge structure which describes the geometric realization on the B-model side of the deformation of the $\mathcal{N}=(2,2)$ superconformal field theory and its chiral ring [50], see also [51] for a review. Schematically the variation takes the form:

$$F^3 \xrightarrow{\nabla_B} F^2 \xrightarrow{\nabla_B} F^1 \xrightarrow{\nabla_B} F^0\,. \tag{3.32}$$

The dimensions of the spaces $(F^3, F^2/F^3, F^1/F^2, F^0/F^1)$ are $(1, h^{2,1}, h^{2,1}, 1)$. Elements in these spaces can be obtained by taking derivatives of $\Omega(x)$ w.r.t the moduli. A section of the filtration can be given by the following vector $w(x)$ which has $2h^{2,1}+2$ components:

$$w(x) = \begin{pmatrix}\Omega(x) & \theta_i\Omega(x) & \theta_*\theta_i\Omega(x) & \theta_*^3\Omega(x)\end{pmatrix}^t\,. \tag{3.33}$$

[11] x is a local coordinate on the moduli space of complex structures $\mathcal{M}$ used to indicate the dependence of the form type on the patch in moduli space.

where $\theta_i = x^i\partial_{x^i}, i = 1, \dots, h^{2,1}(Z^*)$ with x^i some local coordinates and x^* some fixed choice of coordinate such that the $h^{2,1}$ elements $\theta_i\theta_*\Omega(x)$ span F^1/F^2. Using $w(x)$ the period matrix can be defined

$$\Pi(x)_\beta{}^\alpha = \int_{\gamma^\alpha} w(x) , \quad \gamma^\alpha \in H_3(Z^*) , \quad \alpha, \beta = 0, \dots, 2h^{2,1} + 1 , \tag{3.34}$$

the first row of which corresponds to the periods of $\Omega(x)$. The periods are annihilated by the PF operators.

Furthermore Griffiths transversality together with the symplectic pairing given by integrating the wedge product of two three forms over the CY allows one to define the threepoint functions[12]

$$C_{ijk} := -\int_Y \Omega \wedge \nabla_i \nabla_j \nabla_k \Omega . \tag{3.35}$$

3.5.3 Special/flat coordinates

A special set of coordinates on the moduli space of complex structures will be discussed which permit an identification with the physical deformations of the underlying theory. These coordinates are defined within special geometry which was developed studying moduli spaces of $\mathcal{N} = 2$ theories.[13] Choosing a symplectic basis of 3-cycles $A^I, B_J \in H_3(Z^*)$ and a dual basis α_I, β^J of $H^3(Z^*)$ such that

$$\begin{aligned} &A^I \cap B_J = \delta^I_J = -B_J \cap A^I , \quad A^I \cap A^J = B_I \cap B_J = 0 , \\ &\int_{A^I} \alpha_J = \delta^I_J , \quad \int_{B_J} \beta^I = \delta^I_J , \quad I, J = 0, \dots, h^{2,1}(Z^*) , \end{aligned} \tag{3.36}$$

the $(3,0)$ form $\Omega(x)$ can be expanded in the basis α_I, β^J:[14]

$$\Omega(x) = X^I(x)\alpha_I + \mathcal{F}_J(x)\beta^J . \tag{3.37}$$

The periods $X^I(x)$ can be identified with projective coordinates on $\mathcal{M}$ and $\mathcal{F}_J$ with derivatives of a homogeneous function $\mathcal{F}(X^I)$ of weight 2 such that $\mathcal{F}_J = \frac{\partial \mathcal{F}(X^I)}{\partial X^J}$. In a patch where $X^0(x) \neq 0$ a set of special coordinates can be defined

$$t^a = \frac{X^a}{X^0} , \quad a = 1, \dots, h^{2,1}(Z^*).$$

The normalized holomorphic $(3,0)$ $v_0 = (X^0)^{-1}\Omega(t)$ has the expansion:

$$v_0 = \alpha_0 + t^a\alpha_a + \beta^b F_b(t) + (2F_0(t) - t^c F_c(t))\beta^0 , \tag{3.38}$$

[12] The sign here is a choice of convention.

[13] See [5, 12, 50–59].

[14] The relative sign choice is a convention which varies in the literature, this choice identifies $X^I = \int_{A^I} \Omega, F_I = \int_{B_I} \Omega$.

where

$$F_0(t) = (X^0)^{-2}\mathfrak{F} \quad \text{and} \quad F_a(t) := \partial_a F_0(t) = \frac{\partial F_0(t)}{\partial t^a}.$$

$F_0(t)$ is the prepotential. One can define

$$v_a = \alpha_a + \beta^b F_{ab}(t) + (F_a(t) - t^b F_{ab}(t))\beta^0\,, \tag{3.39}$$
$$v^a_D = \beta^a - t^a \beta^0\,, \tag{3.40}$$
$$v^0 = -\beta^0\,. \tag{3.41}$$

The Yukawa coupling in special coordinates is given by

$$C_{abc} := \partial_a\partial_b\partial_c F_0(t) = -\int_{Z^*} v_0 \wedge \partial_a\partial_b\partial_c v_0\,. \tag{3.42}$$

Defining further the vector with $2h^{2,1}+2$ components:

$$v = (v_0\,, \quad v_a\,, \quad v^a_D\,, \quad v^0)^t\,, \tag{3.43}$$

which gives by construction:

$$\partial_a \begin{pmatrix} v_0 \\ v_b \\ v^b_D \\ v^0 \end{pmatrix} = \underbrace{\begin{pmatrix} 0 & \delta^c_a & 0 & 0 \\ 0 & 0 & C_{abc} & 0 \\ 0 & 0 & 0 & \delta^b_a \\ 0 & 0 & 0 & 0 \end{pmatrix}}_{:=C_a} \begin{pmatrix} v_0 \\ v_c \\ v^c_D \\ v^0 \end{pmatrix}, \tag{3.44}$$

which defines the $(2h^{2,1}+2)\times(2h^{2,1}+2)$ matrices C_a, in terms of which the equation can be written in the form:

$$(\partial_a - C_a)\; v = 0\,. \tag{3.45}$$

The entries of v correspond to elements in the different filtration spaces discussed earlier. Equation (3.45) defines the Gauss-Manin connection in special coordinates. The upper triangular structure of the connection matrix reflects the effect of the charge increment of the elements in the chiral ring upon insertion of a marginal operator of unit charge. Since the underlying superconformal field theory is isomorphic for the A- and the B-models, this set of coordinates describing the variation of Hodge structure is the good one for describing mirror symmetry and provide thus the mirror maps. The power of mirror symmetry comes from the fact that the computation on the B-model side are well under control since they are connected to the variation of complex structure. The computations which are involved are classical. The identification with the non-trivial quantum geometry of the A-model is done by finding the right set of mirror maps, which are the special coordinates which are also called flat coordinates. These are the coordinates in terms of which the Gauss-Manin connection has the form of Eq. (3.44). There are several techniques [28, 50, 51, 57, 58, 60–63] to identify the special set of coordinates which allows an identification with the physical parameters and hence with the A-model side. This will be outlined in the following for the example of the quintic.

3.6 Example of mirror geometries

In the following an example of a construction of mirror manifolds as well as finding the appropriate mirror maps will be discussed. The exposition will be rather concise referring to [8, 9, 12, 48, 64] and references therein for further details. The quintic Z denotes the CY manifold defined by

$$Z := \{P(x) = 0\} \subset \mathbb{P}^4 \,, \tag{3.46}$$

where P is a homogeneous polynomial of degree 5 in 5 variables $x_1, \ldots, x_5$. The mirror quintic Z^* can be constructed using the Greene-Plesser construction [11]. Equivalently it may be constructed using Batyrev's dual polyhedra [65] in the toric geometry language[15]. In the Greene-Plesser construction the family of mirror quintics is the one parameter family of quintics defined by

$$\{p(Z^*) = \sum_{i=1}^{5} x_i^5 - \psi \prod_{i=1}^{5} x_i = 0\} \in \mathbb{P}^4 \,, \tag{3.47}$$

after a $(\mathbb{Z}_5)^3$ quotient and resolving the singularities.

In the following, the mirror construction following Batyrev will be outlined. The mirror pair of CY 3-folds (Z, Z^*) is given as hypersurfaces in toric ambient spaces (W, W^*). The mirror symmetry construction of [65] associates the pair (Z, Z^*) to a pair of integral reflexive polyhedra (Δ, Δ^*).

3.6.1 The A-model geometry

The polyhedron Δ is characterized by k relevant integral points ν_i lying in a hyperplane of distance one from the origin in $\mathbb{Z}^5$, ν_0 will denote the origin following the conventions of [65, 66]. The k integral points $\nu_i(\Delta)$ of the polyhedron Δ correspond to homogeneous coordinates u_i on the toric ambient space W and satisfy $n = h^{1,1}(Z)$ linear relations:

$$\sum_{i=0}^{k-1} l_i^a \, \nu_i = 0 \,, \quad a = 1, \ldots, n \,. \tag{3.48}$$

The integral entries of the vectors l^a for fixed a define the weights l_i^a of the coordinates u_i under the $\mathbb{C}^*$ actions

$$u_i \to (\lambda_a)^{l_i^a} u_i \,, \quad \lambda_a \in \mathbb{C}^* \,.$$

[15] For a review of toric geometry see [29, 9].

The l_i^a can also be understood as the $U(1)_a$ charges of the fields of the gauged linear sigma model (GLSM) construction associated with the toric variety [42]. The toric variety W is defined as $W \simeq (\mathbb{C}^k - \Xi)/(\mathbb{C}^*)^n$, where Ξ corresponds to an exceptional subset of degenerate orbits. To construct compact hypersurfaces, W is taken to be the total space of the anti-canonical bundle over a compact toric variety. The compact manifold $Z \subset W$ is defined by introducing a superpotential $\mathcal{W}_Z = u_0 p(u_i)$ in the GLSM, where x_0 is the coordinate on the fiber and $p(u_i)$ a polynomial in the $u_{i>0}$ of degrees $-l_0^a$. At large Kähler volumes, the critical locus is at $u_0 = p(u_i) = 0$ [42].

An example of CY manifold is the quintic which is the compact geometry given by a section of the anti-canonical bundle over $\mathbb{P}^4$. The charge vectors for this geometry are given by:

$$\begin{array}{rl} & \;\; u_0 \;\; u_1 \; u_2 \; u_3 \; u_4 \;\; u_5 \\ l = & (-5 \;\; 1 \;\; 1 \;\; 1 \;\; 1 \;\; 1)\,. \end{array} \tag{3.49}$$

The vertices of the polyhedron Δ are given by:

$$\begin{aligned} &\nu_0 = (0,0,0,0,0)\,, \quad \nu_1 = (1,0,0,0,0)\,, \quad \nu_2 = (0,1,0,0,0)\,, \\ &\nu_3 = (0,0,1,0,0)\,, \quad \nu_4 = (0,0,0,1,0)\,, \quad \nu_5 = (-1,-1,-1,-1,0)\,. \end{aligned} \tag{3.50}$$

3.6.2 The B-model geometry

The B-model geometry $Z^* \subset W^*$ is determined by the mirror symmetry construction of [67, 65] as the vanishing locus of the equation

$$p(Z^*) = \sum_{i=0}^{k-1} a_i y_i = \sum_{\nu_i \in \Delta} a_i X^{\nu_i}\,, \tag{3.51}$$

where a_i parameterize the complex structure of Z^*, y_i are homogeneous coordinates [67] on W^* and $X_m\,, m = 1,\ldots,4$ are inhomogeneous coordinates on an open torus $(\mathbb{C}^*)^4 \subset W^*$ and $X^{\nu_i} := \prod_m X_m^{\nu_{i,m}}$ [46]. The relations (3.48) impose the following relations on the homogeneous coordinates

$$\prod_{i=0}^{k-1} y_i^{l_i^a} = 1\,, \quad a = 1,\ldots,n = h^{2,1}(Z^*) = h^{1,1}(Z)\,. \tag{3.52}$$

The important quantity in the B-model is the holomorphic $(3,0)$ form which is given by:

$$\Omega(a_i) = \mathrm{Res}_{p=0} \frac{1}{p(Z^*)} \prod_{i=1}^{4} \frac{dX_i}{X_i}\,. \tag{3.53}$$

Its periods

$$\pi_\alpha(a_i) = \int_{\gamma^\alpha} \Omega(a_i)\,, \quad \gamma^\alpha \in H_3(Z^*)\,, \quad \alpha = 0,\ldots,2h^{2,1}+1\,, \tag{3.54}$$

are annihilated by an extended system of GKZ [68] differential operators

$$\mathcal{L}(l) = \prod_{l_i>0} \left(\frac{\partial}{\partial a_i}\right)^{l_i} - \prod_{l_i<0} \left(\frac{\partial}{\partial a_i}\right)^{-l_i}, \tag{3.55}$$

$$\mathcal{Z}_k = \sum_{i=0}^{k-1} \nu_{i,j}\theta_i\,, \quad j = 1,\ldots,4. \quad \mathcal{Z}_0 = \sum_{i=0}^{k-1} \theta_i + 1\,, \quad \theta_i = a_i \frac{\partial}{\partial a_i}\,, \tag{3.56}$$

where l can be a positive integral linear combination of the charge vectors l^a. The equation $\mathcal{L}(l)\,\pi_0(a_i) = 0$ follows from the definition (3.53). The equations $\mathcal{Z}_k\,\pi_\alpha(a_i) = 0$ express the invariance of the period integral under the torus action and imply that the period integrals only depend on special combinations of the parameters a_i

$$\pi_\alpha(a_i) \sim \pi_\alpha(z_a)\,, \quad z_a = (-)^{l_0^a} \prod_i a_i^{l_i^a}\,, \tag{3.57}$$

the $z_a\,, a = 1,\ldots,n$ define local coordinates on the moduli space $\mathcal{M}$ of complex structures of Z^*.

The charge vector defining the A-model geometry in Eq. (3.49) gives the mirror geometry defined by:

$$p(Y) = \sum_{i=0}^{5} a_i y_i = 0\,, \tag{3.58}$$

where the coordinates y_i are subject to the relation

$$y_1 y_2 y_3 y_4 y_5 = y_0^5\,. \tag{3.59}$$

Changing the coordinates $y_i = x_i^5$, $i = 1,\ldots,5$ shows the equivalence of (3.47) and (3.58) with

$$\psi^{-5} = -\frac{a_1 a_2 a_3 a_4 a_5}{a_0^5} =: z\,. \tag{3.60}$$

Furthermore, the following Picard-Fuchs (PF) operator annihilating $\tilde{\pi}_\alpha(z_i) = a_0\,\pi_\alpha(a_i)$ is found:

$$\mathcal{L} = \theta^4 - 5z \prod_{i=1}^{4} (5\theta + i)\,, \quad \theta = z\frac{d}{dz}\,. \tag{3.61}$$

The discriminant of this operator is

$$\Delta = 1 - 3125\,z\,, \tag{3.62}$$

and the Yukawa coupling can be computed:

$$C_{zzz} = \frac{5}{z^3\,\Delta}\,. \tag{3.63}$$

The PF operator gives a differential equation which has three regular singular points which correspond to points in the moduli space of the family of quintics where the defining equation becomes singular or acquires additional symmetries, these are the points: [16]

- $z = 0$, the quintic at this value corresponds to the quotient of $\prod_{i=1}^{5} x_i = 0$ which is the most degenerate Calabi-Yau and corresponds to large radius when translated to the A-side.
- $z = 5^{-5}$ this corresponds to a discriminant locus of the differential equation (3.61) and also to the locus where the Jacobian of the defining equation vanishes. This type of singularity is called a *conifold* singularity. Its A-model interpretation is going to be very useful in the context of higher genus amplitudes in at a later stage.
- $z = \infty$, this is known as the orbifold point in the moduli space of the quintic and it corresponds to a non-singular CY threefold with a large automorphism group. This is reflected by a monodromy of order 5.

3.6.3 Finding the special/flat coordinates

The deformations of complex structure of the B-model geometry are typically deformations of the defining equation such as Eq. (3.47). The geometric quantities which correspond to the physical generating functions of interest are computed as functions of these deformation parameters. A physical interpretation of the generating functions in terms of instanton sums using the mirror map to the A-model requires finding the good coordinates to describe the deformation problem on the B-model side. These are the special/flat coordinates discussed earlier. It will be sketched in the following how these can be found in the locus in the moduli space of the mirror quintic which is in the vicinity of $z = 0$. This is the patch in moduli space mirror to the large volume of the A-model, which is the locus which admits an interpretation in terms of quantum cohomology. The logic is to span the Hodge filtration by acting with derivatives on Ω and expressing the PF equation as a connection on the deformation bundle. By a change of normalizations and of coordinates, the connection can be brought to the upper triangular form which reflects the chiral ring property. Traditionally, the singularity structure at large volume also permits an identification of the mirror map and the Yukawa couplings as reviewed in [48] for example. Transforming the connection on the deformation bundle however allows one to find the physical coordinates in various patches in moduli space, which is required for many applications, see for example [28, 61, 62].

Choosing the following section of the Hodge filtration:

$$\vec{\Omega}_z(z) = (\Omega(z), \partial_z\Omega(z), \partial_z^2\Omega(z), \partial_z^3\Omega(z))^t \,, \tag{3.64}$$

[16] For further details see [48], for the physical SCFT interpretation see [69].

one can use the Picard-Fuchs equation (3.61) to read off the following connection matrix

$$\partial_z \vec{\Omega} = \underbrace{\begin{pmatrix} 0 & 1 & 0 & 0 \\ 0 & 0 & 1 & 0 \\ 0 & 0 & 0 & 1 \\ \frac{120}{z^3\Delta} & -\frac{1-15000z}{z^3\Delta} & -\frac{7-45000z}{z^2\Delta} & -\frac{6-25000z}{z\Delta} \end{pmatrix}}_{M_z} \vec{\Omega}\,, \tag{3.65}$$

where $\Delta = 1 - 3125z$. This can be written as

$$\nabla_z \vec{\Omega}_z = (\partial_z - M_z)\vec{\Omega}_z = 0\,. \tag{3.66}$$

To find the basis and the coordinate in terms of which the connection becomes the topological connection, the following ansatz is made

$$\vec{\Omega}_t(z(t)) = \left(\frac{\Omega(z(t))}{S_0(t)}, \partial_t \frac{\Omega(z(t))}{S_0(t)}, C(z(t))^{-1}\partial_t^2 \frac{\Omega(z(t))}{S_0(t)}, \partial_t \left(C(z(t))^{-1}\partial_t^2 \frac{\Omega(z(t))}{S_0(t)} \right) \right)^t, \tag{3.67}$$

the vector transforms as

$$\vec{\Omega}(z(t))_t = A(t)\vec{\Omega}_z\,.$$

Transforming the connection matrix M_z with

$$M_t = (\partial_t A(t) + \frac{\partial z(t)}{\partial t} A(t) \cdot M_z(z(t))) \cdot A^{-1}(t)\,.$$

Requiring the matrix M_t to have the form:

$$M_t = \begin{pmatrix} 0 & 1 & 0 & 0 \\ 0 & 0 & C(z(t)) & 0 \\ 0 & 0 & 0 & 1 \\ 0 & 0 & 0 & 0 \end{pmatrix}, \tag{3.68}$$

gives differential equations for the functions in the ansatz, which are solved by:

$$\begin{aligned} S_0(t) &= 1 + 120q + 21000q^2 + 141150000q^3 + \cdots, \quad q = e^{-t}, \\ z(t) &= q - 770q^2 + 171525q^3 + \cdots, \\ C_{ttt} &= \partial_t^3 F_0 = 5 + 2875q + 4876875q^2 + 8564575000q^3 + \cdots. \end{aligned}$$

4 Higher genus recursion

The central theme in the following are the holomorphic anomaly equations of Bershadsky, Cecotti, Ooguri and Vafa (BCOV) [18, 19]. These relate recursively the topological string amplitudes at genus g, $\mathcal{F}^g$ which were defined in Sect. 3.2 to amplitudes of lower genus.

The derivation of BCOV is based on a world-sheet analysis of the SCFT underlying topological string theory. It was previously shown that one can construct four topological string theories out of the SCFT depending on which finite chiral ring one wants to restrict the attention to. These were the A/*anti*-A-model and the B/anti-B-models corresponding to the $(a,c)/(c,a)$ and $(c,c)/(a,a)$ rings, respectively. The key insight which allows the derivation of the equations is the failure of decoupling of deformations coming from the anti-ring. For instance, deformations coming from the (a,a) ring will have an effect on the amplitudes of the B-model which is supposed to be only deformed by marginal fields coming from the (c,c) ring. The starting point for the analysis are the tt^* equations of Cecotti and Vafa [70]. The setup for these is to reconsider the deformation subring discussed previously and to replace half of the states with states coming from world-sheet CPT conjugate operators. In terms of the variation of Hodge structures the interplay of the two theories leads to a consideration of the non-holomorphic variation of the Hodge split instead of the filtration which varies holomorphically as captured by the Picard-Fuchs operators. The set of equations obtained from the variation are familiar in the language of supergravity where they reflect the special geometry of the manifold of the scalars of the vector multiplets. See [51, 71] for an overview.

4.1 tt^* equations

To state the tt^* equations, half of the states in the deformation bundle are replaced by states coming from the world-sheet CPT conjugate operators. The states created by the charge $(2,2)$ and $(3,3)$ operators that were introduced with upper indices with respect to the topological metric are replaced using the so called tt^* metric which is obtained using the world-sheet CPT operator Θ [17],[18]

$$g_{a\bar{b}} = \langle e_a|\Theta e^b\rangle\,. \tag{4.1}$$

Using this metric a new basis for the dual states can be defined as follows

$$|e_{\bar{i}}\rangle = g_{k\bar{i}}|e^k\rangle\,, \quad |e_{\bar{0}}\rangle = g_{0\bar{0}}|e^0\rangle\,. \tag{4.2}$$

This tt^* metric induces a connection on the bundle $\mathcal{V}$ which is compatible with the holomorphic structure. It is given by

$$D_i(|e_a\rangle) = \partial_i(|e_a\rangle) - (A_i)_a^b(|e_b\rangle)\,, \tag{4.3}$$

[17] The notation here follows [34].

[18] Letters from the beginning of the alphabet are running from $0,\ldots,n$ while letter from the middle of the alphabet are only running over $1,\ldots,n$.

where $(A_i)^a_b = g^{b\bar{c}}\partial_i g_{a\bar{c}}$ and $(|e_a\rangle)$ denotes the vector $(|e_0\rangle, |e_i\rangle, |e_{\bar{i}}\rangle, |e_{\bar{0}}\rangle)^T$, in matrix form the connection reads

$$A_i = \begin{pmatrix} g^{0\bar{0}}\partial_i g_{0\bar{0}} & 0 & 0\,0 \\ 0 & g^{l\bar{j}}\partial_i g_{k\bar{j}} & 0\,0 \\ 0 & 0 & 0\,0 \\ 0 & 0 & 0\,0 \end{pmatrix}, \quad A_{\bar{i}} = \begin{pmatrix} 0\,0 & 0 & 0 \\ 0\,0 & 0 & 0 \\ 0\,0 & g^{l\bar{j}}\partial_{\bar{i}} g_{k\bar{j}} & 0 \\ 0\,0 & 0 & g^{0\bar{0}}\partial_{\bar{i}} g_{0\bar{0}} \end{pmatrix}. \tag{4.4}$$

Moreover the action of the chiral ring operators $\phi_i, \phi_{\bar{i}}$ becomes in this basis

$$\phi_i \begin{pmatrix} |e_0\rangle \\ |e_j\rangle \\ |e_{\bar{j}}\rangle \\ |e_{\bar{0}}\rangle \end{pmatrix} = \underbrace{\begin{pmatrix} 0 & \delta_i^k & 0 & 0 \\ 0 & 0 & C_{ijk}g^{k\bar{k}} & 0 \\ 0 & 0 & 0 & g_{i\bar{j}} \\ 0 & 0 & 0 & 0 \end{pmatrix}}_{:=C_i} \begin{pmatrix} |e_0\rangle \\ |e_k\rangle \\ |e_{\bar{k}}\rangle \\ |e_{\bar{0}}\rangle \end{pmatrix}, \tag{4.5}$$

and

$$\phi_{\bar{i}} \begin{pmatrix} |e_0\rangle \\ |e_j\rangle \\ |e_{\bar{j}}\rangle \\ |e_{\bar{0}}\rangle \end{pmatrix} = \underbrace{\begin{pmatrix} 0 & 0 & 0 & 0 \\ g_{\bar{i}j} & 0 & 0 & 0 \\ 0 & C_{\bar{i}\bar{j}\bar{k}}g^{\bar{k}k} & 0 & 0 \\ 0 & 0 & \delta_{\bar{i}}^{\bar{k}} & \end{pmatrix}}_{:=C_{\bar{i}}} \begin{pmatrix} |e_0\rangle \\ |e_k\rangle \\ |e_{\bar{k}}\rangle \\ |e_{\bar{0}}\rangle \end{pmatrix}. \tag{4.6}$$

As pointed out previously the ground-state $|e_0\rangle$ of the SCFT is unique up to scale. This means that it is a section of a line bundle $\mathcal{L}$ over the moduli space $\mathcal{M}$. Moreover, as the $|e_i\rangle$ states are created by the operators which also parameterize the deformations of the theory, these can be identified with sections of $\mathcal{L} \otimes T\mathcal{M}$. The reason they are also sections of the line bundle is because they inherit the arbitrary scale from the ground-state. The bundle of ground-states can hence be organized as

$$V_{\mathbb{C}} = \mathcal{L} \oplus \mathcal{L} \otimes T\mathcal{M} \oplus \overline{\mathcal{L} \otimes T\mathcal{M}} \oplus \overline{\mathcal{L}}. \tag{4.7}$$

The tt^* equations [70][19] can now be spelled out

$$\begin{aligned} &[D_i, D_j] = 0, && [D_{\bar{i}}, D_{\bar{j}}] = 0, \\ &[D_i, C_j] = [D_j, C_i], && [D_{\bar{i}}, C_{\bar{j}}] = [C_{\bar{i}}, D_{\bar{j}}], \\ &\left[D_i, D_{\bar{j}}\right] = -[C_i, C_{\bar{j}}]. && \end{aligned} \tag{4.8}$$

The last one of these equations is particularly interesting as it allows to define a modified connection which has vanishing curvature

$$\nabla_i = D_i + C_i\,, \quad \nabla_{\bar{i}} = D_{\bar{i}} + C_{\bar{i}}\,, \quad [\nabla_i, \nabla_{\bar{i}}] = 0\,. \tag{4.9}$$

This connection is the Gauss-Manin connection which was already encountered in the discussion of topological field theory and variation of Hodge structures.

[19] For a pedagogical derivation see [9].

4.2 Special geometry of $\mathcal{M}$

The tt^* metric on $V_{\mathbb{C}}$ can further be used to define a metric on $\mathcal{M}$ by restricting to the $T\mathcal{M}$ valued part of the metric and its complex conjugate. The metric obtained in this way is the Zamolodchikov metric which is defined by

$$G_{i\bar{j}} = \frac{g_{i\bar{j}}}{g_{0\bar{0}}}. \tag{4.10}$$

Setting further

$$g_{0\bar{0}} := e^{-K},$$

it is shown that $G_{i\bar{j}} = \partial_i \partial_{\bar{j}} K$ and hence the Zamolodchikov metric is Kähler. Furthermore, the last equation of (4.8) can be translated into the following statement about the curvature of $G_{i\bar{j}}$

$$R_{i\bar{i}\ j}^{\ \ l} = [\bar{\partial}_{\bar{i}}, D_i]^l{}_j = \bar{\partial}_{\bar{i}} \Gamma^l_{ij} = \delta^l_i G_{j\bar{i}} + \delta^l_j G_{i\bar{i}} - C_{ijk} \overline{C}^{kl}_{\bar{i}}, \tag{4.11}$$

where D is the covariant derivative with connection parts which follow from the context

$$\Gamma^k_{ij} = G^{k\bar{k}} \partial_i G_{\bar{k}j}, \quad \text{and} \quad K_i = \partial_i K,$$

for the cotangent bundle and the line bundle respectively and

$$\overline{C}^{jk}_{\bar{i}} := e^{2K} G^{k\bar{k}} G^{l\bar{l}} \overline{C}_{\bar{i}\bar{k}\bar{l}}.$$

The Kähler condition together with the condition on the curvature makes the manifold $\mathcal{M}$ a special Kähler manifold, the geometry of which is called special geometry. (See [51, 56, 59, 71] and references therein.) The Yukawa coupling in this setting is a holomorphic section of $\mathcal{L}^2 \otimes \mathrm{Sym}^3 T^*\mathcal{M}$.

Here it will be shown that the tt^* formulation of the geometry of the deformation bundle corresponds on the B-model side to the non-holomorphic variation of the Hodge bundle

$$V_{\mathbb{C}} = H^3(X, \mathbb{C}) = H^{3,0}(X) \oplus H^{2,1}(X) \oplus H^{1,2}(X) \oplus H^{0,3}(X). \tag{4.12}$$

Taking as section of $\mathcal{L}$ the unique up to scale holomorphic $(3,0)$ form $\Omega \in H^{3,0}(X)$ and the inner product which is given by taking wedge products of forms and integrating over X, the metric on the line bundle is

$$g_{0\bar{0}} = e^{-K} = -i \int_X \Omega \wedge \overline{\Omega}. \tag{4.13}$$

One can define elements in $H^{2,1}$ and $H^{1,2}$ in the following way

$$\chi_i := D_i \Omega = (\partial_i + K_i)\Omega, \quad \overline{\chi_{\bar{i}}} := D_{\bar{i}} \overline{\Omega} = (\partial_{\bar{i}} + K_{\bar{i}}) \overline{\Omega}. \tag{4.14}$$

With inner product

$$g_{i\bar{j}} = i \int_X \chi_i \wedge \overline{\chi_{\bar{j}}} = e^{-K} G_{i\bar{j}} = e^{-K} \partial_i \partial_{\bar{j}} K. \tag{4.15}$$

Defining the Yukawa coupling

$$C_{ijk} := -\int_X \Omega \wedge D_i D_j D_k \Omega = -\int_X \Omega \wedge \partial_i \partial_j \partial_k \Omega\,, \tag{4.16}$$

one can compute

$$D_i D_j \Omega = D_i \chi_j = i C_{ijk} e^K G^{k\bar{k}} \overline{\chi_{\bar{k}}}\,, \quad D_i \overline{\chi_{\bar{\imath}}} = G_{i\bar{\imath}} \overline{\Omega}\,. \tag{4.17}$$

Taking as a section of $V_{\mathbb{C}}$ the following vector

$$\vec{\Omega} = (\Omega, \chi_i, \overline{\chi_{\bar{\imath}}}, \overline{\Omega})\,, \tag{4.18}$$

one recovers the Gauss-Manin connection

$$D_i \begin{pmatrix} \Omega \\ \chi_j \\ \overline{\chi_{\bar{\jmath}}} \\ \overline{\Omega} \end{pmatrix} = \begin{pmatrix} 0 & \delta_i^k & 0 & 0 \\ 0 & 0 & i C_{ijk} e^K G^{k\bar{k}} & 0 \\ 0 & 0 & 0 & G_{i\bar{\jmath}} \\ 0 & 0 & 0 & 0 \end{pmatrix} \begin{pmatrix} \Omega \\ \chi_k \\ \overline{\chi_{\bar{k}}} \\ \overline{\Omega} \end{pmatrix}. \tag{4.19}$$

This matrix equation can be translated into the following fourth order differential equation for Ω

$$D_m D_n (C_{\bar{i}}^{-1})^{lj} D_i D_j \Omega = 0\,, \tag{4.20}$$

which is the Picard-Fuchs equation. In [58, 51] it is shown that this equation is equivalent to the holomorphic Picard-Fuchs equations encountered previously and that choosing the non-holomorphic variation of Hodge structures is merely a choice of gauge for the vector $\vec{\Omega}$.

4.3 Anomaly equations and their recursive solution

It was argued that deformations of the topological string theory corresponds to the addition of terms of the form

$$z^i \int \phi_i^{(2)} + \bar{z}^{\bar{i}} \int \bar{\phi}_{\bar{i}}^{(2)}\,,$$

to the action S. z^i are complex coordinates on $\mathcal{M}$, the notation change from t^i to z^i is to make manifest the difference between the topological/special coordinates t^i for which taking derivatives corresponds to insertions and the more general coordinates z^i for which insertions correspond to taking covariant derivatives. Hence, amplitudes with insertions of n descendants of marginal (c,c) fields (for the B-model) can be defined by [19]

$$\mathcal{F}^{(g)}_{i_1\cdots i_n} = D_{i_n} \mathcal{F}^{(g)}_{i_1\cdots i_{n-1}}\,. \tag{4.21}$$

Taking a derivative with respect to $\bar{z}^{\bar{\imath}}$ on the other hand corresponds to inserting a descendant of marginal (a,a) operator

$$\phi_{\bar{\imath}}^{(2)} = \mathrm{d}z\mathrm{d}\bar{z}\{G^+, [\overline{G}^+, \phi_{\bar{\imath}}]\} = -\frac{1}{2}\mathrm{d}z\mathrm{d}\bar{z}\{G^+ + \overline{G}^+, [G^+ - \overline{G}^+, \phi_{\bar{\imath}}]\}\,. \tag{4.22}$$

Within topological *field* theory such an insertion in a correlator would vanish as the second part of the equation shows that the operator is Q_B exact. The field theory argument for this decoupling is that the operator Q_B is a symmetry of the theory and can be commuted through the correlator until it acts on a ground-state and gives zero. The failure of this argument in the case of topological string theory is outlined in the following. The formula for $\mathcal{F}^g$ is given again

$$\mathcal{F}^g = \int_{\mathcal{M}_g} [dm\, d\bar{m}] \langle \prod_{a=1}^{3g-3} (\int_\Sigma \mu_a G^-)(\int_\Sigma \mu_{\bar{a}} \overline{G}^-) \rangle_{\Sigma_g} , \tag{4.23}$$

one can see that while commuting G^+ and $\overline{G}^+$ through the correlation function defining $\mathcal{F}^g$ that these will hit the G^- and $\overline{G}^-$ which are contracted with the Beltrami differentials, this will give by the superconformal algebra the energy momentum tensor T. The energy momentum tensor in turn measures the response of the theory to a variation of the metric and can hence be traded for derivatives $\frac{\partial^2}{\partial m \partial \bar{m}}$ on the moduli space of Riemann surfaces $\mathcal{M}_g$ as this latter space precisely parameterizes such changes. One is thus finally left with an integral of a derivative with respect to the moduli of Riemann surfaces over the moduli space of Riemann surfaces which is zero up to boundary contributions. It is exactly these contributions coming from the boundary of the moduli space of Riemann surfaces which is arranged in the holomorphic anomaly equations. These are pictured diagrammatically in the following, for the full derivation of the equations reference is given to the original work [18, 19]. The equations read

$$\bar{\partial}_{\bar{i}} \mathcal{F}^{(g)} = \frac{1}{2} \bar{C}_{\bar{i}}^{jk} \left(\sum_{g_1+g_2=g} D_j \mathcal{F}^{(g_1)} D_k \mathcal{F}^{(g_2)} + D_j D_k \mathcal{F}^{(g-1)} \right) , \tag{4.24}$$

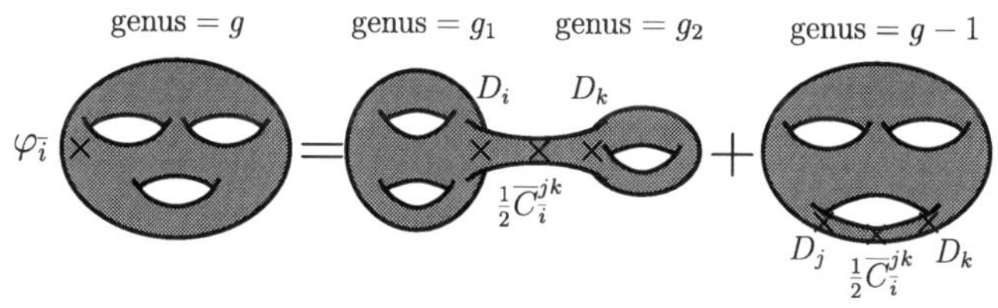

where $g > 1$ and

$$\bar{C}_{\bar{k}}^{ij} = \bar{C}_{\bar{i}\bar{j}\bar{k}} G^{i\bar{i}} G^{j\bar{j}} \, e^{2K}, \qquad \bar{C}_{\bar{i}\bar{j}\bar{k}} = \overline{C_{ijk}} . \tag{4.25}$$

For $g = 1$ the equation is

$$\bar{\partial}_{\bar{i}} \mathcal{F}_j^{(1)} = \frac{1}{2} C_{jkl} C_{\bar{i}}^{kl} + (1 - \frac{\chi}{24}) G_{j\bar{i}} . \tag{4.26}$$

Solution in terms of Feynman diagrams

BCOV proposed together with the anomaly equation [19] a method to recursively solve for the $\mathcal{F}^{(g)}$ s. The idea is to partially integrate the anti-holomorphic derivative appearing in the anomaly equation until one ends up with an expression of the form

$$\partial_{\bar{i}} A = \partial_{\bar{i}} B \Rightarrow A = B + f(z),$$

where $f(z)$ is the holomorphic ambiguity. To be able to do this in practice they first note that the anti-holomorphic Yukawa coupling can locally be written as

$$\overline{C}_{\bar{i}\bar{j}\bar{k}} = e^{-2K} D_{\bar{i}} D_{\bar{j}} \overline{\partial}_{\bar{k}} S,$$

where S is a section of $\mathcal{L}^{-2}$. One can further introduce some objects S^{ij}, S^i which are sections of $\mathcal{L}^{-2} \otimes \mathrm{Sym}^2(T^*\mathcal{M})$ and of $\mathcal{L}^{-2} \otimes T^*(\mathcal{M})$ such that

$$\partial_{\bar{i}} S^{ij} = \bar{C}_{\bar{i}}^{ij}, \qquad \partial_{\bar{i}} S^j = G_{i\bar{i}} S^{ij}, \qquad \partial_{\bar{i}} S = G_{i\bar{i}} S^i. \tag{4.27}$$

Now the procedure is to successively partially integrate the anomaly equation as follows, first write

$$\begin{aligned} &\bar{\partial}_{\bar{i}} \Big(\mathcal{F}^{(g)} - \frac{1}{2} S^{jk} \Big(\sum_{g_1+g_2=g} D_j \mathcal{F}^{(g_1)} D_k \mathcal{F}^{(g_2)} + D_j D_k \mathcal{F}^{(g-1)}\Big)\Big) \\ &\quad = -\frac{1}{2} S^{jk} \bar{\partial}_{\bar{i}} \Big(\sum_{g_1+g_2=g} D_j \mathcal{F}^{(g_1)} D_k \mathcal{F}^{(g_2)} + D_j D_k \mathcal{F}^{(g-1)}\Big), \end{aligned} \tag{4.28}$$

then the repeated use of the expression found for the curvature $[\overline{\partial}_{\bar{i}}, D_j]$ and the knowledge of $\overline{\partial}_{\bar{i}} \mathcal{F}^{\tilde{g}}$ with $\tilde{g} < g$ yields the desired result. They further remarked that the expressions found this way resemble Feynman diagrams if one considers a quantum system with one more degree of freedom than the dimension of the moduli space $\mathcal{M}$, where the propagators are given by

$$K^{ij} = -S^{ij}, \qquad K^{i\phi} = -S^i, \quad \text{and} \quad K^{\phi\phi} = -2S,$$

and vertices are given by the correlation functions with insertions. Furthermore a proof was given for the expansion in terms of Feynman diagrams.

One of the practical shortcomings of this procedure is however that the number of iterations grows very fast, expressions for the $\mathcal{F}^{(g)}$ become very long. On the other hand the recursive information contained in the anomaly equation needs to be supplemented by boundary conditions at every step in order to fix the holomorphic ambiguities.

4.4 Polynomial structure of higher genus amplitudes

The higher genus free energies of the topological string can be interpreted as giving certain amplitudes of the physical string theory.[20] The full topological string partition function conjecturally also encodes the information of $4d$ BPS states [73].

[20] See [72] for a review.

It is thus natural to expect the topological string free energies to be characterized by automorphic forms of the target space duality group. The modularity of the topological string amplitudes was used in [19] to fix the solutions of the anomaly equation. The modularity of the amplitudes is most manifest whenever the modular group is $\mathrm{SL}(2,\mathbb{Z})$ or a subgroup thereof. The higher genus generating functions of the Gromov-Witten invariants for the elliptic curve were expressed as polynomials [74, 75] where the polynomial generators were the elements of the ring of almost holomorphic modular forms E_2, E_4 and E_6 [76]. Polynomials of these generators also appear whenever $\mathrm{SL}(2,\mathbb{Z})$ is a subgroup of the modular group, as for example in [28, 77–81]. The relation of topological strings and almost holomorphic modular forms was further explored in [23, 24, 82]. For most moduli spaces of CY manifolds and their monodromy groups, the appropriate automorphic forms are however not known. It turns out that using special geometry is enough to prove a polynomial structure of the higher genus topological string amplitudes [21, 22], which will be reviewed in the following.

Yamaguchi and Yau proved in [21] that the higher genus amplitudes of the topological string for the mirror quintic and related geometries with a one dimensional space of deformations of complex structures can be expressed as polynomials of degree $3g-3+n$ where n refers to the number of insertions, in a finite number of generators. As generators of the polynomials some elements of special geometry were used which appear on the r.h.s of the holomorphic anomaly equation, namely multi-derivatives of the connections. The non-holomorphic generators of [21] consist of

$$A_p = G^{z\bar{z}}(z\partial_z)^p G_{z\bar{z}} \quad \text{and} \quad B_p = e^K (z\partial_z)^p e^{-K}, \qquad p = 1, 2, 3, \ldots$$

As a holomorphic generator $X \sim \frac{1}{\Delta}$ with Δ discriminant was used. In a next step, relations between the generators are shown such that the infinite number of non-holomorphic generators can be reduced to A_1, B_1, B_2 and B_3. Furthermore from the analysis of the holomorphic anomaly equation it can be proven [21] that only special combinations of the generators appear in the topological string amplitudes and thus the number of non-holomorphic generators gets reduced by one.

The generalization of this construction to CY manifolds with a moduli space of arbitrary dimension was given in [22].[21] This construction will be reviewed in the following along with the necessary boundary conditions to solve the holomorphic ambiguities at each step of the recursion.

In [22] it was proven that the correlation functions $\mathcal{F}^{(g)}_{i_1 \cdots i_n}$ are polynomials of degree $3g-3+n$ in the generators K_i, S^{ij}, S^i, S where a grading $1, 1, 2, 3$ was assigned to these generators, respectively. It was furthermore shown that by making a change of generators [22]

$$\tilde{S}^{ij} = S^{ij},$$

[21]This was solved independently by P. Mayr in an unpublished manuscript.

$$\begin{aligned}\tilde{S}^i &= S^i - S^{ij}K_j,\\ \tilde{S} &= S - S^i K_i + \frac{1}{2}S^{ij}K_iK_j,\\ \tilde{K}_i &= K_i\,,\end{aligned} \tag{4.29}$$

the $\mathcal{F}^{(g)}$ do not depend on $\tilde{K}_i$, i.e., $\partial\mathcal{F}^{(g)}/\partial\tilde{K}_i = 0$. The tilde will be dropped from the modified generators in the following.

The proof relies on expressing the first non-vanishing correlation functions in terms of these generators. At genus zero these are the holomorphic three-point couplings $\mathcal{F}^{(0)}_{ijk} = C_{ijk}$. The holomorphic anomaly equation (4.24) can be integrated using Eq. (4.27) to

$$\mathcal{F}^{(1)}_i = \frac{1}{2}C_{ijk}S^{jk} + (1-\frac{\chi}{24})K_i + f^{(1)}_i, \tag{4.30}$$

with ambiguity $f^{(1)}_i$. As can be seen from this expression, the non-holomorphicity of the correlation functions only comes from the generators. Furthermore the special geometry relation (4.11) can be integrated:

$$\Gamma^l_{ij} = \delta^l_i K_j + \delta^l_j K_i - C_{ijk}S^{kl} + s^l_{ij}\,, \tag{4.31}$$

where s^l_{ij} denote holomorphic functions that are not fixed by the special geometry relation, this can be used to derive the following equations which show the closure of the generators carrying the non-holomorphicity under taking derivatives [22].[22]

$$\begin{aligned}\partial_i S^{jk} &= C_{imn}S^{mj}S^{nk} + \delta^j_i S^k + \delta^k_i S^j - s^j_{im}S^{mk} - s^k_{im}S^{mj} + h^{jk}_i\,,\\ \partial_i S^j &= C_{imn}S^{mj}S^n + 2\delta^j_i S - s^j_{im}S^m - h_{ik}S^{kj} + h^j_i\,,\\ \partial_i S &= \frac{1}{2}C_{imn}S^mS^n - h_{ij}S^j + h_i\,,\\ \partial_i K_j &= K_iK_j - C_{ijn}S^{mn}K_m + s^m_{ij}K_m - C_{ijk}S^k + h_{ij}\,,\end{aligned} \tag{4.32}$$

where h^{jk}_i, h^j_i, h_i and h_{ij} denote holomorphic functions. All these functions together with the functions in Eq. (4.31) are not independent. It was shown in [26] (See also [25]) that the freedom of choosing the holomorphic functions in this ring reduces to functions $\mathcal{E}^{ij}, \mathcal{E}^j, \mathcal{E}$ which can be added to the polynomial generators

$$\begin{aligned}\widehat{S}^{ij} &= S^{ij} + \mathcal{E}^{ij}\,,\\ \widehat{S}^j &= S^j + \mathcal{E}^j\,,\\ \widehat{S} &= S + \mathcal{E}\,.\end{aligned} \tag{4.33}$$

All the holomorphic quantities change accordingly.

[22] These equations are for the tilded generators of Eq. (4.29) and are obtained straightforwardly from the equations in [22]

The topological string amplitudes now satisfy the holomorphic anomaly equations where the $\bar{\partial}_{\bar{i}}$ derivative is replaced by derivatives with respect to the polynomial generators [22, 28].

$$\bar{\partial}_{\bar{i}}\mathcal{F}^{(g)} = \bar{C}_{\bar{i}}^{jk}\left\{\frac{\partial\mathcal{F}^{(g)}}{\partial S^{jk}} - \frac{1}{2}\frac{\partial\mathcal{F}^{(g)}}{\partial S^{k}}K_j - \frac{1}{2}\frac{\partial\mathcal{F}^{(g)}}{\partial S^{j}}K_k + \frac{1}{2}\frac{\partial\mathcal{F}^{(g)}}{\partial S}K_jK_k\right\} + G_{\bar{i}j}\frac{\partial\mathcal{F}^{(g)}}{\partial K_j} \tag{4.34}$$

$$= \frac{1}{2}\bar{C}_{\bar{i}}^{jk}\Big\{\Big(\partial_j\partial_k + (C_{jkm}S^{ml} - s^l_{jk})\partial_l + (4-2g)(h_{jk} - C_{jkl}S^l) \tag{4.35}$$

$$+(3-2g)K_j\,\partial_k + (3-2g)K_k\,\partial_j + (4-2g)(3-2g)K_kK_j\Big)\mathcal{F}^{(g-1)} +$$

$$+\sum_{h=1}^{g-1}\partial_j\mathcal{F}^{(g-h)}\partial_k\mathcal{F}^{(h)} + (2-2h)K_j\mathcal{F}^{(h)}\partial_k\mathcal{F}^{(g-h)}$$

$$+(2-2g+2h)K_k\mathcal{F}^{(g-h)}\partial_j\mathcal{F}^{(h)}$$

$$+K_jK_k(2-2h)(2-2g+2h)\mathcal{F}^{(h)}\,\mathcal{F}^{(g-h)}\Big\}.$$

4.5 Using the polynomial structure

4.5.1 Constructing the generators

The construction of the generators of the polynomial construction has been discussed in [26]. The starting point is to pick a local coordinate z_* on the moduli space such that C_{*ij} is an invertible $n \times n$ matrix in order to rewrite Eq. (4.31) as

$$S^{ij} = (C_*^{-1})^{ik}\left(\delta^j_* K_k + \delta^j_k K_* - \Gamma^j_{*k} + s^j_{*k}\right). \tag{4.36}$$

The freedom in Eq. (4.33) can be used to choose some of the s^k_{ij} [26]. The other generators are then constructed using the equations (4.32) [26]:

$$S^i = \frac{1}{2}\left(\partial_i S^{ii} - C_{imn}S^{mi}S^{ni} + 2s^i_{im}S^{mi} - h^{ii}_i\right), \tag{4.37}$$

$$S = \frac{1}{2}\left(\partial_i S^i - C_{imn}S^m S^{ni} + s^i_{im}S^m + h_{im}S^{mi} - h^i_i\right). \tag{4.38}$$

In both equations the index i is fixed, i.e., there is no summation over that index. The freedom in adding holomorphic functions to the generators of Eq. (4.33) can again be used to make some choice for the functions h^{ii}_i, h^i_i, the other ones are fixed by this choice and can be computed from Eq. (4.32).

4.5.2 Boundary conditions

Genus 1

The holomorphic anomaly equation at genus 1 (4.26) can be integrated to give:

$$\mathcal{F}^{(1)} = \frac{1}{2}\left(3 + n - \frac{\chi}{12}\right)K + \frac{1}{2}\log\det G^{-1} + \sum_i s_i \log z_i + \sum_a r_a \log\Delta_a\,, \tag{4.39}$$

where $i = 1, \ldots, n = h^{1,1}(Z) = h^{2,1}(Z^*)$, and a runs over the number of discriminant components. The coefficients s_i and r_a are fixed by the leading singular behavior of $\mathcal{F}^{(1)}$ which is given by [19]

$$\mathcal{F}^{(1)} \sim -\frac{1}{24}\sum_i \log z_i \int_Z c_2 J_i\,, \tag{4.40}$$

for the algebraic coordinates z_i, for a discriminant Δ corresponding to a conifold singularity the leading behavior is given by

$$\mathcal{F}^{(1)} \sim -\frac{1}{12}\log\Delta\,. \tag{4.41}$$

Higher genus boundary conditions

The holomorphic ambiguity needed to reconstruct the full topological string amplitudes can be fixed by imposing various boundary conditions for $\mathcal{F}^{(g)}$ at the boundary of the moduli space.

The large complex structure limit

The leading behavior of $\mathcal{F}^{(g)}$ at this point (which is mirror to the large volume limit) was computed in [13, 14, 18, 19, 83, 84]. In particular the contribution from constant maps is

$$\mathcal{F}^{(g)}|_{q_a=0} = (-1)^g \frac{\chi}{2} \frac{|B_{2g}B_{2g-2}|}{2g\,(2g-2)\,(2g-2)!}\,, \quad g>1, \tag{4.42}$$

where q_a denote the exponentiated mirror map at this point.

Conifold-like loci

The leading singular behavior of the partition function $\mathcal{F}^{(g)}$ at a conifold locus has been determined in [13, 14, 18, 19, 85, 86]

$$\mathcal{F}^{(g)}(t_c) = b\frac{B_{2g}}{2g(2g-2)t_c^{2g-2}} + O(t_c^0), \qquad g>1\,, \tag{4.43}$$

Here $t_c \sim \Delta^{\frac{1}{m}}$ is the flat coordinate at the discriminant locus $\Delta = 0$. For a conifold singularity $b = 1$ and $m = 1$. In particular the leading singularity in (4.43) as well as the absence of subleading singular terms follows from the Schwinger loop computation of [13, 14], which computes the effect of the extra massless hypermultiplet in the space-time theory [87]. The singular structure and the "gap" of subleading singular terms have been also observed in the dual matrix model [88] and were first used in [89, 90] to fix the holomorphic ambiguity at higher genus. The space-time derivation of [13, 14] is not restricted to the conifold case and applies also to the case $m > 1$ singularities which give rise to a different spectrum of extra massless vector and hypermultiplets in space-time. The coefficient of the Schwinger loop integral is a weighted trace over the spin of the particles [86, 87] leading to the prediction $b = n_H - n_V$ for the coefficient of the leading singular term. The appearance of the prefactor b in the case $m > 1$ has been discussed in [26] for the case of the local $\mathbb{F}_2$ (see also [91]).

The holomorphic ambiguity

The singular behavior of $\mathcal{F}^{(g)}$ is taken into account by the local ansatz

$$\text{hol.ambiguity} \sim \frac{p(\tilde{z}_i)}{\Delta^{(2g-2)}}, \tag{4.44}$$

for the holomorphic ambiguity near $\Delta = 0$, where $p(\tilde{z}_i)$ is a priori a series in the local coordinates $\tilde{z}_i$ near the singularity. Patching together the local information at all the singularities with the boundary divisors with finite monodromy, it follows however that the numerator $p(z_i)$ is generically a polynomial of low degree in the z_i. Here z_i denote the natural coordinates centered at large complex structure, $z_i = 0\ \forall i$. The finite number of coefficients in $p(z_i)$ is constrained by (4.43). The polynomial structure of higher genus amplitudes in combination with the boundary conditions are enough to solve topological string theory perturbatively on local CY manifolds (See for example [26, 92]) and they permit progress in solving for higher genus amplitudes on compact CY manifolds [28, 90, 91].

4.5.3 Example: the quintic

Choice of generators

The choice of the polynomial generators is fixed such that the holomorphic functions appearing in Eqs. (4.31,4.32) are rational expressions in terms of z, the coordinate in the large complex structure patch of the moduli space. For the holomorphic functions in the following lower/upper indices are multiplied/divided by z

$$A_i^j \to \frac{z_i}{z_j} A_i^j .$$

With this convention, all the holomorphic functions appearing in the setup of the polynomial construction can be expressed in terms of polynomials in the local coordinates (See [21, 22, 25]). The holomorphic functions are chosen as in [25]:

$$s_{zz}^z = -\frac{8}{5}, \quad h_z^{zz} = \frac{1}{25}, \quad h_z^z = -\frac{1}{125}, \quad h_z = \frac{2}{3125}, \quad h_{zz} = \frac{2}{25}, \tag{4.45}$$

Genus 1

Using the boundary conditions for genus 1, the equation for $\mathcal{F}^1$ (4.39) can be fixed with:

$$n = 1\,, \quad \chi = -200\,, \quad s = -\frac{31}{12}\,, \quad r = -\frac{1}{12}\,, \tag{4.46}$$

and the starting polynomial expression is fixed to be (with $P = \Delta^{-1}$):

$$z\partial_z \mathcal{F}^{(1)} = \frac{1}{60}\left(-112 + 560 K_z + 5P(1 + 30 S^{zz})\right)\,. \tag{4.47}$$

Higher genus

Using the holomorphic anomaly equation in its polynomial form (4.34,4.36) and the boundary conditions, the higher genus amplitudes can be fixed. For example at genus 2 with the given choice of generators this gives:

$$\begin{aligned}\mathcal{F}^{(2)} = \frac{1}{36000}\Big(&5P^2\left(37500(S^{zz})^3 + 6000(S^{zz})^2 + 325\,S^{zz} + 6\right) \\ &-P\left(25000\,S^z(30\,S^{zz}+1) + 150000(S^{zz})^2 + 4400\,S^{zz} - 23\right) \\ &+16\left(175000\,S + 35000\,S^z + 3500\,S^{zz} - 39\right)\Big)\,.\end{aligned} \tag{4.48}$$

Using the polynomial generators of [21], the boundary conditions reviewed earlier as well as additional counting techniques for invariants as in [93], it was possible in [90] to solve the quintic to very high genus.

Acknowledgments

I would like to thank the Mathematical Sciences Center at Tsinghua University and Prof. Shing-Tung Yau for hospitality and for the invitation to give these lectures. I would also like to thank Michael Hecht, Albrecht Klemm, Dominique Länge, Peter Mayr, Emanuel Scheidegger, Shing-Tung Yau and Jie Zhou for discussions and collaborations on some of the topics covered in these lectures and related subjects. I want furthermore to thank Emanuel Scheidegger and Jie Zhou for comments on the manuscript. This work is supported by the DFG fellowship AL 1407/1-1.

References

[1] N. Seiberg and E. Witten, Electric - magnetic duality, monopole condensation, and confinement in $N = 2$ supersymmetric Yang-Mills theory, *Nucl. Phys.* **B426** (1994) 19–52, `arXiv:hep-th/9407087 [hep-th]`.

[2] A. Klemm, W. Lerche, P. Mayr, C. Vafa, and N. P. Warner, Selfdual strings and N=2 supersymmetric field theory, *Nucl. Phys.* **B477** (1996) 746–766, `arXiv:hep-th/9604034 [hep-th]`.

[3] W. Lerche, Introduction to Seiberg-Witten theory and its stringy origin, *Nucl. Phys. Proc. Suppl.* **55B** (1997) 83–117, `arXiv:hep-th/9611190 [hep-th]`.

[4] L. J. Dixon, Some world-sheet properties of superstring compactifications, on orbifolds and otherwise, in: Superstrings, unified theories and cosmology 1987 (Trieste, 1987), vol. 4 of *ICTP Ser. Theoret. Phys.*, 67–126. World Sci. Publ., Teaneck, NJ, 1988.

[5] W. Lerche, C. Vafa, and N. P. Warner, Chiral rings in $N = 2$ superconformal theories, *Nucl. Phys.* **B324** (1989) 427.

[6] E. Witten, Mirror manifolds and topological field theory, `arXiv:hep-th/9112056 [hep-th]`.

[7] S.-T. Yau, ed., Mirror symmetry. I, vol. 9 of *AMS/IP Studies in Advanced Mathematics*. American Mathematical Society, Providence, RI, 1998. Revised reprint of ıt Essays on mirror manifolds [Internat. Press, Hong Kong, 1992; MR1191418 (94b:32001)].

[8] D. A. Cox and S. Katz, Mirror symmetry and algebraic geometry, vol. 68 of *Mathematical Surveys and Monographs*. American Mathematical Society, Providence, RI, 1999.

[9] K. Hori, S. Katz, A. Klemm, R. Pandharipande, R. Thomas, C. Vafa, R. Vakil, and E. Zaslow, Mirror symmetry, vol. 1 of *Clay Mathematics Monographs*. American Mathematical Society, Providence, RI, 2003. With a preface by Vafa.

[10] P. S. Aspinwall, T. Bridgeland, A. Craw, M. R. Douglas, M. Gross, A. Kapustin, G. W. Moore, G. Segal, B. Szendrői, and P. M. H. Wilson, Dirichlet branes and mirror symmetry, vol. 4 of *Clay Mathematics Monographs*. American Mathematical Society, Providence, RI, 2009.

[11] B. R. Greene and M. Plesser, Duality in Calabi-Yau moduli space, *Nucl. Phys.* **B338** (1990) 15–37.

[12] P. Candelas, X. C. De La Ossa, P. S. Green, and L. Parkes, A Pair of Calabi-Yau manifolds as an exactly soluble superconformal theory, *Nucl. Phys.* **B359** (1991) 21–74.

[13] R. Gopakumar and C. Vafa, M theory and topological strings. 1., `arXiv:hep-th/9809187 [hep-th]`.

[14] R. Gopakumar and C. Vafa, M theory and topological strings. 2., `arXiv:hep-th/9812127 [hep-th]`.

[15] S. H. Katz, A. Klemm, and C. Vafa, Geometric engineering of quantum field theories, *Nucl. Phys.* **B497** (1997) 173–195, `arXiv:hep-th/9609239 [hep-th]`.

[16] S. H. Katz, P. Mayr, and C. Vafa, Mirror symmetry and exact solution of 4-D $N = 2$ gauge theories: 1., *Adv. Theor. Math. Phys.* **1** (1998) 53–114, `arXiv:hep-th/9706110 [hep-th]`.

[17] P. Mayr, Geometric construction of $N = 2$ gauge theories, *Fortsch. Phys.*

47 (1999) 39–63, `arXiv:hep-th/9807096 [hep-th]`.

[18] M. Bershadsky, S. Cecotti, H. Ooguri, and C. Vafa, Holomorphic anomalies in topological field theories, *Nucl. Phys.* **B405** (1993) 279–304, `arXiv:hep-th/9302103 [hep-th]`.

[19] M. Bershadsky, S. Cecotti, H. Ooguri, and C. Vafa, Kodaira-Spencer theory of gravity and exact results for quantum string amplitudes, *Commun. Math. Phys.* **165** (1994) 311–428, `arXiv:hep-th/9309140 [hep-th]`.

[20] E. Witten, Quantum background independence in string theory, `arXiv:hep-th/9306122 [hep-th]`.

[21] S. Yamaguchi and S.-T. Yau, Topological string partition functions as polynomials, *JHEP* **0407** (2004) 047, `arXiv:hep-th/0406078 [hep-th]`.

[22] M. Alim and J. D. Länge, Polynomial structure of the (Open) topological string partition function, *JHEP* **0710** (2007) 045, `arXiv:0708.2886 [hep-th]`.

[23] M. Aganagic, V. Bouchard, and A. Klemm, Topological strings and (almost) modular forms, *Commun. Math. Phys.* **277** (2008) 771–819, `arXiv:hep-th/0607100 [hep-th]`.

[24] T. W. Grimm, A. Klemm, M. Marino, and M. Weiss, Direct integration of the topological string, *JHEP* **0708** (2007) 058, `arXiv:hep-th/0702187 [HEP-TH]`.

[25] S. Hosono, BCOV ring and holomorphic anomaly equation, `arXiv:0810.4795 [math.AG]`.

[26] M. Alim, J. D. Länge, and P. Mayr, Global properties of topological string amplitudes and Orbifold Invariants, *JHEP* **1003** (2010) 113, `arXiv:0809.4253 [hep-th]`.

[27] M. Alim, *Mirror symmetry, toric branes and topological string amplitudes as polynomials.* PhD Thesis. Ludwig-Maximilians-University Library, Munich, 2009. `http://edoc.ub.uni-muenchen.de/10341/`.

[28] M. Alim and E. Scheidegger, Topological Strings on Elliptic Fibrations, `arXiv:1205.1784 [hep-th]`.

[29] B. R. Greene, String theory on Calabi-Yau manifolds, `arXiv:hep-th/9702155 [hep-th]`.

[30] N. Warner, $N = 2$ supersymmetric integrable models and topological field theories, `arXiv:hep-th/9301088 [hep-th]`.

[31] P. S. Aspinwall, D-branes on Calabi-Yau manifolds, `arXiv:hep-th/0403166 [hep-th]`.

[32] P. Fre and P. Soriani, *The $N = 2$ Wonderland: From Calabi-Yau Manifolds to Topological Field Theories*, World Scientific, Singapore, 1995.

[33] R. Dijkgraaf, H. Verlinde, and E. Verlinde, Notes on topological string theory and 2D quantum gravity, in: *String Theory and Quantum Gravity (Trieste, 1990)*, 91–156. World Scientific, River Edge, NJ, 1991.

[34] J. Walcher, Extended holomorphic anomaly and loop amplitudes in open

topological string, *Nucl. Phys.* **B817** (2009) 167–207, arXiv:0705.4098 [hep-th].

[35] E. Witten, Topological Sigma models, *Commun. Math. Phys.* **118** (1988) 411.

[36] M. Vonk, A Mini-course on topological strings, arXiv:hep-th/0504147 [hep-th].

[37] M. Marino, Chern-Simons theory and topological strings, *Rev. Mod. Phys.* **77** (2005) 675–720, arXiv:hep-th/0406005 [hep-th].

[38] M. Marino, Les Houches lectures on matrix models and topological strings, arXiv:hep-th/0410165 [hep-th].

[39] A. Neitzke and C. Vafa, Topological strings and their physical applications, arXiv:hep-th/0410178 [hep-th].

[40] M. Mariño, Chern-Simons theory, matrix models and topological strings, vol. 131 of *International Series of Monographs on Physics*. The Clarendon Press Oxford University Press, Oxford, 2005. http://dx.doi.org/10.1093/acprof:oso/9780198568490.001.0001.

[41] B. Pioline, Lectures on black holes, topological strings and quantum attractors (2.0), *Lect. Notes Phys.* **755** (2008) 283–373.

[42] E. Witten, Phases of $N = 2$ theories in two-dimensions, *Nucl. Phys.* **B403** (1993) 159–222, arXiv:hep-th/9301042 [hep-th].

[43] P. S. Aspinwall, B. R. Greene, and D. R. Morrison, Calabi-Yau moduli space, mirror manifolds and space-time topology change in string theory, *Nucl. Phys.* **B416** (1994) 414–480, arXiv:hep-th/9309097 [hep-th].

[44] P. S. Aspinwall, B. R. Greene, and D. R. Morrison, Measuring small distances in $N = 2$ sigma models, *Nucl. Phys.* **B420** (1994) 184–242, arXiv:hep-th/9311042 [hep-th].

[45] C. Voisin, Mirror symmetry, vol. 1 of *SMF/AMS Texts and Monographs*. American Mathematical Society, Providence, RI, 1999. Translated from the 1996 French original by Roger Cooke.

[46] V. V. Batyrev and D. van Straten, Generalized hypergeometric functions and rational curves on Calabi-Yau complete intersections in toric varieties, *Commun. Math. Phys.* **168** (1995) 493–534, arXiv:alg-geom/9307010 [alg-geom].

[47] V. Bouchard and R. Cavalieri, On the mathematics and physics of high genus invariants of [C**3 / Z(3)], *Adv. Theor. Math. Phys.* **13** (2009) 695–719, arXiv:0709.3805 [math.AG].

[48] M. Gross, Calabi-Yau manifolds and mirror symmetry, in: Calabi-Yau manifolds and related geometries (Nordfjordeid, 2001), *Universitext*, 69–159. Springer, Berlin, 2003. http://dx.doi.org/10.1007/978-3-642-19004-9_2.

[49] C. Voisin, Hodge theory and complex algebraic geometry. I, vol. 76 of *Cambridge Studies in Advanced Mathematics*. Cambridge University Press, Cam-

bridge, 2002. http://dx.doi.org/10.1017/CBO9780511615344. Translated from the French original by Leila Schneps.

[50] W. Lerche, D. Smit, and N. Warner, Differential equations for periods and flat coordinates in two-dimensional topological matter theories, *Nucl. Phys.* **B372** (1992) 87–112, arXiv:hep-th/9108013 [hep-th].

[51] A. Ceresole, R. D'Auria, S. Ferrara, W. Lerche, J. Louis, and T. Regge, Picard-Fuchs equations, special geometry and target space duality, Contribution to second volume of *Essays on Mirror Manifolds*.

[52] B. de Wit and A. Van Proeyen, Potentials and symmetries of general gauged $N = 2$ supergravity: Yang-Mills Models, *Nucl. Phys.* **B245** (1984) 89.

[53] E. Cremmer, C. Kounnas, A. Van Proeyen, J. Derendinger, S. Ferrara, *et al.*, Vector multiplets coupled to $N = 2$ supergravity: SuperHiggs effect, flat potentials and geometric structure, *Nucl. Phys.* **B250** (1985) 385.

[54] B. de Wit, P. Lauwers, and A. Van Proeyen, Lagrangians of $N = 2$ supergravity – matter systems, *Nucl. Phys.* **B255** (1985) 569.

[55] P. Candelas and X. de la Ossa, Moduli space of Calabi-Yau manifolds, *Nucl. Phys.* **B355** (1991) 455–481.

[56] A. Strominger, Special geometry, *Commun. Math. Phys.* **133** (1990) 163–180.

[57] S. Ferrara and J. Louis, Flat holomorphic connections and Picard-Fuchs identities from $N = 2$ supergravity, *Phys. Lett.* **B278** (1992) 240–245, arXiv:hep-th/9112049 [hep-th].

[58] A. Ceresole, R. D'Auria, S. Ferrara, W. Lerche, and J. Louis, Picard-Fuchs equations and special geometry, *Int. J. Mod. Phys.* **A8** (1993) 79–114, arXiv:hep-th/9204035 [hep-th].

[59] D. S. Freed, Special Kähler manifolds, *Comm. Math. Phys.* **203** no. 1, (1999) 31–52. http://dx.doi.org/10.1007/s002200050604.

[60] S. Hosono, B. Lian, and S.-T. Yau, GKZ generalized hypergeometric systems in mirror symmetry of Calabi-Yau hypersurfaces, *Commun. Math. Phys.* **182** (1996) 535–578, arXiv:alg-geom/9511001 [alg-geom].

[61] S. Hosono and B. Lian, GKZ hypergeometric systems and applications to mirror symmetry, arXiv:hep-th/9602147 [hep-th].

[62] M. Noguchi, Mirror symmetry of Calabi-Yau manifolds and flat coordinates, *Int. J. Mod. Phys.* **A12** (1997) 4973–4996, arXiv:hep-th/9609163 [hep-th].

[63] T. Masuda and H. Suzuki, Prepotentials, bilinear forms on periods and enhanced gauge symmetries in Type II strings, *Int. J. Mod. Phys.* **A14** (1999) 1177–1204, arXiv:hep-th/9807062 [hep-th].

[64] A. Klemm, Topological string theory on Calabi-Yau threefolds, *PoS* **RTN2005** (2005) 002.

[65] V. V. Batyrev, Dual polyhedra and mirror symmetry for Calabi-Yau hypersurfaces in toric varieties, arXiv:alg-geom/9310003 [alg-geom].

[66] S. Hosono, A. Klemm, S. Theisen, and S.-T. Yau, Mirror symmetry, mirror map and applications to Calabi-Yau hypersurfaces, *Commun. Math. Phys.* **167** (1995) 301–350, arXiv:hep-th/9308122 [hep-th].
[67] K. Hori and C. Vafa, Mirror symmetry, arXiv:hep-th/0002222 [hep-th].
[68] I. M. Gel'fand, A. V. Zelevinskiĭ, and M. M. Kapranov, Hypergeometric functions and toric varieties, *Funktsional. Anal. i Prilozhen.* **23** no. 2, (1989) 12–26. http://dx.doi.org/10.1007/BF01078777.
[69] P. S. Aspinwall, The Moduli space of $N = 2$ superconformal field theories, arXiv:hep-th/9412115 [hep-th].
[70] S. Cecotti and C. Vafa, Topological anti topological fusion, *Nucl. Phys.* **B367** (1991) 359–461.
[71] A. Ceresole, R. D'Auria, and S. Ferrara, The Symplectic structure of $N = 2$ supergravity and its central extension, *Nucl. Phys. Proc. Suppl.* **46** (1996) 67–74, arXiv:hep-th/9509160 [hep-th].
[72] I. Antoniadis and S. Hohenegger, Topological amplitudes and physical couplings in string theory, *Nucl. Phys. Proc. Suppl.* **171** (2007) 176–195, arXiv:hep-th/0701290 [HEP-TH].
[73] H. Ooguri, A. Strominger, and C. Vafa, Black hole attractors and the topological string, *Phys. Rev.* **D70** (2004) 106007, arXiv:hep-th/0405146 [hep-th].
[74] R. E. Rudd, The String partition function for QCD on the torus, arXiv:hep-th/9407176 [hep-th].
[75] R. Dijkgraaf, Mirror symmetry and elliptic curves, in: The moduli space of curves (Texel Island, 1994), vol. 129 of *Progr. Math.*, 149–163. Birkhäuser Boston, Boston, MA, 1995.
[76] M. Kaneko and D. Zagier, A generalized Jacobi theta function and quasi-modular forms, in: The moduli space of curves (Texel Island, 1994), vol. 129 of *Progr. Math.*, 165–172. Birkhäuser Boston, Boston, MA, 1995.
[77] J. Minahan, D. Nemeschansky, C. Vafa, and N. Warner, E strings and $N = 4$ topological Yang-Mills theories, *Nucl. Phys.* **B527** (1998) 581–623, arXiv:hep-th/9802168 [hep-th].
[78] S. Hosono, M. Saito, and A. Takahashi, Holomorphic anomaly equation and BPS state counting of rational elliptic surface, *Adv. Theor. Math. Phys.* **3** (1999) 177–208, arXiv:hep-th/9901151 [hep-th].
[79] S. Hosono, Counting BPS states via holomorphic anomaly equations, *Fields Inst. Commun.* (2002) 57–86, arXiv:hep-th/0206206 [hep-th].
[80] A. Klemm, M. Kreuzer, E. Riegler, and E. Scheidegger, Topological string amplitudes, complete intersection Calabi-Yau spaces and threshold corrections, *JHEP* **0505** (2005) 023, arXiv:hep-th/0410018 [hep-th].
[81] A. Klemm, J. Manschot, and T. Wotschke, Quantum geometry of elliptic Calabi-Yau manifolds, arXiv:1205.1795 [hep-th].
[82] M. Gunaydin, A. Neitzke, and B. Pioline, Topological wave functions and

heat equations, *JHEP* **0612** (2006) 070, arXiv:hep-th/0607200 [hep-th].

[83] M. Marino and G. W. Moore, Counting higher genus curves in a Calabi-Yau manifold, *Nucl. Phys.* **B543** (1999) 592–614, arXiv:hep-th/9808131 [hep-th].

[84] C. Faber and R. Pandharipande, Hodge integrals and Gromov-Witten theory, *Inventiones Mathematicae* **139** (1999) 173–199, arXiv:98101173 [math.AG].

[85] D. Ghoshal and C. Vafa, $C = 1$ string as the topological theory of the conifold, *Nucl. Phys.* **B453** (1995) 121–128, arXiv:hep-th/9506122 [hep-th].

[86] I. Antoniadis, E. Gava, K. Narain, and T. Taylor, $N = 2$ type II heterotic duality and higher derivative F terms, *Nucl. Phys.* **B455** (1995) 109–130, arXiv:hep-th/9507115 [hep-th].

[87] C. Vafa, A Stringy test of the fate of the conifold, *Nucl. Phys.* **B447** (1995) 252–260, arXiv:hep-th/9505023 [hep-th].

[88] M. Aganagic, A. Klemm, M. Marino, and C. Vafa, Matrix model as a mirror of Chern-Simons theory, *JHEP* **0402** (2004) 010, arXiv:hep-th/0211098 [hep-th].

[89] M.-x. Huang and A. Klemm, Holomorphic Anomaly in Gauge Theories and Matrix Models, *JHEP* **0709** (2007) 054, arXiv:hep-th/0605195 [hep-th].

[90] M.-x. Huang, A. Klemm, and S. Quackenbush, Topological string theory on compact Calabi-Yau: Modularity and boundary conditions, *Lect. Notes Phys.* **757** (2009) 45–102, arXiv:hep-th/0612125 [hep-th].

[91] B. Haghighat and A. Klemm, Solving the topological string on K3 fibrations, *JHEP* **1001** (2010) 009, arXiv:0908.0336 [hep-th]. With an appendix by Sheldon Katz.

[92] B. Haghighat, A. Klemm, and M. Rauch, Integrability of the holomorphic anomaly equations, *JHEP* **0810** (2008) 097, arXiv:0809.1674 [hep-th].

[93] S. H. Katz, A. Klemm, and C. Vafa, M theory, topological strings and spinning black holes, *Adv. Theor. Math. Phys.* **3** (1999) 1445–1537, arXiv:hep-th/9910181 [hep-th].

Open Problems and Surveys of Contemporary Mathematics
SMM 6, pp. 45–97

Introduction to Generalized Geometry and T-Duality

David Baraglia*

Abstract

These notes are an introduction to generalized geometry and T-duality. Both subjects arise in the study of the topology and geometry of string theory compactifications. The theory of bundle gerbes is reviewed and is seen to be a common theme underlying both generalized geometry and T-duality. In bringing together generalized complex geometry and T-duality we arrive at the interchange of complex and symplectic structures characteristic of mirror symmetry.

2000 Mathematics Subject Classification: 53D18, 53C08, 19L50, 81T30

Keywords and Phrases: generalized complex structure, generalized geometry, T-duality, gerbes, twisted K-theory, string theory compactifications

Contents

*School of Mathematics Science, The University of Adelaide, Australia.
Email: david.baraglia@adelaide.edu.au

1 Introduction

These notes are an introduction to generalized geometry and T-duality. The underlying theme which unites these topics and motivates their study is that they arise in the geometry and topology of string theory compactifications. These notes do not assume familiarity with physics and we limit ourselves to occasional discussion of string theory whose purpose is to give some context for the mathematical ideas being explored. In short, the idea is that the physical properties of string theory compactified over a space X correspond to geometric and topological properties of X. However the topology and geometry of X that the string theory "sees" is of an exotic or "stringy" nature. Generalized geometry and T-duality are two interconnected topics within the study of the stringy geometry of spaces.

Aside from the obvious goal of presenting a readable introduction to these topics, we aim to give an indication of how they fit in to an even greater picture which incorporates amongst other things, mirror symmetry. By the end of these notes the reader will see that when generalized geometry and T-duality are brought together we can begin to understand the mysterious interchange of complex and symplectic geometry which lies at the heart of mirror symmetry. The generalized geometry approach to T-duality was developed by Cavalcanti and Gualtieri in [26] following on from [25, 38]. We follow closely this approach to generalized geometry and T-duality. We also consider some of the more topological aspects of T-duality; gerbes, twisted K-theory and topological T-duality following the paper of Bouwknegt, Evslin and Mathai [12].

In Section 1.1 we will give an overview of the idea of string theory compactifications and how it leads to interesting geometry and topology. Following this the notes are divided into four sections which we briefly describe. Section 2 is an introduction to bundle gerbes, their connections and curvature and the relation to the string theory B-field. Gerbes are a hallmark of stringy geometry exhibited in both generalized geometry and T-duality, so they are a crucial part of these notes. Section 3 is on generalized geometry and in particular generalized complex structures which are mixture of complex and symplectic geometry. We also cover twisting by gerbes, twisted cohomology, abstract Courant algebroids and generalized metrics. T-duality is studied in Section 4 starting from the Buscher rules derived in physics, moving to topological T-duality for circle bundles and generalizing to principal torus bundles. We give an introduction to twisted K-theory and show how T-duality gives isomorphisms in twisted K-theory, as well as twisted cohomology. Finally in Section 5 we show how generalized geometry and T-duality can be combined to give a T-duality transformation for generalized complex structures, and argue that this is a simplified version of mirror symmetry.

1.1 String theory compactifications and twisted geometry

The main idea that underlies these notes is that the physical properties of string theories compactified over a space X correspond to interesting and exotic geometric structures on X. The mathematics is self-contained however and we only discuss aspects of string theory to provide some perspective. For more on the mathematics of string theory the reader may consult [31, 42].

For the sake of discussion we will consider only the type II superstring theories throughout the text. Type II superstring theory features a ten-dimensional space-time X_{10} in which the strings propagate. In order to make contact with the four dimensional space-time of our perceptions one considers *compactification*, meaning that we take X_{10} to have the form

$$X_{10} = M_4 \times X_6,$$

where M_4 is four dimensional Minkowski space-time and X_6 is a compact 6-manifold, which is supposed to be very small so that we can not perceive these extra dimensions. The ten dimensional string theory is now approximated by a four dimensional effective theory that lives on M_4. Most importantly, the details of this four dimensional effective theory (field content, coupling constants, etc) are dictated by the topology and geometry of the compactified dimensions X_6. What is exciting about this is that the string theory depends on the geometry and topology of X_6 in a rather exotic way. We may term this the "stringy topology" or "stringy geometry" of X_6. The ideas of stringy topology and geometry that we will consider are not limited to 6-manifolds. They can even be applied to spaces more general than manifolds such as orbifolds and orientifolds. In these notes we will limit ourselves mostly to manifolds however.

A prominent feature of stringy topology and geometry which will be important throughout these notes is the presence of *gerbes*, which we introduce in Section 2. In string theory there is a certain field which goes by a number of names: the Kalb-Ramond field, Neveu-Schwarz field, or simply the B-field. In much the same way that charged particles interact with an electromagnetic field, the strings of string theory are charged with respect to the Neveu-Schwarz field. One way to represent this mathematically is as a bundle gerbe with connection, the subject matter of Section 2. We will adopt this point of view and thus treat the B-field as a gerbe with connection.

For this reason we consider spaces X equipped with a gerbe $\mathcal{G}$. In the presence of a gerbe certain topological invariants of X become "twisted". Instead of the de-Rham cohomology $H^*(X, \mathbb{R})$ of X we will be lead to consider *twisted (de Rham) cohomology* $H^*(X, \mathcal{G})$ of $(X, \mathcal{G})$. Instead of the K-theory $K^*(X)$ of X we will be lead to consider the *twisted K theory* $K^*(X, \mathcal{G})$ of $(X, \mathcal{G})$. The presence of gerbes is an absolutely essential ingredient for T-duality as we will see in Section 4. The underlying idea behind T-duality is that two different spaces with gerbes $(X, \mathcal{G})$,$(\hat{X}, \hat{\mathcal{G}})$ can produce the same string theory and thus have the same stringy topology and geometry. For example T-dual pairs $(X, \mathcal{G})$,$(\hat{X}, \hat{\mathcal{G}})$ have isomorphic twisted cohomology and twisted K-theory. This phenomenon can happen even when X and $\hat{X}$ are topologically distinct. This is similar to the phenomenon of mirror symmetry in which distinct Calabi-Yau manifolds are thought to occur in mirror pairs $X, \hat{X}$ such that invariants on one side correspond to different invariants on the other side. In particular invariants related to the symplectic geometry on one side should correspond to invariants of the complex geometry on the mirror. The famous Strominger-Yau-Zaslow (SYZ) conjecture [61] puts forth the claim that mirror symmetry is really a manifestation of T-duality.

A connection on a gerbe allows one to do differential geometry in the presence of a gerbe. More specifically we will see in Section 3 that generalized geometry serves as a setting for differential geometry which incorporates gerbes with connection. One of the most exciting aspects of generalized geometry is that of *generalized complex structures.* These structures are a mixture of complex and symplectic geometry which already is intriguing since an interchange of complex and symplectic geometry is a hallmark of mirror symmetry. There are other indications that generalized complex manifolds are deeply interconnected with mirror symmetry. One of these is that T-duality can be applied to generalized complex manifolds and under the right circumstances T-duality exchanges complex and symplectic geometries in exactly the way that one would expect under the SYZ conjecture. We will see how this occurs in Section 5.

2 Bundle gerbes and the B-field

Gerbes were originally introduced by Giraud [36] in the pursuit of non-abelian cohomology of degree 2 and further popularized by Brylinski [15]. Roughly speaking gerbes are defined as certain sheaves of groupoids, but rather than delve into the quite complicated specifics of this, we will present abelian gerbes in the form of *bundle gerbes*, introduced by Murray [49]. Not only are bundle gerbes much easier to define, they are also more suited to differential geometry. In Section 2.3 we will offer an explanation for the prevalence of gerbes in string theory.

2.1 Bundle gerbes

Throughout this section we work with paracompact topological spaces for simplicity. References for this section are the original paper [49] on bundle gerbes as well as [60], [11].

To define bundle gerbes we first define a notion of a cover.

Definition 2.1. Let X be a topological space. Then if $\pi : Y \to X$ is a map from another topological space Y to X we say that (Y, π) is a *cover* of X if π admits local sections. By this we mean that for each $x \in X$ there is an open neighborhood $x \in U \subseteq X$ such that $\pi|_{\pi^{-1}(U)} : \pi^{-1}(U) \to U$ admits a (continuous) section.

Example 2.1. To justify the above definition we show that an open cover is a cover. Let $\{U_\alpha\}$ be an open cover of X. We take Y to be the disjoint union $Y = \coprod_\alpha U_\alpha$ and $\pi : Y \to X$ to be the inclusion of each factor $U_\alpha \to X$. Then (Y, π) is a cover.

If $f : Y \to X$ and $r : Z \to Y$ are covers then so is $f \circ r : Z \to X$. The covers of X form a category as follows. Let $(Y, f), (Z, g)$ be covers of X. A morphism $r : (Z, g) \to (Y, f)$ is a map $r : Z \to Y$ such that $g = f \circ r$ and such that $r : Z \to Y$ admits local sections (i.e. r is itself a cover). A *refinement* of a cover (Y, f) is a cover (Z, g) and morphism $r : (Z, g) \to (Y, f)$. Similarly if we are given covers $(Y, f), (Y', f')$, a *common refinement* is a cover (Z, g) and morphisms $r : (Z, g) \to (Y, f)$, $r' : (Z, g) \to (Y', f')$. Observe that if $(Y, f), (Y', f)$ are covers then so is the fiber product $(Y \times_X Y', f \times_X f')$, which is then a common refinement. In fact common refinements of $(Y, f), (Y', f')$ are in bijection with refinements of $(Y \times_X Y', f \times_X f')$, or even more simply with maps $Z \to Y \times_X Y'$ admitting local sections.

If (Y, f) is a cover of X we write $Y^{[k]}$ for the k-fold fiber product. We have an associated topological groupoid with objects Y and morphisms $Y^{[2]}$. Two pairs $(x, y), (y', z) \in Y^{[2]}$ are taken to be composable if and only if $y = y'$ and their composition is taken to be $(x, y)(y, z) = (x, z)$. A refinement $r : Z \to Y$ induces maps $r^{[k]} : Z^{[k]} \to Y^{[k]}$ as well as a functor between associated groupoids (in fact an equivalence of groupoids).

Let us introduce some notation. For $k = 0, 1, \ldots, (n-1)$, let $\partial_k : Y^{[n]} \to Y^{[n-1]}$ denote the map that omits the $(k+1)$-th factor. If $L \to Y^{[k]}$ is a line bundle or more generally a fiber bundle over $Y^{[k]}$ we write $L_{x_1, x_2, \ldots, x_k}$ for the fiber of L over $(x_1, x_2, \ldots, x_k) \in Y^{[k]}$.

Definition 2.2. A *bundle gerbe* on X consists of a cover (Y, f) on X, a Hermitian line bundle $L \to Y^{[2]}$ on $Y^{[2]}$ and for each pair of composable pairs $(x, y), (y, z) \in Y^{[2]}$ an isomorphism $\theta_{x,y,z} : L_{x,y} \otimes L_{y,z} \to L_{x,z}$ of Hermitian vector spaces such that $\theta_{x,y,z}$ depends continuously on the triple $(x, y, z) \in Y^{[3]}$. We will sometimes think of the isomorphisms $\theta_{x,y,z}$ as a kind of "multiplication" since in addition it is required that θ satisfies the following associativity condition. For each quadruple $(x, y, z, w) \in Y^{[4]}$ the following diagram is required to be commutative:

$$\begin{array}{ccc} L_{x,y} \otimes L_{y,z} \otimes L_{z,w} & \xrightarrow{\theta_{x,y,z} \otimes Id} & L_{x,z} \otimes L_{z,w} \\ \Big\downarrow{\scriptstyle Id \otimes \theta_{y,z,w}} & & \Big\downarrow{\scriptstyle \theta_{x,z,w}} \\ L_{x,y} \otimes L_{y,w} & \xrightarrow{\theta_{x,y,w}} & L_{x,w} \end{array}$$

The bundle gerbe defined by the data $f : Y \to X$, $L \to Y^{[2]}$, θ will be denoted by (Y, f, L, θ) or simply (L, θ) if $f : Y \to X$ is understood. If $\mathcal{G} = (Y, f, L, \theta)$ then we say that $\mathcal{G}$ is defined with respect to Y.

Remark 2.1. In the above definition when we say $\theta_{x,y,z}$ is continuous in the triple (x, y, z), we really mean that $(x, y, z) \mapsto \theta_{x,y,z}$ defines a bundle isomorphism $\theta : \partial_2^* L \otimes \partial_0^* L \to \partial_1^* L$ over $Y^{[3]}$. In what follows we will see further occasions where a bundle map is most easily described in terms of the maps on individual fibers.

Continuity refers to the requirement that the fiberwise maps collectively define a continuous bundle mapping.

Remark 2.2. We will often refer to bundle gerbes simply as gerbes as these are the only kind of gerbe we will encounter in these notes.

Example 2.2. The simplest example of a bundle gerbe is the *trivial gerbe* defined with respect to an arbitrary cover $\pi : Y \to X$. Let $L = \mathbb{C} \times Y$ be the trivial line bundle on Y. Then the multiplication is a map $\theta_{x,y,z} : \mathbb{C} \otimes \mathbb{C} \to \mathbb{C}$ for each $(x, y, z) \in Y^{[3]}$. We take $\theta_{x,y,z}$ to be multiplication of complex numbers. The associativity condition on θ follows immediately since multiplication of complex numbers is associative.

Let $f : Y \to X$ be a cover of X. Suppose $\mathcal{G} = (L, \theta)$, $\mathcal{G}' = (L', \theta')$ are gerbes defined with respect to Y. By a *strict isomorphism* we mean an isomorphism $\phi : L \to L'$ of Hermitian line bundles that respects the gerbe products, that is for each $(x, y, z) \in Y^{[3]}$ we have a commutative diagram

$$\begin{array}{ccc} L_{x,y} \otimes L_{y,z} & \xrightarrow{\phi_{x,y} \otimes \phi_{y,z}} & L'_{x,y} \otimes L'_{y,z} \\ \downarrow{\scriptstyle \theta_{x,y,z}} & & \downarrow{\scriptstyle \theta'_{x,y,z}} \\ L_{x,z} & \xrightarrow{\phi_{x,z}} & L'_{x,z} \end{array}$$

The adjective strict is used here to distinguish from the notion of *stable isomorphism* defined below.

Let $M \to Y$ be a Hermitian line bundle on Y. We let $\delta(M) \to Y^{[2]}$ be the line bundle $\delta(M) = \partial_1^*(M) \otimes \partial_0^*(M^*)$ defined on $Y^{[2]}$. That is, for each $(x, y) \in Y^{[2]}$ we have $\delta(M)_{x,y} = M_x \otimes M_y^*$. We claim that $\delta(M)$ admits a natural gerbe product θ^M, indeed we define $\theta^M_{x,y,z} : \delta(M)_{x,y} \otimes \delta(M)_{y,z} \to \delta(M)_{x,z}$ to be the map

$$M_x \otimes M_y^* \otimes M_y \otimes M_z^* \to M_x \otimes M_z^*,$$

obtained by pairing M_y^* and M_y. One checks that θ^M satisfies the associativity condition so that $(\delta(M), \theta^M)$ is a gerbe defined with respect to Y. A gerbe of this form is called a *trivializable gerbe.*

Let $\mathcal{G} = (L, \theta)$, $\mathcal{G}' = (L', \theta')$ be gerbes defined with respect to a cover Y of X. We will define a gerbe $\mathcal{G} \otimes \mathcal{G}'$, the *product* of $\mathcal{G}$ and $\mathcal{G}'$. The underlying line bundle for $\mathcal{G} \otimes \mathcal{G}'$ is simply the tensor product $L \otimes L'$. The gerbe product $(L \otimes L')_{x,y} \otimes (L \otimes L')_{y,z} \to (L \otimes L')_{x,z}$ is given by the composition

$$\begin{array}{ccc} L_{x,y} \otimes L'_{x,y} \otimes L_{y,z} \otimes L'_{y,z} & \xrightarrow{Id \otimes S \otimes Id} & L_{x,y} \otimes L_{y,z} \otimes L'_{x,y} \otimes L'_{y,z} \\ & & \downarrow{\scriptstyle \theta_{x,y,z} \otimes \theta'_{x,y,z}} \\ & & L_{x,z} \otimes L'_{x,z} \end{array}$$

where $S : L'_{x,y} \otimes L_{y,z} \to L_{y,z} \otimes L'_{x,y}$ is the isomorphism that swaps factors. One checks easily that this satisfies the associativity condition and so defines a gerbe.

Let $\mathcal{G} = (Y, f, L, \theta)$ be a gerbe defined with respect to a cover (Y, f) of X. If $r : (Z, g) \to (Y, f)$ is a refinement of Y, then we define a pullback gerbe $r^*\mathcal{G}$ defined with respect to Z. The underlying line bundle is the pullback $r^*(L)$ and the gerbe product is similarly a pullback, that is if $(x, y, z) \in Z^{[3]}$ then $(r(x), r(y), r(z)) \in Y^{[3]}$ and we get $\theta_{r(x),r(y),r(z)} : r^*(L)_{x,y} \otimes r^*(L)_{y,z} \to r^*(L)_{x,z}$.

We can also introduce a notion of pullback that changes the base. Let $\phi : X' \to X$ be a map between spaces and $f : Y \to X$ a cover. Then $\phi^*(Y) = Y \times_X X'$ is a cover of X' and there is a map $r : \phi^*(Y) \to Y$ covering ϕ. Given a gerbe $\mathcal{G}$ defined with respect to Y we can now easily define a pullback $r^*(\mathcal{G})$ which is a gerbe defined with respect to $\phi^*(Y)$.

Definition 2.3. Let $\mathcal{G} = (Y, f, L, \theta)$, $\mathcal{G}' = (Y', f', L', \theta')$ be gerbes on X. A *stable isomorphism* $\mathcal{G} \to \mathcal{G}'$ consists of a common refinement $r : (Z, g) \to (Y, f)$, $r' : (Z, g) \to (Y', f')$, a line bundle $M \to Z$ and a strict isomorphism of gerbes $\phi : r^*(L) \otimes \delta(M) \to r'^*(L')$. Two gerbes $\mathcal{G}, \mathcal{G}'$ are said to be *stably isomorphic* if there exists a stable isomorphism $\mathcal{G} \to \mathcal{G}'$.

A gerbe stably isomorphic to a trivial gerbe is called *trivializable* or *stably trivial*.

Observe that given a gerbe $\mathcal{G}$ defined with respect to some cover Y of X and a refinement $r : Z \to Y$, the pullback $r^*(\mathcal{G})$ is stably isomorphic to $\mathcal{G}$. Thus if we are interested in the classification of gerbes up to stable isomorphism we may choose refinements as desired. In particular, since $f : Y \to X$ admits local sections we may choose an open cover $\{U_\alpha\}_{\alpha \in I}$ of X and local sections $s_\alpha : U_\alpha \to Y$. Setting $Z = \coprod_{\alpha \in I} U_\alpha$ we have a refinement $r : Z \to Y$ given by $r = \coprod_{\alpha \in I} s_\alpha$. Therefore in the classification of gerbes up to stable isomorphism it suffices to consider only those covers arising from open covers of X. If $Y = \coprod_\alpha U_\alpha$ then one sees that $Y^{[2]} = \coprod_{\alpha,\beta} U_{\alpha\beta}$, where as usual $U_{\alpha\beta} = U_\alpha \cap U_\beta$. A gerbe $\mathcal{G}$ defined with respect to Y is then a collection of line bundles $L_{\alpha\beta} \to U_{\alpha\beta}$ together with an associative multiplication $\theta_{\alpha\beta\gamma} : L_{\alpha\beta} \otimes L_{\beta\gamma} \to L_{\alpha\gamma}$.

By refining the cover if necessary we may assume there exists sections $\sigma_{\alpha\beta} : U_{\alpha\beta} \to L_{\alpha\beta}$ of unit norm. Define $g_{\alpha\beta\gamma} : U_{\alpha\beta\gamma} \to \mathrm{U}(1)$ by the relation $\theta_{\alpha\beta\gamma}(\sigma_{\alpha\beta} \otimes \sigma_{\beta\gamma}) = g_{\alpha\beta\gamma}\sigma_{\alpha\gamma}$. Then by the associativity condition on θ we see that $\{g_{\alpha\beta\gamma}\}$ is a Čech 2-cocycle with values in the sheaf $\mathcal{C}(\mathrm{U}(1))$ of continuous $\mathrm{U}(1)$-valued functions. One checks easily that the Čech cohomology class associated to $\{g_{\alpha\beta\gamma}\}$ is independent of the choice of refinements and local sections, so that we get a well defined class $[\mathcal{G}] \in H^2(X, \mathcal{C}(\mathrm{U}(1))) \simeq H^3(X, \mathbb{Z})$. The class $[\mathcal{G}]$ is called the *Dixmier-Douady class* of $\mathcal{G}$.

Example 2.3. We show that conversely every class in $H^3(X, \mathbb{Z})$ is the Dixmier-Douady class of a gerbe. Let $h \in H^3(X, \mathbb{Z})$ and let $\{g_{\alpha\beta\gamma}\}$ be an explicit 2-cocycle for h defined with respect to some open cover $\{U_\alpha\}$. Let $Y = \coprod_\alpha U_\alpha$ be the cover $Y \to X$ associated to the open cover $\{U_\alpha\}$. To define a line bundle $L \to Y^{[2]}$ we must define a line bundle $L_{\alpha\beta}$ for each pair (α, β) with $U_\alpha \cap U_\beta$ non-empty. We set $L_{\alpha\beta} = \mathbb{C}$, the trivial bundle. Finally we need to define a multiplication $\theta_{\alpha\beta\gamma} : \mathbb{C} \otimes \mathbb{C} \to \mathbb{C}$ for each triple intersection. These are defined by $\theta_{\alpha\beta\gamma}(z_1 \otimes z_2) = g_{\alpha\beta\gamma} z_1 z_2$. One sees immediately that this defines a bundle gerbe with Dixmier-Douady class h.

With a little more work we can show that two gerbes have the same Dixmier-Douady class if and only if they are stably isomorphic. From this and the above example we conclude that stable isomorphism classes of gerbes are in bijection with $H^3(X, \mathbb{Z})$.

Remark 2.3. In the same way that isomorphism classes of line bundles on X give a geometric interpretation of the group $H^2(X, \mathbb{Z})$, we can say that bundle gerbes give a geometric interpretation of the group $H^3(X, \mathbb{Z})$.

The multiplication operation on gerbes descends to isomorphism classes and gives $H^3(X, \mathbb{Z})$ an abelian group structure. One sees easily that this group structure coincides with the usual addition on $H^3(X, \mathbb{Z})$.

For the purposes of T-duality and twisted K-theory, the data defining a stable isomorphism carries too much information. The main point is that we do not want to know about the choice of common refinement Z used. Our next step therefore will be to partition stable isomorphisms into suitable equivalence classes.

Let $\mathcal{G} = (Y, f, L, \theta)$, $\mathcal{G}' = (Y', f', L', \theta')$ be stably isomorphic gerbes on X. Consider two stable isomorphisms. Thus we have two common refinements $r : Z \to Y$, $r' : Z \to Y'$, $s : W \to Y$, $s' : W \to Y'$, two line bundles $M \to Z$, $N \to W$ and two strict isomorphisms $\phi : r^*(L) \otimes \delta(M) \to r'^*(L')$ and $\psi : s^*(L) \otimes \delta(N) \to s'^*(L')$. To compare the two stable isomorphisms, choose a common refinement of common refinements, that is let V be a cover which is a refinement of Z and W, $a : V \to Z$, $b : V \to W$ and such that $r \circ a = s \circ b$, $r' \circ a = s' \circ b$. Such refinements are possible to find because the maps $Z \to Y \times_X Y'$ and $W \to Y \times_X Y'$ admit local sections. Indeed we could take V to be $\{(z, w) \in Z \times W \,|\, r(z) = s(w),\ r'(z) = s'(w)\}$. Let $c : V \to Y$ be the composition $c = r \circ a = s \circ b$ and let $c' : V \to Y'$ similarly be given by $c' = r' \circ a = s' \circ b$. We then have stable isomorphisms $a^*(\phi) : c^*(L) \otimes a^*(\delta(M)) \to c'^*(L')$ and $b^*(\psi) : c^*(L) \otimes b^*(\delta(N)) \to c'^*(L')$. Combining the two isomorphisms we get a strict isomorphism $a^*(\delta(M)) \to b^*(\delta(N))$. Alternatively, letting $D \to V$ be the line bundle $D = a^*(M) \otimes b^*(N^*)$ we have a strict isomorphism $\mu : \delta(D) \simeq 1$, where 1 denotes the trivial gerbe with respect to V, that is the gerbe consisting of the trivial line bundle $\mathbb{C} \to V$, and gerbe product simply given by multiplication of complex numbers. From the definition of strict isomorphism this means that for every pair $(x, y) \in V^{[2]}$ there is an isomorphism $\mu_{x,y} : D_y \to D_x$ (continuous in x, y) and satisfying the cocycle condition $\mu_{x,y} \circ \mu_{y,z} = \mu_{x,z}$. Thus μ is a descent isomorphism [15] and there is a uniquely determined line bundle $E \to X$ such that if $t : V \to X$ is the map from the cover V to X then $D \simeq t^*(E)$. It is not hard to see that E is independent of the choice of common refinement V.

We conclude that for any two stable isomorphisms $\alpha, \beta : \mathcal{G} \to \mathcal{G}'$, there is a uniquely determined line bundle $E \to X$ over X. We introduce an equivalence relation $\sim$ on stable isomorphisms, namely $\alpha \sim \beta$ if and only if the corresponding line bundle is trivial. For any two gerbes $\mathcal{G}, \mathcal{G}'$, we let $\mathrm{Hom}(\mathcal{G}, \mathcal{G}')$ denote the set of equivalence classes of stable isomorphisms $\mathcal{G} \to \mathcal{G}'$ (in particular $\mathrm{Hom}(\mathcal{G}, \mathcal{G}')$ is empty if $\mathcal{G}, \mathcal{G}'$ are not stably isomorphic). It is straightforward to see that equivalence classes of stable isomorphisms can be composed and that the composition defines a category $Grb(X)$ of gerbes on X. That is the objects of $Grb(X)$ are gerbes on X and the morphisms are equivalence classes of stable isomorphisms. We state here without proof some useful properties of $Grb(X)$. The proofs left as straightforward (though somewhat lengthy) exercises:

- The only morphisms in $Grb(X)$ are isomorphisms, so $Grb(X)$ is a groupoid.
- $Grb(X)$ is contravariant with respect to X in the sense that for any map $\phi : Y \to X$ there is a pullback functor $\phi^* : Grb(X) \to Grb(Y)$ and if $\psi : Z \to Y$ is another map then $(\phi \circ \psi)^* = \psi^* \circ \phi^*$.
- If $\mathcal{G}, \mathcal{G}' \in Grb(X)$ are isomorphic then $\mathrm{Hom}(\mathcal{G}, \mathcal{G}')$ is naturally a torsor for the group $H^2(X, \mathbb{Z})$ of isomorphism classes of Hermitian line bundles on X. In particular $\mathrm{Hom}(\mathcal{G}, \mathcal{G})$ is canonically the group of Hermitian line bundles on X.
- Think of $\mathrm{Hom}(\mathcal{G}, \mathcal{G}')$ and $\mathrm{Hom}(\mathcal{G}', \mathcal{G}'')$ as torsors for the group of line bundles. Then if $\alpha \in \mathrm{Hom}(\mathcal{G}, \mathcal{G}')$, $\beta \in \mathrm{Hom}(\mathcal{G}', \mathcal{G}'')$ and L, M line bundles, we have $(\beta + M) \circ (\alpha + L) = (\beta \circ \alpha) + (L \otimes M)$.

Finally we will describe a way that gerbes occur as an obstruction to a lifting problem, which is quite useful. For any oriented orthogonal vector bundle $V \to X$ on X we can define a gerbe $L(V)$ called the *lifting gerbe of* V [49, 60], which captures the failure of V to admit a Spin^c-structure. Let $P \to X$ be the $\mathrm{SO}(n)$-frame bundle of V, where n is the rank of V. There is a natural map $\phi : P^{[2]} \to \mathrm{SO}(n)$, defined by letting $\phi(pg, p) = g$ for any $p \in P$ and $g \in \mathrm{SO}(n)$. We can think of $\mathrm{Spin}^c(n) \to \mathrm{SO}(n)$ as a principal circle bundle over $\mathrm{SO}(n)$. Pulling back the associated Hermitian line bundle by ϕ yields a line bundle $L \to P^{[2]}$ which will define the lifting gerbe. The multiplication on L is induced by the group composition on $\mathrm{Spin}^c(n)$. One checks that this defines a gerbe $L(V)$, the lifting gerbe of V. The stable isomorphism class of $L(V)$ is given by $[L(V)] = W_3(V) \in H^3(X, \mathbb{Z})$, where $W_3(V)$ is the third integral Stiefel-Whitney class of V. Recall that $W_3(V)$ is defined by $W_3(V) = \beta(w_2(V))$, where $w_2(V)$ is the second Stiefel-Whitney class and β is the Bockstein homomorphism $\beta : H^2(X, \mathbb{Z}_2) \to H^3(X, \mathbb{Z})$. It can be shown that $W_3(V)$ is the obstruction to a $\mathrm{Spin}^c(n)$-structure on V, so $L(V)$ is trivializable if and only if V admits a $\mathrm{Spin}^c(n)$ structure. In fact the different trivializations of $L(V)$ can be identified with the different possible $\mathrm{Spin}^c(n)$-structures on V.

If V is an oriented rank n vector bundle without metric we can still define a lifting gerbe $L(V, h)$ by first choosing a metric h on V. Although $L(V, h)$ depends on h we have as a consequence of the contractibility of the space of metrics on V, that for any two metrics h_1, h_2 there exists a canonical isomorphism $L(V, h_1) \to L(V, h_2)$ as elements of $Grb(X)$. In this sense the lifting gerbe of V is essentially unique and we continue to denote it by $L(V)$.

2.2 Gerbe connections

In this section we study the notions of connections, curving and curvature for bundle gerbes. These are notions that arise in the interplay of gerbes and differential geometry. Thus we will assume here that all spaces are smooth manifolds. Moreover we assume that all covers $\pi : Y \to X$ are such that π is a surjective submersion. If $Y \to X$, $Y' \to X$ are two surjective submersions then the fiber product $Y \times_X Y'$ is a submanifold of $Y \times Y'$ and the projections $Y \times_X Y \to Y$,

$Y \times_X Y' \to Y'$ are surjective submersions as well. This also applies to the iterated fiber products $Y^{[k]}$ of Y and the projections $\partial_i : Y^{[k]} \to Y^{[k-1]}$ defined in Section 2.1. The whole of Section 2.1 carries over to the smooth setting, so for instance we now take our bundle gerbes to be smooth, meaning they consist of a smooth line bundle $L \to Y^{[2]}$ and the gerbe multiplication $\theta_{x,y,z} : L_{x,y} \otimes L_{y,z} \to L_{x,z}$ is smooth in (x, y, z).

Let $f : Y \to X$ be a surjective submersion. Define maps $\delta : \Omega^p(Y^{[k]}) \to \Omega^p(Y^{[k+1]})$ by

$$\delta = \sum_{i=0}^{k} (-1)^i \partial_i^*.$$

In [49] it is shown that the sequence

$$0 \to \Omega^p(X) \xrightarrow{f^*} \Omega^p(Y^{[1]}) \xrightarrow{\delta^*} \Omega^p(Y^{[2]}) \xrightarrow{\delta^*} \cdots \tag{2.1}$$

is exact for any p.

Definition 2.4. Let $\mathcal{G} = (Y, f, L, \theta)$ be a bundle gerbe defined with respect to Y. A *(gerbe) connection* on $\mathcal{G}$ is a unitary connection ∇ on L such that the gerbe product $\theta : \partial_2^* L \otimes \partial_0^* L \to \partial_1^* L$ preserves the induced connections.

Making use of (2.1), we find that every gerbe admits a gerbe connection. If ∇ is a connection for $\mathcal{G} = (Y, f, L, \theta)$ then the curvature of ∇ is a closed complex 2-form $F \in \Omega^2(Y^{[2]}) \otimes \mathbb{C}$. The fact that ∇ preserves θ ensures that $\delta(F) = 0$. Note also that since ∇ is unitary, $iF/2\pi$ is real.

Definition 2.5. If ∇ is a gerbe connection for the gerbe $\mathcal{G} = (Y, f, L, \theta)$ and $F \in \Omega^2(Y^{[2]}) \otimes \mathbb{C}$ the curvature, then a *curving* for $(\mathcal{G}, \nabla)$ is a real 2-form $B \in \Omega^2(Y)$ such that $\delta(B) = \frac{i}{2\pi} F$.

Note that $\delta(dB) = d\delta(B) = d(iF/2\pi) = 0$, so by (2.1) there exists a unique 3-form $H \in \Omega^3(X)$ such that $dB = f^*(H)$. We call H the *curvature* of the curving B.

Note that for every gerbe $\mathcal{G}$ with connection we can find a curving. Additionally if H is the curvature of such a curving then H is closed and the de Rham cohomology class $[H] \in H^3(X, \mathbb{R})$ is readily seen to be the image in real cohomology of the Dixmier-Douady class of $\mathcal{G}$.

Let $\mathcal{G} = (Y, f, L, \theta)$ be a gerbe, ∇ a connection on $\mathcal{G}$ and B a curving with curvature H. If $B_0 \in \Omega^2(X)$ is any real 2-form on X then $B + f^*(B_0)$ is also a curving. The curvature of $B + f^*(B_0)$ is $H + dB_0$. Such a change in curving will be called a *curving shift* by B_0 or just a *shift* by B_0. Note that any two curvings for $(\mathcal{G}, \nabla)$ are related by a *unique* shift.

Much of Section 2.1 can be adapted to the setting of bundle gerbes with connection and curving. Let us list the main points to be adapted:

- Just as we have a pullback operation on gerbes, we can just as easily pullback connections and curvings.

- There is a notion of strict isomorphism of gerbes with connections and curvings, namely a strict isomorphism of the underlying gerbes that preserves the connections and curvings.
- Given a line bundle $L \to Y$ with connection ∇, the trivializable gerbe $\delta(L)$ has natural choices for a connection and curving. The connection is just the induced connection on $\delta(L) = \partial_1^*(L) \otimes \partial_0^*(L^*)$, while the curving is $-iF/2\pi$, where $F \in \Omega^1(Y) \otimes \mathbb{C}$ is the curvature of L. Notice that the curvature of such a curving is zero.
- There is a natural way to define the product of two gerbes with connections and curvings.

There is also a notion of stable isomorphism for gerbes with connections and curving. One simply adapts Definition 2.3 to the present setting. Thus the line bundle $M \to Z$ in Definition 2.3 now becomes a line bundle with connection and the strict isomorphism $\phi : r^*(L) \otimes \delta(M) \to r'^*(L')$ becomes a strict isomorphism of gerbes with connections and curvings.

Now let us specialize to the case that our cover $Y \to X$ is of the form $Y = \coprod_\alpha U_\alpha$ for an open cover $\{U_\alpha\}$. A gerbe $\mathcal{G}$ defined with respect to the cover Y is a collection $\{L_{\alpha\beta}\}$ of line bundles defined on the double intersections together with the gerbe product $\theta_{\alpha\beta\gamma} : L_{\alpha\beta} \otimes L_{\beta\gamma} \to L_{\alpha\gamma}$ on triple intersections. After refining the cover if necessary we may assume that on each $L_{\alpha\beta}$ admits a unit norm section $\sigma_{\alpha\beta}$. Define $g_{\alpha\beta\gamma}$ by $\theta_{\alpha\beta\gamma}(\sigma_{\alpha\beta} \otimes \sigma_{\beta\gamma}) = g_{\alpha\beta\gamma}\sigma_{\alpha\gamma}$. We have seen that $\{g_{\alpha\beta\gamma}\}$ is a 2-cocycle representing the Dixmier-Douady class of $\mathcal{G}$. Now consider equipping $\mathcal{G}$ with a connection and curving. A connection on $\mathcal{G}$ is a connection on each $L_{\alpha\beta}$ such that each $\theta_{\alpha\beta\gamma}$ is constant. A connection $\nabla^{\alpha\beta}$ on $L_{\alpha\beta}$ corresponds to a 1-form $A_{\alpha\beta}$ such that $\nabla^{\alpha\beta}\sigma_{\alpha\beta} = A_{\alpha\beta}\sigma_{\alpha\beta}$. Note that we have $A_{\alpha\beta} = -A_{\beta\alpha}$ because θ is covariantly constant and θ can be used to induce an isomorphism $L_{\alpha\beta} \simeq L^*_{\beta\alpha}$. Differentiating the equation $\theta_{\alpha\beta\gamma}(\sigma_{\alpha\beta} \otimes \sigma_{\beta\gamma}) = g_{\alpha\beta\gamma}\sigma_{\alpha\gamma}$, we find that the $\{A_{\alpha\beta}\}$ define a gerbe connection if and only if

$$A_{\alpha\beta} + A_{\beta\gamma} + A_{\gamma\alpha} = g^{-1}_{\alpha\beta\gamma} dg_{\alpha\beta\gamma}.$$

The curvature of $\nabla^{\alpha\beta}$ is given by $F_{\alpha\beta} = dA_{\alpha\beta}$, so immediately we have

$$F_{\alpha\beta} + F_{\beta\gamma} + F_{\gamma\alpha} = 0.$$

A curving for $\mathcal{G}$ with respect to such a connection is a collection $\{B_\alpha\}$ of 2-forms such that

$$\frac{i}{2\pi} F_{\alpha\beta} = B_\beta - B_\alpha.$$

Finally the curvature of such a connection and curving is the unique closed 3-form H such that $H|_{U_\alpha} = dB_\alpha$.

2.3 Gerbe holonomy and the B-field

To finish up our discussion of gerbes we will say a few things about how they are related to string theory. The string theory B-field can be represented as a bundle gerbe with connection and curving. To understand how this works let us first consider electromagnetism.

In the theory of electromagnetism the electromagnetic field can be represented by a unitary line bundle L with connection ∇. The curvature 2-form F is the electromagnetic field strength. Over time a charged point particle traces out a curve in space-time. In the quantum theory probabilities are computed by summing over all possible trajectories, keeping account of the relative phases. Suppose γ_1, γ_2 are two such trajectories joining two given points. The electromagnetic interaction contributes a relative phase difference between these paths. The phase difference is the holonomy of ∇ around the path obtained by joining γ_1 and γ_2 at the endpoints.

Turning now to string theory there is a similar story. Here we consider only closed strings. A string has one spacial dimension, so over time traces out a two-dimensional surface in space-time. Closed strings are charged with respect to the Neveu-Schwarz B-field. Just as the electromagnetic field has a 2-form field strength, the Neveu-Schwarz field has a field strength which is a closed 3-form H.

The interaction of strings with the B-field means that in the quantum theory one has to take into account phase factors over different trajectories. Essentially what this means is that for each closed oriented surface $\Sigma \subset X$ in space-time X, we need to be able to assign to Σ a U(1)-valued phase $\chi(\Sigma)$. Furthermore if Σ is the boundary of a 3-manifold $M \subset X$ then it is required that $\chi(\Sigma) = exp(2\pi i \int_M H)$. Such a phase function is called a *Cheeger-Simons character* of degree 2 [27].

Given a gerbe with connection and curving, it is possible to define a holonomy function [35, 8] which assigns to each closed surface a U(1)-valued holonomy in much the same way that a connection on a line bundle assigns a holonomy to closed loops. Let $\mathcal{G}$ be a gerbe with connection and curving on X. The holonomy of $\mathcal{G}$ over a closed oriented surface Σ can be defined in the following manner. Since Σ is a 2-dimensional manifold we have that $H^3(\Sigma, \mathbb{Z}) = 0$, thus $\mathcal{G}|_\Sigma$ is trivializable as a gerbe without connection and curving. We can argue that up to a shift by some 2-form B on Σ that $\mathcal{G}|_\Sigma$ is trivializable as a gerbe with connection and curving. In other words $\mathcal{G}|_\Sigma$ is stably isomorphic to the trivial gerbe with trivial connection, but with curving given by B. Then the holonomy may be defined as $exp(2\pi i \int_\Sigma B)$.

It turns out that every Cheeger-Simons characters of degree 2 arises as the holonomy of a gerbe with connection and curving in this way. Thus gerbes with connections and curving are a way of realizing degree 2 Cheeger-Simons characters.

3 Generalized geometry

3.1 The generalized tangent bundle

Generalized geometry began with Hitchin [40] in which he introduced generalized complex structures and was expanded upon further in the theses of Gualtieri [38] and Cavalcanti [25]. A precedent to this was the development of Courant algebroids by Courant [29], Liu, Weinstein and Xu [50] and others. Generalized geometry rapidly became popular amongst physicists due to various connections to string theory and supersymmetric field theories. Our focus will be on generalized complex structures which unite aspects of complex and symplectic geometry.

Let M be a smooth manifold. We let $E = TM \oplus T^*M$ denote the direct sum of tangent and cotangent bundles and call E the *generalized tangent bundle* of M. The bundle E is in some sense a replacement for the usual tangent bundle. Instead of looking at geometric structures on the tangent bundle like Riemannian metrics or complex structures, we consider structures on E which lead to notions such as generalized metrics and generalized complex structures.

The generalized tangent bundle E is equipped with a triple $(\langle\, ,\, \rangle, \rho, [\, ,\,])$ consisting of:

- A symmetric, non-degenerate bilinear form $\langle\, ,\, \rangle : E \otimes E \to \mathbb{R}$,
- A bundle map $\rho : E \to TM$ called the *anchor*,
- A bracket $\Gamma(E) \otimes_{\mathbb{R}} \Gamma(E) \to \Gamma(E)$ on the sections of E called the *Courant bracket.*

To define these structures let us introduce a convention. We write sections of E as a sum of tangent and cotangent vectors: $X + \xi, Y + \eta, \ldots \in \Gamma(E)$ where $X, Y, \ldots \in \Gamma(TM)$, $\xi, \eta, \ldots \in \Gamma(T^*M)$.

The pairing $\langle\, ,\, \rangle$ is given by

$$\langle X + \xi, Y + \eta \rangle = \frac{1}{2}(\eta(X) + \xi(Y)). \tag{3.1}$$

The anchor $\rho : E \to TM$ is the projection to TM:

$$\rho(X + \xi) = X. \tag{3.2}$$

Finally the Courant bracket (first introduced by Courant in [29]) is given by

$$[X + \xi, Y + \eta] = [X, Y] + \mathcal{L}_X \eta - \mathcal{L}_Y \xi - \frac{1}{2} d(\eta(X) - \xi(Y)), \tag{3.3}$$

where on the right hand side $[X, Y]$ is the usual Lie bracket of vector fields and $\mathcal{L}_X$ is the Lie derivative with respect to X. We will have a lot more to say about the Courant bracket and the motivation behind it in Section 3.2.

If M is n-dimensional then E is a bundle of rank $2n$. One sees that the pairing $\langle\, ,\rangle$ is not positive definite, in fact it has signature (n, n). Thus we may say that the generalized tangent bundle has structure groups $O(n, n)$. Of course the splitting $E = TM \oplus T^*M$ reduces the structure group to $GL(n, \mathbb{R}) \subset O(n, n)$, however part of the general philosophy behind generalized geometry is that it is better to think in terms of the $O(n, n)$-structure than the $GL(n, \mathbb{R})$-structure. Let us also remark that the group $O(n, n)$ is closely related to the T-duality group $O(n, n, \mathbb{Z})$, which occurs in toroidal compactifications of string theory [59].

If $f : M \to M$ is a diffeomorphism there is an induced diffeomorphism $f_* : TM \to TM$ of the total space underlying TM. Moreover for each $x \in M$ we have that f_* restricted to the tangent space $T_x M$ is a linear isomorphism $f_*|_{T_x M} : T_x M \to T_{f(x)} M$. Given a vector field X on M we get a new vector field denoted $f_* X$ which at $x \in M$ takes the value $f_*(X(f^{-1}(x)))$. If X, Y are vector fields on M then one finds that $f_*[X,Y] = [f_*X, f_*Y]$, so that f_* preserves the Lie bracket on $\Gamma(TM)$. If instead of the tangent bundle TM we consider the generalized tangent bundle E, we are lead to consider the following notion of automorphism:

Definition 3.1. An automorphism of the generalized tangent bundle (or *generalized diffeomorphism*) is a pair (f, F) where $f : M \to M$ is a diffeomorphism of M and $F : E \to E$ is a diffeomorphism of E which for each $x \in M$ restricts to a linear isomorphism $F|_{E_x} : E_x \to E_{f(x)}$ and such that the pair (f, F) respects the triple $(\langle\, ,\, \rangle, \rho, [\, ,\,])$ in the sense that for all $a, b \in \Gamma(E)$,

- $\langle Fa, Fb\rangle = \langle a, b\rangle$,
- $\rho(Fa) = f_*(\rho a)$,
- $[Fa, Fb] = F[a, b]$,

where if a is a section of E then we use Fa to denote the section of E which sends $x \in M$ to $F(a(f^{-1}(x)))$.

If $f : M \to M$ is any diffeomorphism of M then we can promote f to a generalized diffeomorphism by setting $F = f_* \oplus (f^{-1})^*$. Such a pair (f, F) is easily seen to be a generalized diffeomorphism.

Let $B \in \Omega^2(M)$ be a 2-form on M. We let B act as a bundle endomorphism of E as follows:

$$B(X + \xi) = i_X B. \tag{3.4}$$

One can check directly that this action defines a section of the $O(n, n)$ adjoint bundle of E, that is for any sections $a, b \in \Gamma(E)$ we have $\langle Ba, b\rangle + \langle a, Bb\rangle = 0$. Let $e^B : E \to E$ denote the exponentiated action, in other words

$$e^B(X + \xi) = X + \xi + i_X B. \tag{3.5}$$

This action clearly preserves the pairing on E. Both the infinitesimal action (3.4) and the exponentiated action (3.5) will be referred to as a *B-shift*. It is a straightforward exercise to show that the pair (Id, e^B) is a generalized diffeomorphism if and only if B is closed as a differential 2-form. Thus the space $\Omega^2_{cl}(M)$ of closed 2-forms generates an abelian group of generalized diffeomorphisms.

Proposition 3.1 ([38]). *Any generalized diffeomorphism is a composition of a generalized diffeomorphism $(f, f_* \oplus (f^{-1})^*)$ associated to an ordinary diffeomorphism $f : M \to M$ and a B-shift (Id, e^B) associated to a closed 2-form. It follows that the group of generalized diffeomorphisms of M is a semi-direct product*

$$Diff(M) \ltimes \Omega^2_{cl}(M),$$

where the action of $f \in Diff(M)$ on $B \in \Omega^2_{cl}(M)$ is by diffeomorphism: $fB = (f^{-1})^*(B)$.

In short, the symmetries of the generalized tangent bundle are generated by diffeomorphisms and closed 2-forms.

We saw that the natural pairing on the generalized tangent bundle gives it an $O(n, n)$-structure. Since the subgroup $GL(n, \mathbb{R})$ of $O(n, n)$ is actually a subgroup of $SO(n,n)$, we find that E carries a canonical orientation and we can consider E to have structure group $SO(n,n)$. In addition it is always possible to lift the structure group from $SO(n,n)$ to $Spin(n,n)$. As shown in [38], there are two homomorphisms $\phi_i : GL(n,\mathbb{R}) \to Spin(n,n)$, $i = 1,2$ covering the inclusion $GL(n,\mathbb{R}) \to SO(n,n)$ and thus two canonical spin structures on E. For our purposes it is not necessary to actually make use of these spin structures. There is a naturally defined bundle of Clifford modules S for E which is easily defined without using a spin structure on E. Let us now describe this bundle S.

The bundle S is defined to be $S = \wedge^* T^*M$, the bundle of differential forms of mixed degree on M. In the context of generalized geometry sections of S may either be referred to as *differential forms* or as *spinors*. The Clifford action of a section $X + \xi \in \Gamma(E)$ on a spinor $\omega \in \Gamma(S)$, denoted $\gamma_{X+\xi}\omega$ is given by

$$\gamma_{X+\xi}\omega = i_X\omega + \xi \wedge \omega, \tag{3.6}$$

where we use i_X to denote the operation on forms of contraction by the vector field X. This is a Clifford action in the sense that

$$\gamma_a\gamma_b\omega + \gamma_b\gamma_a\omega = 2\langle a,b\rangle\omega. \tag{3.7}$$

Note that there is a natural $\mathbb{Z}_2$-grading on S given by differential form parity, that is

$$S = S^0 \oplus S^1 \tag{3.8}$$

where $S^0 = \wedge^{ev}T^*M$ is the bundle of even forms and $S^1 = \wedge^{odd}T^*M$ the odd forms. It will be convenient to use the following notational shortcut: for any $i \in \mathbb{Z}$ let S^i denote S^0 if i is even or S^1 if i is odd. The Clifford action exchanges parity, that is if ω is even (odd) then $\gamma_a\omega$ is odd (even).

Let $d : \Gamma(S^i) \to \Gamma(S^{i+1})$ be the exterior derivative. We now explain how the Clifford action, exterior derivative and Courant bracket are all connected. Let $a, b \in \Gamma(E)$ and $\omega \in \Gamma(S)$. Then one can show the following relation

$$\gamma_{[a,b]}\omega = \frac{1}{2}[[d,\gamma_a],\gamma_b]\omega - \frac{1}{2}[[d,\gamma_b],\gamma_a]\omega. \tag{3.9}$$

The brackets on the right hand side of (3.9) are graded commutators, where d, γ_a, γ_b are viewed as having degree 1. In other words expanding the commutators we get

$$\gamma_{[a,b]}\omega = \frac{1}{2}(d\gamma_a\gamma_b - d\gamma_b\gamma_a + 2\gamma_a d\gamma_b - 2\gamma_b d\gamma_a + \gamma_a\gamma_b d - \gamma_b\gamma_a d)\omega. \tag{3.10}$$

We can simplify this relation by first introducing a new bracket $\{\,,\,\} : \Gamma(E) \otimes_{\mathbb{R}} \Gamma(E) \to \Gamma(E)$ called the *Dorfman bracket* given by

$$\{a,b\} = [a,b] + d\langle a,b\rangle. \tag{3.11}$$

Note that $\{a,b\}$ is not skew-symmetric, in fact the skew-symmetrization of $\{\,,\,\}$ is the Courant bracket. A straightforward computation shows that the Dorfman bracket is given by:

$$\{X+\xi, Y+\eta\} = [X,Y] + \mathcal{L}_X\eta - i_Y d\xi. \tag{3.12}$$

We will say more about the Dorfman bracket in Section 3.2 but for now what is important is that (3.9) follows from the simpler relation

$$\gamma_{\{a,b\}}\omega = [[d,\gamma_a],\gamma_b]\omega, \tag{3.13}$$

or on expanding the graded commutators

$$\gamma_{\{a,b\}}\omega = (d\gamma_a\gamma_b + \gamma_a d\gamma_b - \gamma_b d\gamma_a - \gamma_b\gamma_a d)\omega,$$

which can be verified by a straightforward computation. The significance of Equation (3.13) is that it says the Dorfman bracket is a *derived bracket* [47].

We saw that 2-forms act on the generalized tangent bundle by B-shift: $e^B(X+\xi) = X + \xi + i_X B$. Let us also declare that 2-forms act on the bundle of spinors S by sending a form ω to $e^{-B}\omega$ where

$$e^{-B}\omega = \omega - B\wedge\omega + \frac{1}{2}B\wedge B\wedge\omega + \dots$$

The series is finite because differential form degree is bounded by n. One may wonder why we are using $e^{-B}\omega$ rather than the more obvious $e^B\omega$. The reason is the following easily verified identity:

$$e^{-B}\gamma_a\omega = \gamma_{e^B a}(e^{-B}\omega) \tag{3.14}$$

which shows that the action of B on forms together with the action of B on E respects Clifford multiplication. This will be important when we consider twisting by a gerbe in Section 3.3.

Previously we remarked that the generalized tangent bundle has two canonical spin structures corresponding to homomorphisms $\phi_i : GL(n,\mathbb{R}) \to Spin(n,n)$, $i = 1,2$. Let $\mathcal{S}_1, \mathcal{S}_2$ denote the spinor bundles associated to these spin structures. Then as shown in [38] the relation between $S, \mathcal{S}_1$ and $\mathcal{S}_2$ is as follows:

$$\begin{aligned}\mathcal{S}_1 &= S \otimes det(TM) \otimes |det(TM)|^{-1/2},\\ \mathcal{S}_2 &= S \otimes |det(TM)|^{1/2}.\end{aligned}$$

Thus the spin bundles $\mathcal{S}_1, \mathcal{S}_2$ only differ from S by certain line bundle factors. It is more convenient however to work with S because S is equipped with the exterior derivative, which we have just seen is related to the Dorfman and Courant brackets.

One last piece of structure that we would like to introduce is a bilinear form on spinors valued in the determinant bundle of T^*M, $det(T^*M) = \wedge^n T^*M$, where M is n-dimensional. The pairing, called the *Mukai pairing* is denoted $\langle\, ,\,\rangle : S\otimes S \to det(T^*M)$, although this should not cause any confusion with the bilinear pairing on E. Let $\sigma : S \to S$ be the involution of S which acts in degree k as multiplication by $(-1)^{\frac{1}{2}k(k-1)}$. Alternatively σ can be understood as reversing the order of wedge products since if $\xi_1, \xi_2, \dots, \xi_k$ are 1-forms then $\sigma(\xi_1\wedge\cdots\wedge\xi_k) = \xi_k\wedge\cdots\wedge\xi_1$. Now if $\alpha,\beta\in\Gamma(S)$ we define $\langle\alpha,\beta\rangle$ by[1]

$$\langle\alpha,\beta\rangle = [\alpha\wedge\sigma(\beta)]^n,$$

[1] There is an alternative convention which defines the Mukai pairing by $[\sigma(\alpha)\wedge\beta]^n$. This has the effect of altering some signs in various identities involving the Mukai pairing.

where $[\ \ ]^n$ denotes taking the top degree part. Some properties of the Mukai pairing which are easily verified are:

- $\langle \alpha, \beta \rangle = (-1)^{\frac{1}{2}n(n-1)} \langle \beta, \alpha \rangle$,
- $\langle \gamma_v \alpha, \beta \rangle = (-1)^{n-1} \langle \alpha, \gamma_v \beta \rangle$,
- $\langle \gamma_v \alpha, \gamma_v \beta \rangle = (-1)^{n-1} \langle v, v \rangle \langle \alpha, \beta \rangle$,
- $\langle e^B \alpha, e^B \beta \rangle = \langle \alpha, \beta \rangle$,

where $\alpha, \beta \in \Gamma(S)$, $v \in \Gamma(E)$ and $B \in \Omega^2(M)$. The last of these identities, which follows from $\sigma(e^B\beta) = e^{-B}\sigma(\beta)$ implies that the Mukai pairing is invariant under generalized diffeomorphisms.

3.2 Courant brackets and Dorfman brackets

In Section 3.1 we have seen that the generalized tangent bundle E of a manifold M is equipped with a non-degenerate pairing $\langle\, , \rangle$, an anchor map ρ and a bracket $[\, ,]$. As an exercise, one can show that the triple $(\langle\, , \rangle, \rho, [\, ,])$ satisfies the following useful properties where $a, b, c \in \Gamma(E)$, $f, g \in \mathcal{C}^\infty(M)$:

(C0) $[a, b] = -[b, a]$,
(C1) $\rho[a, b] = [\rho a, \rho b]$,
(C2) $[[a, b], c] + [[b, c], a] + [[c, a], b] = d[a, b, c]$, where

$$[a, b, c] = \frac{1}{3}\left(\langle [a, b], c\rangle + \langle [b, c], a\rangle + \langle [c, a], b\rangle\right), \tag{3.15}$$

(C3) $[a, fb] = f[a, b] + \rho(a)(f)b - \langle a, b\rangle df$,
(C4) $\rho(a)\langle b, c\rangle = \langle [a, b] + d\langle a, b\rangle, c\rangle + \langle b, [a, c] + d\langle a, c\rangle\rangle$,
(C5) $\langle df, dg\rangle = 0$.

The axioms (C0)–(C5) define what is known as a *Courant algebroid*. Courant algebroids were introduced by Liu, Weinstein and Xu in [50] as an abstraction of Courant's bracket. Let us give the definition.

Definition 3.2 (Liu, Weinstein, Xu). Let E be a vector bundle on M equipped with a triple $(\langle\, , \rangle, \rho, [\, ,])$ where $\langle\, , \rangle$ is a symmetric non-degenerate bilinear form on E, ρ a bundle map $\rho : E \to TM$ called the *anchor* and $[\, ,]$ a bracket $[\, ,] : \Gamma(E) \otimes_{\mathbb{R}} \Gamma(E) \to \Gamma(E)$ called the *Courant bracket.*

Define an operator $D : \mathcal{C}^\infty(M) \to \Gamma(E)$ as follows: for a function $f \in \mathcal{C}^\infty(M)$, Df is defined by the equation

$$\langle Df, a\rangle = \frac{1}{2}\rho(a)(f) \tag{3.16}$$

for all $a \in \Gamma(E)$. Then $(E, \langle\, , \rangle, \rho, [\, ,])$ is a *Courant algebroid* if the axioms (C0)-(C5) above hold (where now D is used in place of d).

We see from axiom (C2) that the Courant bracket is not a Lie bracket; it fails to obey the Jacobi identity. Axiom (C2) shows that the Jacobi identity fails in a way that is measured by a triple bracket $[\, , \, ,] : \Gamma(E) \otimes_{\mathbb{R}} \Gamma(E) \otimes_{\mathbb{R}} \Gamma(E) \to \mathcal{C}^\infty(M)$

given in (3.15). Taking this further Roytenberg and Weinstein [56] showed that $\Gamma(E)$ has the structure of an L_∞-algebra, in fact of a special kind known as a Lie 2-algebra (see for instance [4] for the definition of Lie 2-algebras and related notions). Although the higher categorical point of view on Courant algebroids has some applications, we will not make use of it in these notes.

It has been observed (for instance in [50]) that Courant algebroids are equipped with another bracket which is often useful to consider as it can be easier to work with than the Courant bracket. This bracket, called the *Dorfman bracket* will be denoted $\{\,,\,\}$ and is defined in the same way as we did for the generalized tangent bundle:

$$\{a, b\} = [a, b] + D\langle a, b\rangle. \tag{3.17}$$

The Dorfman bracket is not skew-symmetric and from (3.17) it follows that the Courant bracket is the skew-symmetrization of the Dorfman bracket:

$$[a, b] = \frac{1}{2}(\{a, b\} - \{b, a\}).$$

When E is the generalized tangent bundle the Dorfman bracket is given by (3.12).

The axioms for a Courant algebroid in terms of the Dorfman bracket simplify to the following:

(D1) $\rho\{a, b\} = [\rho a, \rho b]$,
(D2) $\{a, \{b, c\}\} = \{\{a, b\}, c\} + \{b, \{a, c\}\}$,
(D3) $\{a, fb\} = f\{a, b\} + \rho(a)(f)b$,
(D4) $\rho(a)\langle b, c\rangle = \langle\{a, b\}, c\rangle + \langle b, \{a, c\}\rangle$,
(D5) $\{a, b\} + \{b, a\} = 2D\langle a, b\rangle$.

These axioms can be proven from axioms (C0)–(C6) together with Definition (3.17) of the Dorfman bracket, see for instance [55] for the proof.

3.3 Twisting by a gerbe

We now show that the Courant algebroid structure on the generalized tangent bundle $E = TM \oplus T^*M$ can be replaced by a twisted construction in the presence of a gerbe with connection. Thus let $(\mathcal{G}, \nabla)$ be a gerbe with connection. For the construction we will assume that the cover $\pi : Y \to M$ over which the gerbe is defined has the form $Y = \coprod_\alpha U_\alpha$ for some open cover $\{U_\alpha\}_\alpha$ of M. Since any gerbe is isomorphic to one of this form, this is not too restrictive. Thus we take $\mathcal{G}$ to be defined by a collection $\{L_{\alpha\beta}\}$ of line bundles $L_{\alpha\beta} \to U_{\alpha\beta} = U_\alpha \cap U_\beta$ and multiplication $\theta_{\alpha\beta\gamma} : L_{\alpha\beta} \otimes L_{\beta\gamma} \to L_{\alpha\gamma}$ over the triple intersections $U_{\alpha\beta\gamma}$. Let $F_{\alpha\beta}$ be the curvature of the connection on $L_{\alpha\beta}$. Thus $\{F_{\alpha\beta}\}$ is a 1-cocycle of closed 2-forms:

$$F_{\alpha\beta} + F_{\beta\gamma} + F_{\gamma\alpha} = 0.$$

Moreover if we set $B_{\alpha\beta} = \frac{i}{2\pi} F_{\alpha\beta}$ and take ∇ to be unitary then the $B_{\alpha\beta}$ are real 2-forms. The twisting construction that follows really only depends on the choice of 1-cocycle $\{B_{\alpha\beta}\}$ of closed 2-forms. As a consequence we will see that up to isomorphism the twisting construction only depends on the image of the Dixmier-Douady class of $\mathcal{G}$ in $H^3(M,\mathbb{R})$. Nevertheless, it is worth emphasizing the relation to gerbes since this ties in well with applications to T-duality.

Since the $B_{\alpha\beta}$ are closed 2-forms they determine locally defined automorphisms $e^{B_{\alpha\beta}}$ of the generalized tangent bundle. We will define a bundle $E_{\mathcal{G}}$ as follows. Over each open set U_α we take $E_{\mathcal{G}}$ to be the generalized tangent bundle $(TM \oplus T^*M)|_{U_\alpha}$. On the intersection $U_{\alpha\beta}$ we use the automorphism $e^{B_{\alpha\beta}}$ as a transition function between U_β and U_α. To be more explicit a section of $E_{\mathcal{G}}$ is a collection of vector fields $\{X_\alpha\}$ and 1-forms $\{\xi_\alpha\}$ such that over $U_{\alpha\beta}$ we have

$$X_\alpha + \xi_\alpha = e^{B_{\alpha\beta}}(X_\beta + \xi_\beta),$$

or on equating components

$$X_\alpha = X_\beta, \tag{3.18}$$
$$\xi_\alpha = \xi_\beta + i_{X_\beta} B_{\alpha\beta} \tag{3.19}$$

Note that the first of these equations says that the $\{X_\alpha\}$ patch together to form a single vector field. The second relation (3.19) is consistent on triple intersections precisely because the $\{B_{\alpha\beta}\}$ satisfy the cocycle condition.

By construction of $E_{\mathcal{G}}$ we have over each U_α a canonical identification $E_{\mathcal{G}}|_{U_\alpha} = (TM \oplus T^*M)|_{U_\alpha}$. Thus $E_{\mathcal{G}}|_{U_\alpha}$ inherits the structure of a Courant algebroid. Moreover the transition functions are by closed 2-forms which we have seen act as automorphisms of the Courant algebroid structure. It follows immediately that $E_{\mathcal{G}}$ inherits the structure of a Courant algebroid. We say that $E_{\mathcal{G}}$ is a *twisted generalized tangent bundle*, twisted by the gerbe with connection $\mathcal{G}$.

In a similar manner one can also construct a twisted spinor bundle $S_{\mathcal{G}}$. The sections of $S_{\mathcal{G}}$ correspond to collections $\{\omega_\alpha\}$ of differential forms which are related on the overlaps as follows:

$$\omega_\alpha = e^{-B_{\alpha\beta}}\omega_\beta.$$

It follows directly from (3.14) that Clifford multiplication survives this construction, so there is a well defined Clifford action $\gamma : E_{\mathcal{G}} \otimes S_{\mathcal{G}} \to S_{\mathcal{G}}$. Denoting a section of $E_{\mathcal{G}}$ by $\{X_\alpha + \xi_\alpha\}$ and a section of $S_{\mathcal{G}}$ by $\{\omega_\alpha\}$, then the Clifford action of $\{X_\alpha + \xi_\alpha\}$ on $\{\omega_\alpha\}$ is given by $\{i_{X_\alpha}\omega_\alpha + \xi_\alpha \wedge \omega_\alpha\}$.

Since the $B_{\alpha\beta}$ are closed 2-forms their action commutes with the exterior derivative. So there is a well defined differential operator $d_{\mathcal{G}} : \Gamma(S_{\mathcal{G}}) \to \Gamma(S_{\mathcal{G}})$, which we call the *twisted differential*. If $u = \{\omega_\alpha\}$ represents a section of $S_{\mathcal{G}}$ then $d_{\mathcal{G}}u$ is represented by $\{d\omega_\alpha\}$. The Dorfman bracket $\{\,,\,\}_{\mathcal{G}}$ on $E_{\mathcal{G}}$ and the twisted differential $d_{\mathcal{G}}$ are related through the identity

$$\gamma_{\{a,b\}_{\mathcal{G}}}\omega = [[d_{\mathcal{G}}, \gamma_a], \gamma_b]\omega, \tag{3.20}$$

which follows immediately since locally it reduces to (3.13). Finally we note that there is also a Mukai pairing $\langle\,,\,\rangle : S_{\mathcal{G}} \otimes S_{\mathcal{G}} \to det(T^*M)$ which follows since the Mukai pairing is invariant under B-shifts.

Unlike the ordinary generalized tangent bundle E, a twisted generalized tangent bundle $E_{\mathcal{G}}$ is not canonically a direct sum. It does however fit into a short exact sequence:

$$0 \to T^*M \xrightarrow{\rho^*} E_{\mathcal{G}} \xrightarrow{\rho} TM \to 0. \tag{3.21}$$

Now since short exact sequences of smooth vector bundles can always be split we find that $E_{\mathcal{G}}$ as a vector bundle is isomorphic to E, but non-canonically. Moreover E and $E_{\mathcal{G}}$ may be non-isomorphic as Courant algebroids, in fact we find this is the case unless the Dixmier-Douady class of $\mathcal{G}$ is trivial in $H^3(M, \mathbb{R})$.

As we now explain a choice of curving for $(\mathcal{G}, \nabla)$ determines a splitting of (3.21). Let $\{B_\alpha\}$ be such a curving. Thus $\{B_\alpha\}$ is a collection of 2-forms such that $B_\beta - B_\alpha = B_{\alpha\beta}$. From this curving we get induced vector bundle isomorphisms $\tau : E_{\mathcal{G}} \to E$, $\psi : S_{\mathcal{G}} \to S$ as follows: $\tau(X_\alpha + \xi_\alpha) = e^{B_\alpha}(X_\alpha + \xi_\alpha)$, $\psi(\omega_\alpha) = e^{-B_\alpha}\omega_\alpha$. One sees that $\{e^{B_\alpha}(X_\alpha + \xi_\alpha)\}$ and $\{e^{-B_\alpha}\omega_\alpha\}$ patch together to form well defined sections of E and S, so τ and ψ are well defined bundle isomorphisms.

Now we can use τ, ψ to transport the structures on $E_{\mathcal{G}}, S_{\mathcal{G}}$ over to E, S. One sees that the pairing $\langle\, , \rangle$ on $E_{\mathcal{G}}$, the anchor $E_{\mathcal{G}} \to TM$, Clifford action $E_{\mathcal{G}} \otimes S_{\mathcal{G}} \to S_{\mathcal{G}}$ and Mukai pairing on $S_{\mathcal{G}}$ map to the standard ones on E and S. We now consider the twisted Dorfman bracket $\{\, , \}_{\mathcal{G}}$, Courant bracket $[\, ,]_{\mathcal{G}}$ and the twisted differential $d_{\mathcal{G}}$. One finds that these do not in general map to the standard Dorfman bracket, Courant bracket and differential d, but to twisted versions. Let H be the curvature of the curving, that is $H|_{U_\alpha} = dB_\alpha$. One finds

$$\tau\{\tau^{-1}(X + \xi), \tau^{-1}(Y + \eta)\}_{\mathcal{G}} = \{X + \xi, Y + \eta\}_H,$$
$$\psi(d_{\mathcal{G}}\psi^{-1}\omega) = d_H\omega,$$

where the *H-twisted Dorfman bracket* $\{\, , \}_H$ is defined by [58]

$$\{X + \xi, Y + \eta\}_H = \{X + \xi, Y + \eta\} + i_X i_Y H,$$

and the *H-twisted differential* d_H is defined by

$$d_H\omega = d\omega + H \wedge \omega.$$

We can also define the *H-twisted Courant bracket* $[\, ,]_H$ to be the Courant bracket corresponding to $\{\, , \}_H$ which is given by

$$[X + \xi, Y + \eta]_H = [X + \xi, Y + \eta] + i_X i_Y H.$$

We have seen that on choosing a curving we can identify $E_{\mathcal{G}}$ and $S_{\mathcal{G}}$ with E and S and the only modification required is that the Dorfman bracket and the differential receive an additional term depending on the curvature 3-form H. An alternative approach which is more straightforward but less motivated is to just start with the curvature form H and simply define the twisted operators $\{\, , \}_H$, d_H as above.

We have seen that the twisting construction gives rise to a twisted differential $d_H : \Gamma(S) \to \Gamma(S)$ on differential forms. That $d_H^2 = 0$ follows since H is a closed 3-form. The cohomology of d_H is called the *H-twisted cohomology* of M or just the twisted cohomology of M. Note that while d_H does not respect differential form degree it does send even (odd) forms to odd (even) forms, so twisted cohomology is naturally $\mathbb{Z}_2$-graded. We use $H^k(M, H)$ to denote H-twisted cohomology in degree k. Twisted cohomology is closely related to twisted K-theory. The relation is much the same as the relation between ordinary K-theory and cohomology, in fact it is possible to define twisted Chern characters from twisted K-theory to twisted cohomology.

There is a sort of converse to the twisting construction. For any Courant algebroid E there is a canonical sequence (3.21). If this sequence is exact we say that E is an *exact Courant algebroid*. It is a theorem [57] that any exact Courant algebroid is of the form $E = TM \oplus T^*M$ with H-twisted Dorfman bracket, for some closed 3-form H. The 3-form is not uniquely determined, only the underlying de Rham cohomology class $[H] \in H^3(M, \mathbb{R})$, called the Ševera class of E is uniquely determined. One sees from this that the set of isomorphisms classes of exact Courant algebroids on M is in bijection with $H^3(M, \mathbb{R})$ by way of the Ševera class.

3.4 Generalized metrics

Let E be an exact Courant algebroid on a manifold M.

Definition 3.3. A *generalized* metric G on E is a bundle endomorphism $G : E \to E$ such that G is compatible with the pairing, that is $\langle Ga, Gb \rangle = \langle a, b \rangle$ for all a, b and G is symmetric, positive definite in the sense that $G(a,b) = \langle Ga, b \rangle$ is a symmetric positive definite bilinear form on E.

If G is a generalized metric we see from the definition that $\langle G^2 a, b \rangle = G(Ga, b) = G(a, Gb) = \langle Ga, Gb \rangle = \langle a, b \rangle$ for all a, b. Thus $G^2 = 1$. We then have a decomposition $E = V_+ \oplus V_-$ into the ± 1 eigenbundles of G. From the properties of G it is clear that V_+ is a maximal positive definite subbundle of E with respect to $\langle\, ,\, \rangle$. Similarly V_- is maximal negative definite and in fact V_- is the orthogonal complement of V_+ with respect to $\langle\, ,\, \rangle$. Conversely we see that a choice $V_+ \subset E$ of maximal positive definite subbundle of E determines an orthogonal decomposition $E = V_+ \oplus V_-$ and hence a generalized metric G by letting $V_\pm$ be the ± 1 eigenspaces of G.

Suppose now that $E = TM \oplus T^*M$ is the generalized tangent bundle of M. A maximal positive definite subbundle V_+ of E can be expressed as a graph of a map $C : TM \to T^*M$ as follows:

$$V_+ = \{X + CX \,|\, X \in TM\}.$$

Since C maps from TM to T^*M we can also view C as a bilinear form C' on TM by setting $C'(X,Y) = (CX)(Y)$. We can then decompose C' into symmetric and skew-symmetric parts: $C' = g - B$ where g is a symmetric bilinear form and B is a 2-form[2]. In what follows we will also think of g as a map $g : TM \to T^*M$ which sends a vector field X to the 1-form $g(X, \,)$. Similarly we can view B as a map $B : TM \to T^*M$ which sends X to $i_X B = B(X, \,)$. We need to determine when V_+ is positive definite. One sees that $\langle X + CX, Y + CY \rangle = g(X,Y)$, so V_+ is

[2]The convention that is normally used is to write $C' = g + B$, but it fits better with other conventions we are using to instead write $C' = g - B$.

positive definite if and only if g is positive definite, i.e. a Riemannian metric. We conclude that generalized metrics on $TM \oplus T^*M$ correspond to pairs (g, B) where g is a Riemannian metric and B is a 2-form. We observe also that the orthogonal complement of V_+ can now be written as

$$V_- = \{X + (-g - B)X \mid X \in TM\}.$$

Generalized metrics are related to gerbe curvings. To see this let $\mathcal{G}$ be a gerbe $\{L_{\alpha\beta}\}$ defined with respect to some open cover and choose also a gerbe connection with curvature 2-forms $F_{\alpha\beta}$ and set $B_{\alpha\beta} = \frac{i}{2\pi} F_{\alpha\beta}$.

Let $E_{\mathcal{G}}$ be the twisted generalized tangent bundle defined by $\mathcal{G}$. Suppose that G is a generalized metric on $E_{\mathcal{G}}$. Then G determines a curving for $\mathcal{G}$. Indeed the exact sequence $T^*M \to E_{\mathcal{G}} \to TM$ can be split, since we can lift TM to the subspace of $E_{\mathcal{G}}$ which is the G-orthogonal complement of T^*M. We can also show this explicitly. Over each U_α we have an identification $E_{\mathcal{G}}|_{U_\alpha} \simeq TM \oplus T^*M|_{U_\alpha}$. So over U_α the generalized metric G is determined by the graph of $g_\alpha - B_\alpha$, where g_α is a Riemannian metric and B_α a 2-form. Now recall that the transition functions for $E_{\mathcal{G}}$ are shifts by the $B_{\alpha\beta}$, that is if $\{X_\alpha + \xi_\alpha\}$ is a section of $E_{\mathcal{G}}$ then we have relations $X_\alpha = X_\beta$, $\xi_\alpha = \xi_\beta + i_{X_\beta} B_{\alpha\beta}$. Now assume that $\{X_\alpha + \xi_\alpha\}$ is a section of V_+ so that $\xi_\alpha = (g_\alpha - B_\alpha)X_\alpha$. It follows that on $U_{\alpha\beta}$ we have

$$\begin{aligned} g_\alpha &= g_\beta, \\ B_\alpha &= B_\beta - B_{\alpha\beta}. \end{aligned}$$

We see that the collection of 2-forms $\{B_\alpha\}$ is precisely a curving for $\mathcal{G}$, while the collection of metrics $\{g_\alpha\}$ patch together to form a single Riemannian metric on M. A short calculation shows that the G-orthogonal complement of T^*M over U_α is given by $\{X - B_\alpha X \mid X \in TM\}$. So the splitting $E_{\mathcal{G}} \simeq TM \oplus T^*M$ is given over U_α by sending $X_\alpha + \xi_\alpha$ to $X_\alpha + \xi_\alpha + B_\alpha X_\alpha = e^{B_\alpha}(X_\alpha + \xi_\alpha)$. This is precisely the map τ associated to the curving $\{B_\alpha\}$ as in Section 3.3.

3.5 Generalized complex structures

Let M be a manifold and $E = TM \oplus T^*M$ the generalized tangent bundle equipped with H-twisted Dorfman bracket for some closed 3-form $H \in \Omega^3(M)$.

Definition 3.4. An *almost generalized complex structure* is an endomorphism $J : E \to E$ such that

- $J^2 = -1$,
- $\langle Ja, Jb \rangle = \langle a, b \rangle$ for all a, b.

In terms of structure groups an almost generalized complex structure is a reduction of structure from $O(m, m)$ to $U(m/2, m/2)$ where m is the dimension of M. In particular, this requires M to be even dimensional and we write $m = 2n$.

If J is an almost generalized complex structure then we have a decomposition

$$E_{\mathbb{C}} = L \oplus \overline{L}$$

where $E_{\mathbb{C}} = E \otimes \mathbb{C}$ is the complexification of E, L is the $+i$ eigenspace of J and $\overline{L}$ the $-i$ eigenspace. We note that from the properties of J we have that L is a maximal isotropic complex subbundle of $E_{\mathbb{C}}$. Conversely a maximal isotropic complex subbundle L of $E_{\mathbb{C}}$ such that $L \cap \overline{L} = 0$ defines a decomposition $E_{\mathbb{C}} = L \oplus \overline{L}$ and thus an almost generalized complex structure. We observe also that by use of the pairing on $E_{\mathbb{C}}$ there is a complex linear isomorphism $\overline{L} = L^*$.

Definition 3.5. Let J be an almost generalized complex structure and L the corresponding $+i$ eigenbundle. We say that J is *H-integrable* or J is a *generalized complex structure* if the space of sections of L is closed under the H-twisted Courant bracket. If the 3-form H is understood we may simply say integrable.

We note that since L is isotropic the Courant bracket and Dorfman bracket coincide on sections of L. Thus the space of sections of L is closed under the Courant bracket if and only if it is closed under the Dorfman bracket. Moreover when J is integrable we find that the induced bracket on the space of sections of L makes L into a complex Lie algebroid. Let us recall the definition

Definition 3.6. A *Lie algebroid* on M is a vector bundle $L \to M$ equipped with a bundle map $\rho : L \to TM$ called the *anchor* and a bracket $[\,,\,] : \Gamma(L) \otimes \Gamma(L) \to \Gamma(L)$ which makes $\Gamma(L)$ a Lie algebra and such that the following holds:

- $\rho[a, b] = [\rho a, \rho b]$,
- $[a, fb] = f[a, b] + \rho(a)(f)b$,

for any $a, b \in \Gamma(L)$, $f \in \mathcal{C}^\infty(M)$. The notion of a complex Lie algebroid is essentially the same.

Example 3.1. Let I be an almost complex structure on M, that is an endomorphism $I : TM \to TM$ of the tangent bundle such that $I^2 = -1$. Then from I we may construct the following almost generalized complex structure $J_I : TM \oplus T^*M \to TM \oplus T^*M$:

$$J_I = \begin{bmatrix} -I & 0 \\ 0 & I^* \end{bmatrix}$$

where I^* is the dual endomorphism $I^* : T^*M \to T^*M$ of I. It is straightforward to see that $J_I^2 = -1$ and that J_I is compatible with the pairing, so J_I is indeed an almost generalized complex structure.

For an almost complex structure I we let $T^{1,0}M$ denote the i eigenspace of I on $TM_{\mathbb{C}}$ and $T^{0,1}M$ the $-i$ eigenspace. Similarly we let $T^{*1,0}M$ and $T^{*0,1}M$ be the corresponding dual spaces. The corresponding i eigenspace of J_I is $L = T^{0,1}M \oplus T^{*1,0}M$. As an exercise one can show that J_I is integrable as an almost generalized complex structure if and only if I is integrable as an ordinary almost complex structure.

Example 3.2. Let ω be an almost symplectic structure on M, meaning a 2-form $\omega \in \Omega^2(M)$ which is non-degenerate in the sense that the contraction map $\omega : TM \to T^*M$ is an isomorphism. If in addition ω is closed then it is a symplectic structure on M. From any such ω we may construct an almost generalized complex structure $J_\omega : TM \oplus T^*M \to TM \oplus T^*M$:

$$J_\omega = \begin{bmatrix} 0 & -\omega^{-1} \\ \omega & 0 \end{bmatrix},$$

where we are viewing ω as a map $\omega : TM \to T^*M$. The i eigenspace of J_ω is given by

$$L = \{X - i\omega X \,|\, X \in TM_{\mathbb{C}}\}.$$

By a straightforward calculation we find that J_ω is integrable if and only if $d\omega = 0$.

Let φ be a complex spinor, that is a section of the bundle $\Gamma(S_{\mathbb{C}})$ of complex differential forms. We define the *annihilator* $Ann(\varphi)_x$ of φ at $x \in M$ to be

$$Ann(\varphi)_x = \{a \in (E_{\mathbb{C}})_x \,|\, \gamma_a \varphi(x) = 0\},$$

in other words $Ann(\varphi)_x$ is the subspace of $(E_{\mathbb{C}})_x$ for which the Clifford action on φ is trivial. If $\varphi(x)$ is non-zero then $Ann(\varphi)_x$ is an isotropic subspace of $(E_{\mathbb{C}})_x$ for if $a, b \in Ann(\varphi)_x$ we have

$$0 = \gamma_a \gamma_b \varphi(x) + \gamma_b \gamma_a \varphi(x) = 2\langle a, b\rangle \varphi(x)$$

so if $\varphi(x) \neq 0$ then $\langle a, b\rangle = 0$.

Let $Ann(\varphi)$ denote the disjoint union of the subspaces $Ann(\varphi)_x$ as x ranges over $x \in M$. In general $Ann(\varphi)$ is not a subbundle of $E_{\mathbb{C}}$ because the dimension of $Ann(\varphi)_x$ will vary with x. However for suitable spinors $Ann(\varphi)$ will be a subbundle. Let V be a complex vector space of complex dimension $2k$. An isotropic subspace of V is said to be *maximal* if it has dimension k (the largest possible dimension for an isotropic subspace in V).

Definition 3.7. Let $\varphi \in \Gamma(S_{\mathbb{C}})$ be a spinor. We say that φ is a *pure spinor* if φ is non-vanishing and for each $x \in M$ the subspace $Ann(\varphi)_x$ is a maximal isotropic subspace of $(E_{\mathbb{C}})_x$. In this case $Ann(\varphi)$ defines a bundle of maximal isotropic subspaces of $E_{\mathbb{C}}$.

Note that for any spinor φ and any non-vanishing function f, we have $Ann(\varphi) = Ann(f\varphi)$, so φ is pure if and only of $f\varphi$ is pure.

To understand pure spinors better let us consider the underlying linear algebra. Let V be a finite dimensional real or complex vector space and consider $W = V \oplus V^*$ with the natural bilinear pairing $\langle\, , \rangle$. We will find a description for the maximal isotropic subspaces of W.

Let $U \subseteq V$ be a subspace of V. Let $Ann(U)$ denote the subspace of V^* which annihilates U, that is $Ann(U) = \{\xi \in V^* \,|\, \xi|_U = 0\}$. Now set $L(U) = U \oplus Ann(U) \subset V \oplus V^*$. It is straightforward to see that $L(U)$ is a maximal isotropic subspace of W. Let $\xi_1, \xi_2, \ldots, \xi_k$ be a basis for $Ann(U)$. Then $\varphi = \xi_1 \wedge \xi_2 \wedge \cdots \wedge \xi_k \in \wedge^* V^*$ is easily seen to be a pure spinor for $L(U)$. One sees that every pure spinor for $L(U)$ has the form $c\varphi$ for some non-zero constant c.

Let $\epsilon \in \wedge^2 U^*$ and set

$$L(U, \epsilon) = \{X + \xi \,|\, X \in U,\, \xi|_U = i_X \epsilon\}.$$

For every (U, ϵ), the space $L(U, \epsilon)$ is maximal isotropic. The space $L(U)$ is a special case of this construction since $L(U) = L(U, 0)$. Now we would like to find a pure spinor for $L(U, \epsilon)$. For this we observe the following trick. Given $B \in \wedge^2 V^*$ we let B act as a B-shift in the usual manner:

$$e^B(X + \xi) = X + \xi + i_X B,$$

where $X \in V$, $\xi \in V^*$. Choose B such that $B|_U = \epsilon$ which is always possible. We then find $L(U, \epsilon) = e^B L(U)$. It follows that every pure spinor φ for $L(U, \epsilon)$ has the form

$$\varphi = ce^{-B} \wedge \Omega, \tag{3.22}$$

where $\Omega = \xi_1 \wedge \cdots \wedge \xi_k$ with $\xi_1, \ldots, \xi_k$ a basis for $Ann(U)$.

Proposition 3.2. *Every maximal isotropic subspace of $W = V \oplus V^*$ has the form $L(U, \epsilon)$ for some (U, ϵ).*

Proof. Let L be a maximal isotropic subspace of W. Let $\rho : W \to V$ be the projection to V and set $U = \rho(L)$. For each $X \in U$ we can choose a lift of X to an element $X + \xi \in L$. If we do this for a basis of U we get a linear map $\tau : U \to V^*$ such that $\{X + \tau(X) \,|\, X \in U\}$ is a subspace of L. Let K be the kernel of $\rho|_L$. Since L is isotropic we see that $K \subseteq Ann(U)$ and since L is maximal isotropic we find by dimensional reasons that $K = Ann(U)$. We find therefore that

$$L = \{X + \tau(X) + \xi' \,|\, X \in U,\, \xi' \in Ann(U)\}. \tag{3.23}$$

Let $\epsilon : U \to U^*$ be the composition of $\tau : U \to V^*$ with the restriction map $V^* \to U^*$. We see immediately from (3.23) that L can also be expressed as

$$L = \{X + \xi \,|\, X \in U,\, \xi|_U = \epsilon(X)\}.$$

To complete the proof it suffices to show that $\epsilon : U \to U^*$ is skew-symmetric, that is ϵ is an element of $\wedge^2 U^*$. This follows almost immediately since L is isotropic. □

Corollary 3.1. *Every maximal isotropic subspace of $V \oplus V^*$ has a pure spinor, unique up to scale.*

Proof. This follows since we already determined the pure spinors for each $L(U, \epsilon)$ as given in (3.22). □

Let J be an almost generalized complex structure on $E = TM \oplus T^*M$. The i-eigenbundle L is a bundle of maximal isotropics, so at every point there is a pure spinor for L unique up to scale. We see from this that J determines a complex line bundle $K \subset S$ such that the pure spinors of J are precisely the non-vanishing sections of K. It follows that any almost generalized complex structure pure spinors exist locally, but not necessarily globally. The bundle K will be called the *canonical bundle* of J.

We have seen that every almost generalized complex structure locally admits a pure spinor. Consider now the converse, that is let φ be a (locally defined) complex pure spinor and let $L = Ann(\varphi)$ be the corresponding bundle of maximal isotropics. We ask when is L the i-eigenbundle of an almost generalized complex structure. It is clear that this is the case if and only if $L \cap \overline{L} = 0$, for then we have a decomposition $E_{\mathbb{C}} = L \oplus \overline{L}$ and we can define an almost generalized complex structure J by letting it act as multiplication by i on L and by $-i$ on $\overline{L}$.

So if φ is a complex pure spinor we are interested in determining when the condition $Ann(\varphi) \cap Ann(\overline{\varphi}) = 0$ holds.

Proposition 3.3 ([28], III.2.4.). *Let φ, ψ be real or complex pure spinors. Then $Ann(\varphi) \cap Ann(\psi) = 0$ if and only if $\langle \varphi, \psi \rangle \neq 0$, where $\langle\, , \rangle$ is the Mukai pairing on forms.*

Thus a pure spinor φ determines an almost generalized complex structure if and only if $\langle \varphi, \overline{\varphi} \rangle \neq 0$.

Let V be a real vector space, $W = V \oplus V^*$ and let φ be a complex pure spinor. We have seen that every such pure spinor has the form $\varphi = ce^A \wedge \Omega$ where $\Omega = \xi_1 \wedge \cdots \wedge \xi_k$ for a collection $\xi_1, \ldots, \xi_k$ of linearly independent elements of $V^*_{\mathbb{C}}$, c is a non-zero constant and $A \in \wedge^2 V^*_{\mathbb{C}}$. Let us further write $A = B + i\omega$ where $B, \omega \in \wedge^2 V^*$ are real. Suppose also that V has real dimension m. Then one finds

$$\langle \varphi, \overline{\varphi} \rangle = |c|^2 (2i\omega)^{n-k} \wedge \Omega \wedge \overline{\Omega}.$$

So φ determines a generalized complex structure on $V \oplus V^*$ if and only if

$$\omega^{n-k} \wedge \Omega \wedge \overline{\Omega} \neq 0.$$

For instance when $k = 0$ we are left with $\omega^n \neq 0$, so ω is a symplectic form and when $k = n$ we are left with $\Omega \wedge \overline{\Omega} \neq 0$, so Ω is a complex volume form.

Definition 3.8. Let φ be a complex pure spinor for $(V \oplus V^*)_{\mathbb{C}}$ such that $\langle \varphi, \overline{\varphi} \rangle \neq 0$. Then we have seen that φ has the form $\varphi = ce^{B+i\omega} \wedge \Omega$, where $\Omega = \xi_1 \wedge \cdots \wedge \xi_k$ for some linearly independent elements $\xi_1, \ldots, \xi_k \in V^*_{\mathbb{C}}$. The *type* of φ is the integer k.

If J is an almost generalized complex structure on M we have at each point a type, so we get a type function $t : M \to \mathbb{Z}$. Loosely speaking the type function measures the number of complex dimensions of M. More precisely for each $m \in M$, J_m induces a complex structure on $(T^*M)_m \cap J_m(T^*M)_m$ and $t(m)$ is the complex dimension of this space. If M has real dimension $m = 2n$, then the type at any point is an integer between 0 and n. Type 0 means that there are no complex dimensions and M has an almost symplectic structure. Type n means that all dimensions of M are complex and M has an almost complex structure.

It is important to note that the type function is generally not continuous. However the type of φ is the degree of the lowest non-zero component of φ, so the type is upper semi-continuous. In other words for each $x \in M$ there is a neighborhood of x in which type can only decrease. In particular if J is of type 0 at a point x, then it is type 0 throughout a neighborhood of x. In this sense type 0 (symplectic type) represents the most generic type of generalized complex structure, while maximal type (complex type) the least generic.

Note further that every pure spinor is either of even or odd degree, which follows immediately from the general form (3.22). If φ is a pure spinor on a manifold M, it is clear also that the parity of φ has to be locally constant. We see that the type of M has the same parity as φ, so in particular type can only change by even amounts. For example a generalized almost complex structure on a 2-dimensional manifold M has a type function $t : M \to \{0, 1\}$ of definite parity. So in the 2-dimensional case, the type actually has to be locally constant. In dimension 4 or higher however the type function will generally experience jumps.

We have seen that an almost generalized complex structure J has a canonical bundle $K \subset S$ the non-vanishing sections of which are the pure spinors for J. By considering the Clifford action of $E_{\mathbb{C}}$ we can extend K to a full decomposition of S. Let M have dimension $m = 2n$ and set $S_{-n} = K$. Let S_{k-n} be the subbundle of S spanned by elements of the form $\gamma_{a_1}\gamma_{a_2}\cdots\gamma_{a_k}\varphi$, where $a_1, \ldots, a_k$ are sections of L^* and φ is a pure spinor. Since L^* is maximal isotropic and L is the annihilator of K one finds that $S_{k-n} \simeq \wedge^k L^* \otimes K$ and we get a decomposition

$$S = S_{-n} \oplus S_{1-n} \oplus \cdots \oplus S_{n-1} \oplus S_n. \tag{3.24}$$

In particular this decomposition gives a grading on S in which Clifford multiplication by L^* has degree 1 and Clifford multiplication by L has degree -1. It follows that $S_n = \overline{K}$ is the canonical bundle for $L^* = \overline{L}$ and from this it further follows more generally that $\overline{S_j} = S_{-j}$.

So far we have not said much about integrability and how it relates to pure spinors. Let H be a closed 3-form. As usual we have a twisted differential $d_H = d + H\wedge : \Gamma(S) \to \Gamma(S)$. Now if J is a generalized almost complex structure then we have the decomposition (3.24) and we may define operators

$$\begin{aligned} \partial &: \Gamma(S_i) \to \Gamma(S_{i+1}), \\ \overline{\partial} &: \Gamma(S_i) \to \Gamma(S_{i-1}) \end{aligned}$$

by letting ∂ be the component of d_H of degree 1 and $\overline{\partial}$ the component of degree -1. As the notation suggests, $\overline{\partial}$ is the complex conjugate of ∂.

Proposition 3.4. *The almost generalized complex structure J is H-integrable if and only if we have an equality*

$$d_H = \partial + \overline{\partial},$$

in other words if and only if d_H sends $\Gamma(S_i)$ to $\Gamma(S_{i-1}) \oplus \Gamma(S_{i+1})$ for all i.

Proof. Let us first show that J is H-integrable if and only if d_H sends $\Gamma(S_i)$ to $\Gamma(S_{i-1}) \oplus \Gamma(S_{i+1})$ for $i = -n$. Then we show that the cases $i > -n$ follow from the $i = -n$ case.

Since integrability is a local condition we may assume that φ is a pure spinor for J, so φ is a non-vanishing section of $K = S_{-n}$. J is H-integrable if and only if for all sections $a, b \in \Gamma(L)$ we have $\gamma_{\{a,b\}_H}\varphi = 0$. Now we use the derived bracket property (3.20) to see that

$$\begin{aligned} \gamma_{\{a,b\}_H}\varphi &= [[d_H, \gamma_a], \gamma_b]\varphi \\ &= (d_H\gamma_a\gamma_b + \gamma_a d_H \gamma_b - \gamma_b d_H \gamma_a - \gamma_b\gamma_a d_H)\varphi \end{aligned}$$

$$= -\gamma_b \gamma_a d_H \varphi$$
$$= \gamma_a \gamma_b d_H \varphi.$$

Now $d_H\varphi$ has opposite parity to φ, so we know that $d_H\varphi$ is a section of $S_{-n+1} \oplus S_{-n+3} \oplus \cdots$. But from the above calculation we see that $\gamma_{\{a,b\}_H}\varphi = 0$ for all $a, b \in \Gamma(L)$ if and only if $d_H\varphi \in \Gamma(S_{-n+1})$. Thus J is H-integrable if and only if $d_H\varphi = \overline{\partial}\varphi = (\partial + \overline{\partial})\varphi$, since $\partial\varphi = 0$ on the grounds that $S_{-n-1} = 0$. We find therefore that φ is H-integrable if and only if d_H and $(\partial + \overline{\partial})$ agree on all non-vanishing sections of S_{-n}. But since d_H and $\partial + \overline{\partial}$ are first order differential operators this is enough to see that d_H and $(\partial + \overline{\partial})$ agree on all sections of S_{-n}.

Now we need only show that if J is integrable then d_H sends $\Gamma(S_i)$ to $\Gamma(S_{i-1}) \oplus \Gamma(S_{i+1})$ for all i and we already know the case $i = -n$ holds. Assume by induction the result holds for $i < j$ where $j \geq -n$. To prove the $i = j$ case it suffices to consider sections of S_j of the form $\omega' = \gamma_b\omega$ where $\omega \in \Gamma(S_{j-1})$ and $b \in \Gamma(L^*)$. To show that $d_H\omega'$ is a section of $S_{j-1} \oplus S_{j+1}$ we will show that for any $a \in \Gamma(L)$ we have that $\gamma_a d_H \omega'$ is a section of $S_{j-2} \oplus S_j$.

From (3.20) we have

$$\gamma_a d_H \gamma_b \omega = \gamma_{\{a,b\}_H}\omega - d_H \gamma_a \gamma_b \omega + \gamma_b d_H \gamma_a \omega + \gamma_b \gamma_a d_H \omega.$$

By induction we see that each terms on the right hand side is a section of $S_{j-2} \oplus S_j$ as required. □

Remark 3.1. In the proof of Proposition 3.4 we argued that if d_H and $(\partial+\overline{\partial})$ agree on all non-vanishing pure spinors then J is integrable. In fact, it suffices for d_H and $(\partial+\overline{\partial})$ to coincide on a single non-vanishing pure spinor since d_H and $(\partial+\overline{\partial})$ are first order differential operators and one can check that their first order terms coincide.

Suppose that J is H-integrable. We have just shown that $d_H = \partial + \overline{\partial}$. Now since $d_H^2 = 0$ we find, by comparing degrees that $\partial^2 = \overline{\partial}^2 = \partial\overline{\partial} + \overline{\partial}\partial = 0$. In particular, $\overline{\partial}$ is a differential. We may therefore define corresponding cohomology groups

$$H^k_{\overline{\partial}}(M) = \frac{Ker(\overline{\partial} : \Gamma(S_k) \to \Gamma(S_{k+1}))}{Im(\overline{\partial} : \Gamma(S_{k-1}) \to \Gamma(S_k))}.$$

The groups $H^k_{\overline{\partial}}(M)$ are a generalized complex analogue of Dolbeault cohomology. One may ask what is the relation between the $\overline{\partial}$ cohomology and the twisted cohomology. In general they are related by a spectral sequence [25] which is a generalization of the Frölicher spectral sequence [62] for complex manifolds. There is however a special case in which the relation is more direct:

Definition 3.9. A generalized complex manifold is said to satisfy the $\partial\overline{\partial}$-lemma if

$$Ker(\partial) \cap Im(\overline{\partial}) = Im(\partial) \cap Ker(\overline{\partial}) = Im(\partial\overline{\partial}).$$

This is a generalization of the usual $\partial\overline{\partial}$-lemma in complex geometry that holds for compact Kähler manifolds [30]. A generalization of the argument used in [30] shows the following:

Proposition 3.5. *Let M, J be a generalized complex manifold where J is H-integrable and such that the $\partial\overline{\partial}$-lemma holds. There are natural inclusions of $\overline{\partial}$ cohomology into twisted cohomology such that*

$$H^k(M,H)_{\mathbb{C}} = \bigoplus_{j=k\,(\mathrm{mod}\,2)} H^{j+p-n}_{\overline{\partial}}(M),$$

where M has dimension $2n$ and $p = 0$ or 1 according to whether pure spinors for J are even or odd degree forms. Moreover $H^j_{\overline{\partial}}(M)$ and $H^{-j}_{\overline{\partial}}(M)$ are complex conjugates. Therefore we have a natural Hodge structure on twisted cohomology.

The inclusions $H^j_{\overline{\partial}}(M) \to H^{j+p-n}(M,H)_{\mathbb{C}}$ of Proposition 3.5 arise as follows. For each $\overline{\partial}$ cohomology class one can find a representative that is d_H-closed. Moreover any two such representatives define the same twisted cohomology class.

Example 3.3. We give here an example from [38] of a generalized complex manifold which exhibits the type change phenomenon.

Let $M = \mathbb{R}^4 = \mathbb{C}^2$ with complex coordinates z_1, z_2 and we define a generalized complex structure on M by giving a pure spinor

$$\varphi = z_1 + dz_1 \wedge dz_2.$$

At points where $z_1 \neq 0$, φ has type 0 (symplectic type). In fact we can rewrite φ as $\varphi = z_1 e^{\frac{dz_1\wedge dz_2}{z_1}}$ and we see that the corresponding symplectic form is $\omega = Im(\frac{dz_1\wedge dz_2}{z_1})$.

At points where $z_1 = 0$ we have $\varphi = dz_1 \wedge dz_2$ so φ has type 2 (complex type) and the holomorphic volume form is $\Omega = dz_1 \wedge dz_2$. This is the usual holomorphic volume form on $\mathbb{C}^2$, so the underlying complex structure is just the usual complex structure (restricted to the tangent spaces over the submanifold $z_1 = 0$).

We see that φ has type 0 everywhere except on the submanifold $z_1 = 0$ where the type is 2. It remains to check that the almost generalized complex structure corresponding to φ is integrable (with H=0). For this we observe that $d\varphi = dz_1 = -i_{\partial_2}\varphi = \gamma_{-\partial_2}\varphi$ and that $\partial_2 \in \Gamma(L^*)$ since $\overline{\partial}_2$ annihilates φ. By Proposition 3.4 and Remark 3.1 we have established integrability.

In the previous example we showed the phenomenon of type change. Now we consider the structure of generalized complex manifolds away from points where the type changes. A point $x \in M$ in a generalized complex manifold is called *regular* if there is a neighborhood of x in which the type is constant. For a regular point the local geometry is completely determined by the type:

Proposition 3.6 ([38], 4.35 (Generalized Darboux Theorem)). *Let x be a regular point of type k in a generalized complex manifold M of dimension $m = 2n$. Equip $\mathbb{R}^{2n-2k}$ with the standard symplectic structure and $\mathbb{C}^k$ with the standard complex structure so that $\mathbb{R}^{2n-2k} \times \mathbb{C}^k$ is naturally a generalized complex manifold of dimension $2n$ and of type k. There exists a neighborhood U of x and an inclusion $U \to \mathbb{R}^{2n-2k} \times \mathbb{C}^k$ so that the generalized complex structures on U induced from M and from $\mathbb{R}^{2n-2k} \times \mathbb{C}^k$ coincide, up to a B-shift by some 2 form on U.*

In other words if one takes into account diffeomorphisms and B-shifts then every regular point in a generalized complex manifold has a neighborhood that looks like an open set in $\mathbb{R}^{2n-2k} \times \mathbb{C}^k$.

4 T-duality

Our goal is to introduce T-duality from the perspectives of topology and geometry. However the idea of T-duality has its origins in physics, so let us begin with some historical remarks. It had been observed at least as far back as 1984 that there is a duality between string theory compactified over a circle of radius R and of radius $1/R$ (in suitable units of length) [46]. Many subsequent papers further studied string theory compactifications over a circle or multiple circles (toroidal compactification). The transformation on the bosonic string theory fields underlying this duality are known as the *Buscher rules* [20, 21]. In the simplest kind of toroidal compactification one considers string theory on a product $X = M \times T^d$, where X represents the space-time in the theory and T^d is a flat d-dimensional torus representing d compactified dimensions. One thinks of T^d as being small, so that space-time is effectively replaced by the lower dimensional space M.

One can also consider more complicated compactifications, for instance instead of a product X could be a principal torus bundle over M or something yet more complicated. For the purposes of these notes we will always take X to be a principal torus bundle over a space M. The Buscher rules provide transformation laws that can be applied over a local trivialization of X. Actually they also require a local trivialization of the Neveu-Schwarz B-field of string theory. Recall that the NS B-field can be thought of as a gerbe with connection. So the Buscher rules are local transformation rules that apply only when the torus bundle and B-field have been trivialized. This leads to a question of whether the Buscher rules can be understood globally. In 2004 Bouwknegt, Evslin and Mathai put forth an approach to T-duality which would make global sense of the Buscher rules for topologically non-trivial torus bundles and B-fields [12]. This approach to T-duality has been subsequently expanded upon considerably [1, 5, 6, 7, 13, 14, 16–19, 26, 45, 52, 53] leading to new perspectives and generalizations. This approach to T-duality is known as *topological T-duality*, since it can be expressed as a purely topological duality which does not require local data such a metric. Thus topological T-duality strips away the local field content of string theory leaving only a global topological remainder. However it is possible to refine topological T-duality to a duality between different geometric structures. We will see in Section 5 that such a refinement is possible using Courant algebroids, leading to T-duality transformations between generalized complex structures and generalized metrics. The relationship between T-duality and Courant algebroids was formulated by Cavalcanti and Gualtieri in [25, 26, 38] and has also been considered in [8, 39, 41, 43, 44].

4.1 T-duality from the Buscher rules

The Buscher rules are a set of equations which describe locally how the bosonic fields in string theory transform under T-duality. More precisely let X be a manifold which will be the spacetime of the string theory. If we consider either bosonic string theory or the NSNS-sector of the type II superstring theories the field content is as follows:

- A Riemannian metric g,
- The Neveu-Schwarz B-field B,
- A function φ called the *dilaton.*

Locally the B-field can be thought of as just a 2-form. As we have previously discussed a better way to interpret the B-field globally is that it is the curving for a gerbe with connection. If the gerbe is defined by a collection of line bundles with connections $\{L_{\alpha\beta}\}$, then the curving is a collection of locally defined 2-forms $\{B_\alpha\}$. The field strength of the B-field is the globally defined closed 3-form H such that $H|_{U_\alpha} = dB_\alpha$. The dilaton field φ turns out to be not important as far as the topological aspects of T-duality are concerned, so we will say nothing more about it. Notice that the fields g and $\{B_\alpha\}$ can be thought of as a generalized metric on a twisted generalized tangent bundle.

The Buscher rules apply to spacetimes in which one or more dimensions are circles. In other words let us assume that X is a locally trivial circle bundle $\pi : X \to M$ over a space M. The Buscher rules are also local in nature so we restrict to a coordinate chart on the base M over which X is trivializable. Let $U \subset M$ be a coordinate chart with coordinates $x^1, \ldots, x^m$ and identify the restriction of X over U with the product $U \times S^1$, where we view the circle S^1 as $\mathbb{R}/\mathbb{Z}$ with the standard coordinate $t = x^0$. The local field content consists of a metric g and a 2-form B on $U \times S^1$. Let $\hat{S}^1$ denote the dual circle $\hat{S}^1 = \mathbb{R}^*/\mathbb{Z}^*$ with coordinate $\hat{x}^0$. We assume also that g and B are S^1-invariant. The Buscher rules are a set of equations giving a metric $\hat{g}$ and 2-form $\hat{B}$ on $U \times \hat{S}^1$ in terms of g, B:

$$\begin{aligned}
\hat{g}_{\mu\nu} &= g_{\mu\nu} - \frac{1}{g_{00}}(g_{\mu 0}g_{\nu 0} - B_{\mu 0}B_{\nu 0}),\\
\hat{B}_{\mu\nu} &= B_{\mu\nu} - \frac{1}{g_{00}}(g_{\mu 0}B_{\nu 0} - g_{\nu 0}B_{\mu 0}),\\
\hat{g}_{\mu 0} &= \frac{B_{\mu 0}}{g_{00}},\\
\hat{B}_{\mu 0} &= \frac{g_{\mu 0}}{g_{00}},\\
\hat{g}_{00} &= \frac{1}{g_{00}}.
\end{aligned}$$

Notice in particular that g_{00} represents the radius of S^1 and $\hat{g}_{00}$ the radius of the dual circle $\hat{S}^1$, so in particular the Buscher rules implement the operation of inverting the radius which is a feature of T-duality. These equations may look very mysterious, but we will see that they make a lot more sense once the global nature of T-duality is properly understood.

Let us attempt to simplify the Buscher rules and try to understand them more globally. First let us assume for simplicity that $g_{00} = 1$, so that $\hat{g}_{00} = 1$. Since $g, \hat{g}$ are invariant metrics they determine connections $A, \hat{A}$, which have the form $A = dx^0 + A_\mu dx^u = dx^0 + g_{\mu 0} dx^\mu$ and $\hat{A} = d\hat{x}^0 + \hat{A}_\mu dx^\mu = d\hat{x}^0 + \hat{g}_{\mu 0} dx^\mu$. We can then use the connection A to decompose g and B. Since A is the connection determined by g we have that g is of the form $g = g_M + A \otimes A$, for some metric g_M on the base M. Likewise the decomposition of B has the form $B = B_2 + B_1 \wedge A$, where B_2 is a 2-form on the base and B_1 a 1-form on the base. If we similarly write $\hat{g} = \hat{g}_M + \hat{A} \otimes \hat{A}$, $\hat{B} = \hat{B}_2 + \hat{B}_1 \wedge \hat{A}$, then the Buscher rules take the remarkably simpler form

$$
\begin{aligned}
\hat{g}_M &= g_M, \\
\hat{B}_2 &= B_2 + A_\mu dx^\mu \wedge \hat{A}_\nu dx^\nu, \\
\hat{B}_1 &= A_\mu dx^\mu, \\
\hat{A}_\mu dx^\mu &= B_1.
\end{aligned}
$$

Expressed in this form we see that the main feature of the Buscher rules is that it exchanges the 1-form part of the connection on one side with the 1-form part of the B-field on the other side. Let $F = dA$, $\hat{F} = d\hat{A}$ be the curvature 2-forms of the connections and let $H = dB$, $\hat{H} = d\hat{B}$ be the curvature 3-forms of the B-fields. Let us define a 3-form H_3 on the base by $H_3 = dB_2 - (\hat{A}_\mu dx^\mu) \wedge F = d\hat{B}_2 - (A_\mu dx^\mu) \wedge \hat{F}$. Then we have

$$
\begin{aligned}
H &= H_3 + \hat{F} \wedge A, \\
\hat{H} &= H_3 + F \wedge \hat{A}.
\end{aligned}
$$

In particular this means that if we integrate H over the circle S^1, we get the curvature $\hat{F}$ of the dual circle bundle and if we integrate $\hat{H}$ over the dual circle $\hat{S}^1$ we get the curvature F of the circle bundle corresponding to S^1. Lastly we also observe that $\hat{H} - H = d(A \wedge \hat{A})$, so that H and $\hat{H}$ differ by an exact term on the fiber product space $U \times S^1 \times \hat{S}^1$. The topological implications of these relations will form the basis for topological T-duality in the next section.

4.2 Topological T-duality for circle bundles

Let us begin with topological T-duality for circle bundles, along the lines of Bouwknegt, Evslin and Mathai. Although we are primarily interested in smooth manifolds, topological T-duality applies to much more general spaces, for instance assuming all spaces involved to be locally contractible will suffice.

We are given a space X which could be thought of as string theory space-time. The space X is a principal circle bundle $\pi : X \to M$ over a base space M and there is a bundle gerbe $\mathcal{G}$ on X which has a Dixmier-Douady class $[\mathcal{G}] = h \in H^3(X, \mathbb{Z})$. For this section it is possible to work directly with the cohomology class h and forget about the gerbe $\mathcal{G}$. When we proceed to higher rank T-duality we will see that it is much better to think in terms of gerbes rather than their Dixmier-Douady classes. T-duality is a relation between a pair of such spaces $X, \hat{X}$. That is to say $X, \hat{X}$ are both circle bundles equipped with gerbes. Shortly we will give a definition of the T-duality relation, but first let us establish some conventions.

Let $X, \hat{X}$ be spaces, $\mathcal{G}$ a gerbe on X and $\hat{\mathcal{G}}$ a gerbe on $\hat{X}$. Assume that $X, \hat{X}$ are principal circle bundles over a common base space M. Let $\pi : X \to M$, $\hat{\pi} : \hat{X} \to M$ be the projections. Let $C = X \times_M \hat{X}$ be the fiber product, which will also be called the *correspondence space*. We may define various projection maps $p, \hat{p}, q$ forming a commutative diagram:

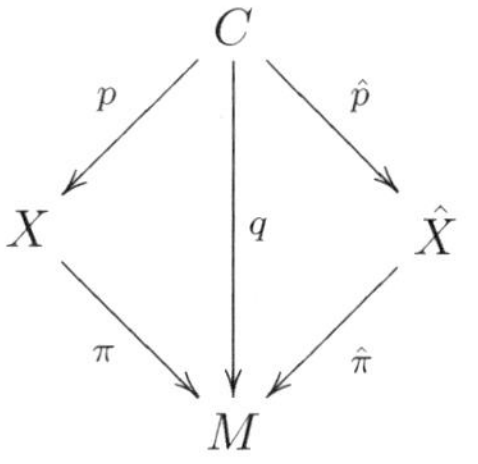

Let $[\mathcal{G}] = h \in H^3(X, \mathbb{Z})$ and $[\hat{\mathcal{G}}] = \hat{h} \in H^3(\hat{X}, \mathbb{Z})$ be the Dixmier-Douady classes of $\mathcal{G}, \hat{\mathcal{G}}$.

The cohomology of X and M are related through the *Gysin sequence* [10] which is a long exact sequence:

$$\cdots \to H^k(M, \mathbb{Z}) \xrightarrow{\pi^*} H^k(X, \mathbb{Z}) \xrightarrow{\pi_*} H^{k-1}(M, \mathbb{Z}) \xrightarrow{c\smile} H^{k+1}(M, \mathbb{Z}) \to \cdots$$

In the above sequence the map $c \smile$ denotes the cup product with the first Chern class $c = c_1(X) \in H^2(M, \mathbb{Z})$. The map $\pi_* : H^k(X, \mathbb{Z}) \to H^{k-1}(M, \mathbb{Z})$ is known as the *Gysin homomorphism* or *push-forward map*. There are a number of ways it can be defined. If X and M are compact oriented manifolds then π_* can be obtained from the push-forward map in homology together with Poincaré duality. If X, M are smooth manifolds and we use de Rham cohomology instead of singular cohomology then π_* is integration over the fiber. Perhaps the definition most appropriate to the Gysin sequence is to define π_* using the Serre spectral sequence, which is the approach used in [10].

Definition 4.1. Let $c = c_1(X) \in H^2(M, \mathbb{Z})$ and $\hat{c} = c_1(\hat{X}) \in H^2(M, \mathbb{Z})$ be the Chern classes of $X, \hat{X}$. We say that $(X, \mathcal{G})$ and $(\hat{X}, \hat{\mathcal{G}})$ are *T-dual* if:

(T1) $\pi_*(h) = \hat{c}$,
(T2) $\hat{\pi}_*(\hat{h}) = c$,
(T3) $p^*(h) = \hat{p}^*(\hat{h})$.

The axioms (T1)–(T3) are succinct, but arguably not very enlightening. Later we will revisit the connection between these axioms and the Buscher rules. Axioms (T1),(T2) state that there is an interchange between the topology of the torus bundle on one side and the Dixmier-Douady class of the gerbe on the dual side. Axiom (T3) states that when the gerbes $\mathcal{G}, \hat{\mathcal{G}}$ are pulled back to the correspondence space C, they become isomorphic.

Proposition 4.1 ([5, 12, 16]). *For any principal circle bundle $\pi : X \to M$ and gerbe $\mathcal{G}$ on X there exists a unique T-dual pair $(\hat{X}, \hat{\mathcal{G}})$. Uniqueness means that if $(\hat{X}', \mathcal{G}')$ is another T-dual pair then there is an isomorphism $\phi : \hat{X} \to \hat{X}'$ of principal circle bundles over M such that $\hat{\mathcal{G}}$ and $\phi^*(\hat{\mathcal{G}}')$ are stably isomorphic.*

Proof (sketch). Let $c = c_1(X)$ be the first Chern class of X and $h = [\mathcal{G}]$ the Dixmier-Douady class of $\mathcal{G}$. To construct the T-dual circle bundle $\hat{X} \to M$, we need only specify what the Chern class of $\hat{X}$ is. But by axiom (T1) we must have $\hat{c} = \pi_*(h)$, so this determines the T-dual space $\hat{\pi} : \hat{X} \to M$ up to principal bundle isomorphism and by construction (T1) is satisfied.

The construction of the T-dual gerbe $\hat{\mathcal{G}}$ is more complicated. First we observe that by the Gysin sequence for $\pi : X \to M$ we have $c \smile \pi_*(h) = c \smile \hat{c} = 0$. Now if we feed this into the Gysin sequence for $\hat{\pi} : \hat{X} \to M$, we find that there exists a class $h' \in H^3(\hat{X}, \mathbb{Z})$ such that $\hat{\pi}_*(h') = c$. Now while h' satisfies (T2), it need not satisfy (T3). However, by the Gysin sequence again we find that h' is only unique up to the image of π^*. In other words we want to find a class $\alpha \in H^3(M, \mathbb{Z})$ such that $\hat{h} = h' + \pi^*(\alpha)$ satisfies (T3). This will prove existence. To find α we use a spectral sequence argument. Set $d = p^*(h) - \hat{p}^*(h') \in H^3(C, \mathbb{Z})$. Then by considering the Serre spectral sequence on C we find (see [5]) that d is a pullback $d = q^*(\alpha)$ of some $\alpha \in H^3(M, \mathbb{Z})$. Therefore (T3) is satisfied by taking $\hat{h} = h' + \hat{\pi}^*(\alpha)$.

For uniqueness suppose $\hat{h}, \hat{h}'$ are such that (T2),(T3) both hold and set $d = \hat{h}' - \hat{h}$. Then we have that $\hat{\pi}_*(d) = 0$ and $\hat{p}^*(d) = 0$. Another spectral sequence argument [5] shows that d must have the form $d = \hat{\pi}^*(c \smile e)$ for some class $e \in H^1(M, \mathbb{Z})$. So $\hat{h}' = \hat{h} + \hat{\pi}^*(c \smile e)$. Recall that $H^1(M, \mathbb{Z})$ describes homotopy classes of maps $M \to S^1$. Let $f : M \to S^1$ be such a map representing the class e. Then f determines a principal bundle automorphism $\phi_f : \hat{X} \to \hat{X}$. One shows [16] that the action of ϕ_f on cohomology is as follows:

$$\phi_f^*(x) = x + \hat{\pi}^*(\hat{\pi}_*(x) \smile e)$$

for any $x \in H^*(\hat{X}, \mathbb{Z})$. From this and (T2) we immediately see that $\phi_f^*(\hat{h}') = \hat{h}$ proving uniqueness. □

4.3 T-duality and differential forms

We will assume the spaces involved are smooth manifolds and show how T-duality can be viewed at the level of differential forms.

Let us begin with a pair of T-dual circle bundles with gerbes $(X, \mathcal{G}), (\hat{X}, \hat{\mathcal{G}})$. We let $h, \hat{h}$ be the Dixmier-Douday classes of $\mathcal{G}, \hat{\mathcal{G}}$ and $h_{\mathbb{R}} \in H^3(X, \mathbb{R})$, $\hat{h}_{\mathbb{R}} \in H^3(\hat{X}, \mathbb{R})$ the images of $h, \hat{h}$ in real cohomology. Next choose closed 3-forms $H \in \Omega^3(X)$, $\hat{H} \in \Omega^3(\hat{X})$ representing $h_{\mathbb{R}}, \hat{h}_{\mathbb{R}}$. We may also assume by averaging that $H, \hat{H}$ are invariant under the S^1-actions.

Let A be a connection for the principal circle bundle $X \to M$, that is A is an invariant 1-form on X such that if v is the vertical vector field generating the

S^1-action[3], then $i_v A = 1$. There is a closed 2-form on M such that $dA = \pi^*(F)$, the curvature of A. Note that the cohomology class of F is the image of the Chern class of X in real cohomology.

Proposition 4.2. *For any choice of representative H and connection A it is possible to choose also a representative $\hat{H}$ and connection $\hat{A}$ for $\hat{X}$ such that*

$$H = \pi^*(H_3) + A \wedge \pi^*(\hat{F}), \tag{4.1}$$
$$\hat{H} = \hat{\pi}^*(H_3) + \hat{A} \wedge \hat{\pi}^*(F). \tag{4.2}$$

Where H_3 is a 3-form on M and $\hat{F}$ is the curvature of $\hat{A}$.

We claim that this point of view helps to clarify the T-duality axioms (T1)–(T3). For instance if we take H and integrate over the fibers we find from (4.1) that

$$\int_{S^1} H = \hat{F},$$

where we use $\int_{S^1} : \Omega^*(X) \to \Omega^{*-1}(M)$ to denote the operation of integrating differential forms on X over the fibers. If we pass to de Rham cohomology we find that

$$\pi_*(h_{\mathbb{R}}) = \hat{c}_{\mathbb{R}}$$

which is axiom (T1), except at the level of real cohomology. Similarly from (4.2) we find $\int_{S^1} \hat{H} = F$ and thus $\hat{\pi}_* \hat{h}_{\mathbb{R}} = c_{\mathbb{R}}$, which is (T2) at the level of real cohomology. Finally let us define $\mathcal{B} \in \Omega^2(C)$ by

$$\mathcal{B} = p^*(A) \wedge \hat{p}^*(\hat{A}).$$

Combining (4.1) and (4.2) we find

$$\hat{p}^*(\hat{H}) - p^*(H) = d\mathcal{B}.$$

Passing to cohomology we find $p^*(h_{\mathbb{R}}) = \hat{p}^*(\hat{h}_{\mathbb{R}})$, which of course is (T3) at the level of real cohomology. So we see that the equations (4.1), (4.2) imply the axioms (T1)–(T3) for T-duality, except only at the level of real cohomology. In Equations (4.1) and (4.2) we see that $\hat{H}$ and H are related by exchanging the connections $A, \hat{A}$. We have already seen that these relations are very closely related to the Buscher rules.

We saw in Section 3.3 that in the presence of a closed 3-form H on X, we may consider the twisted differential $d_H = d + H\wedge$ and twisted cohomology groups $H^*(X, H)$. We may do the same for $(\hat{X}, \hat{H})$ leading to the twisted cohomology $H^*(\hat{X}, \hat{H})$. A fundamental aspect of T-duality is that it entails an isomorphism of twisted cohomology, with an appropriate degree shift. Let $S = \wedge^*(T^*X)$ be the bundle of forms on X and $\hat{S} = \wedge^*(T^*\hat{X})$ the bundle of forms on $\hat{X}$. We define the *T-duality transformation* $T : \Gamma(S) \to \Gamma(\hat{S})$ by

$$T(\omega) = \int_{S^1} e^{-\mathcal{B}} \wedge p^*(\omega), \tag{4.3}$$

[3] That is, we think of S^1 as $\mathbb{R}/\mathbb{Z}$, then ∂_t defines an element of the Lie algebra of S^1. We then take v to be the vector field on X associated to ∂_t through the principal bundle circle action

where the integration is over the fibers of $C \to \hat{X}$. Note the similarity of (4.3) to the Fourier transform, or to the Fourier-Mukai transform [48]. Notice also that due to the fiber integration, T sends even forms to odd forms and odd forms to even.

Proposition 4.3. *The T-duality transformation satisfies $d_{\hat{H}}T\omega = Td_H\omega$ for all $\omega \in \Gamma(S)$, so induces a morphism in cohomology $T_* : H^i(X,H) \to H^{i-1}(\hat{X},\hat{H})$. We have that T_* is an isomorphism, in particular the twisted cohomologies of $X, \hat{X}$ are isomorphic up to a shift in degree.*

Proof. We have

$$\begin{aligned} d_{\hat{H}}T\omega &= d_{\hat{H}} \int_{S^1} e^{-\mathcal{B}} \wedge p^*(\omega) \\ &= \int_{S^1} d_{\hat{p}^*(\hat{H})} e^{-\mathcal{B}} \wedge p^*(\omega) \\ &= \int_{S^1} e^{-\mathcal{B}} (d_{\hat{p}^*(\hat{H})} - d\mathcal{B}) \wedge p^*\omega \\ &= \int_{S^1} e^{-\mathcal{B}} \wedge p^*(d_H\omega) \\ &= Td_H\omega. \end{aligned}$$

So as claimed there is an induced map $T_* : H^i(X,H) \to H^{i-1}(\hat{X},\hat{H})$. To show that this is an isomorphism we make two claims. First, every twisted cohomology class on X or $\hat{X}$ has an invariant representative and secondly T restricted to invariant forms is an isomorphism $T : \Gamma(S)^{S^1} \to \Gamma(\hat{S})^{\hat{S}^1}$. The latter claim is easy to show by a direct calculation. The former claim is a little more complicated and can be shown by a spectral sequence argument [5]. Combining these arguments we see that T_* has to be both injective and surjective. □

4.4 Topological T-duality for principal torus bundles

So far we have seen topological T-duality for circle bundles. There are multiple directions in which this most basic form of T-duality can be expanded. The most obvious extension is to generalize from principal circle bundles to principal torus bundles which is what we will do in this section. For technical reasons we assume our spaces to be locally contractible and have the property that every subspace is paracompact.

There are complications in generalizing Definition 4.1 to the case of principal torus bundles. If $\pi : X \to M$, $\hat{\pi} : \hat{X} \to M$ are principal T^n-bundles, where $T^n = \mathbb{R}^n/\mathbb{Z}^n$ is the n-torus and $n > 1$, then axioms (T1),(T2) as written no longer make sense. It is possible to rephrase the two axioms (T1),(T2) in a way that still makes sense for higher rank torus bundles [17]. This is done by appealing to the Leray-Serre spectral sequences for $\pi : X \to M$, $\hat{\pi} : \hat{X} \to M$. However even then it turns out that the right definition of T-duality requires an additional axiom. In [7] we obtained a new definition of T-duality which is equivalent to the previous approach but which is a lot easier to state. In particular it does away with the need to involve spectral sequences in the definition. We will adopt the definition of T-duality for torus bundles in [7].

Let $\pi : X \to M$, $\hat{\pi} : \hat{X} \to M$ be principal T^n-bundles, define the correspondence space $C = X \times_M \hat{X}$ and maps $p, \hat{p}, q$ as in Section 4.2. Given a point $m \in M$ we let $T_m = \pi^{-1}(m)$, $\hat{T}_m = \hat{\pi}^{-1}(m)$ denote the fibers of $X, \hat{X}$ over m. In addition it follows that $q^{-1}(m) = T_m \times \hat{T}_m$.

Let V denote a real n-dimensional vector space, Λ a rank n lattice in V and $T^n = V/\Lambda$ the associated torus. For example we could take $V = \mathbb{R}^n$, $\Lambda = \mathbb{Z}^n$. Let $\hat{V}$ be another n-dimensional vector space, $\hat{\Lambda}$ a rank n lattice in $\hat{V}$ and $\hat{T}^n = \hat{V}/\hat{\Lambda}$ the associated torus. Of course T^n, $\hat{T}^n$ are isomorphic as Lie groups, but it will be convenient to think of T^n and $\hat{T}^n$ as distinct tori. Moreover we will think of $\pi : X \to M$ as a principal T^n-bundle and $\hat{\pi} : \hat{X} \to M$ as a principal $\hat{T}^n$-bundle.

Definition 4.2. Let $\pi : X \to M$, $\hat{\pi} : \hat{X} \to M$ be rank n principal torus bundles on M, $\mathcal{G}, \hat{\mathcal{G}}$ gerbes on $X, \hat{X}$ respectively. We say that the pairs $(X, \mathcal{G})$ and $(\hat{X}, \hat{\mathcal{G}})$ are *(topologically) T-dual* if the following conditions hold:

(TT1) For all $m \in M$ the restrictions $\mathcal{G}|_{T_m}$, $\hat{\mathcal{G}}|_{\hat{T}_m}$ are trivial.
(TT2) There exists an isomorphism $\gamma : p^*(\mathcal{G}) \to \hat{p}^*(\hat{\mathcal{G}})$.
(TT3) The isomorphism γ can be chosen to satisfy the Poincaré property, defined below.

Remark 4.1. Nowhere in the definition of topological T-duality do we actually use the fact that $X, \hat{X}$ are principal torus bundles (this includes the Poincaré property below). With a small modification this definition of T-duality can be applied to arbitrary torus bundles [7].

To explain the Poincaré property we assume axioms (TT1), (TT2) of Definition 4.2 hold. Choose a point $m \in M$. Then by (TT1) there exist trivializations $\tau_m : 1 \to \mathcal{G}|_{T_m}$ and $\hat{\tau}_m : 1 \to \hat{\mathcal{G}}|_{\hat{T}_m}$, where 1 denotes the trivial gerbe. On $q^{-1}(m) = T_m \times \hat{T}_m$ we have two isomorphisms $p^*(\mathcal{G}|_{T_m}) \to \hat{p}^*(\hat{\mathcal{G}}|_{\hat{T}_m})$. The first is given by the restriction of γ:

$$p^*(\mathcal{G}|_{T_m}) \xrightarrow{\gamma|_{T_m \times \hat{T}_m}} \hat{p}^*(\hat{\mathcal{G}}|_{\hat{T}_m}),$$

and the second by the composition

$$p^*(\mathcal{G}|_{T_m}) \xrightarrow{p^*(\tau_m^{-1})} 1 \xrightarrow{\hat{p}^*(\hat{\tau}_m)} \hat{p}^*(\hat{\mathcal{G}}|_{\hat{T}_m}).$$

Recall from Section 2.1 that whenever we have two isomorphisms between two given gerbes, we get a line bundle which measures the difference between the two isomorphisms. So we get a line bundle on $T_m \times \hat{T}_m$ which has a Chern class $d \in H^2(T_m \times \hat{T}_m, \mathbb{Z})$. However the class d is not unique because we had to choose trivializations $\tau_m, \hat{\tau}_m$. Changing these trivializations we find that the change in d corresponds to pulling back line bundles on T_m and $\hat{T}_m$ to $T_m \times \hat{T}_m$, namely the line bundles corresponding to change in trivializations. Factoring out this ambiguity and using the Künneth theorem we have for each $m \in M$ a well defined class $\delta_m \in H^1(T_m, \mathbb{Z}) \otimes H^1(\hat{T}_m, \mathbb{Z})$. Alternatively we can rewrite this as $\delta_m \in Hom(H_1(T_m, \mathbb{Z}), H^1(\hat{T}_m, \mathbb{Z}))$. We say that γ satisfies the *Poincaré property at* $m \in M$ if $\delta_m \in Hom(H_1(T_m, \mathbb{Z}), H^1(\hat{T}_m, \mathbb{Z}))$ is an isomorphism. If γ satisfies the Poincaré property at all points $m \in M$ we simply say that γ satisfies the Poincaré property.

The groups $H^1(T_m, \mathbb{Z})$, $H^1(\hat{T}_m, \mathbb{Z})$ are the stalks at m of the locally constant sheaves $R^1\pi_*\mathbb{Z}$, $R^1\hat{\pi}_*\mathbb{Z}$. Recall that $R^1\pi_*\mathbb{Z}$ is the sheaf associated to the presheaf which assigns to an open set $U \subseteq M$ the group $H^1(\pi^{-1}(U), \mathbb{Z})$. The locally constant sheaf $R^1\hat{\pi}_*\mathbb{Z}$ is similarly defined. Note however that since $X, \hat{X}$ are principal the locally constant sheaves $R^1\pi_*\mathbb{Z}$, $R^1\hat{\pi}_*\mathbb{Z}$ are actually constant sheaves. Indeed for each $m \in M$ we can identify the group $H^1(T_m, \mathbb{Z})$ with the dual lattice $\Lambda^* = Hom_{\mathbb{Z}}(\Lambda, \mathbb{Z})$. This gives an isomorphism $R^1\pi_*\mathbb{Z} = \Lambda^*$ and similarly $R^1\hat{\pi}_*\mathbb{Z} = \hat{\Lambda}^*$. It is straightforward to see that the assignment $m \mapsto \delta_m$ is actually a global section δ of $Hom(R^1\pi_*\mathbb{Z}, (R^1\hat{\pi}_*\mathbb{Z})^*) = Hom(\Lambda^*, \hat{\Lambda}^*)$. If the Poincaré property holds we find that δ is an isomorphism between Λ and the dual of $\hat{\Lambda}$. We can therefore view $\Lambda, \hat{\Lambda}$ as dual lattices. It follows immediately that the corresponding vector spaces $V = \Lambda \otimes \mathbb{R}$, $\hat{V} = \hat{\Lambda} \otimes \mathbb{R}$ are dual vector spaces. Finally the tori $T^n = V/\Lambda$, $\hat{T}^n = \hat{V}/\hat{\Lambda}$ are dual tori. In short we can say that if $X, \hat{X}$ are T-dual then the fibers of $X, \hat{X}$ are dual tori. This is one of the characteristic properties of T-duality.

For circle bundles Definition 4.2 can be shown to be equivalent to the previously used definition of T-duality, Definition 4.1. We saw that for circle bundles a T-dual always exists and is unique. In the higher rank case neither of these statements hold. Not every pair $(X, \mathcal{G})$ where X is a principal torus bundle and $\mathcal{G}$ a gerbe on X admit a T-dual and even when they do it is generally not unique. An alternative point of view is that a T-dual still exists, but exists as a more exotic kind of structure like a bundle of non-commutative tori. We will not explore such possibilities here however.

Given a pair $(X, \mathcal{G})$, where X is a principal torus bundle and $\mathcal{G}$ is a gerbe on X it can be shown [7] that a T-dual according to Definition 4.2 exists if and only if the Dixmier-Douady class $[\mathcal{G}] \in H^3(X, \mathbb{Z})$ lies in the image of a natural map $\phi : H^2(M, \pi_*(\mathcal{U})) \to H^3(X, \mathbb{Z})$. Here $\mathcal{U}$ denotes the sheaf of $U(1)$-valued functions on X and the map ϕ is the composition of the natural pullback $\pi^* : H^2(M, \pi_*(\mathcal{U})) \to H^2(X, \mathcal{U})$ with the coboundary $\delta : H^2(X, \mathcal{U}) \to H^3(X, \mathbb{Z})$ for the exponential sequence $0 \to \mathbb{Z} \to \mathcal{C} \to \mathcal{U} \to 0$ (where $\mathcal{C}$ is the sheaf of real valued continuous functions).

When a T-dual for a pair $(X, \mathcal{G})$ exists, it will generally not be unique. We will not attempt a classification of all possible T-duals here.

One may ask what motivation is there for Definition 4.2. We will see in Section 4.5 that Definition 4.2 is exactly the right definition needed to construct an isomorphism in twisted K-theory which along the fibers looks like the Fourier-Mukai transform. Some further insight will be gained by expressing T-duality at the level of differential forms as we did in Section 4.3.

Let $\pi : X \to M$ be a principal $T^n = V/\Lambda$-bundle. The vertical tangent bundle $Ker(\pi_*) \subset TX$ canonically identifies with V. A connection on X can be thought of as an invariant V-valued 1-form A such that on restriction to the vertical bundle $A|_V : V \to V$ is the identity. The curvature of A is the V-valued 2-form F on M such that $dA = \pi^*(F)$. Now let $(X, \mathcal{G}), (\hat{X}, \hat{\mathcal{G}})$ be a T-dual pair. Choose connections $A, \hat{A}$ for $X, \hat{X}$. Proposition 4.2 extends as follows:

Proposition 4.4. *Given a T-dual pair $(X, \mathcal{G}), (\hat{X}, \hat{\mathcal{G}})$ and connections $A, \hat{A}$ there exists a 3-form $H_3 \in \Omega^3(M)$ such that the Dixmier-Douady classes $[\mathcal{G}] \in H^3(X, \mathbb{Z})$, $[\hat{\mathcal{G}}] \in H^3(\hat{X}, \mathbb{Z})$ are represented at the level of de Rham cohomology by*

$$H = \pi^*(H_3) + \langle A \wedge \pi^*(\hat{F}) \rangle, \tag{4.4}$$
$$\hat{H} = \hat{\pi}^*(H_3) + \langle \hat{A} \wedge \hat{\pi}^*(F) \rangle, \tag{4.5}$$

where $F, \hat{F}$ are the curvatures of $A, \hat{A}$ and $\langle\ \_ \wedge \_\ \rangle$ denotes the pairing of the dual spaces V, V^* extended to V and V^*-valued differential forms by the wedge product.

It follow from this that we also have the relation

$$\hat{p}^*(\hat{H}) - p^*(H) = d\mathcal{B},$$

where

$$\mathcal{B} = \langle p^*(A) \wedge \hat{p}^*(\hat{A}) \rangle.$$

We can now define a T-duality transformation between twisted cohomology on X and $\hat{X}$ exactly as we did for the circle bundle case. Define $T : \Gamma(S) \to \Gamma(\hat{S})$, where $S = \wedge^* T^* X$ and $\hat{S} = \wedge^* T^* \hat{X}$ by

$$T\omega = \int_{T^n} e^{-\mathcal{B}} \wedge p^*(\omega), \tag{4.6}$$

where the integration is over the T^n fibers of $C \to \hat{X}$. One sees again that T commutes with the twisted differential, so defines a map $T_* : H^k(X, H) \to H^{k-n}(\hat{X}, \hat{H})$. A straightforward extension of the proof of Proposition 4.3 shows that T_* is an isomorphism, thus the twisted cohomology groups of X and $\hat{X}$ are isomorphic up to a shift of degree n.

4.5 T-duality and twisted K-theory

In the presence of a gerbe $\mathcal{G}$ on X the K-theory groups $K^*(X)$ of X can be replaced by the *twisted K-theory* $K^*(X, \mathcal{G})$. From the string theory point of view twisted K-theory plays a role of describing D-brane charges and Ramond-Ramond fluxes. Now since T-duals are supposed to give rise to equivalent string theories we expect that T-duals should have the same twisted K-theory, up to a degree shift. This turns out to be the case, in fact the requirement of an isomorphism of twisted K-theories is the very basis for our definition of T-duality.

To get a feel for the twisted K-theory groups $K^*(X, \mathcal{G})$, let us consider the case that $\mathcal{G}$ has torsion Dixmier-Douady class. To further simplify suppose that $\mathcal{G}$ is defined with respect to an open cover $\{U_\alpha\}$ of X. Thus $\mathcal{G}$ consists of line bundles $\{L_{\alpha\beta}\}$ and an associative multiplication $\theta_{\alpha\beta\gamma} : L_{\alpha\beta} \otimes L_{\beta\gamma} \to L_{\alpha\gamma}$. The usual K-theory group $K^0(X)$ is defined using vector bundles on X and the other K-theory group $K^1(X)$ can just be defined as $K^0(X \times \mathbb{R})$. In the case of a gerbe of torsion class the twisted K-theory is constructed from *twisted vector bundles*.

Definition 4.3. A *twisted vector bundle* $(E_\alpha, m_{\alpha\beta})$ for the gerbe $(L_{\alpha\beta}, \theta_{\alpha\beta\gamma})$ is a collection $\{E_\alpha\}$ of vector bundles where E_α is defined over U_α together with a collection of multiplication maps $m_{\alpha\beta} : L_{\alpha\beta} \otimes E_\beta \to E_\alpha$ defined on each non-trivial double intersection such that the multiplication is an action in the sense that the following diagram commutes:

$$\begin{array}{ccc} L_{\alpha\beta} \otimes L_{\beta\gamma} \otimes E_\gamma & \xrightarrow{\theta_{\alpha\beta\gamma} \otimes Id} & L_{\alpha\gamma} \otimes E_\gamma \\ \Big\downarrow{\scriptstyle Id \otimes m_{\beta\gamma}} & & \Big\downarrow{\scriptstyle m_{\alpha\gamma}} \\ L_{\alpha\beta} \otimes E_\beta & \xrightarrow{m_{\alpha\beta}} & E_\alpha \end{array}$$

Two bundle gerbe modules $(E_\alpha, m_{\alpha\beta})$, $(F_\alpha, n_{\alpha\beta})$ are isomorphic if there are bundle isomorphisms $\phi_\alpha : E_\alpha \to F_\alpha$ which commute with the actions in the sense that the following diagram commutes:

$$\begin{array}{ccc} L_{\alpha\beta} \otimes E_\beta & \xrightarrow{Id \otimes \phi_\beta} & L_{\alpha\beta} \otimes F_\beta \\ \Big\downarrow{\scriptstyle m_{\alpha\beta}} & & \Big\downarrow{\scriptstyle n_{\alpha\beta}} \\ E_\alpha & \xrightarrow{\phi_\alpha} & F_\alpha \end{array}$$

Observe that a twisted vector bundle $(E_\alpha, m_{\alpha\beta})$ for a gerbe $\mathcal{G} = (L_{\alpha\beta}, \theta_{\alpha\beta\gamma})$ such that each E_α has rank 1 gives rise to a trivialization of $\mathcal{G}$. Indeed the multiplications $m_{\alpha\beta} : L_{\alpha\beta} \otimes E_\beta \to E_\alpha$ can be rearranged as isomorphisms $m_{\alpha\beta} : L_{\alpha\beta} \to E_\alpha \otimes E_\beta^*$ which yield the trivialization.

Suppose that X is connected. Under this assumption one can see that if $(E_\alpha, m_{\alpha\beta})$ is a twisted vector bundle for $\mathcal{G}$ then the ranks of the E_α are all the same value r. If we take determinants the maps $m_{\alpha\beta} : L_{\alpha\beta} \otimes E_\beta \to E_\alpha$ give rise to maps $m_{\alpha\beta}^r : L_{\alpha\beta}^r \otimes \wedge^r E_\beta \to \wedge^r E_\alpha$. This gives a trivialization of $\mathcal{G}^{\otimes r}$. Thus we see that a gerbe $\mathcal{G}$ has a non-trivial twisted vector bundle only if $\mathcal{G}$ is of torsion class. This is why we have restricted to gerbes of torsion class here.

There is an obvious way to take the direct sum of two twisted vector bundles $(E_\alpha, m_{\alpha\beta})$, $(F_\alpha, n_{\alpha\beta})$ for $\mathcal{G}$ by simply taking the direct sum $(E_\alpha \oplus F_\alpha, m_{\alpha\beta} \oplus n_{\alpha\beta})$. Moreover the direct sum is defined at the level of isomorphism classes of twisted vector bundles. We see that the set of isomorphism classes of twisted vector bundles is a semi-group. The associated Grothendieck group $K^0(X, \mathcal{G})$ is the twisted K-theory of X of degree 0. When $\mathcal{G}$ is non-torsion this construction does not work because as we have seen there are no twisted vector bundles for $\mathcal{G}$. To define the twisted K-theory group $K^0(X, \mathcal{G})$ in the general case, one has to take a different arguably less intuitive approach. The twisted K-theory can be defined as homotopy classes of a bundle of Fredholm operators over X constructed from $\mathcal{G}$. We omit the details since they will not be particularly useful to us here, but the interested reader can consult various references such as [2, 3, 11, 33, 54].

Once the twisted K-theory groups $K^0(X, \mathcal{G})$ have been defined we get the twisted K^1-groups simply by setting it equal to $K^0(X \times \mathbb{R}, \mathcal{G})$. We will use the notation $K^i(X, \mathcal{G})$ where i is any integer to denote $K^0(X, \mathcal{G})$ if i is even or $K^1(X, \mathcal{G})$ if i is odd.

If $\mathcal{G}, \mathcal{H}$ are gerbes of torsion class then $\mathcal{G} \otimes \mathcal{H}$ is also of torsion class. Suppose $(E_\alpha, m_{\alpha\beta})$ is a twisted vector bundle for $\mathcal{G}$ and $(F_\alpha, n_{\alpha\beta})$ is a twisted vector bundle for $\mathcal{H}$. It is easy to see that the tensor product $(E_\alpha \otimes F_\alpha, m_{\alpha\beta} \otimes n_{\alpha\beta})$ is a twisted vector bundle for $\mathcal{G} \otimes \mathcal{H}$. Moreover this tensor product descends to a product of twisted K-theory groups $\otimes : K^0(X, \mathcal{G}) \otimes K^0(X, \mathcal{H}) \to K^0(X, \mathcal{G} \otimes \mathcal{H})$. In fact, we can extend this to twisted K-theory for arbitrary gerbes and in arbitrary degree, so there is always a graded commutative multiplication $K^i(X, \mathcal{G}) \otimes K^j(X, \mathcal{H}) \to K^{i+j}(X, \mathcal{G} \otimes \mathcal{H})$.

If $(E_\alpha, m_{\alpha\beta})$ is a twisted vector bundle for a gerbe $\mathcal{H}$ and L is a line bundle on X then $(E_\alpha \otimes L, m_{\alpha\beta} \otimes Id)$ is also a twisted vector bundle for $\mathcal{H}$. From this we see that the group of line bundles on X acts as automorphisms $L\otimes : K^0(X, \mathcal{H}) \to K^0(X, \mathcal{H})$. This can be extended to the non-torsion case: for any line bundle L and any gerbe $\mathcal{H}$ on X we get an action of L on the K-theory groups $K^*(X, \mathcal{H})$ by an automorphism $L\otimes : K^*(X, \mathcal{H}) \to K^*(X, \mathcal{H})$. When X is compact we can think of L as a class in $K^0(X)$ and the action $L\otimes$ reduces to the product in twisted K-theory.

Recall from Section 2.1 that we defined a category $Grb(X)$ whose objects are gerbes on X and whose morphisms are equivalence classes of stable isomorphisms. Twisted K-theory becomes a functor $K^*(X, \;\;) : Grb(X) \to Gr_{\mathbb{Z}_2}\mathrm{Ab}$ from the category of gerbes on X defined to the category of $\mathbb{Z}_2$-graded abelian groups. In particular, suppose $\mathcal{G}, \mathcal{H}$ are two gerbes and $f, g : \mathcal{G} \to \mathcal{H}$ are two stable isomorphisms from $\mathcal{G}$ to $\mathcal{H}$. Then as morphisms in $Grb(X)$ we have seen that f and g differ by a line bundle L on X. The induced isomorphisms $f, g : K^*(X, \mathcal{G}) \to K^*(X, \mathcal{H})$ differ by the action of L, that is we have a commutative diagram of the form:

$$\begin{array}{ccc} K^*(X,\mathcal{G}) & \xrightarrow{\;f\;} & K^*(X,\mathcal{H}) \\ & \searrow^{g} & \downarrow{\scriptstyle \otimes L} \\ & & K^*(X,\mathcal{H}) \end{array}$$

Twisted K-theory is also functorial with respect to the underlying space. We have chosen to work with compactly supported twisted K-theory, so the twisted K-theory is functorial with respect to proper maps. That is for any proper map $\phi : Y \to X$ there are pullback maps $\phi^* : K^*(X, \mathcal{G}) \to K^*(Y, \phi^*(\mathcal{G}))$ which are contravariant in ϕ.

For application to T-duality we need to make use of a pushforward operation in twisted K-theory [33],[23],[24]. For any smooth map $\phi : Y \to X$ there is a pushforward map $\phi_* : K^i(Y, \phi^*(\mathcal{G}) \otimes L(\phi)) \to K^{i-n}(Y, \mathcal{G})$, where $L(\phi)$ is the lifting gerbe of $TY \oplus \phi^*(TX)$ as in Section 2.1 and n is the rank of $TY \oplus \phi^*(TX)$. In the case where $\phi : Y \to X$ is a fiber bundle we can replace $L(\phi)$ by $L(V)$ where $V = \mathrm{Ker}(\phi_*)$ is the vertical tangent bundle [5].

There is a Mayer-Vietoris sequence in twisted K-theory [33],[23]. If U_1, U_2 are open sets covering X and $\mathcal{G}$ is a gerbe on X, there is a natural long exact sequence

$$\begin{array}{ccccc} K^0(X,\mathcal{G}) & \longleftarrow & K^0(U_1,\mathcal{G}) \oplus K^0(U_2,\mathcal{G}) & \longleftarrow & K^0(U_1 \cap U_2,\mathcal{G}) \\ \downarrow & & & & \uparrow \\ K^1(U_1 \cap U_2,\mathcal{G}) & \longrightarrow & K^1(U_1,\mathcal{G}) \oplus K^1(U_2,\mathcal{G}) & \longrightarrow & K^1(X,\mathcal{G}) \end{array}$$

where $\mathcal{G}$ defines twists on $U_1, U_2, U_1 \cap U_2$ by restriction. The Mayer-vietoris sequence is natural is the sense that pullback, pushforward and isomorphism of twists yield commuting Mayer-Vietoris sequences.

One last property of twisted K-theory that we would like to say a few words about is the existence of a twisted Chern character. Suppose $\mathcal{G}$ is a gerbe on X so that the twisted K-theory group $K^*(X,\mathcal{G})$ is defined. If we choose a connection and curving with curvature 3-form H then one can define a twisted Chern character $Ch_{\mathcal{G}} : K^*(X,\mathcal{G}) \to H^*(X,H)$ [11],[51]. When X is non-compact one needs to use compactly supported twisted cohomology. When X is compact, the twisted Chern character yields an isomorphism $Ch_{\mathcal{G}} : K^*(X,\mathcal{G}) \otimes \mathbb{R} \to H^*(X,H)$ [34].

Given a T-dual pair $(X,\mathcal{G}), (\hat{X},\hat{\mathcal{G}})$, choose an isomorphism $\gamma : p^*(\mathcal{G}) \to \hat{p}^*(\hat{\mathcal{G}})$ satisfying the Poincaré property. From the choice of isomorphism γ we may define a T-duality transform $T : K^*(X,\mathcal{G}) \to K^{*-n}(\hat{X},\hat{\mathcal{G}})$ in twisted K-theory, where n is the rank of the fibers of $X, \hat{X}$. We can show that T is an isomorphism, at least when the base M admits a finite good cover. The map T is defined to make the following diagram commute:

$$\begin{array}{ccc} K^*(C,p^*(\mathcal{G})) & \xrightarrow{\gamma} & K^*(C,\hat{\mathcal{G}}) \\ {\scriptstyle p^*}\uparrow & & \downarrow{\scriptstyle \hat{p}_*} \\ K^*(X,\mathcal{G}) & \xrightarrow{T} & K^{*-n}(\hat{X},\hat{\mathcal{G}}) \end{array}$$

that is, $T = \hat{p}_* \circ \gamma \circ p^*$. Notice that this is quite similar to the Fourier-Mukai transform and to the T-duality transform in twisted cohomology defined in Equation (4.6).

We say that an open cover $\mathcal{U}$ of M is a *good cover* if all non-empty finite intersections in $\mathcal{U}$ are contractible. If M is compact then M admits a finite good cover [10, Theorem 5.1].

Proposition 4.5. *Suppose that M admits a finite good cover. Then for any T-dual pairs $(X, \mathcal{G}), (\hat{X}, \hat{\mathcal{G}})$ on M and isomorphism $\gamma : p^*(\mathcal{G}) \to \hat{p}^*(\hat{\mathcal{G}})$ satisfying the Poincaré property, the map $T : K^*(X, \mathcal{G}) \to K^{*-n}(\hat{X}, \hat{\mathcal{G}})$ is an isomorphism.*

Proof. We prove this result by using induction on the number of elements in a finite good cover. To start the induction we need to show the result is true if the base is contractible. We will prove this shortly.

Now assume the result holds is whenever the base has a good cover with at most k elements. Suppose M has a good cover $\mathfrak{U}$ with $k+1$ elements, say $\mathfrak{U} = \{U_0, U_1, \ldots, U_k\}$. Then $N = U_1 \cup U_2 \cup \cdots \cup U_k$ has a finite good cover with k-elements and so does $N \cap U_0$, namely $\{U_0 \cap U_1, U_0 \cap U_2, \ldots U_0 \cap U_k\}$. For an open subset $W \subset M$ we let $X|_W$ denote $\pi^{-1}(W)$ and $\mathcal{G}|_W$ denote the restriction of $\mathcal{G}$ to $X|_W$. Similarly define $\hat{X}|_W, C|_W$ and $\hat{\mathcal{G}}|_W$. By restriction γ defines isomorphisms $\gamma|_W : p^*(\mathcal{G}|_W) \to \hat{p}^*(\hat{\mathcal{G}}|_W)$ satisfying the Poincaré property. Using $\gamma|_W$ we may construct over W a T-duality transformation which we denote by $T|_W : K^*(X|_W, \mathcal{G}|_W) \to K^{*-n}(\hat{X}|_W, \hat{\mathcal{G}}|_W)$. We obtain a commutative diagram with rows given by Mayer-Vietoris sequences:

$$\begin{array}{ccccccc}
\cdots \to & K^*(X|_{N\cap U_0}, \mathcal{G}) & \to & K^*(X|_N, \mathcal{G}) \oplus K^*(X|_{U_0}, \mathcal{G}) & \to & K^*(X, \mathcal{G}) & \to \cdots \\
& \downarrow{\scriptstyle T|_{N\cap U_0}} & & \downarrow{\scriptstyle T|_N \oplus T_{U_0}} & & \downarrow{\scriptstyle T} & \\
\cdots \to & K^{*-n}(\hat{X}|_{N\cap U_0}, \hat{\mathcal{G}}) & \to & K^{*-n}(\hat{X}|_N, \hat{\mathcal{G}}) \oplus K^{*-n}(\hat{X}|_{U_0}, \hat{\mathcal{G}}) & \to & K^{*-n}(\hat{X}, \hat{\mathcal{G}}) & \to \cdots
\end{array}$$

where we have dropped the restriction notation for the gerbes $\mathcal{G}, \hat{\mathcal{G}}$. Now since N, U_0 and $N \cap U_0$ all admit good covers with no more than k elements we have by induction that $T|_N, T|_{U_0}$ and $T|_{N\cap U_0}$ are isomorphisms. Thus by an application of the five lemma we find that T is also an isomorphism.

All that remains is to show the result when M is contractible. In this case the torus bundles $X, \hat{X}$ are both trivial, so we can identify $X, \hat{X}$ with the trivial bundles $M \times T^n$, $M \times \hat{T}^n$ (here $T^n = \mathbb{R}^n/\mathbb{Z}^n$ and $\hat{T}^n = (\mathbb{R}^n)^*/(\mathbb{Z}^n)^*$ is the dual torus). The gerbes $\mathcal{G}, \hat{\mathcal{G}}$ are by definition of T-duality trivial on the fibers, so in this case we have that $\mathcal{G}, \hat{\mathcal{G}}$ are trivial. Choosing trivializations of $\mathcal{G}, \hat{\mathcal{G}}$, we can then identify $K^*(X, \mathcal{G})$ with $K^*(M \times T^n)$, which itself is isomorphic to $K^{*-b}(T^n)$, where b is the dimension of the base. To see this one can apply the Atiyah-Hirzebruch spectral sequence [2] to the fiber bundle $M \times T^n \to M$. We can similarly identify $K^*(\hat{X}, \hat{\mathcal{G}})$ with $K^{*-b}(\hat{T}^n)$. The T-duality transform then reduces to a map $T' : K^{*-b}(T^n) \to K^{*-b-n}(\hat{T}^n)$ defined by a commutative diagram

$$\begin{array}{ccc}
K^*(T^n \times \hat{T}^n) & \xrightarrow{\otimes P} & K^*(T^n \times \hat{T}^n) \\
{\scriptstyle p^*}\uparrow & & \downarrow{\scriptstyle \hat{p}_*} \\
K^*(T^n) & \xrightarrow{T'} & K^{*-n}(\hat{T}^n)
\end{array} \tag{4.7}$$

where P is a certain line bundle on $T^n \times \hat{T}^n$. In fact, since γ satisfies the Poincaré property we can choose the trivializations of $\mathcal{G}, \hat{\mathcal{G}}$ so that the map $K^*(T^n \times \hat{T}^n) \to$

$K^*(T^n \times \hat{T}^n)$ is given by tensoring with the Poincaré line bundle. Thus T' is a K-theoretic version of the Fourier-Mukai transform and a straightforward computation shows that T' is an isomorphism. □

Remark 4.2. In the last step it is crucial to know that the line bundle in (4.7) P is the Poincaré line bundle. Indeed not all line bundles on $T^n \times \hat{T}^n$ give rise to an isomorphism in (4.7). Thus the Poincaré property is critical to the construction of an isomorphism of twisted K-theories. Looking back at Definition 4.2, we see that the axioms (TT1)-(TT3) are precisely what is needed to construct the isomorphism $T : K^*(X, \mathcal{G}) \to K^{*-n}(\hat{X}, \hat{\mathcal{G}})$.

5 T-duality and generalized geometry

Having studied both generalized geometry and T-duality we now see how the two can be combined leading to a T-duality transformation of Courant algebroids and accompanying structures. In particular this will lead to a T-duality transformation for generalized complex structures. We will see moreover that in the right circumstances the transformation exchanges generalized complex structures of complex and symplectic type suggesting a connection to mirror symmetry. This section follows [26] quite closely as well as [25, 38].

Let $\pi : X \to M$ be a principal T^n-bundle. Let H be a closed invariant 3-form on X. It follows that the T^n-action on X defines a T^n-action on the generalized tangent bundle $E = TX \oplus T^*X$ and since H is invariant the action preserves the H-twisted Dorfman bracket on E. In particular this implies that the space $\Gamma(E)^{T^n}$ of T^n-invariant sections of E is closed under the Dorfman bracket. Moreover the space $\Gamma(E)^{T^n}$ can be identified with the space of sections of a vector bundle on M, namely the quotient E/T^n, which we denote by E^r. Under the bijection $\Gamma(E^r) = \Gamma(E)^{T^n}$ we see that the Dorfman bracket $\{\,,\,\}_H$ induces a bracket $\{\,,\,\}^r_H$ on the space of sections of E^r. It is not hard to see that this defines a Courant algebroid structure on E^r. For instance the pairing $\langle\,,\,\rangle$ on E^r is induced by the pairing $\langle\,,\,\rangle$ on E, since the T^n-action preserves the pairing on E and the anchor $\rho^r : E^r \to TM$ is obtained from the composition of the anchor $\rho : E \to TX$ with $\pi_* : TX \to TM$. The resulting Courant algebroid $(E^r, \{\,,\,\}^r_H, \langle\,,\,\rangle, \rho^r)$ will be called the *dimensional reduction* of $(E, \{\,,\,\}_H, \langle\,,\,\rangle, \rho)$.

Remark 5.1. Suppose that H, H' are closed invariant 3-forms representing the same cohomology class on X. Then we can find a 2-form B such that $H' = H + dB$. By averaging we may assume B is also invariant. By considering the B-shift e^B we find that there is a T^n-invariant isomorphism between the Courant algebroids $(E, \{\,,\,\}_H, \langle\,,\,\rangle, \rho)$ and $(E, \{\,,\,\}_{H'}, \langle\,,\,\rangle, \rho)$. It follows that the dimensional reductions of $(E, \{\,,\,\}_H, \langle\,,\,\rangle, \rho)$ and $(E, \{\,,\,\}_{H'}, \langle\,,\,\rangle, \rho)$ are isomorphic as well. Thus up to isomorphism the dimensional reduction of $(E, \{\,,\,\}_H, \langle\,,\,\rangle, \rho)$ depends only on the cohomology class of H in $H^3(X, \mathbb{R})$ (i.e. the Ševera class) and thus only depends on the isomorphism class of the original exact Courant algebroid.

Now suppose $(X, \mathcal{G}), (\hat{X}, \hat{\mathcal{G}})$ are a pair of T-dual principal torus bundles with gerbes. The gerbes have Dixmier-Douady classes $h = [\mathcal{G}] \in H^3(X, \mathbb{Z})$, $\hat{h} = [\hat{\mathcal{G}}] \in H^3(\hat{X}, \mathbb{Z})$. Let $h_{\mathbb{R}}, \hat{h}_{\mathbb{R}}$ be the images of the Dixmier-Douady classes in real cohomology. Then we have exact Courant algebroids $E, \hat{E}$ on $X, \hat{X}$ such that $h_{\mathbb{R}}, \hat{h}_{\mathbb{R}}$ are the Ševera classes of $E, \hat{E}$ respectively. By the above dimensional reduction procedure we get two Courant algebroids $E^r, \hat{E}^r$ on M.

Proposition 5.1. *The Courant algebroids $E^r, \hat{E}^r$ are isomorphic.*

Proof. Let $H, \hat{H}$ be closed invariant 3-forms on $X, \hat{X}$ representing the cohomology classes $h_{\mathbb{R}}, \hat{h}_{\mathbb{R}}$. We may identify E with the generalized tangent bundle $E = TX \oplus T^*X$ with H-twisted Dorfman bracket. Similarly $\hat{E} = T\hat{X} \oplus T^*\hat{X}$ with $\hat{H}$-twisted Dorfman bracket. Since the isomorphism classes of $E^r, \hat{E}^r$ are independent of the choice of invariant 3-forms, we may choose $H, \hat{H}$ to have the form given by Equations (4.1) and (4.2), for some connections $A, \hat{A}$ on $X, \hat{X}$ and some 3-form H_3 on M.

Using the connection A we get an isomorphism $TX = \pi^*(TM) \oplus V$, where we have used V to denote the trivial vector bundle corresponding to the vector space V. Moreover since the connection is T^n-invariant this decomposition is preserved by the T^n-action so on factoring out the action we get $TX/T^n = TM \oplus V$ as vector bundles on M. Dually we have $T^*X/T^n = T^*M \oplus V^*$. Similarly using $\hat{A}$ we have $T\hat{X}/T^n = TM \oplus \hat{V}$ and $T^*\hat{X}/T^n = T^*M \oplus \hat{V}^*$. So the connections $A, \hat{A}$ induce isomorphisms

$$E^r = TM \oplus V \oplus V^* \oplus T^*M, \tag{5.1}$$

$$\hat{E}^r = TM \oplus \hat{V} \oplus \hat{V}^* \oplus T^*M. \tag{5.2}$$

Recall that $V, \hat{V}$ are dual vector spaces. Let $\phi : E^r \to \hat{E}^r$ be the isomorphism which identifies V^* with $\hat{V}$ and V with $\hat{V}^*$. We will show that ϕ is an isomorphism of Courant algebroids.

Let $S = \wedge^* T^*X$ and $\hat{S} = \wedge^* T^*\hat{X}$ be the bundles of differential forms on X and $\hat{X}$ and write $S^r = S/T^n$, $\hat{S}^r = \hat{S}/T^n$ for the quotients by the T^n-action which are bundles over M. The Clifford action of E on S is T^n-invariant so descends to a Clifford action of E^r on S^r. Similarly $\hat{E}^r$ has a Clifford action on $\hat{S}^r$. Recall also the T-duality transformation on differential forms $T : \Gamma(S) \to \Gamma(\hat{S})$ given by $T\omega = \int_{T^n} e^{-\mathcal{B}} \wedge p^*(\omega)$. Since T sends invariant forms to invariant forms it defines also a transformation $T : \Gamma(S^r) \to \Gamma(\hat{S}^r)$ which in fact is a bundle isomorphism between S^r and $\hat{S}^r$ as can be seen by a direct computation. Next observe that since $H, \hat{H}$ are invariant the differentials $d_H, d_{\hat{H}}$ preserve the subspace of invariant forms so defines differentials on $\Gamma(S^r), \Gamma(\hat{S}^r)$. We have seen that T interwinds the differentials, that is $Td_H = d_{\hat{H}}T$. The key identity we need is that T also intertwines the Clifford actions:

$$T(\gamma_a(\omega)) = \gamma_{\phi(a)}(T\omega),$$

where a is a section of E^r and ω a section of S^r. This is a matter of straightforward computation which we leave as an exercise.

The final ingredient is to recall Equation (3.20), which states that the Dorfman bracket is a derived bracket:

$$\gamma_{\{a,b\}_H}\omega = [[d_H, \gamma_a], \gamma_b]\omega,$$

and similarly for $\{\,,\,\}_{\hat{H}}$. Now letting a, b be invariant sections of E and ω an invariant section of S we get

$$\begin{aligned}
\gamma_{\phi(\{a,b\}_H)}T\omega &= T(\gamma_{\{a,b\}_H}\omega),\\
&= T([[d_H, \gamma_a], \gamma_b]\omega),\\
&= [[d_{\hat{H}}, \gamma_{\phi(a)}], \gamma_{\phi(b)}]T\omega,\\
&= \gamma_{\{\phi(a),\phi(b)\}_{\hat{H}}}T\omega.
\end{aligned}$$

Therefore $\phi(\{a,b\}_H) = \{\phi(a), \phi(b)\}_{\hat{H}}$, in other words ϕ interchanges the brackets on E^r and $\hat{E}^r$. This almost completes the proof. It remains only to show that ϕ also exchanges the anchors on $E^r, \hat{E}^r$ and the natural pairings. For the anchors this follows immediately from the definition of ϕ. For the pairings we observe that is a, b are invariant sections of E and ω an invariant section of S then

$$\begin{aligned}
2\langle\phi(a), \phi(b)\rangle T\omega &= (\gamma_{\phi(a)}\gamma_{\phi(b)} + \gamma_{\phi(b)}\gamma_{\phi(a)})T\omega,\\
&= T(\gamma_a\gamma_b + \gamma_b\gamma_a)\omega,\\
&= T(2\langle a, b\rangle\omega),\\
&= 2\langle a, b\rangle T\omega.
\end{aligned}$$

So we must have $\langle a, b\rangle = \langle\phi(a), \phi(b)\rangle$ for all a, b and the proof is complete. □

We have just seen that if $(X, \mathcal{G}), (\hat{X}, \hat{\mathcal{G}})$ are T-dual then on choosing connections $A, \hat{A}$ and representative 3-forms $H, \hat{H}$ satisfying (4.1) and (4.2) we get an isomorphism ϕ between invariant sections of the generalized tangent bundles of X and $\hat{X}$, which preserves the Courant algebroid structure. Now suppose the generalized tangent bundle is equipped with some additional structure, say a generalized complex structure J. If J is also T^n-invariant then it induces an endomorphism J^r on E^r. Using ϕ we get a corresponding endomorphism on $\hat{E}^r$ and thus an endomorphism $\hat{J}$ on the generalized tangent bundle of $\hat{X}$. Moreover, since ϕ preserves Courant brackets we see that $\hat{J}$ is integrable. In this way we see that invariant generalized complex structures on X can be transformed to invariant generalized complex structures on $\hat{X}$. Of course this is not limited to generalized complex structures. Other invariant structures on the generalized tangent bundle of X are transported over to $\hat{X}$ in a similar manner.

Let G be a T^n-invariant generalized metric on E. Then G defines an endomorphism of E^r and by the isomorphism ϕ an endomorphism of $\hat{E}^r$. One sees easily that this is an invariant generalized metric $\hat{G}$ on $\hat{E}$. Recall that a generalized metric G on $TX \oplus T^*X$ corresponds to a Riemannian metric g and a

2-form B. Similarly the generalized metric $\hat{G}$ on $T\hat{X} \oplus T^*\hat{X}$ corresponds to a metric $\hat{g}$ and 2-form $\hat{B}$. The transformation $G \mapsto \hat{G}$ corresponds to a transformation $(g, B) \mapsto (\hat{g}, \hat{B})$. One can show that this gives a global reformulation of the Buscher rules. Thus we can say that the Buscher rules are nothing more than the transformation of generalized metrics on isomorphic Courant algebroids.

Since generalized complex structures have been the focus of the generalized geometry section, let us examine in further detail the behavior of the T-duality transform on generalized complex structures. More specifically we will be interested in how T-duality can change the type of generalized complex structures and thus exchange complex and symplectic geometries.

Since the type of a generalized complex structure J on $E = TX \oplus T^*X$ at a given point $x \in X$ depends only on the endomorphism $J_x : E_x \to E_x$ the problem of determining type change reduces to a problem in linear algebra. Instead of an isomorphism $\phi : E^r \to \hat{E}^r$ where $E^r, \hat{E}^r$ are given by (5.1),(5.2) it will suffice to consider an isomorphism of vector spaces of the form $\psi : F \to \hat{F}$ where

$$F = U \oplus V \oplus V^* \oplus U^*,$$
$$\hat{F} = U \oplus V^* \oplus V \oplus U^*,$$

and ψ is the map that interchanges the V and V^* factors. Set $W = U \oplus V$ and $\hat{W} = U \oplus V^*$. Thus $F = W \oplus W^*$ and $\hat{F} = \hat{W} \oplus \hat{W}^*$. The duality pairing of W with W^* and $\hat{W}$ with $\hat{W}^*$ give F and $\hat{F}$ natural pairings. Then ψ is an isometry. Let $\rho : F \to W$ and $\hat{\rho} : \hat{F} \to \hat{W}$ be the projections. We can think of $\rho, \hat{\rho}$ as being anchors. Now suppose we have a generalized complex structure J on F, that is J is an endomorphism such that $J^2 = -1$ and J is compatible with the pairing on F. We can define the type of J in the usual manner. Using ψ we get an induced generalized complex structure $\hat{J}$ on $\hat{F}$. We would like to compare the types of J and $\hat{J}$. In general there are many different cases to consider so rather than work out the general situation, let us consider the most interesting special cases.

Example 5.1 (Complex fibrations). Let J have complex type, that is under the decomposition $F = W \oplus W^*$, J has the form

$$J = \begin{bmatrix} -I & 0 \\ 0 & I^* \end{bmatrix}$$

where I is a complex structure on W. In addition we suppose that the decomposition $W = U \oplus V$ is preserved by I, so that U, V are complex subspaces of W. This situation corresponds to the case where X, M are complex manifolds and $\pi : X \to M$ is a complex fibration, a holomorphic family of complex tori. If we let I_U, I_V denote the complex structures on U, V then under the decomposition $F = U \oplus V \oplus V^* \oplus U^*$ we find that

$$J = \begin{bmatrix} -I_U & 0 & 0 & 0 \\ 0 & -I_V & 0 & 0 \\ 0 & 0 & I_V^* & 0 \\ 0 & 0 & 0 & I_U^* \end{bmatrix}.$$

Now under the isomorphism ψ that swaps V and V^* we get

$$\hat{J} = \begin{bmatrix} -I_U & 0 & 0 & 0 \\ 0 & I_V^* & 0 & 0 \\ 0 & 0 & -I_V & 0 \\ 0 & 0 & 0 & I_U^* \end{bmatrix}.$$

In particular we see that $\hat{J}$ has the same form as J, in other words it is induced by a complex structure on $\hat{W}$ such that U and V^* are complex subspaces. Geometrically this means that the dual generalized complex manifold $\hat{X}$ is the holomorphic family $\hat{\pi} : \hat{X} \to M$ of dual tori.

Example 5.2 (Symplectic fibrations). Let J have symplectic type, that is under $F = W \oplus W^*$ we have

$$J = \begin{bmatrix} 0 & -\omega^{-1} \\ \omega & 0 \end{bmatrix},$$

for some symplectic form ω on W. Now suppose that $W = U \oplus V$ is a decomposition into symplectic subspaces and such that U, V are orthogonal with respect to ω. So there are symplectic forms ω_U, ω_V on U, V such that $\omega = \omega_U + \omega_V$. Thus under $F = U \oplus V \oplus V^* \oplus U^*$ we have

$$J = \begin{bmatrix} 0 & 0 & 0 & -\omega_U^{-1} \\ 0 & 0 & -\omega_V^{-1} & 0 \\ 0 & \omega_V & 0 & 0 \\ \omega_U & 0 & 0 & 0 \end{bmatrix}.$$

Geometrically J corresponds to a symplectic fibration $\pi : X \to M$.

Now under the isomorphism ψ the corresponding generalized complex structure on $\hat{F} = U \oplus V \oplus V^* \oplus U^*$ is

$$\hat{J} = \begin{bmatrix} 0 & 0 & 0 & -\omega_U^{-1} \\ 0 & 0 & \omega_V & 0 \\ 0 & -\omega_V^{-1} & 0 & 0 \\ \omega_U & 0 & 0 & 0 \end{bmatrix}.$$

We see that this is again of the same form as J. Geometrically the dual generalized complex manifold $\hat{\pi} : \hat{X} \to M$ is a symplectic fibration by dual tori which inherits a dual symplectic structure.

Example 5.3 (Lagrangian fibrations and real structures). Let J again be of symplectic type, that is let ω be a symplectic form on W such that under the decomposition $F = W \oplus W^*$ we have

$$J = \begin{bmatrix} 0 & -\omega^{-1} \\ \omega & 0 \end{bmatrix}.$$

But this time suppose that U, V are Lagrangian subspaces of W. Thus ω determines a duality pairing $(\, , \,) : U \otimes V \to \mathbb{R}$ by setting $(u, v) = \omega(u, v)$. Now view ω as a map $\omega : W \to W^*$ and use the isomorphisms $W = U \oplus V$, $W^* = V^* \oplus U^* = U \oplus V$. Thus ω can be viewed as a map $\omega : U \oplus V \to U \oplus V$ of the form

$$\omega = \begin{bmatrix} Id & 0 \\ 0 & -Id \end{bmatrix}.$$

So now under the decomposition $F = U \oplus V \oplus U \oplus V$ we have

$$J = \begin{bmatrix} 0 & 0 & -Id & 0 \\ 0 & 0 & 0 & Id \\ Id & 0 & 0 & 0 \\ 0 & -Id & 0 & 0 \end{bmatrix}.$$

To apply the isomorphism ψ we swap the inner two factors and arrive at

$$\hat{J} = \begin{bmatrix} 0 & -Id & 0 & 0 \\ Id & 0 & 0 & 0 \\ 0 & 0 & 0 & Id \\ 0 & 0 & -Id & 0 \end{bmatrix}.$$

Now we see that $\hat{J}$ is of the form

$$\hat{J} = \begin{bmatrix} -I & 0 \\ 0 & I^* \end{bmatrix}$$

where I is a complex structure on $U \oplus U \simeq U \oplus V^* \simeq \hat{W}$. Thus the T-dual $\hat{\pi} : \hat{X} \to M$ of a Lagrangian fibration is a complex manifold, but the fibers of $\hat{\pi}$ are not complex submanifolds. In fact we see that the vertical bundle of $\hat{X} \to M$ induces a real structure on every tangent space of $T\hat{X}$. In other words if $V \subset TX$ is the vertical bundle of $\pi : X \to M$ then we have a decomposition $TX = V \oplus IV$. Let us say that any such fibration of a complex manifold is a *real fibration*. We have just shown that T-duality exchanges Lagrangian fibrations with real fibrations.

Recall that the Strominger-Yau-Zaslow conjecture of [61] claims that mirror symmetry of Calabi-Yau manifolds is a form of T-duality. More specifically it is claimed that mirror Calabi-Yau manifolds $X, \hat{X}$ are special Lagrangian torus fibrations over a common base M and that $\hat{X}$ is obtained from X by dualizing the fibers. Special Lagrangians are a particular kind of Lagrangian submanifold which are also minimal submanifolds. For this discussion however the point we want to highlight is that $X, \hat{X}$ are Lagrangian fibrations of Kähler manifolds. Now it is easy to see that a Lagrangian fibration of a Kähler manifold is also a real fibration. Therefore Example 5.3 predicts that the T-dual of a Lagrangian fibration of a Kähler manifold is again a Lagrangian fibration of a Kähler manifold, a statement that is clearly compatible with the SYZ conjecture. Note also that $X, \hat{X}$ being Kähler have both a complex and symplectic structure. Example 5.3 shows that T-duality in such a situation relates the symplectic structure on one side to the complex structure on the other. The exchange of complex and symplectic is also a characteristic feature of mirror symmetry. We conclude that Example 5.3 is compatible with the SYZ picture of mirror symmetry.

Example 5.3 is still a far cry from mirror symmetry. Since in the above we have assumed the torus fibration is a non-singular principal torus bundle and the generalized complex structures are T^n-invariant. There are many compact Calabi-Yau manifolds for which any special Lagrangian fibration must necessarily admit singular fibers. As an indication of this observe that the total space of a principal torus bundle over a simply connected base always has vanishing Euler characteristic, yet there are many compact simply connected Calabi-Yau manifolds with non-zero Euler characteristic. There arises the question as to whether T-duality can accommodate singularities. There is at least some evidence in favor of this. The work of Gross on topological mirror symmetry [37] puts forward a construction of mirror manifolds at the topological level and appears to be essentially a kind of T-duality construction. In works of Căldăraru [22], Donagi and Pantev [32] and Ben-Bassat [9] it is seen that the duality of holomorphic torus fibrations extends to fibrations with singularities. This gives further indication for how a theory of T-duality with singularities could proceed. These works can be considered a T-duality of generalized complex structures of complex type, a more sophisticated version of Example 5.1.

References

[1] M. Aldi, R. Heluani, *Dilogarithms, OPE and twisted T-duality*, arXiv:1105.4280 (2011).

[2] M. Atiyah, G. Segal, Twisted K-theory. *Ukr. Math. Bull.* **1** (2004), no. 3, 291–334.

[3] M. Atiyah, G. Segal, Twisted K-theory and cohomology. Inspired by S. S. Chern, 5–43, Nankai Tracts Math., 11, World Sci. Publ., Hackensack, NJ, (2006).

[4] J. C. Baez, A. S. Crans, Higher-dimensional algebra. VI. Lie 2-algebras. *Theory Appl. Categ.* **12** (2004), 492–538.

[5] D. Baraglia, Topological T-duality for general circle bundles, arXiv:1105.0290v2 (2011).

[6] D. Baraglia, Conformal Courant algebroids and orientifold T-duality, arXiv:1109.0875v2, (2011).

[7] D. Baraglia, Topological T-duality for torus bundles with monodromy, arXiv:1201.1731v1, (2012).

[8] D. Belov, C. Hull, R. Minasian, T-duality, gerbes and loop spaces, arXiv:07105151v1, (2007).

[9] O. Ben-Bassat, Twisting derived equivalences. *Trans. Amer. Math. Soc.* **361** (2009), no. 10, 5469–5504.

[10] R. Bott, L. W. Tu, *Differential forms in algebraic topology.* Graduate Texts in Mathematics, 82. Springer-Verlag, New York-Berlin, (1982).

[11] P. Bouwknegt, A. Carey, V. Mathai, M. Murray, D. Stevenson, Twisted K-theory and K-theory of bundle gerbes. *Comm. Math. Phys.* **228** (2002), no. 1, 17–45.

[12] P. Bouwknegt, J. Evslin, V. Mathai, T-duality: topology change from H-flux. *Comm. Math. Phys.* **249** (2004), no. 2, 383–415.
[13] P. Bouwknegt, K. Hannabuss, V. Mathai, T-duality for principal torus bundles and dimensionally reduced Gysin sequences. *Adv. Theor. Math. Phys.* **9** (2005), no. 5, 749–773.
[14] P. Bouwknegt, K. Hannabuss, V. Mathai, Nonassociative tori and applications to T-duality. *Comm. Math. Phys.* **264** (2006), no. 1, 41–69.
[15] J.-L. Brylinski, Loop spaces, characteristic classes and geometric quantization. Progress in Mathematics, 107. Birkhäuser Boston, Inc., Boston, MA, (1993) 300 pp.
[16] U. Bunke, T. Schick, On the topology of T-duality. *Rev. Math. Phys.* **17** (2005), no. 1, 77–112.
[17] U. Bunke, P. Rumpf, T. Schick, The topology of T-duality for T^n-bundles. *Rev. Math. Phys.* **18** (2006), no. 10, 1103–1154.
[18] U. Bunke, T. Schick, T-duality for non-free circle actions. Analysis, geometry and topology of elliptic operators, 429–466, World Sci. Publ., Hackensack, NJ, (2006).
[19] U. Bunke, T. Schick, M. Spitzweck, Periodic twisted cohomology and T-duality. *Astérisque* No. **337** (2011), 134 pp.
[20] T. H. Buscher, A symmetry of the string background field equations. *Phys. Lett.* B **194** (1987), no. 1, 59–62.
[21] T. H. Buscher, Path-integral derivation of quantum duality in nonlinear sigma-models. *Phys. Lett. B* **201** (1988), no. 4, 466–472.
[22] A. Căldăraru, Derived categories of twisted sheaves on elliptic threefolds. *J. Reine Angew. Math.* **544** (2002), 161–179.
[23] A. L. Carey, B.-L. Wang, Thom isomorphism and push-forward map in twisted K-theory. *J. K-Theory* **1** (2008), no. 2, 357–393.
[24] A. L. Carey, B.-L. Wang, Riemann-Roch and index formulae in twisted K-theory. Superstrings, geometry, topology, and C^*-algebras, 95–131, *Proc. Sympos. Pure Math.*, **81**, Amer. Math. Soc., Providence, RI, (2010).
[25] G. Cavalcanti, *New aspects of the dd^c-lemma.* D.Phil. thesis, Oxford University, (2004). math.DG/0501406.
[26] G. Cavalcanti, M. Gualtieri, Generalized complex geometry and T-duality, A celebration of the mathematical legacy of Raoul Bott, AMS (2010), 341–366.
[27] J. Cheeger, J. Simons, Differential characters and geometric invariants. Geometry and topology, Lecture Notes in Math., 1167, Springer, Berlin, (1985), 50–800.
[28] C. Chevalley, The algebraic theory of spinors and Clifford algebras. Collected works. Vol. 2. Springer-Verlag, Berlin, (1997), 214 pp.
[29] T. J. Courant, Dirac manifolds. *Trans. Amer. Math. Soc.* **319** (1990), no. 2, 631–661.
[30] P. Deligne, P. Griffiths, J. Morgan, D. Sullivan, Real homotopy theory of Kähler manifolds. *Invent. Math.* **29** (1975), no. 3, 245–274.
[31] P. Deligne et. al., *Quantum Fields and Strings: A Course for Mathematicians.* Vol. 1, 2. American Mathematical Society, Providence, (1999).
[32] R. Donagi, T. Pantev, Torus fibrations, gerbes, and duality. With an ap-

pendix by D. Arinkin. *Mem. Amer. Math. Soc.* **193** (2008), no. 901, 90 pp.

[33] D. S. Freed, M. J. Hopkins, C. Teleman, Loop groups and twisted K-theory I, arXiv:0711.1906v1 (2007).

[34] D. S. Freed, M. J. Hopkins, C. Teleman, Twisted equivariant K-theory with complex coefficients. *J. Topol.* **1** (2008), no. 1, 16–44.

[35] J. Fuchs, T. Nikolaus, C. Schweigert, K. Waldorf, Bundle gerbes and surface holonomy. European Congress of Mathematics, 167–195, *Eur. Math. Soc.*, Zürich, (2010).

[36] J. Giraud, Cohomologie non ablienne. Die Grundlehren der mathematischen Wissenschaften, Band 179. Springer-Verlag, Berlin-New York, (1971). 467 pp.

[37] M. Gross, Topological mirror symmetry. *Invent. Math.* **144** (2001), no. 1, 75–137.

[38] M. Gualtieri. *Generalized complex geometry.* D. Phil. thesis, Oxford University, (2003). math.DG/0401221.

[39] P. Hekmati, V. Mathai, *T-duality of current algebras and their quantization*, arXiv:1203.1709v2 (2012).

[40] N. Hitchin, Generalized Calabi-Yau manifolds. *Q. J. Math.* **54**, (2003), no. 3, 281–308.

[41] O. Hohm, C. Hull, B. Zwiebach, Generalized metric formulation of double field theory. *J. High Energy Phys.* (2010), no. 8, **008**, 35 pp.

[42] K. Hori et. al., *Mirror symmetry.* Clay Mathematics Monographs, 1. American Mathematical Society, Providence, (2003).

[43] C. M. Hull, R. A. Reid-Edwards, Non-geometric backgrounds, doubled geometry and generalised T-duality. *J. High Energy Phys.* (2009), no. 9, **014**, 79 pp.

[44] C. Hull, B. Zwiebach, The gauge algebra of double field theory and Courant brackets. *J. High Energy Phys.* (2009), no. 9, **090**, 26 pp.

[45] A. Kahle, A. Valentino, T-duality and Differential K-Theory, arXiv:0912.2516v2 (2009).

[46] K. Kikkawa, M. Yamasaki, Casimir effects in superstring theories. *Phys. Lett.* B **149** (1984), no. 4, 5, 357.

[47] Y. Kosmann-Schwarzbach, Derived brackets. *Lett. Math. Phys.* **69** (2004), 61–87.

[48] S. Mukai, Duality between $D(X)$ and $D(\hat{X})$ with its application to Picard sheaves. *Nagoya Math. J.* **81** (1981), 153–175.

[49] M. K. Murray, Bundle gerbes. *J. London Math. Soc.* (2) **54** (1996), no. 2, 403–416.

[50] Z.-J. Liu, A. Weinstein, P. Xu, Manin triples for Lie bialgebroids. *J. Differential Geom.* **45** (1997), no. 3, 547–574.

[51] V. Mathai, D. Stevenson, Chern character in twisted K-theory: equivariant and holomorphic cases. *Comm. Math. Phys.* **236** (2003), no. 1, 161–186.

[52] V. Mathai, J. Rosenberg, T-duality for torus bundles with H-fluxes via noncommutative topology. *Comm. Math. Phys.* **253** (2005), no. 3, 705–721.

[53] V. Mathai, J. Rosenberg, T-duality for torus bundles with H-fluxes via noncommutative topology. II. The high-dimensional case and the T-duality

group. *Adv. Theor. Math. Phys.* **10** (2006), no. 1, 123–158.

[54] J. Rosenberg, Continuous-trace algebras from the bundle theoretic point of view. *J. Austral. Math. Soc. Ser. A* **47** (1989), no. 3, 368–381.

[55] D. Roytenberg, *Courant algebroids, derived brackets and even symplectic supermanifolds.* PhD thesis, Berkeley (1993). arXiv:math/9910078v1.

[56] D. Roytenberg, A. Weinstein, Courant algebroids and strongly homotopy Lie algebras. *Lett. Math. Phys.* **46** (1998), no. 1, 81–93.

[57] P. Ševera, Letters to Alan Weinstein, http://sophia.dtp.fmph.uniba.sk/~severa/letters/.

[58] P. Ševera, A. Weinstein, Poisson geometry with a 3-form background. Noncommutative geometry and string theory, *Progr. Theoret. Phys. Suppl.* No. **144** (2001), 145-154.

[59] A. Shapere, F. Wilczek, Self-dual models with theta terms. *Nuclear Phys.* B **320** (1989), no. 3, 669–695.

[60] D. Stevenson, The Geometry of Bundle Gerbes, PhD thesis, University of Adelaide (2000). math.DG/0004117.

[61] A. Strominger, S.-T. Yau, E. Zaslow, Mirror symmetry is T-duality. *Nuclear Phys.* B **479** (1996), no. 1–2, 243–259.

[62] C. Voisin, Hodge theory and complex algebraic geometry. I. Cambridge Studies in Advanced Mathematics, 76. Cambridge University Press, Cambridge, (2002) 322 pp.

Open Problems and Surveys of Contemporary Mathematics
SMM 6, pp. 99–132

Theta Functions, Old and New

Arnaud Beauville*

Abstract

We first recall the classical theory of theta functions: in modern terms, they are sections of a line bundle on a complex torus. We discuss a fundamental example of such a complex torus, the *Jacobian* JC of an algebraic curve C, which parametrizes topologically trivial line bundles on C. Then we define generalized theta functions by replacing JC by a moduli space of higher rank vector bundles. We discuss the similarities and differences with the classical theory, the known results and the open problems.

2000 Mathematics Subject Classification: 14K25; 14H40, 14H60

Keywords and Phrases: Complex torus, polarizations, theta functions, theta divisor, Jacobians, vector bundles, generalized theta functions, theta map

Contents

*Laboratoire J.-A. Dieudonné, Université de Nice, Parc Valrose 06108, Nice cedex 2, France. Email: beauville@unice.fr

Introduction

Theta functions are holomorphic functions on $\mathbb{C}^g$, quasi-periodic with respect to a lattice. For $g = 1$ they have been introduced by Jacobi; in the general case they have been thoroughly studied by Riemann and his followers. From a modern point of view they are sections of line bundles on certain complex tori; in particular, the theta functions associated to an algebraic curve C are viewed as sections of a natural line bundle (and of its tensor powers) on a complex torus associated to C, the *Jacobian*, which parametrizes topologically trivial line bundles on C.

Around 1980, under the impulsion of mathematical physics, the idea emerged gradually that one could replace in this definition line bundles by higher rank vector bundles. The resulting sections are called *generalized* (or *non-abelian*) *theta functions*; they turn out to share some (but not all) of the beautiful properties of classical theta functions.

The goal of these lectures is to develop first the modern theory of classical theta functions (complex tori, line bundles, Jacobians), then to explain how it can be generalized by considering higher rank vector bundles. We have tried to make them accessible for students with a minimal background in complex geometry: Chapter 0 of [13] should be more than enough. At a few places, especially in the last chapters, we had to use some more advanced results. Also we have not tried to be exhaustive: sometimes we just give a sketch of proof, or we prove a particular case, or we just admit the result.

These notes come from a series of lectures given at Tsinghua University in April 2011. I am grateful to Tsinghua University and Professor S.-T. Yau for the generous invitation.

1 The cohomology of a torus

1.1 Real tori

Let V be a real vector space, of dimension n. A *lattice* in V is a $\mathbb{Z}$-module $\Gamma \subset V$ such that the induced map $\Gamma \otimes_{\mathbb{Z}} \mathbb{R} \to V$ is an isomorphism; equivalently, any basis of Γ over $\mathbb{Z}$ is a basis of V. In particular $\Gamma \cong \mathbb{Z}^n$.

The quotient $T := V/\Gamma$ is a smooth, compact Lie group, isomorphic to $(\mathbb{S}^1)^n$. The quotient homomorphism $\pi : V \to V/\Gamma$ is the *universal covering* of T. Thus Γ is identified with the fundamental group $\pi_1(T)$.

We want to consider the cohomology algebra $H^*(T, \mathbb{C})$. We think of it as being de Rham cohomology: recall that a smooth p-form ω on T is *closed* if $d\omega = 0$, *exact* if $\omega = d\eta$ for some $(p-1)$-form η. Then

$$H^p(T, \mathbb{C}) = \frac{\{\text{closed } p\text{-forms}\}}{\{\text{exact } p\text{-forms}\}} .$$

Let ℓ be a linear form on V. The 1-form $d\ell$ on V is invariant by translation, hence is the pullback by π of a 1 form on T that we will still denote $d\ell$. Let $(x_1, \ldots, x_n)$ be a system of coordinates on V. The forms $(dx_1, \ldots, dx_n)$ form a basis of the cotangent space $T_a^*(T)$ at each point $a \in T$; thus a p-form ω on T can be written in a unique way

$$\omega = \sum_{i_1 < \cdots < i_p} \omega_{i_1 \cdots i_p}(x)\, dx_{i_1} \wedge \cdots \wedge dx_{i_p},$$

where the $\omega_{i_1 \cdots i_p}$ are smooth functions on T (with complex values).

An important role in what follows will be played by the translations $t_a : x \mapsto x + a$ of T. We say that a p-form ω is *constant* if it is invariant by translation, that is, $t_a^* \omega = \omega$ for all $a \in T$; in terms of the above expression for ω, it means that the functions $\omega_{i_1 \ldots i_p}$ are constant. Such a form is determined by its value at 0, which is a skew-symmetric p-linear form on $V = T_0(T)$. We will denote by $\mathrm{Alt}^p(V, \mathbb{C})$ the space of such forms, and identify it to the space of constant p-forms. A constant form is closed, hence we have a linear map $\delta^p : \mathrm{Alt}^p(V, \mathbb{C}) \to H^p(T, \mathbb{C})$. Note that $\mathrm{Alt}^1(V, \mathbb{C})$ is simply $\mathrm{Hom}_{\mathbb{R}}(V, \mathbb{C})$, and δ^1 maps a linear form ℓ to the class of $d\ell$.

Proposition 1.1. *The map* $\delta^p : \mathrm{Alt}^p(V,\mathbb{C}) \to H^p(T,\mathbb{C})$ *is an isomorphism.*

Proof . There are various elementary proofs of this, see for instance [8], III.4. To save time we will use the Künneth formula. We choose our coordinates $(x_1,\ldots,x_n)$ so that $V=\mathbb{R}^n$, $\Gamma=\mathbb{Z}^n$. Then $T=T_1\times\cdots\times T_n$, with $T_i\cong\mathbb{S}^1$ for each i, and dx_i is a 1-form on T_i, which generates $H^1(T_i,\mathbb{C})$. The Künneth formula gives an isomorphism of graded algebras $H^*(T,\mathbb{C}) \xrightarrow{\sim} \bigotimes_i H^*(T_i,\mathbb{C})$. This means that $H^*(T,\mathbb{C})$ is the exterior algebra on the vector space with basis $(dx_1,\ldots,dx_n)$, and this is equivalent to the assertion of the Proposition. □

What about $H^*(T,\mathbb{Z})$? The Künneth isomorphism shows that it is torsion free, so it can be considered as a subgroup of $H^*(T,\mathbb{C})$. By definition of the de Rham isomorphism the image of $H^p(T,\mathbb{Z})$ in $H^p(T,\mathbb{C})$ is spanned by the closed p-forms ω such that $\int_\sigma \omega \in \mathbb{Z}$ for each p-cycle σ in $H_p(T,\mathbb{Z})$. Write again $T=\mathbb{R}^n/\mathbb{Z}^n$; the closed paths $\gamma_i : t\mapsto te_i$, for $t\in[0,1]$, form a basis of $H_1(T,\mathbb{Z})$, and we have $\int_{\gamma_i} d\ell = \ell(e_i)$. Thus $H^1(T,\mathbb{Z})$ is identified with the subgroup of $H^1(T,\mathbb{C}) = \mathrm{Hom}_{\mathbb{R}}(V,\mathbb{C})$ consisting of linear forms $V\to\mathbb{C}$ which take integral values on Γ; it is isomorphic to $\mathrm{Hom}_{\mathbb{Z}}(\Gamma,\mathbb{Z})$. Applying again the Künneth formula gives:

Proposition 1.2. *For each* p*, the image of* $H^p(T,\mathbb{Z})$ *in* $H^p(T,\mathbb{C})\cong \mathrm{Alt}^p(V,\mathbb{C})$ *is the subgroup of forms which take integral values on* Γ*; it is isomorphic to* $\mathrm{Alt}^p(\Gamma,\mathbb{Z})$. □

1.2 Complex tori

From now on we assume that V has a complex structure, that is, V is a complex vector space, of dimension g. Thus $V\cong\mathbb{C}^g$ and $\Gamma\cong\mathbb{Z}^{2g}$. Then $T:=V/\Gamma$ is a complex manifold, of dimension g, in fact a complex Lie group; the covering map $\pi : V\to V/\Gamma$ is holomorphic. We say that T is a *complex torus*. **Beware:** while all real tori of dimension n are diffeomorphic to $(\mathbb{S}^1)^n$, there are many non-isomorphic complex tori of dimension g – more about that in Sect. 3.3 below.

The complex structure of V provides a natural decomposition

$$\mathrm{Hom}_{\mathbb{R}}(V,\mathbb{C}) = V^*\oplus \overline{V}^* ,$$

where $V^* := \mathrm{Hom}_{\mathbb{C}}(V,\mathbb{C})$ and $\overline{V}^* = \mathrm{Hom}_{\mathbb{C}}(\overline{V},\mathbb{C})$ are the subspaces of $\mathbb{C}$-linear and $\mathbb{C}$-antilinear forms respectively. We write the corresponding decomposition of $H^1(T,\mathbb{C})$

$$H^1(T,\mathbb{C}) = H^{1,0}(T)\oplus H^{0,1}(T) .$$

If $(z_1,\ldots,z_g)$ is a coordinate system on V, $H^{1,0}(T)$ is the subspace spanned by the classes of $dz_1,\ldots,dz_g$, while $H^{1,0}(T)$ is spanned by the classes of $d\bar z_1,\ldots,d\bar z_g$.

The decomposition $\mathrm{Hom}_{\mathbb{R}}(V,\mathbb{C}) = V^*\oplus\overline{V}^*$ gives rise to a decomposition

$$\mathrm{Alt}^p(V,\mathbb{C}) \cong \wedge^pV^*\oplus(\wedge^{p-1}V^*\otimes\overline{V}^*)\oplus\cdots\oplus\wedge^p\overline{V}^*$$

which we write

$$H^p(T,\mathbb{C}) = H^{p,0}(T)\oplus\cdots\oplus H^{0,p}(T) .$$

The forms in $\mathrm{Alt}^p(V,\mathbb{C})$ which belong to $H^{p,0}(T)$ (resp. $H^{0,p}(T)$) are those which are $\mathbb{C}$-linear (resp. $\mathbb{C}$-antilinear) in each variable. It is not immediate to characterize those which belong to $H^{q,r}(T)$ for $q,r>0$; for $p=2$ we have:

Proposition 1.3. *Via the identification $H^2(T,\mathbb{C})=\mathrm{Alt}^2(V,\mathbb{C})$, $H^{2,0}$ is the space of $\mathbb{C}$-bilinear forms, $H^{0,2}$ the space of $\mathbb{C}$-biantilinear forms, and $H^{1,1}$ is the space of $\mathbb{R}$-bilinear forms E such that $E(ix,iy)=E(x,y)$.*

Proof. We have only to prove the last assertion. For $\varepsilon\in\{\pm1\}$, let $\mathcal{E}_\varepsilon$ be the space of forms $E\in\mathrm{Alt}^2(V,\mathbb{C})$ satisfying $E(ix,iy)=\varepsilon E(x,y)$. We have $\mathrm{Alt}^2(V,\mathbb{C})=\mathcal{E}_1\oplus\mathcal{E}_{-1}$, and $H^{2,0}$ and $H^{0,2}$ are contained in $\mathcal{E}_{-1}$.

For $\ell\in V^*$, $\ell'\in\overline{V}^*$, $v,w\in V$, we have

$$(\ell\wedge\ell')(iv,iw)=\ell(iv)\ell'(iw)-\ell(iw)\ell'(iv)=(\ell\wedge\ell')(v,w)\ ,$$

hence $H^{1,1}$ is contained in $\mathcal{E}_1$; it follows that $H^{2,0}\oplus H^{0,2}=\mathcal{E}_{-1}$ and $H^{1,1}=\mathcal{E}_1$. □

2 Line bundles on complex tori

2.1 The Picard group of a manifold

Our next goal is to describe all holomorphic line bundles on our complex torus T. Recall that line bundles on a complex manifold M form a group, the *Picard group* $\mathrm{Pic}(M)$ (the group structure is given by the tensor product of line bundles). It is canonically isomorphic to the first cohomology group $H^1(M,\mathcal{O}_M^*)$ of the sheaf $\mathcal{O}_M^*$ of invertible holomorphic functions on M. To compute this group a standard tool is the *exponential exact sequence* of sheaves

$$0\to\mathbb{Z}_M\to\mathcal{O}_M\xrightarrow{\ \mathbb{e}\ }\mathcal{O}_M^*\to1$$

where $\mathbb{e}(f):=\exp(2\pi if)$, and $\mathbb{Z}_M$ denotes the sheaf of locally constant functions on M with integral values. This gives a long exact sequence in cohomology

$$H^1(M,\mathbb{Z})\longrightarrow H^1(M,\mathcal{O}_M)\longrightarrow\mathrm{Pic}(M)\xrightarrow{\ c_1\ }H^2(M,\mathbb{Z})\longrightarrow H^2(M,\mathcal{O}_M)\qquad(2.1)$$

For $L\in\mathrm{Pic}(M)$, the class $c_1(L)\in H^2(M,\mathbb{Z})$ is the *first Chern class* of L. It is a topological invariant, which depends only on L as a topological complex line bundle (this is easily seen by replacing holomorphic functions by continuous ones in the exponential exact sequence).

When M is a projective (or compact Kähler) manifold, Hodge theory provides more information on this exact sequence.[1] The image of c_1 is the kernel of the natural map $H^2(M,\mathbb{Z})\to H^2(M,\mathcal{O}_M)$. This map is the composition of the maps $H^2(M,\mathbb{Z})\to H^2(M,\mathbb{C})\to H^2(M,\mathcal{O}_M)$ deduced from the injections of

[1] In this section and the following we use standard Hodge theory, as explained in [13], 0.6. Note that Hodge theory is much easier in the two cases of interest for us, namely complex tori and algebraic curves.

sheaves $\mathbb{Z}_M \hookrightarrow \mathbb{C}_M \hookrightarrow \mathcal{O}_M$. Now the map $H^2(M,\mathbb{C}) \to H^2(M,\mathcal{O}_M) \cong H^{0,2}$ is the projection onto the last summand of the Hodge decomposition

$$H^2(M,\mathbb{C}) = H^{2,0} \oplus H^{1,1} \oplus H^{0,2}$$

(for the experts: this can be seen by comparing the de Rham complex with the Dolbeault complex.)

Thus the image of c_1 consists of classes $\alpha \in H^2(M,\mathbb{Z})$ whose image $\alpha_{\mathbb{C}} = \alpha^{0,2} + \alpha^{1,1} + \alpha^{0,2}$ in $H^2(M,\mathbb{C})$ satisfies $\alpha^{0,2} = 0$. But since $\alpha_{\mathbb{C}}$ comes from $H^2(M,\mathbb{R})$ we have $\alpha^{2,0} = \overline{\alpha^{0,2}} = 0$: *the image of* c_1 *consists of the classes in* $H^2(M,\mathbb{Z})$ *whose image in* $H^2(M,\mathbb{C})$ *belongs to* $H^{1,1}$ ("Lefschetz theorem").

The kernel of c_1, denoted $\operatorname{Pic}^{\circ}(M)$, is the group of topologically trivial line bundles. The exact sequence (2.1) shows that it is isomorphic to the quotient of $H^1(M,\mathcal{O}_M)$ by the image of $H^1(M,\mathbb{Z})$. We claim that this image is a lattice in $H^1(M,\mathcal{O}_M)$: this is equivalent to saying that the natural map $H^1(M,\mathbb{R}) \to H^1(M,\mathcal{O}_M)$ is bijective. By Hodge theory, this map is identified with the restriction to $H^1(M,\mathbb{R})$ of the projection of $H^1(M,\mathbb{C}) = H^{1,0} \oplus H^{0,1}$ onto $H^{0,1}$. Since $H^1(M,\mathbb{R})$ is the subspace of classes $\alpha + \bar{\alpha}$ in $H^1(M,\mathbb{C})$, the projection $H^1(M,\mathbb{R}) \to H^{0,1}$ is indeed bijective. Thus $\operatorname{Pic}^{\circ}(M)$ is naturally identified with the complex torus $H^1(M,\mathcal{O}_M)/H^1(M,\mathbb{Z})$.

2.2 Flat line bundles

There is another description of $\operatorname{Pic}^{\circ}(M)$ which will be of interest for us. Instead of holomorphic line bundles, defined by holomorphic transition functions, we can consider *flat* line bundles, defined by locally constant transition functions; they are parametrized by $H^1(M,\mathbb{C}^*)$.

More important for us will be the *unitary flat* line bundles, defined by locally constant transition functions with values in the unit circle $\mathbb{S}^1$. Let us assume for simplicity that $H^2(M,\mathbb{Z})$ *is torsion free*. In that case the diagram of exact sequences of sheaves

$$\begin{array}{ccccccccc}
0 & \longrightarrow & \mathbb{Z}_M & \longrightarrow & \mathbb{R}_M & \xrightarrow{\ \mathrm{e}\ } & \mathbb{S}^1_M & \longrightarrow & 1 \\
 & & \downarrow & & \downarrow & & \downarrow & & \\
0 & \longrightarrow & \mathbb{Z}_M & \longrightarrow & \mathcal{O}_M & \xrightarrow{\ \mathrm{e}\ } & \mathcal{O}^*_M & \longrightarrow & 1
\end{array}$$

gives rise to a diagram in cohomology

$$\begin{array}{ccccccccc}
0 & \longrightarrow & H^1(M,\mathbb{Z}) & \longrightarrow & H^1(M,\mathbb{R}) & \xrightarrow{\ \varepsilon\ } & H^1(M,\mathbb{S}^1) & \longrightarrow & 0 \\
 & & \| & & \downarrow{\scriptstyle \pi} & & \downarrow & & \\
0 & \longrightarrow & H^1(M,\mathbb{Z}) & \longrightarrow & H^1(M,\mathcal{O}_M) & \longrightarrow & \operatorname{Pic}^{\circ}(M) & \longrightarrow & 0
\end{array}$$

(the homomorphism ε is surjective because $H^2(M,\mathbb{Z})$ is torsion free and therefore injects into $H^2(M,\mathbb{R})$). We have seen in the previous section that π is bijective, so *the map* $H^1(M,\mathbb{S}^1) \to \operatorname{Pic}^{\circ}(M)$ *is an isomorphism.* In other words, every line bundle $L \in \operatorname{Pic}^{\circ}(M)$ admits a unique unitary flat structure.

2.3 Systems of multipliers

We go back to our complex torus $T = V/\Gamma$.

Lemma 2.1. *Every line bundle on V is trivial.*

Proof. We have $H^2(V, \mathbb{Z}) = 0$ and $H^1(V, \mathcal{O}_V) = 0$ (see [13], p. 46), hence $\mathrm{Pic}(V) = 0$ by the exact sequence (2.1). □

Let L be a line bundle on T. We consider the diagram

$$\begin{array}{ccc} \pi^* L & \longrightarrow & L \\ \big\downarrow & & \big\downarrow \\ V & \xrightarrow{\ \pi\ } & T \end{array}$$

The action of Γ on V lifts to an action on $\pi^* L = V \times_T L$. We know that $\pi^* L$ is trivial; we choose a trivialization $\pi^* L \xrightarrow{\sim} V \times \mathbb{C}$. We obtain an action of Γ on $V \times \mathbb{C}$, so that L is the quotient of $V \times \mathbb{C}$ by this action. An element γ of Γ acts linearly on the fibers, hence by

$$\gamma \cdot (z, t) = (z + \gamma, e_\gamma(z)\, t) \quad \text{for } z \in V,\ t \in \mathbb{C}$$

where e_γ is a holomorphic invertible function on V. This formula defines a group action of Γ on $V \times \mathbb{C}$ if and only if the functions e_γ satisfy

$$e_{\gamma+\delta}(z) = e_\gamma(z + \delta)\, e_\delta(z) \quad (\text{“cocycle condition”}).$$

A family $(e_\gamma)_{\gamma \in \Gamma}$ of holomorphic invertible functions on V satisfying this condition is called a system of multipliers. Every line bundle on T is defined by such a system.

A *theta function* for the system $(e_\gamma)_{\gamma \in \Gamma}$ is a holomorphic function $V \to \mathbb{C}$ satisfying

$$\theta(z + \gamma) = e_\gamma(z)\theta(z) \quad \text{for all } \gamma \in \Gamma, z \in V.$$

Proposition 2.2. *Let $(e_\gamma)_{\gamma \in \Gamma}$ be a system of multipliers, and L the associated line bundle. The space $H^0(T, L)$ is canonically identified with the space of theta functions for $(e_\gamma)_{\gamma \in \Gamma}$.*

Proof. Any global section s of L lifts to a section $\hat{s} = \pi^* s$ of $\pi^* L = V \times_T L$ over V, defined by $\hat{s}(z) = (z, s(\pi z))$; it is Γ-invariant in the sense that $\hat{s}(z + \gamma) = \gamma \cdot \hat{s}(z)$. Conversely, a Γ-invariant section of $\pi^* L$ comes from a section of L. Now a section of $\pi^* L \cong V \times \mathbb{C}$ is of the form $z \mapsto (z, \theta(z))$, where $\theta : V \to \mathbb{C}$ is holomorphic. It is Γ-invariant if and only if θ is a theta function for $(e_\gamma)_{\gamma \in \Gamma}$. □

Let $(e_\gamma)_{\gamma \in \Gamma}$ and $(e'_\gamma)_{\gamma \in \Gamma}$ be two systems of multipliers, defining line bundles L and L'. The line bundle $L \otimes L'$ is the quotient of the trivial line bundle $V \times (\mathbb{C} \otimes \mathbb{C})$ by the tensor product action $\gamma \cdot (z, t \otimes t') = (z + \gamma, e_\gamma(z)\, t \otimes e'_\gamma(z)\, t')$; therefore it is defined by the system of multipliers $(e_\gamma e'_\gamma)_{\gamma \in \Gamma}$. In other words, multiplication defines a group structure on the set of systems of multipliers, and we have a surjective group homomorphism

$$\{\text{systems of multipliers}\} \longrightarrow \operatorname{Pic}(T) .$$

A system of multipliers $(e_\gamma)_{\gamma\in\Gamma}$ lies in the kernel if and only if the associated line bundle admits a section which is everywhere $\neq 0$; in view of Proposition 2.2, this means that *there exists a holomorphic function* $h : V \to \mathbb{C}^*$ *such that* $e_\gamma(z) = \dfrac{h(z+\gamma)}{h(z)}$. We will call such systems of multipliers trivial. Thus we can always multiply a given system (e_γ) by a trivial one without changing the associated line bundle.

Remark 2.3. (only for the readers who know group cohomology) Put $H^* := H^0(V, \mathcal{O}_V^*)$. The system of multipliers are exactly the 1-cocycles of Γ with values in H^*, and the trivial systems are the coboundaries. Thus we get a group isomorphism $H^1(\Gamma, H^*) \xrightarrow{\sim} \operatorname{Pic}(T)$ (see [20], §2 for a more conceptual explanation of this isomorphism).

Remark 2.4. The argument in this section apply equally well to *flat* line bundles, since obviously $H^1(V, \mathbb{C}^*) = 0$. The corresponding systems of multipliers are of the form $a(\gamma)_{\gamma\in\Gamma}$, where $a : \Gamma \to \mathbb{C}^*$ is a homomorphism. Similarly, unitary flat line bundles correspond to homomorphisms $\Gamma \to \mathbb{S}^1$. This is nothing but the classical isomorphism $H^1(T, A) \xrightarrow{\sim} \operatorname{Hom}(\pi_1(T), A)$ for $A = \mathbb{C}^*$ or $\mathbb{S}^1$.

2.4 Interlude: hermitian forms

There are many holomorphic invertible functions on V, hence many systems of multipliers giving rise to the same line bundle. Our next goal will be to find a subset of such systems such that each line bundle corresponds exactly to one system of multipliers in that subset. This will involve hermitian forms on V, so let us fix our conventions.

A *hermitian form* H on V will be $\mathbb{C}$-linear in the second variable, $\mathbb{C}$-antilinear in the first. We put $S(x,y) = \operatorname{Re} H(x,y)$ and $E(x,y) = \operatorname{Im} H(x,y)$. S and E are $\mathbb{R}$-bilinear forms on V, S is symmetric, E is skew-symmetric; they satisfy:

$$S(x,y) = S(ix,iy), \quad E(x,y) = E(ix,iy), \quad S(x,y) = E(x,iy)$$

Using these relations one checks easily that the following data are equivalent:

- The hermitian form H;
- The symmetric $\mathbb{R}$-bilinear form S with $S(x,y) = S(ix,iy)$;
- The skew-symmetric $\mathbb{R}$-bilinear form E with $E(x,y) = E(ix,iy)$.

Moreover,

$$H \text{ non-degenerate} \iff E \text{ non-degenerate} \iff S \text{ non-degenerate.}$$

2.5 Systems of multipliers associated to hermitian forms

We denote by $\mathcal{P}$ the set of pairs (H,α), where H is a hermitian form on V, α a map from Γ to $\mathbb{S}^1$, satisfying:

$$E := \operatorname{Im}(H) \text{ takes integral values on } \Gamma, \text{ and } \alpha \text{ satisfies}$$

$$\alpha(\gamma+\delta) = \alpha(\gamma)\alpha(\delta)(-1)^{E(\gamma,\delta)} .$$

The law $(H,\alpha)\cdot(H',\alpha') = (H+H',\alpha\alpha')$ defines a group structure on $\mathcal{P}$. For $(H,\alpha)\in\mathcal{P}$, we put

$$e_\gamma(z) = \alpha(\gamma)e^{\pi[H(\gamma,z)+\frac{1}{2}H(\gamma,\gamma)]} .$$

We leave as an (easy) exercise to check that this defines a system of multipliers. The corresponding line bundle will be denoted $L(H,\alpha)$. The map $(H,\alpha)\mapsto L(H,\alpha)$ from $\mathcal{P}$ onto $\operatorname{Pic}(T)$ is a group homomorphism; we want to prove that it is an isomorphism.

Proposition 2.5. *The first Chern class $c_1(L(H,\alpha))$ is equal to $E\in \operatorname{Alt}^2(\Gamma,\mathbb{Z})\cong H^2(T,\mathbb{Z})$.*

Proof. We will use Chern's original definition of the first Chern class of a line bundle L on a compact manifold M (see [13], p. 141). One chooses a $\mathcal{C}^\infty$ metric h on L; this is nothing but a $\mathcal{C}^\infty$ function $L\to\mathbb{R}_+$, which is positive outside the zero section and satisfies $h(\lambda x) = |\lambda|^2h(x)$ for $x\in L$, $\lambda\in\mathbb{C}$. If s is a local non-vanishing holomorphic section of L, the 2-form $\omega_{L,h} := \frac{1}{2\pi i}\partial\bar\partial\log h(s)$ does not depend on the choice of s; thus $\omega_{L,h}$ is a globally defined closed 2-form, whose class in $H^2(M,\mathbb{C})$ represents $c_1(L)$.

To apply this in our situation, we observe that the metric $\tilde h$ on $V\times\mathbb{C}$ defined by $\tilde h(z,t) = e^{-\pi H(z,z)}|t|^2$ is invariant under Γ; hence it is the pullback of a metric h on $L(H,\alpha)$. The form $\pi^*\omega_{L,h}$ is equal to $\omega_{V\times\mathbb{C},\tilde h}$; to compute it we apply our formula to the section $s: z\mapsto(z,1)$ of $V\times\mathbb{C}$. We find

$$\omega_{V\times\mathbb{C},\tilde h} = \frac{1}{2\pi i}\partial\bar\partial\log e^{-\pi H(z,z)} = \frac{i}{2}\partial\bar\partial\, H(z,z) .$$

It remains to prove that $\frac{i}{2}\partial\bar\partial\, H(z,z)$ is the constant 2-form defined by E. It suffices to prove this when $H(x,y) = \bar x_j y_j$; then $\frac{i}{2}\partial\bar\partial\, H(z,z) = \frac{i}{2}dz_j\wedge d\bar z_j$. Let $v=(v_1,\dots,v_g)$, $w=(w_1,\dots,w_g)$ two vectors of V; we have

$$(dz_j\wedge d\bar z_j)(v,w) = z_j(v)\bar z_j(w) - z_j(w)\bar z_j(v) = v_j\bar w_j - w_j\bar v_j = -2i\operatorname{Im}H(v,w) ,$$

hence our assertion. □

(See [20], §2 for a proof in terms of group cohomology.)

Theorem 2.6. *The map $(H,\alpha)\mapsto L(H,\alpha)$ defines a group isomorphism $\mathcal{P}\xrightarrow{\sim}\operatorname{Pic}(T)$.*

Proof . Let $\mathcal{Q}$ be the group of hermitian forms H on V such that $\mathrm{Im}(H)$ is integral on Γ. By Proposition 2.5 and Section 2.1 we have a commutative diagram

$$\begin{array}{ccccccccc}
0 & \longrightarrow & \mathrm{Hom}(\Gamma, \mathbb{S}^1) & \longrightarrow & \mathcal{P} & \longrightarrow & \mathcal{Q} & \longrightarrow & 0 \\
 & & \downarrow{\scriptstyle L^o} & & \downarrow{\scriptstyle L} & & \downarrow{\scriptstyle \iota} & & \\
0 & \longrightarrow & \mathrm{Pic}^o(T) & \longrightarrow & \mathrm{Pic}(T) & \xrightarrow{c_1} & H^2(T,\mathbb{Z}) & &
\end{array}$$

with $\iota(H) = \mathrm{Im}(H) \in \mathrm{Alt}^2(\Gamma, \mathbb{Z}) \cong H^2(T, \mathbb{Z})$.

Let us first prove that ι is bijective onto $\mathrm{Im}(c_1)$. Let $E \in \mathrm{Alt}^2(\Gamma,\mathbb{Z}) \cong H^2(T,\mathbb{Z})$; we have seen in Section 2.1 that E belongs to $\mathrm{Im}(c_1)$ if and only if it belongs to $H^{1,1}$, that is satisfies $E(ix, iy) = E(x, y)$ (Proposition 1.3). By Section 2.4 this is equivalent to $E = \mathrm{Im}(H)$ for a hermitian form $H \in \mathcal{Q}$; moreover H is uniquely determined by E, hence our assertion.

The map L^o associates to a unitary character $\alpha : \Gamma \to \mathbb{S}^1$ the unitary flat bundle $L(0, \alpha)$; we have already seen that it is bijective (Section 2.2 and Remark 2.4). Thus the map $(H, \alpha) \mapsto L(H, \alpha)$ is bijective. □

2.6 The theorem of the square

This section is devoted to an important result, Theorem 2.8 below, which is actually an easy consequence of our description of line bundles on T (we encourage the reader to have a look at the much more elaborate proof in [20], §6, valid over any algebraically closed field).

Lemma 2.7. *Let $a \in V$. We have $t^*_{\pi(a)}L(H, \alpha) = L(H, \alpha')$ with $\alpha'(\gamma) = \alpha(\gamma)\mathbb{e}(E(\gamma, a))$.*

Proof . In general, let L be a line bundle on T defined by a system of multipliers $(e_\gamma)_{\gamma\in\Gamma}$. Then $(e_\gamma(z + a))_{\gamma\in\Gamma}$ is a system of multipliers, defining a line bundle L'; the self-map $(z, t) \mapsto (z + a, t)$ of $V \times \mathbb{C}$ is equivariant w.r.t. the actions of Γ defined by $(e_\gamma(z + a))$ on the source and $(e_\gamma(z))$ on the target, so it induces an isomorphism $L' \xrightarrow{\sim} t^*_{\pi(a)}L$.

We apply this to the multiplier $e_\gamma(z) = a(\gamma)e^{\pi[H(\gamma,z)+\frac{1}{2}H(\gamma,\gamma)]}$; we find $e_\gamma(z+a) = e_\gamma(z)\ e^{\pi H(\gamma,a)}$. Recall that we are free to multiply $e_\gamma(z)$ by $\frac{h(z+\gamma)}{h(z)}$ for some holomorphic invertible function h; taking $h(z) = e^{-\pi H(a,z)}$, our multiplier becomes $e_\gamma\ e^{\pi[H(\gamma,a)-H(a,\gamma)]} = e_\gamma\ e^{2\pi i E(\gamma,a)}$. □

Theorem 2.8 (Theorem of the square). *Let L be a line bundle on T.*

1) *The map*

$$\lambda_L : T \to \mathrm{Pic}^o(T), \quad \lambda_L(a) = t_a^* L \otimes L^{-1}$$

is a group homomorphism.

2) *Let $E \in \mathrm{Alt}^2(\Gamma, \mathbb{Z})$ be the first Chern class of L. We have*

$$\mathrm{Ker}\,\lambda_L = \Gamma^\perp/\Gamma\ , \quad \textit{with}\ \ \Gamma^\perp := \{z \in V \mid E(z, \gamma) \in \mathbb{Z}\ \ \textit{for all}\ \gamma \in \Gamma\}\,.$$

3) *If E is non-degenerate, λ_L is surjective and has finite kernel.*

4) *If E is unimodular, λ_L is a group isomorphism.*

Proof . By the Lemma, λ_L is the composition

$$T \xrightarrow{\varepsilon} \mathrm{Hom}(\Gamma, \mathbb{S}^1) \xrightarrow{L^o} \mathrm{Pic}^o(T) ,$$

where $\varepsilon(a)$, for $a = \pi(\tilde{a}) \in T$, is the map $\gamma \mapsto \mathbb{e}(E(\gamma, \tilde{a})$, and L^o is the isomorphism $\alpha \mapsto L(0, \alpha)$ (Theorem 2.6). Therefore we can replace λ_L by ε in the proof. Then 1) and 2) become obvious.

Assume that E is non-degenerate. Let $\chi \in \mathrm{Hom}(\Gamma, \mathbb{S}^1)$. Since Γ is a free $\mathbb{Z}$-module, we can find a homomorphism $u : \Gamma \to \mathbb{R}$ such that $\chi(\gamma) = \mathbb{e}(u(\gamma))$ for each $\gamma \in \Gamma$. Extend u to a $\mathbb{R}$-linear form $V \to \mathbb{R}$; since E is non-degenerate, there exists $a \in V$ such that $u(z) = E(z, a)$, hence $\varepsilon(\pi(a)) = \chi$. Thus ε is surjective.

Let us denote by $e : V \to \mathrm{Hom}_{\mathbb{R}}(V, \mathbb{R})$ the $\mathbb{R}$-linear isomorphism associated to E. The dual $\Gamma^* := \mathrm{Hom}_{\mathbb{Z}}(\Gamma, \mathbb{Z})$ embeds naturally in $\mathrm{Hom}_{\mathbb{R}}(V, \mathbb{R})$, and $\Gamma^\perp$ is by definition $e^{-1}(\Gamma^*)$; then e identifies $\Gamma^\perp$ with Γ^*, so that the inclusion $\Gamma \subset \Gamma^\perp$ corresponds to the map $\Gamma \to \Gamma^*$ associated to $E_{|\Gamma}$. This map has finite cokernel, and it is bijective if E is unimodular; this achieves the proof. □

Remark 2.9. We have seen in Section 2.1 that $\mathrm{Pic}^o(T)$ has a natural structure of complex torus; it is not difficult to prove that the map λ_L is holomorphic. In particular, when E is unimodular, λ_L is an isomorphism of complex tori.

Corollary 2.10. *Assume that $c_1(L)$ is non-degenerate. Any line bundle L' with $c_1(L') = c_1(L)$ is isomorphic to t_a^*L for some a in T.*

Proof . $L' \otimes L^{-1}$ belongs to $\mathrm{Pic}^o(T)$, hence is isomorphic to $t_a^*L \otimes L^{-1}$ for some a in T by 3). □

The following immediate consequence of 1) will be very useful:

Corollary 2.11. *Let $a_1, \ldots, a_r$ in T with $\sum a_i = 0$. Then $t_{a_1}^*L \otimes \cdots \otimes t_{a_r}^*L \cong L^{\otimes r}$.* □

3 Polarizations

In this section we will consider a line bundle $L = L(H, \alpha)$ on our complex torus T such that the hermitian form H is *positive definite*. We will first look for a concrete expression of the situation using an appropriate basis.

3.1 Frobenius lemma

The following easy result goes back to Frobenius:

Proposition 3.1. *Let Γ be a free finitely generated $\mathbb{Z}$-module, and $E : \Gamma \times \Gamma \to \mathbb{Z}$ a skew-symmetric, non-degenerate form. There exist positive integers $d_1, \ldots, d_g$ with $d_1 \,|\, d_2 \,|\, \cdots \,|\, d_g$ and a basis $(\gamma_1, \ldots, \gamma_g; \delta_1, \ldots, \delta_g)$ of Γ such that the matrix of E in this basis is $\begin{pmatrix} 0 & \mathbf{d} \\ -\mathbf{d} & 0 \end{pmatrix}$, where $\mathbf{d}$ is the diagonal matrix with entries $(d_1, \ldots, d_g)$.*

As a consequence we see that the determinant of E is the square of the integer $d_1 \cdots d_g$, called the *Pfaffian* of E and denoted $\mathrm{Pf}(E)$. The most important case for us will be when $d_1 = \cdots = d_g = 1$, or equivalently $\det(E) = 1$; in that case one says that E is *unimodular*, and that $(\gamma_1, \ldots, \gamma_g; \delta_1, \ldots, \delta_g)$ is a *symplectic basis* of Γ.

Proof. Let d_1 be the minimum of the numbers $E(\alpha, \beta)$ for $\alpha, \beta \in \Gamma$, $E(\alpha, \beta) > 0$; choose γ, δ such that $E(\gamma, \delta) = d_1$. For any $\varepsilon \in \Gamma$, $E(\gamma, \varepsilon)$ is divisible by d_1 – otherwise using Euclidean division we would find ε with $0 < E(\gamma, \varepsilon) < d_1$. Likewise $E(\delta, \varepsilon)$ is divisible by d_1. Put $U = \mathbb{Z}\gamma \oplus \mathbb{Z}\delta$; we claim that $\Gamma = U \oplus U^\perp$. Indeed, for $x \in \Gamma$, we have

$$x = \frac{E(x,\delta)}{d_1}\gamma + \frac{E(\gamma,x)}{d_1}\delta + (x - \frac{E(x,\delta)}{d_1}\gamma - \frac{E(\gamma,x)}{d_1}\delta) \,.$$

Reasoning by induction on the rank of Γ, we find integers $d_2 \,|\, d_3 \,|\, \cdots \,|\, d_g$ and a basis $(\gamma, \gamma_2, \ldots, \gamma_g;\ \delta, \delta_2, \ldots, \delta_g)$ of Γ, such that the matrix of E is $\begin{pmatrix} 0 & \mathbf{d} \\ -\mathbf{d} & 0 \end{pmatrix}$. It remains to prove that d_1 divides d_2; otherwise, using Euclidean division again, we can find $k \in \mathbb{Z}$ such that $0 < E(\gamma + \gamma_2, k\delta + \delta_2) < d_1$, a contradiction. □

3.2 Polarizations and the period matrix

Going back to our complex torus $T = V/\Gamma$, we assume given a *positive definite* hermitian form H on V, such that $E := \mathrm{Im}(H)$ takes integral values on Γ. Such a form is called a *polarization* of T; if E is unimodular, we say that H is a *principal polarization.* A complex torus which admits a polarization is classically called a (polarized) *abelian variety*; we will see below that it is actually a projective manifold. It is common to use the abbreviation *p.p.a.v.* for "principally polarized abelian variety".

We choose a basis $(\gamma_1, \ldots, \gamma_g; \delta_1, \ldots, \delta_g)$ as in Proposition 3.1 (note that E is non-degenerate by Section 2.4); we put $\gamma'_j := \frac{\gamma_j}{d_j}$ for $j = 1, \ldots, g$.

Lemma 3.2. *$(\gamma'_1, \ldots, \gamma'_g)$ is a basis of V over $\mathbb{C}$.*

Proof. Let $W = \mathbb{R}\gamma'_1 \oplus \cdots \oplus \mathbb{R}\gamma'_g$. Our statement is equivalent to $V = W \oplus iW$. But if $x \in W \cap iW$, we have $H(x,x) = E(x, ix) = 0$ since $E_{|W} = 0$, hence $x = 0$. □

Expressing the δ_j's in this basis gives a matrix $\tau \in M_g(\mathbb{C})$ with $\delta_j = \sum_i \tau_{ij}\gamma'_i$. In the corresponding coordinates, we have

$$\Gamma = \mathbf{d}\mathbb{Z}^g \oplus \tau\mathbb{Z}^g$$

in other words, the elements of Γ are the column vectors $\mathbf{d}p+\tau q$ with $p,q\in\mathbb{Z}^g$. The matrix τ is often called the *period matrix.*

Note that in case the polarization is principal we have $\gamma'_i=\gamma_i$ and $\Gamma=\mathbb{Z}^g\oplus\tau\mathbb{Z}^g$.

Proposition 3.3. *The matrix τ is symmetric, and* $\operatorname{Im}(\tau)$ *is positive definite.*

Proof. Put $\tau=A+iB$, with $A,B\in M_g(\mathbb{R})$. We will compare the bases $(\gamma_1,\ldots,\gamma_g;\delta_1,\ldots,\delta_g)$ and $(\gamma'_1,\ldots,\gamma'_g;i\gamma'_1,\ldots,i\gamma'_g)$ of V over $\mathbb{R}$. The change of basis matrix (expressing the vectors of the first basis in the second one) is $P=\begin{pmatrix}\mathbf{d} & A\\ 0 & B\end{pmatrix}$. Therefore the matrix of E in the second basis is

$$ {}^tP^{-1}\begin{pmatrix}0 & \mathbf{d}\\ -\mathbf{d} & 0\end{pmatrix}P^{-1}=\begin{pmatrix}0 & B^{-1}\\ -{}^tB^{-1} & {}^tB^{-1}(A-{}^tA)B^{-1}\end{pmatrix} $$

Now the condition $E(ix,iy)=E(x,y)$, expressed in the basis $(\gamma'_1,\ldots,\gamma'_g;i\gamma'_1,\ldots,i\gamma'_g)$, is equivalent to $A={}^tA$ and $B={}^tB$; we have $H(\gamma'_j,\gamma'_k)=E(\gamma'_j,i\gamma'_k)$, so the matrix of H in the basis $(\gamma'_1,\ldots,\gamma'_g)$ (over $\mathbb{C}$) is B^{-1}, and the positivity of H is equivalent to that of B. $\square$

3.3 The moduli space of p.p.a.v.

In this section we restrict for simplicity to the case where the polarization is *principal*; we encourage the reader to adapt the argument to the general case (see for instance [8], VII.1).

We have seen that the choice of a symplectic basis determines the matrix τ, which in turn completely determines T and H: we have $V=\mathbb{C}^g$ and $\Gamma=\Gamma_\tau:=\mathbb{Z}^g\oplus\tau\mathbb{Z}^g$; the hermitian form H is given by the matrix $\operatorname{Im}(\tau)^{-1}$, and its imaginary part E by $E(p+\tau q,p'+\tau q')={}^tpq'-{}^tqp'$.

The space of symmetric matrices $\tau\in M_g(\mathbb{C})$ with $\operatorname{Im}(\tau)$ positive definite is denoted $\mathbb{H}_g$, and called the *Siegel upper half space.* It is an open subset of the vector space of complex symmetric matrices. From what we have seen it follows that $\mathbb{H}_g$ parametrizes p.p.a.v. $(V/\Gamma,H)$ endowed with a symplectic basis of the lattice Γ.

Now we want to get rid of the choice of the symplectic basis. We have associated to a symplectic basis an isomorphism $V\xrightarrow{\sim}\mathbb{C}^g$ which maps Γ to the lattice Γ_τ. A change of the basis amounts to a linear automorphism M of $\mathbb{C}^g$, inducing a symplectic isomorphism $\Gamma_\tau\xrightarrow{\sim}\Gamma_{\tau'}$. Such an isomorphism is given by $\begin{pmatrix}p'\\ q'\end{pmatrix}=P\begin{pmatrix}p\\ q\end{pmatrix}$, where P belongs to the symplectic group $\mathrm{Sp}(2g,\mathbb{Z})$, that is, $P\in M_{2g}(\mathbb{Z})$ and ${}^tPJP=J$, with $J=\begin{pmatrix}0 & \mathbf{1}\\ -\mathbf{1} & 0\end{pmatrix}$.

We have $M(p+\tau q)=p'+\tau' q'$, hence

$$ (\mathbf{1}\ \tau')=M\,(\mathbf{1}\ \tau)\,P^{-1}\quad\text{or equivalently}\quad \begin{pmatrix}\mathbf{1}\\ \tau'\end{pmatrix}={}^tP^{-1}\begin{pmatrix}\mathbf{1}\\ \tau\end{pmatrix}{}^tM\,. $$

If $P = \begin{pmatrix} a & b \\ c & d \end{pmatrix}$, with $a, b, c, d \in M_g(\mathbb{Z})$, we have ${}^tP^{-1} = -JPJ = \begin{pmatrix} d & -c \\ -b & a \end{pmatrix}$. We find

$$\mathbf{1} = (d - c\tau){}^tM \ , \quad \tau' = (-b + a\tau){}^tM \ , \quad \text{hence} \quad \tau' = (a\tau - b)(-c\tau + d)^{-1} \ .$$

Thus the group $\mathrm{Sp}(2g, \mathbb{Z})$ acts on $\mathbb{H}_g$ by $(P, \tau) \mapsto (a\tau - b)(-c\tau + d)^{-1}$, and two matrices τ, τ' correspond to the same p.p.a.v. with different symplectic bases iff they are conjugated under this action. To get a nicer formula, we observe that

$$\begin{pmatrix} a & -b \\ -c & d \end{pmatrix} = tPt \ , \ \text{with} \ t = \begin{pmatrix} \mathbf{1} & 0 \\ 0 & -\mathbf{1} \end{pmatrix} \ ;$$

since $tJt = -J$, the map $P \mapsto tPt$ is an automorphism of $\mathrm{Sp}(2g, \mathbb{Z})$. Composing our action with this automorphism, we obtain:

Proposition 3.4. *The group* $\mathrm{Sp}(2g, \mathbb{Z})$ *acts on* $\mathbb{H}_g$ *by* $\begin{pmatrix} a & b \\ c & d \end{pmatrix} \cdot \tau = (a\tau + b)(c\tau + d)^{-1}$. *The quotient* $\mathcal{A}_g := \mathbb{H}_g / \mathrm{Sp}(2g, \mathbb{Z})$ *parametrizes isomorphism classes of* g*-dimensional p.p.a.v.*

It is not difficult to show that the action of $\mathrm{Sp}(2g, \mathbb{Z})$ on $\mathbb{H}_g$ is nice ("properly discontinuous"), so that $\mathcal{A}_g$ is an analytic space ([8], VII.1). A much more subtle result is that $\mathcal{A}_g$ is Zariski open in a projective variety, the *Satake compactification* $\overline{\mathcal{A}_g}$.

We have not made precise in which sense $\mathcal{A}_g$ parametrizes p.p.a.v. It is actually what is called a moduli space; we will give a precise definition in the case of vector bundles (see Section 4.2 below), which can be adapted without difficulty to this case.

3.4 Theta functions

Let H be a polarization on T; we keep the notation of the previous sections. Let $\alpha : \Gamma \to \mathbb{S}^1$ be any map satisfying $\alpha(\gamma + \delta) = \alpha(\gamma)\alpha(\delta)(-1)^{E(\gamma, \delta)}$.

Theorem 3.5. $\dim H^0(T, L(H, \alpha)) = d_1 \cdots d_g = \mathrm{Pf}(E)$.

Proof. We first treat the case $d_1 = \cdots = d_g = 1$. According to Prop. 2.2, we are looking for theta functions satisfying

$$\theta(z + \gamma) = \alpha(\gamma) e^{\pi[H(\gamma, z) + \frac{1}{2}H(\gamma, \gamma)]} \theta(z) \ .$$

Recall that we are free to multiply $e_\gamma(z)$ by $\frac{h(z+\gamma)}{h(z)}$ for some $h \in H^0(V, \mathcal{O}_V^*)$ (this amounts to multiply θ by h). We will use this to make θ *periodic* with respect to the basis elements $\gamma_1, \ldots, \gamma_g$ of Γ.

As before we put $W = \mathbb{R}\gamma_1 \oplus \cdots \oplus \mathbb{R}\gamma_g$. Since $E_{|W} = 0$, the form $H_{|W}$ is a real symmetric form; since $V = W \oplus iW$ (Lemma 3.2), it extends as a $\mathbb{C}$*-bilinear* symmetric form B on V. We put $h(z) = e^{-\frac{\pi}{2}B(z,z)}$: this amounts to replace H in $e_\gamma(z)$ by $H' := H - B$. We have

Lemma 3.6. $H'(p+\tau q, z) = -2i\,{}^tqz$.

Proof. Let $w \in W$. We have $H'(w,y) = 0$ for $y \in W$, hence also for any $y \in V$ because H' is $\mathbb{C}$-linear in y. On the other hand for $z \in V$ we have $H'(z,w) = (H-B)(z,w) = (\bar{H}-B)(w,z) = (\bar{H}-H)(w,z) = 2iE(z,w)$. Thus for $z = \sum z_i\gamma_i \in V$ we have $H'(\gamma_j, z) = 0$ and $H'(\delta_j, z) = \sum_k z_k H'(\delta_j, \gamma_k) = -2iz_j$, hence the lemma. □

Put $L = L(H,\alpha)$. By Cor. 2.10, changing α amounts to replace L by t_a^*L for some $a \in T$. Since the pullback map $t_a^* : H^0(T,L) \to H^0(T, t_a^*L)$ is an isomorphism, it suffices to prove the theorem for a particular value of α; we choose $\alpha(p+\tau q) = (-1)^{{}^tpq}$. Indeed we have mod. 2:

$$ {}^t(p+p')(q+q') \equiv {}^tpq + {}^tp'q' + ({}^tpq' - {}^tp'q) = {}^tpq + {}^tp'q' + E(p+\tau q, p'+\tau q') $$

$$ \text{for } p,q,p',q' \in \mathbb{Z}^g . $$

Thus our theta functions must satisfy the quasi-periodicity condition

$$ \theta(z+p+\tau q) = \theta(z)\,\mathrm{e}(-{}^tqz - \frac{1}{2}{}^tq\tau q) \qquad \text{for } z \in \mathbb{C}^g,\ p,q \in \mathbb{Z}^g . $$

In particular, they are periodic with respect to the subgroup $\mathbb{Z}^g \subset \mathbb{C}^g$. This implies that they admit a Fourier expansion of the form $\theta(z) = \sum_{m\in\mathbb{Z}^g} c(m)\mathrm{e}({}^tmz)$. Now let us express the quasi-periodicity condition; we have:

$$ \theta(z+p+\tau q) = \sum_{m\in\mathbb{Z}^g} c(m)\mathrm{e}({}^tm\tau q)\mathrm{e}({}^tmz) $$

and

$$ \theta(z)\mathrm{e}(-{}^tqz - \frac{1}{2}{}^tq\tau q) = \sum_{m\in\mathbb{Z}^g} c(m)\mathrm{e}({}^t(m-q)z - \frac{1}{2}{}^tq\tau q) $$

$$ = \sum_{m\in\mathbb{Z}^g} c(m+q)\mathrm{e}(-\frac{1}{2}{}^tq\tau q)\mathrm{e}({}^tmz) . $$

Comparing we find $c(m+q) = c(m)\mathrm{e}({}^t(m+\frac{q}{2})\tau q)$. Taking $m=0$ gives $c(q) = c(0)\,\mathrm{e}(\frac{1}{2}{}^tq\tau q)$. Thus our theta functions, if they exist, are all proportional to

$$ \theta(z) = \sum_{m\in\mathbb{Z}^g} \mathrm{e}({}^tmz + \frac{1}{2}{}^tm\tau m) . $$

It remains to prove that this function indeed exists, that is that the series converges. But the coefficients $c(m)$ of the Fourier series satisfy $|c(m)| = e^{-q(m)}$, where q is a positive definite quadratic form, and therefore they decrease very fast as $m \to \infty$.

Now we treat the case $d_1 = \cdots = d_g = d$. In this case the form $\frac{1}{d}H$ is a principal polarization, so we can take $L = L_0^{\otimes d}$, where L_0 is the line bundle considered above. The corresponding theta functions satisfy

$$ \theta(z+p+\tau q) = \theta(z)\mathrm{e}(-d{}^tqz - \frac{d}{2}{}^tq\tau q) $$

for $z \in \mathbb{C}^g,\ p, q \in \mathbb{Z}^g$ ("theta functions of order d").

We write again $\theta(z) = \sum_{m\in\mathbb{Z}^g} c(m)\mathbb{e}({}^t\!mz)$; the quasi-periodicity condition gives

$$c(m+dq) = c(m)\mathbb{e}({}^t(m+\frac{d}{2})\tau q) = c(m)\mathbb{e}(\frac{-1}{d}{}^t\!m\tau m)\mathbb{e}(\frac{1}{2d}{}^t(m+dq)\tau(m+dq))$$

This determines up to a constant all coefficients $c(m)$ for m in a given coset ε of $\mathbb{Z}^g$ modulo $d\mathbb{Z}^g$; the corresponding theta function is

$$\theta[\varepsilon](z) = \sum_{m\in\varepsilon} \mathbb{e}({}^t\!mz + \frac{1}{2d}{}^t\!m\tau m)\ . \tag{3.1}$$

By what we have seen the functions $\theta[\varepsilon]$, where ε runs through $\mathbb{Z}^g/d\mathbb{Z}^g$, form a basis of the space of theta functions of order d; in particular, the dimension of this space is d^g.

The proof of the general case is completely analogous but requires more complicated notations. We will not need it in these lectures, so we leave it as an exercise for the reader. □

3.5 Comments

The proof of the theorem gives much more than the dimension of the space of theta functions, namely an explicit basis $(\theta[\varepsilon])_{\varepsilon\in\mathbb{Z}^g/d\mathbb{Z}^g}$ of this space given by formula (3.1). In particular, when the polarization H is principal, the line bundles $L(H,\alpha)$ have a unique non-zero section (up to a scalar); the divisor of this section is called a *theta divisor* of the p.p.a.v. (T,H). By Corollary 2.10 it is well-defined up to translation, so one speaks sometimes of "the" theta divisor. The choice of a symplectic basis gives a particular theta divisor Θ_τ, defined by the celebrated Riemann theta function

$$\theta(z) = \sum_{m\in\mathbb{Z}^g} \mathbb{e}({}^t\!mz + \frac{1}{2}{}^t\!m\tau m)\ .$$

It is quite remarkable that starting from a linear algebra data (a lattice Γ in V and a polarization), we get a hypersurface $\Theta \subset T = V/\Gamma$. When the p.p.a.v. comes from a geometric construction (Jacobians, Prym varieties, intermediate Jacobians), this divisor has a rich geometry, which reflects the objects we started with. In particular it is often possible to recover these objects from the data (T,Θ) ("Torelli theorem"), or to characterize the p.p.a.v. obtained in this way ("Schottky problem").

3.6 Reminder: line bundles and maps into projective space

Let M be a projective variety, and L a line bundle on M. The *linear system* $|L|$ is by definition $\mathbb{P}(H^0(M,L))$. Sending a nonzero section to its divisors identifies $|L|$ with the set of effective divisors E on M such that $\mathcal{O}_M(E) \cong L$.

The *base locus* $B(L)$ of L is the intersection of the divisors in $|L|$. Assume that L has no base point, that is, $B(L) = \varnothing$. Then the divisors of $|L|$ passing through a point $m \in M$ form a hyperplane in $|L|$, corresponding to a point $\varphi_L(m)$ in the dual projective space $|L|^*$. This defines a morphism $\varphi_L : M \to |L|^*$. Choosing a basis $(s_0, \ldots, s_n)$ of $H^0(M, L)$ identifies $|L|$, hence also its dual $|L|^*$, to $\mathbb{P}^n$; then $\varphi_L(m) = (s_0(m), \ldots, s_n(m))$, where we have fixed an isomorphism $L_m \xrightarrow{\sim} \mathbb{C}$ to evaluate the s_i at m.

If $E \in |L|$, we also denote the linear system $|L|$ by $|E|$, and the map φ_L by φ_E. Thus $|E|$ is the set of effective divisors linearly equivalent to E.

3.7 The Lefschetz theorem

Theorem 3.7 (Lefschetz). *Let L be a line bundle on T.*

1) *Assume $H^0(T, L) \neq 0$. For $k \geq 2$, the linear system $|L^{\otimes k}|$ has no base points.*

2) *Assume that the hermitian form associated to L is positive definite. For $k \geq 3$, the map $\varphi_{L^{\otimes k}} : T \to |L^{\otimes k}|^*$ is an embedding.*

Proof. Let us prove 1) in the case $k = 2$ – the proof in the general case is identical. Let $D \in |L|$. A simple but crucial observation is that

$$x \in t_a^* D \iff a \in t_x^* D\ .$$

By Corollary 2.11 we have $t_a^* D + t_{-a}^* D \in |L^{\otimes 2}|$ for all a in T. Given $x \in T$, we choose a outside the divisors $t_x^* D$ and $t_{-x}^* D^-$, where D^- denotes the image of D by the involution $z \mapsto -z$; then $x \notin t_a^* D + t_{-a}^* D$, which proves 1).

We will prove only a part of 2), namely the *injectivity* of $\varphi_{L^{\otimes k}}$; the proof that its tangent map at each point is injective is analogous but requires some more preparation. We will do it for $k = 3$ (the same proof works for all k) and assume moreover that the polarization is *principal* – again the general case requires more work, see [20], §17.

Let x, y in T such that $\varphi_{L^{\otimes 3}}(x) = \varphi_{L^{\otimes 3}}(y)$. This means that any divisor $E \in |L^{\otimes 3}|$ passing through x passes through y. Let Θ be the unique element of $|L|$. Let $a \in t_x^* \Theta$; we choose $b \in T$ outside $t_y^* \Theta$ and $t_{a-y}^* \Theta^-$, and take $E = t_a^* \Theta + t_b^* \Theta + t_{-a-b}^* \Theta$. We have $x \in E$, hence $y \in E$, but $y \notin t_b^* \Theta$ and $y \notin t_{-a-b}^* \Theta$, so $y \in t_a^* \Theta$, that is, $a \in t_y^* \Theta$. We conclude that the divisors $t_x^* \Theta$ and $t_y^* \Theta$ have the same support. But Θ has no multiple component, since by 1) this would imply $\dim H^0(T, L) > 1$. Thus $t_x^* \Theta = t_y^* \Theta$, and by Theorem 2.8.4) this implies $x = y$. □

Remark 3.8. A line bundle L such that $\varphi_{L^{\otimes k}}$ is an embedding for k large enough is said to be *ample.* The celebrated (and difficult) Kodaira embedding theorem states that this is the case if and only if the class $c_1(L)$ can be represented by a $(1,1)$-form which is everywhere positive definite (see [13], Section I.4, for a precise statement and a proof). The Lefschetz theorem gives a much more elementary version for complex tori. It is also more precise, since it says that $k \geq 3$ is enough for $L^{\otimes k}$ to give an embedding. We are now going to discuss the map defined by $L^{\otimes 2}$ in the case of a principal polarization.

3.8 The linear system $|2\Theta|$

Let us again focus on the case of a principal polarization. The Riemann theta function is even, so its divisor Θ_τ is symmetric – that is, $\Theta_\tau^- = \Theta_\tau$. From Theorem 2.8.4) one deduces that the symmetric theta divisors are the translates $t_a^*\Theta_\tau$ where a runs through the 2^{2g} points of order 2 of T. Note that by Theorem 2.8.1), all divisors 2Θ, where Θ is a symmetric theta divisor, are linearly equivalent. Thus the linear system $|2\Theta|$ is canonically associated to the principal polarization.

We will denote by i_T the involution $z \mapsto -z$. The quotient $K := T/i_T$ is called the *Kummer variety* of T; it has 2^{2g} singular points, which are the images of the points of order 2 in T.

Proposition 3.9. *Let Θ be an* irreducible *symmetric theta divisor on T. The map $\varphi_{2\Theta} : T \to |2\Theta|^*$ factors through i_T and embeds $K = T/i_T$ into $|2\Theta|^*$.*

(See Remark 3.10 below for the irreducibility hypothesis.)

Proof. By Theorem 3.7 the map is everywhere defined. Recall that a basis of $H^0(T, \mathcal{O}_T(2\Theta))$ is given by the theta functions

$$\theta[\varepsilon](z) = \sum_{m\in\varepsilon} \mathrm{e}({}^t mz + \frac{1}{4}{}^t m\tau m) \ ,$$

where ε runs through the cosets of $\mathbb{Z}^g/2\mathbb{Z}^g$. Thus $\varphi_{2\Theta}$ maps $\pi(z) \in T$ to $\big(\theta[\varepsilon](z)\big)_{\varepsilon\in\mathbb{Z}^g/2\mathbb{Z}^g}$ in $\mathbb{P}^{2^g-1}$. Since each $\varepsilon \in \mathbb{Z}^g/2\mathbb{Z}^g$ is stable under the involution $m \mapsto -m$, the functions $\theta[\varepsilon]$ are even; therefore $\varphi_{2\Theta}$ factors through i_T, and induces a map $K \to |2\Theta|^*$.

Let us prove that this map is injective. Let $x \neq y$ in T with $\varphi_{2\Theta}(x) = \varphi_{2\Theta}(y)$. Let a be a general point of $t_x^*\Theta$. The divisor $t_a^*\Theta + t_{-a}^*\Theta$ belongs to $|2\Theta|$ and contains x, hence also y.

Since $t_x^*\Theta \neq t_y^*\Theta$, a does not belong to $t_y^*\Theta$; thus $y \notin t_a^*\Theta$, and therefore $y \in t_{-a}^*\Theta$. This means $y - a \in \Theta$, and since Θ is symmetric $a \in t_{-y}^*\Theta$. We conclude that $t_x^*\Theta = t_{-y}^*\Theta$, hence $x = -y$, which proves our assertion.

The injectivity of the tangent map at the smooth points of K is proved in the same way; the analysis at the singular points is more delicate, see [18]. □

Remark 3.10. What if Θ is reducible? It is not difficult to show that T must be a product of lower-dimensional p.p.a.v.; that is, $T = T_1 \times \cdots \times T_p$ and $\Theta = \Theta_1 \times T_2 \times \cdots \times T_p + \cdots + T_1 \times \cdots \times T_{p-1} \times \Theta_p$. In that case the geometry of (T, Θ) is determined by that of the (T_i, Θ_i).

Example 3.11. Suppose $g = 2$. Then $\varphi_{2\Theta}$ embeds $K = T/i_T$ in $\mathbb{P}^3$. It is easy to see that K has degree 4 (hint: use $K_T = \mathcal{O}_T = \varphi_{2\Theta}^*\mathcal{O}_{\mathbb{P}^3}(\deg(K) - 4)$); it has 16 double points corresponding to the 16 points of order 2 in T. This is the celebrated Kummer quartic surface, found by Kummer in 1864.

The following remarkable formula explains (in part) the particular role of the linear system $|2\Theta|$. We use the notations of the proof of Theorem 3.5.

Proposition 3.12 (Addition formula).

$$\theta(z+a)\,\theta(z-a) = \sum_{\varepsilon\in\mathbb{Z}^g/2\mathbb{Z}^g} \theta[\varepsilon](z)\,\theta[\varepsilon](a) \qquad \textit{for } z, a \textit{ in } \mathbb{C}^g\ .$$

Proof . We have

$$\theta(z+a)\,\theta(z-a) = \sum_{p,q\in\mathbb{Z}^g} \mathbb{e}\big({}^t(p+q)z + {}^t(p-q)a + \frac{1}{2}({}^tp\tau p + {}^tq\tau q)\big)\ .$$

Putting $r = p+q$, $s = p-q$ defines a bijection of $\mathbb{Z}^g\times\mathbb{Z}^g$ onto the set of pairs (r,s) in $\mathbb{Z}^g\times\mathbb{Z}^g$ with $r\equiv s$ (mod. $2\mathbb{Z}^g$). This set is the union of the subsets $\varepsilon\times\varepsilon\subset\mathbb{Z}^g\times\mathbb{Z}^g$ for $\varepsilon\in\mathbb{Z}^g/2\mathbb{Z}^g$; thus

$$\begin{aligned}\theta(z+a)\,\theta(z-a) &= \sum_{\varepsilon\in\mathbb{Z}^g/2\mathbb{Z}^g}\ \sum_{r,s\in\varepsilon} \mathbb{e}({}^trz + \frac{1}{4}{}^tr\tau r)\,\mathbb{e}({}^tsa + \frac{1}{4}{}^ts\tau s) \\ &= \sum_{\varepsilon\in\mathbb{Z}^g/2\mathbb{Z}^g} \theta[\varepsilon](z)\,\theta[\varepsilon](a)\ . \qquad \square\end{aligned}$$

The addition formula has the following geometric interpretation:

Corollary 3.13. *Let* Θ *be a symmetric theta divisor on* T. *For* $a\in T$, *put* $\kappa(a) := t_a^*\Theta + t_{-a}^*\Theta \in |2\Theta|$. *There is a commutative diagram:*

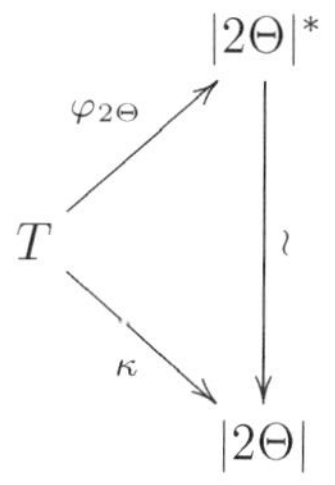

Proof . After a translation by a point of order 2 we can assume $\Theta = \Theta_\tau$ for some symplectic basis of Γ. We identify both $|2\Theta_\tau|$ and its dual to $\mathbb{P}^{2^g-1}$ using the basis $(\theta[\varepsilon])_{\varepsilon\in\mathbb{Z}^g/2\mathbb{Z}^g}$. Then, by the addition formula,

$$\kappa(a) = \big(\theta[\varepsilon](a)\big)_{\varepsilon\in\mathbb{Z}^g/2\mathbb{Z}^g} = \varphi_{2\Theta}(a)\ . \quad \square$$

(For a more intrinsic description of the isomorphism $|2\Theta| \xrightarrow{\sim} |2\Theta|^*$, see [21], p. 555.)

4 Curves and their Jacobians

In this section we denote by C a smooth projective curve (= compact Riemann surface) of genus g.

4.1 Hodge theory for curves

We first recall briefly Hodge theory for curves, which is much easier than in the general case. We start from the exact sequence of sheaves

$$0 \to \mathbb{C}_C \longrightarrow \mathcal{O}_C \xrightarrow{d} K_C \to 0 ,$$

where $\mathbb{C}_C$ is the sheaf of locally constant complex functions, and K_C (also denoted Ω^1_C or ω_C) is the sheaf of holomorphic 1-forms. Taking into account $H^0(C, \mathcal{O}_C) = \mathbb{C}$ and $H^1(C, K_C) \cong \mathbb{C}$ (Serre duality), we obtain an exact sequence

$$0 \to H^0(C, K_C) \xrightarrow{\partial} H^1(C, \mathbb{C}) \xrightarrow{p} H^1(C, \mathcal{O}_C) \to 0 .$$

By definition $g = \dim H^0(C, K_C)$; by Serre duality we have also $\dim H^1(C, \mathcal{O}_C) = g$, hence $\dim H^1(C, \mathbb{C}) = 2g$.

We put $H^{1,0} := \operatorname{Im} \partial$ and $H^{0,1} := \overline{H^{1,0}}$; $H^{1,0}$ is the subspace of classes in $H^1(C, \mathbb{C})$ which can be represented by holomorphic forms, and $H^{0,1}$ by antiholomorphic forms.

Lemma 4.1. *Let $\alpha \neq 0$ in $H^0(C, K_C)$; then $i \int_C \alpha \wedge \bar{\alpha} > 0$.*

Proof. Let $z = x + iy$ be a local coordinate in an open subset U of C. We can write $\alpha = f(z)dz$ in U, so that

$$i \int_U \alpha \wedge \bar{\alpha} = \int_U |f(z)|^2 \, i\, dz \wedge d\bar{z} = \int_U |f(z)|^2 \, 2 \, dx \wedge dy > 0 . \quad \square$$

Proposition 4.2. *$H^1(C, \mathbb{C}) = H^{1,0} \oplus H^{0,1}$; the map p induces an isomorphism $H^{0,1} \to H^1(C, \mathcal{O}_C)$.*

Proof. The second assertion follows from the first and from the above exact sequence. For dimension reasons it suffices to prove that $H^{1,0} \cap H^{0,1} = (0)$. Let $x \in H^{1,0} \cap H^{0,1}$. There exists $\alpha, \beta \in H^0(C, K_C)$ such that $x = [\alpha] = [\bar{\beta}]$, hence $\alpha - \bar{\beta} = df$ for some C^∞ function f on C. Then $\beta \wedge \bar{\beta} = df \wedge \beta = d(f\beta)$, hence $\int_C \beta \wedge \bar{\beta} = 0$ by Stokes theorem. By the Lemma this implies $\beta = 0$ hence $x = 0$. $\square$

Proposition 4.3. *$p(H^1(C, \mathbb{Z}))$ is a lattice in $H^{0,1}$; the hermitian form H on $H^{0,1}$ defined by $H(\alpha, \beta) := 2i \int_C \bar{\alpha} \wedge \beta$ induces a principal polarization on the complex torus $H^{0,1}/p(H^1(C, \mathbb{Z}))$.*

Proof. The first assertion has already been proved (Section 2.1). Lemma 4.1 shows that the form H is positive definite on $H^{0,1} = \overline{H^{1,0}}$. Let $a, b \in H^1(C, \mathbb{Z})$; we have

$$a = \bar{\alpha} + \alpha , \quad b = \bar{\beta} + \beta \quad \text{with} \quad \alpha = p(a) , \ \beta = p(b) .$$

Their cup-product in $H^2(C, \mathbb{Z}) = \mathbb{Z}$ is given by

$$a \cdot b = \int_C (\bar{\alpha} + \alpha) \wedge (\bar{\beta} + \beta) = \frac{1}{2i} \big(H(\alpha, \beta) - H(\beta, \alpha) \big) = \operatorname{Im}(H)(\alpha, \beta) ;$$

thus $\mathrm{Im}(H)$ induces on $H^1(C,\mathbb{Z})$ the cup-product, which is unimodular by Poincaré duality. □

The g-dimensional abelian variety $JC := H^{0,1}/p(H^1(C,\mathbb{Z}))$ with the principal polarization H is called the *Jacobian* of C; it plays an essential role in the study of the curve.

4.2 Line bundles on C

To study line bundles on C we use again the exact sequence (2.1):

$$0 \to H^1(C,\mathbb{Z}) \xrightarrow{i} H^1(C,\mathcal{O}_C) \longrightarrow \mathrm{Pic}(C) \xrightarrow{c_1} H^2(C,\mathbb{Z}) \cong \mathbb{Z} \to 0 .$$

Here for a line bundle L on C, $c_1(L)$ is simply the degree $\deg(L)$ (through the canonical isomorphism $H^2(C,\mathbb{Z}) \cong \mathbb{Z}$): $\deg(L) = \deg(D)$ for any divisor D such that $\mathcal{O}_C(D) \cong L$.

Note that i is the composition of the maps $H^1(C,\mathbb{Z}) \to H^1(C,\mathbb{C}) \xrightarrow{p} H^1(C,\mathcal{O}_C)$ deduced from the inclusions of sheaves $\mathbb{Z}_C \subset \mathbb{C}_C \subset \mathcal{O}_C$. Hence:

Proposition 4.4. *We have an exact sequence* $0 \to JC \longrightarrow \mathrm{Pic}(C) \xrightarrow{\deg} \mathbb{Z} \to 0$. □

Thus JC is identified with $\mathrm{Pic}^{\mathrm{o}}(C)$, the group of isomorphism classes of degree 0 line bundles on C – or the group of degree 0 divisors modulo linear equivalence. More precisely, one can show that JC is a *moduli space* for degree 0 line bundles on C. This means the following. Let S be a complex manifold (or analytic space), and let $\mathcal{L}$ be a line bundle on $C \times S$. For $s \in S$, put $\mathcal{L}_s := \mathcal{L}_{C\times\{s\}}$. We say that $(\mathcal{L}_s)_{s\in S}$ is a *holomorphic family* of line bundles on C parametrized by S. If the line bundles $\mathcal{L}_s$ have degree 0, we get a map $S \to JC$; we want this map to be holomorphic.

In fact we have even more: the line bundles L in JC form a holomorphic family. Namely, there exists a line bundle $\mathcal{P}$ on $C \times JC$ such that $\mathcal{P}_L = L$ for each $L \in JC$. Such a line bundle is called a *Poincaré line bundle*. It is unique up to tensor product by the pullback of a line bundle on JC.

4.3 The Abel-Jacobi maps

As an illustration, choose a divisor D_1 of degree 1 on C and define $\alpha : C \to JC$ by $\alpha(p) = \mathcal{O}_C(p - D_1)$. It is holomorphic (hence algebraic, since both C and JC are projective manifolds): indeed it is defined by the line bundle $\mathcal{O}_{C\times C}(\Delta - p^*D_1)$ on $C \times C$, where Δ is the diagonal and p the first projection.

More generally, let $C^{(d)}$ denote the d-th symmetric power of the curve C, that is, the quotient of C^d by the symmetric group $\mathfrak{S}_d$. This is a *smooth* variety: indeed since this a local question it suffices to prove it for the affine line $\mathbb{C}$; but the map $(z_1,\ldots,z_d) \mapsto (s_1,\ldots,s_d)$, where s_i is the i-th elementary symmetric function of $z_1,\ldots,z_d$, identifies $\mathbb{C}^{(d)}$ to $\mathbb{C}^d$. Using the map $(p_1,\ldots,p_d) \mapsto p_1 + \cdots + p_d$ we will view the elements of $C^{(d)}$ as effective divisors of degree d on C.

Now we choose a divisor D_d of degree d and define a map $\alpha_d : C^{(d)} \to JC$ by $\alpha_d(E) = \mathcal{O}_C(E - D_d)$. Again this is holomorphic, for instance because

$\alpha_d(p_1 + \cdots + p_d) = \alpha(p_1) + \cdots + \alpha(p_d)$ up to a constant. For $E \in C^{(d)}$, the fiber $\alpha_d^{-1}(\alpha_d(E))$ is the linear system $|E|$ (Sect. 3.6).

Proposition 4.5. *For $d \leq g$ the map α_d is generically injective.*

Proof. By the observation preceding the Proposition we must prove[2] $h^0(E) = 1$ for a general $E \in C^{(d)}$. If D is an effective divisor, we have $h^0(D-p) = h^0(D)-1$ for p general in C, hence by induction $h^0(D-E) = h^0(D) - d$ for E general in $C^{(d)}$ with $d \leq h^0(D)$. Taking $D = K$ gives $h^0(K-E) = g - d$ for $d \leq g$, hence $h^0(E) = 1$ by Riemann-Roch. □

Corollary 4.6. *$\alpha_g : C^{(g)} \to JC$ is birational.*

Another consequence is that the image of $\alpha_{g-1} : C^{(g-1)} \to JC$ is a divisor in JC. In fact:

Theorem 4.7 (Riemann). *The image of $\alpha_{g-1} : C^{(g-1)} \to JC$ is a theta divisor of JC.*

We have to refer to [1], p. 23 for the proof.

Remark 4.8. 1) Recall that the map α_{g-1} depends on the choice of a divisor D, or equivalently of the line bundle $L = \mathcal{O}_C(D)$, of degree $g-1$. We will denote by Θ_L the corresponding theta divisor; explicitly:

$$\Theta_L = \{M \in JC \mid H^0(M \otimes L) \neq 0\} .$$

2) There is a way to avoid the inelegant choice of a divisor D_d in the definition of α_d. Let J^d denote the set of isomorphism classes of line bundles of degree d on C. Choosing a line bundle L_d of degree d defines a bijection $JC \to J^d$ (by $M \mapsto M \otimes L_d$). This provides a structure of projective variety on J^d which does not depend on the choice of L_d. By construction J^d is isomorphic to JC, but there is no canonical isomorphism.

Now we have a canonical map $\alpha_d : C^{(d)} \to J^d$ defined simply by $\alpha_d(E) = \mathcal{O}_C(E)$. In particular we have a canonical divisor $\Theta \subset J^{g-1}$, which is the locus of the line bundles L in J^{g-1} with $H^0(L) \neq 0$.

3) A consequence of the Riemann theorem is that the theta divisor is *irreducible*, so a Jacobian cannot be a product of non-trivial p.p.a.v. (Remark 3.10).

5 Vector bundles on curves

As explained in the introduction, generalized theta functions appear when we replace JC, the moduli space of degree 0 line bundles on C, by the analogous moduli spaces for higher rank vector bundles. We will now explain what this means.

[2] We use the standard notations $h^0(\mathcal{F}) := \dim H^0(C, \mathcal{F})$ for a sheaf $\mathcal{F}$ on C, and $h^0(D) := h^0(\mathcal{O}_C(D))$ for a divisor D.

5.1 Elementary properties

Let E be a vector bundle on C, of rank r. The maximum wedge power $\wedge^r E$ is a line bundle on C, denoted $\det(E)$. Its degree is denoted by $\deg(E)$. It has the following properties:

- In an exact sequence $0 \to F \to E \to G \to 0$ we have $\det(E) \cong \det(F) \otimes \det(G)$;
- For any line bundle L on C, we have $\det(E \otimes L) = \det(E) \otimes L^{\otimes r}$.
- (Riemann-Roch) $h^0(E) - h^1(E) = \deg(E) + r(1-g)$.

It will be convenient to introduce the *slope* $\mu(E) := \dfrac{\deg(E)}{r} \in \mathbb{Q}$. Thus Riemann-Roch can be written $h^0(E) - h^1(E) = r(\mu(E) + 1 - g)$.

5.2 Moduli spaces

We have seen that the Jacobian of C parametrizes line bundles of degree 0, in the sense that for any holomorphic family $(\mathcal{L}_s)_{s\in S}$ the corresponding map $S \to JC$ is holomorphic. Unfortunately such a nice moduli space does not exist in higher rank. Indeed we will show the following:

Lemma 5.1. *Let L be a non-trivial line bundle on C with no base point (Sect. 3.6). There exists a holomorphic family of vector bundles $(\mathcal{E}_t)_{t\in\mathbb{C}}$ on C such that:*

$$\mathcal{E}_t \cong \mathcal{O}_C \oplus \mathcal{O}_C \quad \textit{for } t \neq 0 \qquad \mathcal{E}_0 \cong L \oplus L^{-1} .$$

Proof. We first take $C = \mathbb{P}^1$ and $L = \mathcal{O}_{\mathbb{P}^1}(2)$. Put $F := \mathbb{P}^1 \times \mathbb{C}$, $\mathcal{O}_F(k) := \mathrm{pr}_1^*\mathcal{O}_{\mathbb{P}^1}(k)$. Consider the homomorphism

$$u : \mathcal{O}_F^{\oplus 3} \longrightarrow \mathcal{O}_F(2) \quad \text{given by} \quad (X^2, Y^2, tXY) ,$$

where t is the coordinate on $\mathbb{C}$ and (X,Y) the homogeneous coordinates on $\mathbb{P}^1$. The map u is surjective, so its kernel is a rank 2 bundle $\mathcal{F}$ on $\mathbb{P}^1 \times \mathbb{C}$. We claim that

$$\mathcal{F}_t \cong \mathcal{O}_{\mathbb{P}^1}(-1)^{\oplus 2} \quad \text{for } t \neq 0 \qquad \mathcal{F}_0 \cong \mathcal{O}_{\mathbb{P}^1} \oplus \mathcal{O}_{\mathbb{P}^1}(-2) .$$

There is a variety of ways to prove this. Perhaps the easiest is to observe that any vector bundle on $\mathbb{P}^1$ is a direct sum of line bundles. Since $\mathcal{F}_t$ is a sub-bundle of $\mathcal{O}_{\mathbb{P}^1}^{\oplus 3}$ with determinant $\mathcal{O}_{\mathbb{P}^1}(-2)$, it is either $\mathcal{O}_{\mathbb{P}^1}(-1)^{\oplus 2}$ or $\mathcal{O}_{\mathbb{P}^1} \oplus \mathcal{O}_{\mathbb{P}^1}(-2)$; the first case occurs if and only if $h^0(\mathcal{F}_t) = 0$, that is, if and only if $t \neq 0$. Then the vector bundle $\mathcal{F}(1)$ on F has the required properties.

Now we consider the general case. Let s be a nonzero section of L; we can find a section t of L which does no vanish at the zeroes of s. Then (s,t) defines a map $u : C \to \mathbb{P}^1$ such that $u^*\mathcal{O}_{\mathbb{P}^1}(1) = L$. The vector bundle $\mathcal{E} := (u, \mathrm{Id})^*\mathcal{F}(1)$ on $C \times \mathbb{C}$ has the required properties. □

This implies that there is no reasonable moduli space $\mathcal{M}$ containing both $\mathcal{O}_C^{\oplus 2}$ and $L \oplus L^{-1}$: the family constructed in the lemma would give rise to a holomorphic map $\mathbb{C} \to \mathcal{M}$ mapping $\mathbb{C} \smallsetminus \{0\}$ to a point, and 0 to a different point.

There are two ways to deal with this problem. The sophisticated one, which we will not discuss here, replaces moduli spaces by a more elaborate notion called *moduli stacks*. The reader interested by this point of view may look at [12].

Instead we will follow the classical (by now) approach, which eliminates certain vector bundles, for instance those of the form $L \oplus L^{-1}$ which appear in the lemma; this is done as follows:

Definition 5.2. *A vector bundle E on C is* stable *if $\mu(F) < \mu(E)$ for every sub-bundle $0 \subsetneq F \subsetneq E$. It is* polystable *if it is a direct sum of stable sub-bundles of slope $\mu(E)$.*

Theorem 5.3. *There exists a moduli space $\mathcal{M}^s(r,d)$ for stable vector bundles of rank r and degree d. It is a smooth connected quasi-projective manifold; it admits a projective compactification $\mathcal{M}(r,d)$ whose points correspond to isomorphism classes of polystable bundles.*

Note that we do not claim that $\mathcal{M}(r,d)$ is a moduli space for polystable bundles; the situation is more complicated. We refer to [19] for a precise statement as well as the proof.

An important by-product of the proof is the fact that stability is an open condition: if S is a variety and $\mathcal{E}$ a vector bundle on $C \times S$, the set of $s \in S$ such that $\mathcal{E}_s$ is stable is Zariski open in S ([19], Prop. 7.2.6).

5.3 The moduli space $\mathcal{M}(r)$

We will in fact focus on a slightly different moduli space. The map $\det : \mathcal{M}(r,d) \to J^d$ which associates to a vector bundle its determinant is holomorphic. Let L be a line bundle of degree d; the fiber $\det^{-1}(L)$ is denoted $\mathcal{M}(r,L)$. We denote by J_r the subgroup (isomorphic to $(\mathbb{Z}/r\mathbb{Z})^{2g}$) of line bundles $\alpha \in JC$ with $\alpha^{\otimes r} \cong \mathcal{O}_C$.

Proposition 5.4. *The map $\mathcal{M}(r,L) \times JC \to \mathcal{M}(r,d)$ given by $(E,\lambda) \mapsto E \otimes \lambda$ identifies $\mathcal{M}(r,d)$ with the quotient of $\mathcal{M}(r,L) \times JC$ by J_r acting by $\alpha \cdot (E,\lambda) = (E \otimes \alpha, \lambda \otimes \alpha^{-1})$.*

Proof. Let E in $\mathcal{M}(r,d)$. The pairs (F,λ) with $F \in \mathcal{M}(r,L)$, $\lambda \in JC$ and $E \cong F \otimes \lambda$ are obtained by taking $\lambda \in JC$ with $\lambda^{\otimes r} = \det(E) \otimes L^{-1}$ and $F = E \otimes \lambda^{-1}$. We can always find such a λ, hence a pair (F,λ), and two such pairs differ by the action of J_r. Thus $\mathcal{M}(r,d)$ is determined by JC and $\mathcal{M}(r,L)$; from now on we will focus on the latter space. Note that for $N \in \mathrm{Pic}(C)$ the map $E \mapsto E \otimes N$ induces an isomorphism $\mathcal{M}(r,L) \xrightarrow{\sim} \mathcal{M}(r, L \otimes N^{\otimes r})$; thus up to isomorphism, $\mathcal{M}(r,L)$ depends only of the degree d of L (mod. r). When r and d are coprime $\mathcal{M}(r,L) = \mathcal{M}^s(r,L)$ is smooth, and is a nice moduli space; however the most interesting case for us will be $d = 0$, and the moduli space $\mathcal{M}(r,\mathcal{O}_C)$, which we will denote simply $\mathcal{M}(r)$. This is also the moduli space of principal $\mathrm{SL}(r)$-bundles, so its study fits into the more general theory of principal G-bundles for a semisimple group G.

Let us summarize in the next Proposition some elementary properties of $\mathcal{M}(r)$, which follow from its construction (see [19]). From now on we will assume that the genus g of C is ≥ 2 (for $g \leq 1$ there are no stable bundles of degree 0 and rank > 1).

Proposition 5.5. *$\mathcal{M}(r)$ is a projective normal irreducible variety, of dimension $(r^2-1)(g-1)$, with mild singularities (so-called* rational *singularities). Except when $r = g = 2$, its singular locus is the locus of non-stable bundles.*

As algebraic varieties, the moduli spaces $\mathcal{M}(r, L)$ are very different from complex tori:

Proposition 5.6. *The moduli space $\mathcal{M}(r, L)$ is* unirational*; that is, there exists a rational dominant map*[3] $\mathbb{P}^N \dashrightarrow \mathcal{M}(r, L)$.

Proof. Using the isomorphism $\mathcal{M}(r, L) \xrightarrow{\sim} \mathcal{M}(r, L \otimes N^{\otimes r})$ we may assume $\deg(L) > r(2g-1)$, so $\mu(E) > 2g-1$ for $E \in \mathcal{M}(r, L)$. Since E is polystable this implies $H^0(E^* \otimes K_C(p)) = 0$ for any $p \in C$, hence by Serre duality $H^1(E(-p)) = 0$. Then the exact sequence

$$0 \to E(-p) \to E \to E_p \to 0$$

gives for each p a surjection $\mathrm{ev}_p : H^0(E) \twoheadrightarrow E_p$; that is, the global sections of E generate E at p.

Now we claim that a general subspace of dimension $r+1$ of $H^0(E)$ still generates E at each point. For $p \in C$, let Z_p be the subvariety of the Grassmannian $\mathbb{G}(r+1, H^0(E))$ consisting of subspaces V which do not span E_p. This is equivalent to $\dim V \cap \mathrm{Ker}(\mathrm{ev}_p) \geq 2$, so Z_p has codimension 2 (exercise!). Thus $Z = \cup_{p \in C} Z_p$ has codimension 1 in the Grassmannian; any V in the complement of Z generates E at each point. For such a V the evaluation map $V \otimes \mathcal{O}_C \to E$ is surjective. Its kernel is a line bundle; taking determinants we see that it is L^{-1}. Thus E^* is the kernel of a surjective map $V^* \otimes \mathcal{O}_C \to L$.

Conversely, let $\mathbb{G}_0$ be the open subset of the Grassmannian $\mathbb{G}(r+1, H^0(L))$ parametrizing subspaces which span L at each point. For $W \in \mathbb{G}_0$, we have an exact sequence

$$0 \to F_W \longrightarrow W \otimes_{\mathbb{C}} \mathcal{O}_C \xrightarrow{\mathrm{ev}} L \to 0 ;$$

The dual $E_W := F_W^*$ is a rank r vector bundle with determinant L; we obtain in this way an algebraic family of such bundles, parametrized by $\mathbb{G}_0$, such that every element of $\mathcal{M}^s(r, L)$ appears in the family. The subspaces $W \in \mathbb{G}_0$ such that E_W is stable form a Zariski open subset $\mathbb{G}_1 \subset \mathbb{G}_0$ (Sect. 5.2), and we have a surjective map $f : \mathbb{G}_1 \to \mathcal{M}^s(r, L)$ such that $f(W) = E_W$. Since Grassmannians are rational varieties, composing f with a birational map $\mathbb{P}^N \dashrightarrow \mathbb{G}_1$ gives the required rational dominant map. □

Corollary 5.7. *Any rational map from $\mathcal{M}(r, L)$ to a complex torus is constant.*

[3] In the rest of this section we assume some familiarity with the notion of rational maps – see e.g. [13], p. 490.

Proof . Let $T = V/\Gamma$ be a complex torus. In view of the proposition, it suffices to show that any rational map $\varphi : \mathbb{P}^N \dashrightarrow T$ is constant. Let p, q be two general points of $\mathbb{P}^N$. The restriction of φ to the line $\langle p, q\rangle$ defines a map $\mathbb{P}^1 \to T$, which factors through V since $\mathbb{P}^1$ is simply connected, hence is constant. Thus $\varphi(p) = \varphi(q)$. □

5.4 Rationality

The *Lüroth problem* asks whether an unirational variety X is necessarily rational. The answer is positive when X is a curve (Lüroth, 1876) or a surface (Castelnuovo, 1895), but not in higher dimension (see for instance [9]).

While $\mathcal{M}(r, L)$ is known to be rational when $\deg(L)$ is prime to r [17], the rationality of $\mathcal{M}(r)$ is an open problem, already for $r = 2$ and $g = 3$ – despite the fact that in this case we have an explicit description of $\mathcal{M}(2)$ as a quartic hypersurface in $\mathbb{P}^7$ (Sect. 6.5).

6 Generalized theta functions

6.1 The theta divisor

Since $\mathcal{M}(r)$ is simply connected, there is no hope to describe its line bundles by systems of multipliers as for complex tori. However we may try to mimic the definition of the theta divisor: for $L \in J^{g-1}$, we put

$$\Delta_L := \{E \in \mathcal{M}(r) \mid H^0(E \otimes L) \neq 0\}.$$

Theorem 6.1 ([10]). 1) *Δ_L is a Cartier divisor on $\mathcal{M}(r)$.*
2) *The line bundle $\mathcal{L} = \mathcal{O}(\Delta_L)$ is independent of L, and* $\mathrm{Pic}(\mathcal{M}(r)) = \mathbb{Z}[\mathcal{L}]$.

Recall that an effective Cartier divisor is a subvariety locally defined by an equation – or, globally, as the zero locus of a section of a line bundle. On a singular variety (as is $\mathcal{M}(r)$) this is stronger than having codimension 1.

Proof . We will only show why Δ_L is a divisor on the stable locus $\mathcal{M}^s(r)$, referring to [10] for the rest of the proof. It is a consequence of the following lemma:

Lemma 6.2. *Let S be a complex variety, $(\mathcal{E}_s)_{s\in S}$ a family of vector bundles on C, with $\mu(\mathcal{E}_s) = g - 1$ for all $s \in S$. Then the locus*

$$\{s \in S \mid H^0(C, \mathcal{E}_s) \neq 0\}$$

is defined locally by one equation (possibly trivial).

Proof . We will use a general fact about cohomology of coherent sheaves (see [20], §5) : locally on S there exist vector bundles F, G and a homomorphism $u : F \to G$ such that we have for each s in S an exact sequence

$$0 \to H^0(C, \mathcal{E}_s) \longrightarrow F(s) \xrightarrow{u(s)} G(s) \longrightarrow H^1(C, \mathcal{E}_s) \to 0\ .$$

By Riemann-Roch we have $h^0(\mathcal{E}_s) = h^1(\mathcal{E}_s)$, hence F and G have the same rank. We see that $H^0(C, \mathcal{E}_s) \neq 0$ if and only if $\det(u(s)) = 0$, that is, the section $\det(u)$ of $\det(G) \otimes \det(F)^{-1}$ vanishes at s, hence the lemma. □

Coming back to $\mathcal{M}(r)$, the construction of the moduli space implies that locally for the complex topology, there is a "Poincaré bundle", that is a rank r vector bundle $\mathcal{E}$ on $C \times V$ such that $\mathcal{E}_{|C\times\{E\}} \cong E$ for E in V. Applying the lemma to $\mathcal{E} \otimes L$ shows that Δ_L is a divisor on $\mathcal{M}^s(r)$, unless $\Delta_L = \mathcal{M}(r)$. But this cannot hold: if α is a general element of JC, we have $H^0(L \otimes \alpha) = 0$, hence $\alpha^{\oplus r} \notin \Delta_L$. □

6.2 Generalized theta functions

By analogy with the case of Jacobians, the sections of $H^0(\mathcal{M}(r), \mathcal{L}^{\otimes k})$ are called *generalized* (or *non-abelian*) *theta functions* of order k. They are associated to the group $\mathrm{SL}(r)$ (there are more general theta functions associated to each complex reductive group, but we will not discuss them in these notes).

Like for complex tori, the first question we can ask about these theta functions is the dimension of the space $H^0(\mathcal{M}(r), \mathcal{L}^{\otimes k})$. The answer, much more intricate than Theorem 3.5 for complex tori, is known as the *Verlinde formula*; it has been first found by E. Verlinde using physics arguments, then proved mathematically in many different ways – see e.g. [29]. The formula is as follows:

$$\dim H^0(\mathcal{M}(r), \mathcal{L}^{\otimes k}) = \Big(\frac{r}{r+k}\Big)^g \sum_{\substack{S \amalg T = [1, r+k] \\ |S| = r}} \prod_{\substack{s \in S \\ t \in T}} \Big|2 \sin \pi \frac{s-t}{r+k}\Big|^{g-1} \tag{6.1}$$

For $r = 2$ it reduces (after some trigonometric manipulations) to:

$$\dim H^0(\mathcal{M}(2), \mathcal{L}^{\otimes k}) = (\frac{k}{2}+1)^{g-1} \sum_{i=1}^{k+1} \frac{1}{(\sin \frac{i\pi}{k+2})^{2g-2}} \cdot$$

Even in rank 2, it is not at all obvious that the right hand side is an integer!

6.3 Linear systems and rational maps in $\mathbb{P}^n$

This section is the logical continuation of Section 3.6; we again assume some familiarity with the notion of rational map ([13], p. 490). We keep our projective variety M and a line bundle L on M; we do not assume $B(L) = \varnothing$. We still have a map $M \smallsetminus B(L) \to |L|^*$, which we see as a rational map $\varphi_L : M \dashrightarrow |L|^*$.

Conversely, suppose given a rational map φ of M to a projective space $\mathbb{P}(V)$. We assume that M is *normal*; then the indeterminacy locus B of φ has codimension ≥ 2. We assume moreover that the line bundle $\varphi^*\mathcal{O}_{\mathbb{P}(V)}(1)$ on $M \smallsetminus B$ extends to a line bundle L on M. By Hartogs theorem the restriction map $H^0(M, L) \to H^0(M \smallsetminus B, L)$ is bijective, so we get a pullback homomorphism

$\varphi^* : V^* \to H^0(M, L)$. We have a commutative diagram

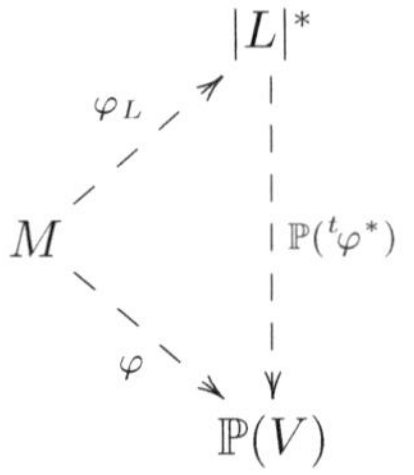

Indeed for m general in M, $\varphi_L(m)$ is the hyperplane of $|L|$ formed by the divisors passing through m; its image under $\mathbb{P}({}^t\varphi^*)$ is the hyperplane of $\mathbb{P}(V)^*$ formed by the hyperplanes of $\mathbb{P}(V)$ passing through $\varphi(m)$, and this corresponds by duality to the point $\varphi(m) \in \mathbb{P}(V)$.

6.4 The theta map

We go back to our moduli space $\mathcal{M}(r)$ and the generator $\mathcal{L}$ of its Picard group. The next step is to ask for the map defined by the linear systems $|\mathcal{L}^{\otimes k}|$. In fact we will concentrate on the simplest one, namely $\varphi_{\mathcal{L}}$. Our task will be to give a geometric description of this map. In order to do this we associate to each vector bundle $E \in \mathcal{M}(r)$ the locus

$$\theta(E) := \{L \in J^{g-1} \mid H^0(E \otimes L) \neq 0\}$$

Proposition 6.3. *$\theta(E)$ is either equal to J^{g-1}, or is a divisor in J^{g-1}, belonging to the linear system $|r\Theta|$.*

Proof. Consider the vector bundle $E \otimes \mathcal{P}$ on $C \times J^{g-1}$, where $\mathcal{P}$ is a Poincaré line bundle (Section 4.2). It defines the family of vector bundles $(E \otimes L)_{L \in J^{g-1}}$ on C. These bundles have slope $g-1$, hence we can apply Lemma 6.2, which shows that $\theta(E)$ is defined locally by one (possibly trivial) equation.

Let S be an irreducible variety, and $\mathcal{E}$ a vector bundle on $C \times S$, with $\deg(\mathcal{E}_s) = 0$ for each s. Lemma 6.2, applied to the vector bundle $\mathcal{E} \otimes \mathcal{P}$ on $C \times S \times J^{g-1}$, gives a line bundle $\mathcal{N}$ on $J^{g-1} \times S$ and a section τ of $\mathcal{N}$ with zero locus $Z = \cup_s \theta(\mathcal{E}_s)$. Put

$$S^{\mathrm{o}} := \{s \in S \mid \theta(\mathcal{E}_s) \neq J^{g-1}\} ;$$

S^{o} is the projection on S of the complement of Z in $J^{g-1} \times S$, so it is a Zariski open subset of S.

Applying this locally to our moduli space $\mathcal{M}(r)$, we see that the vector bundles $E \in \mathcal{M}(r)$ with $\theta(E) = J^{g-1}$ form a closed analytic (and therefore algebraic) subset of $\mathcal{M}(r)$. Let $\mathcal{M}(r)^{\mathrm{o}}$ be the complement of this subset. When E runs through $\mathcal{M}(r)^{\mathrm{o}}$, the Chern class $c_1(\theta(E))$ is constant. So if we fix $E_0 \in \mathcal{M}(r)^{\mathrm{o}}$, we have a rational map $\mathcal{M}(r) \dashrightarrow \mathrm{Pic}^{\mathrm{o}}(J^{g-1})$ mapping $E \in \mathcal{M}(r)^{\mathrm{o}}$ to $\mathcal{O}_J(\theta(E) - \theta(E_0))$. By Corollary 5.7 this map is constant, hence $\mathcal{O}_J(\theta(E))$ is independent of E.

Let $\alpha_1, \ldots, \alpha_r$ be distinct elements of JC. We have

$$\theta(\alpha_1 \oplus \cdots \oplus \alpha_r) = t^*_{\alpha_1}\Theta + \cdots + t^*_{\alpha_r}\Theta \in |r\Theta|$$

(see Corollary 2.11). Thus whenever $\theta(E) \neq J^{g-1}$ we have $\theta(E) \in |r\Theta|$. □

Thus we have a rational map $\theta : \mathcal{M}(r) \dashrightarrow |r\Theta|$.

Theorem 6.4 ([6]). *There is a natural isomorphism*

$$H^0(\mathcal{M}(r), \mathcal{L}) \xrightarrow{\sim} H^0(J^{g-1}, \mathcal{O}(r\Theta))^*$$

making the following diagram commutative:

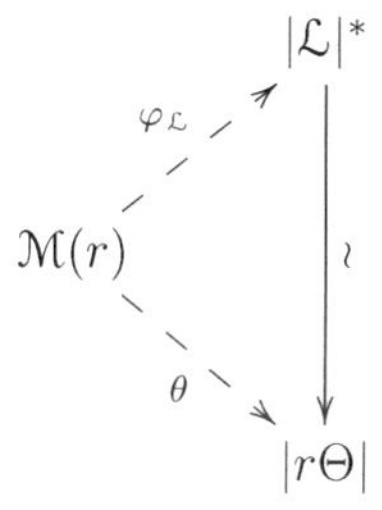

Sketch of proof : For $L \in J^{g-1}$, let H_L be the hyperplane in $|r\Theta|$ consisting of the divisors passing through L. By definition the pullback of H_L under θ is the divisor Δ_L. Thus, as explained in Section 6.3, we get a commutative diagram

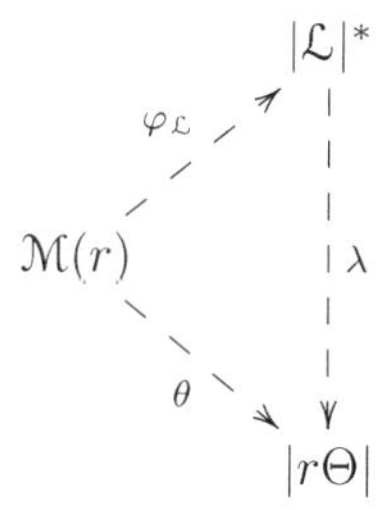

with $\lambda := \mathbb{P}({}^t\theta^*)$. It remains to prove that λ is bijective. Surjectivity is not difficult, let us prove it in the case $r = 2$. If λ is not surjective, the image of θ is contained in a hyperplane of $|2\Theta|$. But this image contains all the divisors $\theta(\alpha \oplus \alpha^{-1}) = t^*_\alpha\Theta + t^*_{-\alpha}\Theta$ for $\alpha \in JC$; and Corollary 3.13 implies that these divisors span $|2\Theta|$ (otherwise the image of $\varphi_{2\Theta}$ would be contained in a hyperplane).

We have $\dim|r\Theta| = r^g - 1$ by Theorem 3.5, so the crucial point is to prove the same equality for $\dim|\mathcal{L}|$. Of course this follows (in a non-trivial way) from the Verlinde formula (6.1); in [6], since the Verlinde formula was not yet available, we constructed a rational dominant map from a certain abelian variety to the moduli space, and applied Theorem 3.5 to get the result. □

Corollary 6.5. *The base locus of the linear system $|\mathcal{L}|$ on $\mathcal{M}(r)$ is the set of vector bundles $E \in \mathcal{M}(r)$ such that $\theta(E) = J^{g-1}$.* □

Thus the rather mysterious map $\varphi_{\mathcal{L}}$ is identified with the more concrete map θ; one usually refers to θ, or $\varphi_{\mathcal{L}}$, as the *theta map.* We will now see that this explicit description allows a good understanding of the theta map in the rank 2 case.

6.5 Rank 2

In rank 2 the theta map is by now fairly well understood. We summarize what is known in one theorem:

Theorem 6.6. 1) *The theta map* $\theta : \mathcal{M}(2) \to |2\Theta|$ *is a morphism.*

2) *If* C *is not hyperelliptic or* $g = 2$, θ *is an embedding.*

3) *If* C *is hyperelliptic of genus* ≥ 3, θ *is* 2*-to-*1 *onto its image in* $|2\Theta|$, *and this image admits an explicit description.*

This is the conjunction of various results. Part 1) is due to Raynaud [28], part 3) to Bhosle-Ramanan [11]. In case 2), the fact that θ is generically injective was proved in [2]; from this Brivio and Verra deduced that θ embeds $\mathcal{M}^s(2)$, and this was extended to $\mathcal{M}(2)$ in [14].

Recall that $\mathcal{M}(2)$ has dimension $3g-3$. In particular:

Corollary 6.7 ([23]). *For* $g = 2$, $\theta : \mathcal{M}(2) \to |2\Theta| \cong \mathbb{P}^3$ *is an isomorphism.* □

Consider the map $k : JC \to \mathcal{M}(2)$ given by $k(L) = L \oplus L^{-1}$. The composition $\theta \circ k$ is the map κ studied in Section 3.8; thus k embeds the Kummer variety K of JC into $\mathcal{M}(2)$, and the restriction of θ to K is the natural embedding of K into $|2\Theta|$. For $g > 2$ K is the singular locus of $\mathcal{M}(2)$ (Proposition 5.5); when C is not hyperelliptic, we obtain a variety in $|2\Theta|$ which is singular along the Kummer variety.

For $g = 3$ and C not hyperelliptic, a very nice application appears in [24]. In that case $\dim \mathcal{M}(2) = 6$, so θ embeds $\mathcal{M}(2)$ as a hypersurface in $|2\Theta| \cong \mathbb{P}^7$. It is not difficult to prove that it has degree 4 (for instance by computing its canonical bundle). Now Coble had found long ago that there is a *unique* quartic hypersurface in $|2\Theta|$ which is singular along the Kummer variety, for which he had written down an explicit equation (see [3] for a modern account). Therefore this hypersurface is $\mathcal{M}(2)$.

We will illustrate the methods used to prove the above results by giving the proof of 1).

6.6 Raynaud's theorem

We will prove part 1) of Theorem 6.6 in the following form:

Proposition 6.8. *Let* $E \in \mathcal{M}(2)$. *Then* $\theta(E) \neq J^{g-1}$.

Proof . If $E = L \oplus L^{-1}$, we have $\theta(E) = \Theta_L + \Theta_{L^{-1}} \neq J^{g-1}$. Therefore we may assume that E is stable.

Suppose $\theta(E) = J^{g-1}$. Put $F = E \otimes L$ for some L in J^{g-1}; our hypothesis becomes $h^0(F \otimes \alpha) > 0$ for all $\alpha \in JC$. Put $h := \min_{\alpha \in JC} h^0(F \otimes \alpha)$. Replacing F by $F \otimes \alpha$ for an appropriate α we may assume $h^0(F) = h$. We will use the semi-continuity theorem in cohomology, which implies that there is a Zariski open subset $U \subset JC$ (containing 0) such that $h^0(F \otimes \alpha) = h$ for $\alpha \in U$ ([20], §5).

Put $F' := F^* \otimes K_C$; let $p \in C$. The Riemann-Roch theorem gives

$$\big(h^0(F(p)) - h^0(F)\big) + \big(h^0(F') - h^0(F'(-p)\big) = 2 \, .$$

For p general we have $h^0(F') - h^0(F'(-p) \geq 1$, hence $h^0(F(p)) - h^0(F) \leq 1$. But $h^0(F(p)) = h$ would imply $h^0(F(p-q)) < h$ for q general, contradicting the definition of h. Thus $h^0(F(p)) = h + 1$.

Put $G := F(p)$. We have $h^0(G) = h+1$, and $h^0(G(-q)) = h^0(F(p-q)) = h$ for q general in C. From the exact sequence

$$0 \to G(-q) \to G \to G_q \to 0$$

we see that the global sections of G generate a rank 1 subsheaf L_0 of G. This is not necessarily a sub-line bundle because the quotient G/L_0 may have torsion; but it is contained in a unique sub-line bundle L, the kernel of the projection from G to the torsion-free quotient of G/L (sometimes called the *saturation* of L_0 in G). This sub-line bundle L has the property that *any rank* 1 *subsheaf* $M \subset G$ *with* $h^0(M) > 0$ *is contained in* L. Indeed if s is a nonzero section of M, at a general point x of C $s(x)$ generates M_x and L_x; therefore the map $M \to G/L$ is zero generically, hence everywhere.

Now take q, r general in C and consider $G(q-r)$. As before its global sections generate a sub-line bundle $L' \subset G(q-r)$. But we have $L(q-r) \subset G(q-r)$, and $h^0(L(q-r)) > 0$ since $h^0(L) = h^0(G) \geq 2$. Hence $L(q-r) \subset L'$. But symmetrically we have $L'(r-q) \subset L$, hence $L(q-r) = L'$. In particular we find $h^0(L(q-r)) = h = h^0(L)$ for q, r general in C. This implies $h^0(L(q)) = h^0(L)+1$, hence by Riemann-Roch $h^0(K \otimes L^{-1}) = h^0(K \otimes L^{-1}(-q))$ for q general; this is possible only if $h^0(K \otimes L^{-1}) = 0$. Applying again Riemann-Roch and using $h^0(L) \geq 2$, we get $\deg(L) \geq g+1 > \mu(G)$, a contradiction. □

6.7 Higher rank

In contrast with the rank 2 case, not much is known in higher rank. It is known since [28] that there exist stable bundles E with $\theta(E) = J^{g-1}$ – that is, base points for the linear system $|\mathcal{L}|$; in fact, they exist as soon as $r \geq g+2$, and even $r \geq 4$ if C is hyperelliptic [26]. On the other hand, in rank 3 there are no base points for $g = 2$ [28], $g = 3$ [4], or if C is general enough [28].

The situation is somewhat particular when $g = 2$, since $\dim \mathcal{M}(r) = \dim |r\Theta| = r^2 - 1$.

Proposition 6.9. *Let* $g = 2$.

1) $\theta : \mathcal{M}(r) \dashrightarrow |r\Theta|$ *is generically finite.*

2) *Its degree is* 1 *for* $r = 2$, 2 *for* $r = 3$, 30 *for* $r = 4$.

Part 1) is proved in [4]. The rank 2 case has been discussed in Corollary 6.7. In rank three $\theta : \mathcal{M}(3) \to |3\Theta| \cong \mathbb{P}^8$ is a double covering, branched along a sextic hypersurface which can be explicitly described [25]. The case $r = 4$ is due to Pauly [27].

Let us conclude with a

Conjecture 6.10. *For $g \geq 3$, the theta map $\theta : \mathcal{M}(r) \dashrightarrow |r\Theta|$ is generically 2-to-1 onto its image if C is hyperelliptic, and generically injective otherwise.*

This is unknown even for $r = g = 3$.

6.8 Further reading

There are a number of topics which I would have liked to cover in these lectures but could not by lack of time. Here are a few of them, with references to the literature:

- The beautiful interplay between curves and their Jacobians: Torelli theorem, Schottky problem, etc. A nice overview can be found in [22]; some of the topics are developed in [1].
- The heat equation and its extension to generalized theta functions. The original paper [16] of Hitchin is of course somewhat advanced, but still quite readable.
- Higgs bundles. Though not directly related to generalized theta functions, this is an important subject with many applications. Here again one can look at the original paper [15] of Hitchin; see also [7] for a short introduction.
- Principal bundles. This amounts to replace the group $\mathrm{SL}(r)$ by any semi-simple group. Essentially all we have said extends to this set-up. There are few results on the theta map, see [5] for the orthogonal and symplectic groups.

References

[1] E. Arbarello, M. Cornalba, P. Griffiths, J. Harris: *Geometry of algebraic curves*. Vol. I. Grund. Math. Wiss. **267**. Springer-Verlag, New York, 1985.

[2] A. Beauville: *Fibrés de rang 2 sur les courbes, fibré déterminant et fonctions thêta*. Bull. Soc. Math. France **116** (1988), 431–448.

[3] A. Beauville: *The Coble hypersurfaces*. C. R. Math. Acad. Sci. Paris **337** (2003), no. 3, 189–194.

[4] A. Beauville: *Vector bundles and theta functions on curves of genus 2 and 3*. Amer. J. of Math. **128** (2006), 607–618.

[5] A. Beauville: *Orthogonal bundles on curves and theta functions*. Ann. Inst. Fourier **56** (2006), 1405–1418.

[6] A. Beauville, M.S. Narasimhan, S. Ramanan: *Spectral curves and the generalised theta divisor*. J. Reine Angew. Math. **398** (1989), 169–179.

[7] S. Bradlow, O. García-Prada, and P. Gothen: *What is ... a Higgs bundle?* Notices of the AMS **54**, no. 8 (2007).

[8] O. Debarre: *Complex Tori and Abelian Varieties*. SMF/AMS Texts and Monographs **11**. AMS, Providence, 2005.

[9] P. Deligne: *Variétés unirationnelles non rationnelles*. Séminaire Bourbaki (1971/1972), Exp. No. 402, pp. 45-57. Lecture Notes in Math. **317**, Springer, Berlin, 1973.

[10] J.M. Drezet, M.S. Narasimhan: *Groupe de Picard des variétés de modules de fibrés semi-stables sur les courbes algébriques*. Invent. math. **97** (1989), 53–94.

[11] U. Desale, S. Ramanan: *Classification of vector bundles of rank* 2 *on hyperelliptic curves*. Invent. Math. **38** (1976), 161–185.

[12] T. Gómez: *Algebraic stacks*. Proc. Indian Acad. Sci. Math. Sci. **111** (2001), no. 1, 1–31.

[13] P. Griffiths, J. Harris: *Principles of Algebraic Geometry*. Wiley-Interscience, New York, 1978.

[14] B. Van Geemen, E. Izadi: *The tangent space to the moduli space of vector bundles on a curve and the singular locus of the theta divisor of the Jacobian*. J. Algebraic Geom. **10** (2001), 133–177.

[15] N. Hitchin: *The self-duality equations on a Riemann surface*. Proc. London Math. Soc. (3) **55** (1987) 59–126.

[16] N. Hitchin: *Flat connections and geometric quantization*. Comm. Math. Phys. **131** (1990), 347–380.

[17] A. King, A. Schofield: *Rationality of moduli of vector bundles on curves*. Indag. Math. (N.S.) **10** (1999), no. 4, 519–535.

[18] H. Lange, M.S. Narasimhan: *Squares of ample line bundles on abelian varieties*. Exposition. Math. **7** (1989), no. 3, 275–287.

[19] J. Le Potier: *Lectures on vector bundles*. Cambridge Studies in Advanced Mathematics, **54**. Cambridge University Press, Cambridge, 1997.

[20] D. Mumford: *Abelian Varieties*. Oxford University Press, London, 1970.

[21] D. Mumford: *Prym Varieties* I. Contributions to analysis, 325–350. Academic Press, New York, 1974.

[22] D. Mumford: *Curves and Their Jacobians*. Appendix to The Red Book of Varieties and Schemes, LN **1358**, Springer-Verlag, Berlin, 1999.

[23] M.S. Narasimhan, S. Ramanan: *Moduli of vector bundles on a compact Riemann surface*. Ann. of Math. **89** (1969), 19–51.

[24] M.S. Narasimhan, S. Ramanan: *2θ-linear systems on Abelian varieties*. Vector bundles on algebraic varieties, 415–427; Oxford University Press, London, 1987.

[25] A. Ortega: *On the moduli space of rank* 3 *vector bundles on a genus* 2 *curve and the Coble cubic*. J. Algebraic Geom. **14** (2005), no. 2, 327–356.

[26] C. Pauly: *On the base locus of the linear system of generalized theta functions*. Math. Res. Lett. **15** (2008), no. 4, 699–703.

[27] C. Pauly: *Rank four vector bundles without theta divisor over a curve of*

genus two. Adv. Geom. **10** (2010), no. 4, 647–657.

[28] M. Raynaud: *Sections des fibrés vectoriels sur une courbe*. Bull. Soc. math. France **110** (1982), 103–125.

[29] C. Sorger: *La formule de Verlinde*. Séminaire Bourbaki, Exp. 794, 1994/95. Astérisque **237** (1996), 87–114.

Open Problems and Surveys of Contemporary Mathematics
SMM 6, pp. 133–193

Structure-Preserving Doubling Algorithms — Rediscovery, Redevelopment and Applications

Eric King-wah Chu*, Wen-Wei Lin†

Abstract

In this paper, we attempt to tell the interesting story about the recent rediscovery and revival of the once faded and almost forgotten doubling algorithm for the discrete-time algebraic Riccati equation. Armed with some new insight and theoretical redevelopment, the method, now called the structure-preserving doubling algorithm (SDA), is first linked to palindromic eigenvalue problems. Generalizations were then developed for continuous-time, nonsymmetric and PCP algebraic Riccati equations and nonlinear matrix equations, for a vast array of applications in vibration analysis, surface acoustic wave simulation, optimal control, stochastic systems, transport theory, time-delayed systems, nano research, etc. We shall present some key theoretical results associated with selected important applications, and a summary of the recent discovery that the SDA can be adapted to solve large-scale problems of dimension n, with an $O(n)$ computational complexity and memory requirement.

2000 Mathematics Subject Classification: 15A18, 15A22, 15A24, 15A30, 65F15, 65F50

Keywords and Phrases: algebraic Riccati equation, continuous-time system, discrete-time system, doubling algorithm, eigenvalue, invariant subspace, large-scale problem, Lyapunov equation, matrix pencil, nano research, Newton's method, nonlinear matrix equation, palindromic eigenvalue problem, singular system, Stein equation, surface acoustic wave, time-delayed system, vibration

*School of Mathematical Sciences, Building 28, Monash University 3800. Email: `eric.chu@monash.edu`

†Department of Applied Mathematics, Chiao Tung University, Hsinchu 300. Email: `wwlin@am.nctu.edu.tw`

Contents

1 Introduction

Our story starts from the discrete-time linear system:

$$x_{j+1} = Ax_j + Bu_j$$

with the state $x \in \mathbb{R}^n$, the control $u \in \mathbb{R}^m$, and the system and control matrices $A \in \mathbb{R}^{n\times n}$ and $B \in \mathbb{R}^{n\times m}$, respectively. In optimal control with an infinite time horizon, we choose u_j such that

$$\min_{u_j} \mathcal{J} = \frac{1}{2}\sum_{j=1}^{\infty}\left(x_j^\top H x_j + u_j^\top R u_j\right).$$

with both $H \in \mathbb{R}^{n\times n}$ and $R \in \mathbb{R}^{m\times m}$ beng symmetric. With the stabilizing solution X from the discrete algebraic Riccati equation (DARE)

$$X = A^\top X(I+GX)^{-1}A + H, \quad G = BR^{-1}B^\top, \tag{1.1}$$

the optimal control has the form $u_j = -(R + B^\top XB)^{-1}B^\top XAx_j$.

One way to solve the DARE (1.1) is to use the structure-preserving doubling algorithm (SDA):

$$\begin{aligned} A_{k+1} &= A_k(I_n + G_kH_k)^{-1}A_k, \\ G_{k+1} &= G_k + A_kG_k(I_n + H_kG_k)^{-1}A_k^\top, \\ H_{k+1} &= H_k + A_k^\top(I_n + H_kG_k)^{-1}H_kA_k; \end{aligned} \tag{1.2}$$

with $A_0 = A$, $H_0 = H$ and $G_0 = G$. A second type of structure-preserving algorithm (SDA_2) will be introduced later, thus the iteration in (1.2) is referred to as the SDA of type 1, or SDA_1. Under favourable conditions, the SDA_1 converges quadratically, with $A_k \to 0$, $H_k \to X$ and $G_k \to Y$ (the adjoint solution) as $k \to \infty$.

The SDA_1 was well-known in the seventies, but faded until its revival by Chu, Fan, Lin and Wang in 2004 [17] (actually for the more general periodic DARE). The origin of the SDA is, interestingly, unknown. In [2, p. 159], we have the following quotation by Anderson and Moore:

> Doubling algorithms have been part of the folklore associated with Riccati equations in linear systems problems for some time. We are unable to give any original reference containing material close to that presented here. However, more recent references include
> [31] Bierman, G.J., Steady-state covariance computation for discrete linear systems, *Proc. 1971 JACC*, paper No. 8-C3.
> [32] Friedlander, B., T. Kailath and L. Ljung, Scattering theory and linear least squares estimation – II. Discrete-time problems, *J. Franklin Inst.*, Vol. 301, Nos. 1 and 2, 1976, pp. 71–82.
> [33] Anderson, B.D.O., Second-order convergent algorithms for the steady-state Riccati equations, *Int. J. Control*, Vol. 28, No. 2, pp. 295–306, 1978.

Many attributed the SDA to Anderson (and his 1978 paper above), but that is wrong according to his quotation above.

In this paper, we shall discuss what is new in the recent rediscovery and redevelopment of the SDA in Section 2. The algorithms and theoretical results for selected important applications, in vibration analysis of fast trains, surface acoustic wave simulation, Green's function computation in nano research, sensitivity analysis of time-delayed systems, and the linear palindromic eigenvalue problem $A^*x = \lambda Ax$, with applications such as the optimal control for singular descriptor linear systems, occupy most of Section 3. A brief summary of the recent generalizations of the SDA for large-scale problems is contained in Section 3.6, before the conclusions in Section 4.

To limit the length of the paper, we neglect some minor results and all numerical experiments. Readers should consult the quoted references for details if so desired. For an earlier brief survey paper on the SDA, see [19].

2 What is new?

We shall sketch the big picture on how the SDA has been revived and redeveloped, before the details for selected applications in the next section. We shall describe how the SDA_1 (1.2) is generalized for several seemingly very different problems. So how these generalizations of the SDA and the associated array of diverse applications are linked?

2.1 SDA_1 and SSF_1 for DAREs; Palindromic eigenvalue problems

The original SDA borrowed from the physical phenomenon of scattering, also acted as a convergence acceleration technique. Subsequently, only a small set of techniques can be employed in its analysis, limiting its understanding, growth and applicability.

The first major modern step is to link the DARE (1.1) to an equivalent eigenvalue problem:

$$\mathcal{M}Z = \mathcal{L}Z\Lambda, \quad \mathcal{M} \equiv \begin{bmatrix} A & 0 \\ -H & I_n \end{bmatrix}, \quad \mathcal{L} \equiv \begin{bmatrix} I_n & G \\ 0 & A^\top \end{bmatrix}, \quad Z = \begin{bmatrix} I_n \\ X \end{bmatrix}. \tag{2.1}$$

The eigenvalue problem (2.1) is described as symplectic because the pencil $(\mathcal{M}, \mathcal{L})$ satisfies

$$\mathcal{M}\mathcal{J}\mathcal{M}^\top = \mathcal{L}\mathcal{J}\mathcal{L}^\top, \quad \mathcal{J} \equiv \begin{bmatrix} 0 & I_n \\ -I_n & 0 \end{bmatrix}.$$

The canonical form $(\mathcal{M}, \mathcal{L})$ in (2.1) is called a standard symplectic form of type 1 (SSF_1). It is easy to show that eigenvalues for $(\mathcal{M}, \mathcal{L})$ occur in reciprocal pairs; i.e., with $\sigma(\cdot)$ denoting the spectrum of matrices, matrix pencils or polynomials, $\lambda \in \sigma(\mathcal{M}, \mathcal{L})$ implies $\lambda \in \sigma(\mathcal{L}, \mathcal{M})$.

To obtain the stabilizing solution X for the DARE (1.1), the more general "doubling" action on the matrix pencil $(\mathcal{M}, \mathcal{L})$ is applied — find $[\widetilde{M}, \widetilde{L}]$ such that

$$\left[\widetilde{\mathcal{M}}, \widetilde{\mathcal{L}}\right] \begin{bmatrix} \mathcal{L} \\ -\mathcal{M} \end{bmatrix} = 0, \tag{2.2}$$

which is equivalent to $\widetilde{\mathcal{M}}\mathcal{L} = \widetilde{\mathcal{L}}\mathcal{M}$, and then define

$$\left(\widehat{\mathcal{M}}, \widehat{\mathcal{L}}\right) \equiv \left(\widetilde{\mathcal{M}}\mathcal{M}, \widetilde{\mathcal{L}}\mathcal{L}\right).$$

The doubling action $(\mathcal{M}, \mathcal{L}) \to \left(\widehat{\mathcal{M}}, \widehat{\mathcal{L}}\right)$ preserves symplecticity. More importantly, the stronger SSF_1 form in $(\mathcal{M}, \mathcal{L})$ is passed onto $\left(\widehat{\mathcal{M}}, \widehat{\mathcal{L}}\right)$ when $\left(\widetilde{M}, \widetilde{L}\right)$ in (2.2) is selected as in [17], where some generalized elementary row-block operations are applied. The formulae in (1.2) are then retrieved.

Assume that $\mathcal{M}x = \lambda\mathcal{L}x$, after the doubling action, we have

$$\widehat{\mathcal{M}}x = \widetilde{\mathcal{M}}\mathcal{M}x = \lambda\widetilde{\mathcal{M}}\mathcal{L}x = \lambda\widetilde{\mathcal{L}}\mathcal{M}x = \lambda^2\widetilde{\mathcal{L}}\mathcal{L}x = \lambda^2\widehat{\mathcal{L}}x.$$

In other words, the power of λ is doubled, thus the term "doubling". Assuming no unimodular eigenvalues exists, several doubling actions reduce the problem to finding the null space of the converged matrix pencil.

In the vibration analysis of fast trains and the simulation of surface acoustic waves (SAW), we are required to solve the palindromic eigenvalue problem

$$\left(\lambda^2 A_{d1}^\top + \lambda A_{d0} + A_{d1}\right) x = 0, \quad A_{d0}^\top = A_{d0}, \quad x \neq 0. \tag{2.3}$$

After a deflation process, (2.3) can be written in SSF_1 form, and the SDA_1 can then be applied for its solution; see [18] or Section 3.1 for the details.

2.2 SDA_2 and SSF_2 for NMEs

Alternatively, and more efficiently, the palindromic eigenvalue problem (2.3) can be solved via the solvent approach. With the factorization

$$\lambda^2 A_{d1}^\top + \lambda A_{d0} + A_{d1} = (\lambda Y - A_{d1})Y^{-1}(\lambda A_{d1} - Y)^\top$$

provided that the nonlinear matrix equation (NME) holds:

$$Y + AY^{-1}A^\top = Q, \tag{2.4}$$

with $Q = -A_{d0}$ and $A = A_{d1}$. This can be shown to be equivalent to the eigenvalue problem

$$\mathcal{M}Z = \mathcal{L}Z\Lambda, \quad \mathcal{M} \equiv \begin{bmatrix} A & 0 \\ Q & -I_n \end{bmatrix}, \quad \mathcal{L} \equiv \begin{bmatrix} 0 & I_n \\ A^\top & 0 \end{bmatrix}, \quad Z \equiv \begin{bmatrix} I_n \\ Y \end{bmatrix}. \tag{2.5}$$

Again, the matrix pencil $(\mathcal{M}, \mathcal{L})$ in (2.5) is symplectic and is called a standard symplectic form of type 2 (SSF_2). A specially designed doubling action can then be applied to $(\mathcal{M}, \mathcal{L})$, producing the SDA_2 for the solution of the NME (2.4):

$$\begin{aligned} A_{k+1} &= A_k(Q_k - P_k)^{-1}A_k, \\ P_{k+1} &= P_k + A_k(Q_k - P_k)^{-1}A_k^\top, \\ Q_{k+1} &= Q_k - A_k^\top(Q_k - P_k)^{-1}A_k^\top; \end{aligned} \tag{2.6}$$

with $A_0 = A$, $P_0 = 0$ and $Q_0 = Q$. Under favourable conditions, we have $A_k \to 0$ and $Q_k \to X$ as $k \to \infty$.

2.3 Advantages of SDA

The first advantage of the SDA lies in its wide applicability. From the brief discussion in Sections 2.1–2.2, we now have most of the ingredients for all the generalizations of the SDA. For other matrix equations, the continuous-time algebraic Riccati equation (CARE) can be treated after the Cayley transform, and Stein and Lyapunov equations are special cases without nonlinear terms. Further generalizations were also found for the nonsymmetric algebraic Riccati equation (NARE).

Algebraic Riccati equations arise in many important applications, including the total least squares problems with or without symmetric constraints [26], the spectral factorizations of rational matrix functions [30], the linear and nonlinear optimal controls [8], the contractive rational matrix functions [55], the structured complex stability radius [39], the transport theory [20, 52, 58], the Wiener-Hopf factorization of Markov chains [83], and the optimal solutions of linear differential systems [55]. Symmetric algebraic Riccati equations have been the topic of extensive research, and the theory, applications and numerical solution of these equations are the subject of [13, 15, 17, 19] as well as the monographs [55, 70]. For NAREs, see [9, 20, 32, 36, 58]. Importantly, different applications possess different structures and solution requirement, throwing up new theoretical and computational challenges.

The palindromic quadratic eigenvalue problem can be generalized to the PCP-palindromic eigenvalue problem, for the sensitivity analysis of time-delayed systems. Furthermore, a SDA was discovered for the linear palindromic eigenvalue problem $A^*x = \lambda Ax$, from a specific doubling action for (A^*, A). When the problems are large, doubling can also be adapted, under additional sparseness and low-rank assumptions. In addition to the palindromic eigenvalue problems and the NMEs mentioned before, the SDA can now be applied to a wide range of interesting problems from important applications. This is the biggest advantage of the SDA over the other methods. In fact, there are no comparable unifying suite of methods for such as large collection of problems and applications.

The next advantage of the SDA lies in its simplicity. With our new understanding of the method related to symplectic pencils, proofs can be constructed using established tools such as the Kronecker canonical form. The convergence of the SDA is now well understood, with fast quadratic convergence when the corresponding pencil $(\mathcal{M}, \mathcal{L})$ has no unimodular eigenvalues. When unimodular eigenvalues with even partial multiplicity exist, linear convergence of factor 1/2 can be achieved.

Lastly, for large-scale problems, additional new insight of the SDA is obtained. When the size of the problem n approaches infinity, the iterates in the SDA differed from the previous ones by diminishing low-rank updates. With this new insight and under some minor additional assumptions, we can show, for the first time, that the SDA can be implemented with an $O(n)$ computational complexity and memory requirement. Consequently, the SDA will be viable alternative for large-scale problems; for details, see Section 3.6 and [60, 61, 62].

3 Applications

3.1 Vibration of fast trains

Quoted from [18], we present here the study of the resonance phenomena of the track under high frequent excitation forces. Research in this area not only contributes to the safety of the operations of high-speed trains but also new designs of train bridges, embedded rail structures (ERS) and train suspension systems. Recently, the dynamic response of the vehicle-rails-bridge interaction system under different train speeds has been studied by Wu, Yang, and Yau [84] and a procedure for designing an optimal ERS is proposed by Markine, de Man, Jovanovic and Esveld [69]. An accurate numerical estimation to the resonance frequencies of the rail plays an important role in both works. However, as mentioned by Ipsen [46], the classic finite element packages fail to deliver correct resonance frequencies for such problems [46]. Here, we would like to compare the method proposed by Mackey, Mackey, Mehl and Mehrmann [67] with the generalized SDA methods proposed by Chu, Fan, Lin and Wang [17] and Lin and Xu [64], in solving the palindromic eigenvalue problems arising from spectral modal analysis of the resonance of the rail under a periodic excitation force.

We assume that the rail sections between consecutive sleeper bays are identical, distances between consecutive wheels are the same and the wheel loads are equal. Figure 3.1 shows an example of the rail section we consider here.

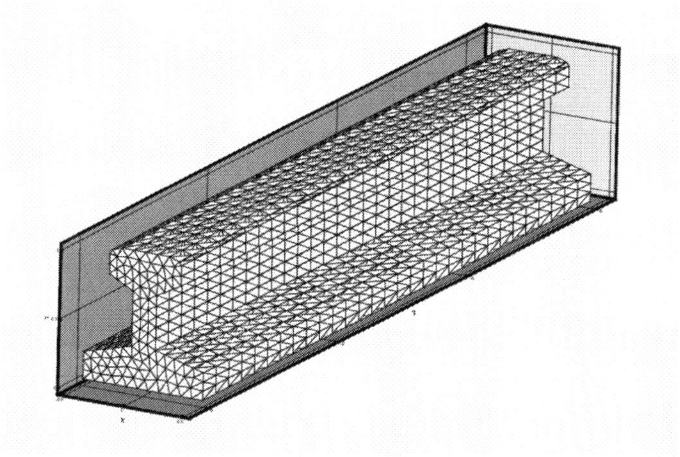

Figure 3.1 A 3D rail model.

Base on our assumptions, we model the rail under cargo wheel loads by a section of rail between two sleepers. The external force is assumed to be periodic and the displacements of two boundary cross sections of the modelled rail are assumed to have a ratio κ, which is dependent on the excitation frequency of the external force. In the following, we consider the rail as a three dimensional isotropic elastic solid and a 3D finite element model of the solid with linear isoparametric tetrahedron elements is introduced.

From the element virtual work principle, the equilibrium state of the solid element e under external body forces satisfies the following equation:

$$\int_e (\delta\epsilon^\top) C\epsilon \, dV + \int_e (\delta q^\top)\rho\ddot{q} \, dV = \int_e (\delta q^\top) f \, dV. \tag{3.1}$$

Here, ρ is the mass density, f is the time-dependent body force, $q = [u, v, w]$ is the displacement vector, $\epsilon = [\frac{\partial u}{\partial x}, \frac{\partial v}{\partial y}, \frac{\partial w}{\partial z}, \frac{\partial u}{\partial y} + \frac{\partial v}{\partial x}, \frac{\partial w}{\partial x} + \frac{\partial u}{\partial z}, \frac{\partial v}{\partial z} + \frac{\partial w}{\partial y}]$, and $\delta q^\top$ and $\delta\epsilon^\top$ are the virtual displacement and the corresponding virtual strain vectors, and

$$C = \frac{E}{(1+\upsilon)(1-2\upsilon)} \operatorname{diag}(C_1, \ C_2)$$

is the well-known strain-stress relationship, where E is the Young's modulus, υ is the Poisson ratio and

$$C_1 = \begin{bmatrix} 1-\upsilon & \upsilon & \upsilon \\ \upsilon & 1-\upsilon & \upsilon \\ \upsilon & \upsilon & 1-\upsilon \end{bmatrix}, \quad C_2 = \left(\frac{1-2\upsilon}{2}\right) I_3.$$

Let ϕ_i and $[u_i, v_i, w_i]^\top$ $(i = 1, \ldots, 4)$ be the linear nodal basis function and the nodal displacement vector associated with the i-th node of the element e, respectively, and let $X_e = [X_1^\top, X_2^\top, X_3^\top, X_4^\top]^\top$, $B_e = [B_1, B_2, B_3, B_4]$ and $N_e =$

$[N_1, N_2, N_3, N_4]$, where

$$X_i = \begin{bmatrix} u_i \\ v_i \\ w_i \end{bmatrix}, \quad N_i = \begin{bmatrix} \phi_i & 0 & 0 \\ 0 & \phi_i & 0 \\ 0 & 0 & \phi_i \end{bmatrix}, \quad B_i = \begin{bmatrix} \frac{\partial \phi_i}{\partial x} & 0 & 0 \\ 0 & \frac{\partial \phi_i}{\partial y} & 0 \\ 0 & 0 & \frac{\partial \phi_i}{\partial z} \\ \frac{\partial \phi_i}{\partial y} & \frac{\partial \phi_i}{\partial x} & 0 \\ \frac{\partial \phi_i}{\partial z} & 0 & \frac{\partial \phi_i}{\partial x} \\ 0 & \frac{\partial \phi_i}{\partial z} & \frac{\partial \phi_i}{\partial y} \end{bmatrix}.$$

Equation (3.1) can now be discretized into the following linear equations

$$\sum_e \left(\int_e B_e^\top C B_e \, dV \right) X_e + \rho \left(\int_e N_e^\top N_e \, dV \right) \ddot{X}_e = \sum_e \left(\int_e N_e^\top N_e \, dV \right) F_e, \quad (3.2)$$

where $F_e = [F_1^\top, F_2^\top, F_3^\top, F_4^\top]^\top$ and F_i $(i = 1, \ldots, 4)$ is the i-th nodal force vector acting on element e. In the following, we denote $K = \sum\limits_e \int_e B_e^\top C B_e \, dV$, and $M = \sum\limits_e \rho \int_e N_e^\top N_e \, dV$. Equation (3.2) can now be written as

$$KX + M\ddot{X} = \rho^{-1} MF.$$

When considering the dynamic response of the solid, dissipative forces such as the force due to frictions have to be considered. Their effect is introduced in the form of the so-called viscous damping $D\dot{X}$ where D is the damping matrix. For the lack of better alternatives, proportional damping proposed by Strutt (Lord Rayleigh) [79] is employed where D is a linear combination of K and M. The equation of motion involving viscous damping can now be written as

$$KX + D\dot{X} + M\ddot{X} = \rho^{-1} MF.$$

Due to the given boundary conditions on a uniform mesh, K, D and M have the following form

$$\begin{bmatrix} G_{11} & G_{12} & 0 & \cdots & 0 & \frac{1}{\kappa} G_{m,m+1}^\top \\ G_{12}^\top & G_{22} & G_{23} & 0 & & 0 \\ 0 & \ddots & \ddots & \ddots & \ddots & \vdots \\ \vdots & 0 & \ddots & \ddots & \ddots & 0 \\ 0 & & \ddots & G_{m-2,m-1}^\top & G_{m-1,m-1} & G_{m-1,m} \\ \kappa G_{m,m+1} & 0 & \cdots & 0 & G_{m-1,m}^\top & G_{m,m} \end{bmatrix}$$

with $G_{ii} \in \mathbb{C}^{n_i \times n_i}$ for $i = 1, \ldots, m$. Furthermore, from the spectral modal analysis, one considers $X = \hat{X} e^{i\omega t}$ where ω is the frequency of the external excitation force and $\hat{X}$ is the corresponding eigenmode. Consequently, we arrive to a palindromic eigenvalue problem

$$\kappa^{-1} \mathcal{P}(\kappa) \hat{X} = 0, \quad \mathcal{P}(\kappa) \equiv \left(\kappa^2 A_1 + \kappa A_0 + A_1^\top \right),$$

where $A_0, A_1 \in \mathbb{C}^{n\times n}$ with $n = n_1 + \cdots + n_m$ and

$$[A_1]_{ij} = \begin{cases} K_{m,m+1} + i\omega D_{m,m+1} - \omega^2 M_{m,m+1} & (\text{if } i = m \text{ and } j = 1), \\ 0 & (\text{otherwise}), \end{cases} \tag{3.3}$$
$$[A_0]_{ij} = \begin{cases} K_{i,j} + i\omega D_{i,j} - \omega^2 M_{i,j} & (\text{if } i-1 \le j \le i+1), \\ 0 & (\text{otherwise}). \end{cases}$$

3.1.1 Deflation

We shall consider the deflation of zero and infinite eigenvalues in this section. For the deflation of $\lambda = \pm 1$, consult [67].

From their definitions in (3.3), $A_1, A_0 \in \mathbb{C}^{n\times n}$ can be partitioned as

$$A_1 = \begin{bmatrix} 0 & 0 & 0 \\ 0 & 0 & 0 \\ L & 0 & 0 \end{bmatrix}, \qquad A_0 = \begin{bmatrix} C_{11} & C_{12} & 0 \\ C_{12}^\top & C_{22} & C_{23} \\ 0 & C_{23}^\top & C_{33} \end{bmatrix}$$

where $L \in \mathbb{C}^{n_m\times n_1}$, $C_{11} = C_{11}^\top \in \mathbb{C}^{n_1\times n_1}$, $C_{33} = C_{33}^\top \in \mathbb{C}^{n_m\times n_m}$ and $C_{22} = C_{22}^\top \in \mathbb{C}^{\ell\times\ell}$ with $\ell = n - n_1 - n_m$. Assume that C_{22} is nonsingular. We have observed that this assumption is generically valid from the numerical examples we have encountered. Otherwise, the preprocessing procedure in [67] should be applied.

Let

$$\Theta = \begin{bmatrix} I_{n_1} & -C_{12}C_{22}^{-1} & 0 \\ 0 & I_\ell & 0 \\ 0 & -C_{23}^\top C_{22}^{-1} & I_{n_m} \end{bmatrix}, \quad \Pi = \begin{bmatrix} I_{n_1} & 0 & 0 \\ 0 & 0 & I_{n_m} \\ 0 & I_\ell & 0 \end{bmatrix}.$$

Then, using a similarity transformation, $\mathcal{P}(\lambda)$ can be transferred to the following form

$$\begin{aligned} \Pi\Theta\mathcal{P}(\lambda)\Theta^\top\Pi^\top &= \begin{bmatrix} \lambda(C_{11} - C_{12}C_{22}^{-1}C_{12}^\top) & L^\top - \lambda C_{12}C_{22}^{-1}C_{23} & 0 \\ \lambda(\lambda L - C_{23}^\top C_{22}^{-1}C_{12}^\top) & \lambda(C_{33} - C_{23}^\top C_{22}^{-1}C_{23}) & 0 \\ 0 & 0 & \lambda C_{22} \end{bmatrix} \\ &= \operatorname{diag}(I_{n_1}, \lambda I_{n_m}, I_\ell) \begin{bmatrix} \mathcal{S}(\lambda) & 0 \\ 0 & \lambda C_{22} \end{bmatrix} \end{aligned}$$

where

$$\mathcal{S}(\lambda) = \begin{bmatrix} \lambda\tilde{C}_{11} & L^\top - \lambda\tilde{C}_{12} \\ \lambda L - \tilde{C}_{12}^\top & \tilde{C}_{22} \end{bmatrix}$$

with $\tilde{C}_{11} \equiv C_{11} - C_{12}C_{22}^{-1}C_{12}^\top$, $\tilde{C}_{12} \equiv C_{12}C_{22}^{-1}C_{23}$ and $\tilde{C}_{22} \equiv C_{33} - C_{23}^\top C_{22}^{-1}C_{23}$.

3.1.2 Structure-Preserving Algorithms

SDA_1

After swapping the row-blocks, the pencil $\mathcal{S}(\lambda)$ is equivalent to

$$\lambda \begin{bmatrix} L & 0 \\ \tilde{C}_{11} & -\tilde{C}_{12} \end{bmatrix} + \begin{bmatrix} -\tilde{C}_{12}^\top & \tilde{C}_{22} \\ 0 & L^\top \end{bmatrix},$$

which is in a generalized standard symplectic form (GSSF) [44]. The structure-preserving doubling algorithm (SDA_1) in [44] can then be applied to solve the corresponding eigenvalue problem.

In terms of accuracy and speed of convergence, SDA_1 behaves similarly as SDA_2 below. However, the operation count for SDA_1 doubles that of SDA_2. As a result, we shall not discuss SDA_1 further. However, SDA_1 is a weapon in reserve against difficult palindromic eigenvalue problems, when some assumptions for SDA_2 are not satisfied.

Assume that $\tilde{C}_{22}$ is invertible. Define a new λ-matrix $\tilde{\mathcal{S}}(\lambda)$ as follows:

$$\begin{aligned}\tilde{\mathcal{S}}(\lambda) &\equiv \begin{bmatrix} I_{n_1} & -L^\top \tilde{C}_{22}^{-1} \\ 0 & I_{n_3} \end{bmatrix} \mathcal{S}(\lambda) \begin{bmatrix} I_{n_1} & 0 \\ \tilde{C}_{22}^{-1}\tilde{C}_{12}^\top & I_{n_3} \end{bmatrix} \\ &= \begin{bmatrix} \lambda(\tilde{C}_{11} - L^\top \tilde{C}_{22}^{-1} L - \tilde{C}_{12}\tilde{C}_{22}^{-1}\tilde{C}_{12}^\top) + L^\top \tilde{C}_{22}^{-1}\tilde{C}_{12}^\top & -\lambda \tilde{C}_{12} \\ \lambda L & \tilde{C}_{22} \end{bmatrix}\end{aligned}$$

and let $\left[\tilde{x}^\top, \tilde{y}^\top\right]^\top$ be an eigenvector of $\tilde{\mathcal{S}}(\lambda)$; i.e.,

$$\lambda\left[(\tilde{C}_{11} - L^\top \tilde{C}_{22}^{-1} L - \tilde{C}_{12}\tilde{C}_{22}^{-1}\tilde{C}_{12}^\top)\tilde{x} - \tilde{C}_{12}\tilde{y}\right] + L^\top \tilde{C}_{22}^{-1}\tilde{C}_{12}^\top \tilde{x} = 0, \tag{3.4a}$$

$$\lambda L\tilde{x} + \tilde{C}_{22}\tilde{y} = 0. \tag{3.4b}$$

Since $\tilde{C}_{22}$ is invertible, from (3.4b) $\tilde{y}$ can be represented as

$$\tilde{y} = -\lambda \tilde{C}_{22}^{-1} L\tilde{x}. \tag{3.5}$$

Substituting (3.5) into (3.4a), we get the following new small size palindromic quadratic eigenvalue problem:

$$\mathcal{P}_d(\lambda)\tilde{x} \equiv (\lambda^2 A_{d1} + \lambda A_{d0} + A_{d1}^\top)\tilde{x} = 0, \tag{3.6}$$

where

$$A_{d1} = \tilde{C}_{12}\tilde{C}_{22}^{-1} L, \quad A_{d0} = \tilde{C}_{11} - L^\top \tilde{C}_{22}^{-1} L - \tilde{C}_{12}\tilde{C}_{22}^{-1}\tilde{C}_{12}^\top.$$

The SDA_1 can then be applied to the palindromic eigenvalue problem (3.6), which is in the same form as (2.3), written in the SSF_1 as in (2.1).

SDA_2

Suppose that X is nonsingular. Rewrite $\mathcal{P}_d(\lambda)$ in (3.6) as

$$\mathcal{P}_d(\lambda) = (\lambda A_{d1} - X)X^{-1}(\lambda X - A_{d1}^\top) + \lambda(A_{d1}X^{-1}A_{d1}^\top + X + A_{d0}).$$

It follows that $\mathcal{P}_d(\lambda)$ can be factorized (or square-rooted) as

$$\mathcal{P}_d(\lambda) = (\lambda A_{d1} - X)X^{-1}(\lambda X - A_{d1}^\top)$$

for some nonsingular X if and only if X is satisfied following nonlinear matrix equation with the plus sign (NME):

$$A_{d1}X^{-1}A_{d1}^\top + X + A_{d0} = 0. \tag{3.7}$$

We can easily prove the following lemma on the existence of the solutions of the NME:

Lemma 3.1.1. *Let* $(\Lambda_1 \oplus \Lambda_2, [Y_1, Y_2])$ *be an eigenpair of* $\mathcal{P}_d(\lambda)$ *in the sense that*

$$A_{d1} Y_i \Lambda_i^2 + A_{d0} Y_i \Lambda_i + A_{d1}^\top Y_i = 0 \quad (i = 1, 2),$$

where $Y_i \in \mathbb{C}^{n_1 \times n_1}$ *for* $i = 1, 2$. *Suppose that* A_{d1} *and* Y_i $(i = 1, 2)$ *are invertible. Then the corresponding NME* (3.7) *has the solutions* $X = A_{d1}^\top Y_i \Lambda_i^{-1} Y_i^{-1}$ $(i = 1, 2)$.

Evidently, there are many solutions to the NME, each will facilitate the factorization of $\mathcal{P}_d(\lambda)$ we aim for. Assume that there are no eigenvalues on the unit circle. Consequently, we can partition the spectrum into $\Lambda_s \oplus \Lambda_s^{-1}$, with Λ_s containing the stable eigenvalues (inside the unit circle). The SDA will seek a stable solution $X_s \equiv A_{d1}^\top Y_s \Lambda_s^{-1} Y_s^{-1}$, where Y_s contains the eigenvectors corresponding to Λ_s. Note that X_s is unique as it is independent of the order of the eigenvalues in Λ_s.

The structure-preserving doubling algorithm SDA_2 (2.6) in [64] can then be applied to solve the NME, and subsequently the palindromic eigenvalue problem.

In the SDA_2 for fast trains, we assume the invertibility of the matrices $Q_k - P_k$. This is the case for large values of k, as indicated by Corollary 3.1.4. A more involved analysis [34] removes the assumption.

Finally, equations similar to (3.7), for real matrices or with the transpose replaced by Hermitian, have been studied and existing results do not apply. The lack of positivity of matrices or an associated inner product make our NME difficult to study.

3.1.3 Convergence of Algorithms

The behavior of the structure-preserving doubling algorithms is well-documented in [15, 17, 36, 44, 45, 64]. However, these results are mostly written for real problem with real variables and have to be modified for our situation. Following the development in [64], let $\mathcal{M} - \lambda\mathcal{L} \in \mathbb{C}^{2n \times 2n}$ be a *T-symplectic* pencil, in the sense that

$$\mathcal{M} J \mathcal{M}^\top = \mathcal{L} J \mathcal{L}^\top, \quad J = \begin{bmatrix} 0 & I \\ -I & 0 \end{bmatrix}. \tag{3.8}$$

Define the nonempty null set

$$\mathcal{N}(\mathcal{M}, \mathcal{L}) \\ \equiv \left\{ [\mathcal{M}_*, \mathcal{L}_*] : \mathcal{M}_*, \mathcal{L}_* \in \mathbb{C}^{2n \times 2n}, \text{ rank}\,[\mathcal{M}_*, \mathcal{L}_*] = 2n, \; [\mathcal{M}_*, \mathcal{L}_*] \begin{bmatrix} \mathcal{L} \\ -\mathcal{M} \end{bmatrix} = 0 \right\}.$$

For any given $[\mathcal{M}_*, \mathcal{L}_*] \in \mathcal{N}(\mathcal{M}, \mathcal{L})$, define $\widehat{\mathcal{M}} = \mathcal{M}_* \mathcal{M}$, $\widehat{\mathcal{L}} = \mathcal{L}_* \mathcal{L}$. For the *doubling transformation* $\mathcal{M} - \lambda\mathcal{L} \to \widehat{\mathcal{M}} - \lambda\widehat{\mathcal{L}}$, we have the adaptation of [64, Theorem 2.1]:

Theorem 3.1.2. *Let* $\widehat{\mathcal{M}} - \lambda\widehat{\mathcal{L}}$ *be a doubling transformation of a T-symplectic pencil* $\mathcal{M} - \lambda\mathcal{L}$. *Then we have:*

(a) *The pencil* $\widehat{\mathcal{M}} - \lambda\widehat{\mathcal{L}}$ *is still T-symplectic.*

(b) *If* $\mathcal{M} \begin{bmatrix} U \\ V \end{bmatrix} = \mathcal{L} \begin{bmatrix} U \\ V \end{bmatrix} S$, *where* $U, V \in \mathbb{C}^{n \times m}$ *and* $S \in \mathbb{C}^{m \times m}$, *then*

$$\widehat{\mathcal{M}} \begin{bmatrix} U \\ V \end{bmatrix} = \widehat{\mathcal{L}} \begin{bmatrix} U \\ V \end{bmatrix} S^2.$$

(c) *If the pencil $\mathcal{M} - \lambda\mathcal{L}$ has the Kronecker canonical form*

$$W\mathcal{M}Z = \begin{bmatrix} J_r & 0 \\ 0 & I_{2n-r} \end{bmatrix}, \quad W\mathcal{L}Z = \begin{bmatrix} I_r & 0 \\ 0 & N_{2n-r} \end{bmatrix} \tag{3.9}$$

where W, Z are nonsingular, J_r is a Jordan matrix corresponding to the finite eigenvalues of $\mathcal{M} - \lambda\mathcal{L}$ and N_{2n-r} is a nilpotent Jordan matrix corresponding to the infinite eigenvalues of $\mathcal{M} - \lambda\mathcal{L}$, then there exists a nonsingular matrix $\widehat{W}$ such that

$$\widehat{W}\widehat{\mathcal{M}}Z = \begin{bmatrix} J_r^2 & 0 \\ 0 & I_{2n-r} \end{bmatrix}, \quad \widehat{W}\widehat{\mathcal{L}}Z = \begin{bmatrix} I_r & 0 \\ 0 & N_{2n-r}^2 \end{bmatrix}. \tag{3.10}$$

Proof. (a) Since $[\mathcal{M}_*, \mathcal{L}_*] \in \mathcal{N}(\mathcal{M}, \mathcal{L})$ implies that $\mathcal{M}_*\mathcal{L} = \mathcal{L}_*\mathcal{M}$, it follows from (3.8) that

$$\begin{aligned} \widehat{\mathcal{M}}J\widehat{\mathcal{M}}^\top &= \mathcal{M}_*\mathcal{M}J\mathcal{M}^\top\mathcal{M}_*^\top = \mathcal{M}_*\mathcal{L}J\mathcal{L}^\top\mathcal{M}_*^\top \\ &= (\mathcal{L}_*\mathcal{M})J(\mathcal{M}^\top\mathcal{L}_*^\top) = \mathcal{L}_*\mathcal{L}J\mathcal{L}^\top\mathcal{L}_*^\top = \widehat{\mathcal{L}}J\widehat{\mathcal{L}}^\top, \end{aligned}$$

implying that $\widehat{\mathcal{M}} - \lambda\widehat{\mathcal{L}}$ is T-symplectic.

(b) Again using $\mathcal{M}_*\mathcal{L} = \mathcal{L}_*\mathcal{M}$, we have

$$\mathcal{M}_*\mathcal{L}\begin{bmatrix} U \\ V \end{bmatrix} = \mathcal{L}_*\mathcal{M}\begin{bmatrix} U \\ V \end{bmatrix} S,$$

and hence

$$\begin{aligned} \widehat{\mathcal{M}}\begin{bmatrix} U \\ V \end{bmatrix} &= \mathcal{M}_*\mathcal{M}\begin{bmatrix} U \\ V \end{bmatrix} = \mathcal{M}_*\mathcal{L}\begin{bmatrix} U \\ V \end{bmatrix} S = \mathcal{L}_*\mathcal{M}\begin{bmatrix} U \\ V \end{bmatrix} S \\ &= \mathcal{L}_*\mathcal{L}\begin{bmatrix} U \\ V \end{bmatrix} S^2 = \widehat{\mathcal{L}}\begin{bmatrix} U \\ V \end{bmatrix} S^2. \end{aligned}$$

(c) Let

$$\widetilde{\mathcal{M}}_* = \begin{bmatrix} J_r & 0 \\ 0 & I_{2n-r} \end{bmatrix} W, \quad \widetilde{\mathcal{L}}_* = \begin{bmatrix} I_r & 0 \\ 0 & N_{2n-r} \end{bmatrix} W.$$

We then have

$$\begin{aligned} \widetilde{\mathcal{M}}_*\mathcal{L}Z &= \begin{bmatrix} J_r & 0 \\ 0 & I_{2n-r} \end{bmatrix} W\mathcal{L}Z = \begin{bmatrix} J_r & 0 \\ 0 & I_{2n-r} \end{bmatrix}\begin{bmatrix} I_r & 0 \\ 0 & N_{2n-r} \end{bmatrix} = \begin{bmatrix} J_r & 0 \\ 0 & N_{2n-r} \end{bmatrix}, \\ \widetilde{\mathcal{L}}_*\mathcal{M}Z &= \begin{bmatrix} I_r & 0 \\ 0 & N_{2n-r} \end{bmatrix} W\mathcal{M}Z = \begin{bmatrix} I_r & 0 \\ 0 & N_{2n-r} \end{bmatrix}\begin{bmatrix} J_r & 0 \\ 0 & I_{2n-r} \end{bmatrix} = \begin{bmatrix} J_r & 0 \\ 0 & N_{2n-r} \end{bmatrix}. \end{aligned}$$

As Z is nonsingular, this implies that $\left[\widetilde{\mathcal{M}}_*, \widetilde{\mathcal{L}}_*\right] \in \mathcal{N}(\mathcal{M}, \mathcal{L})$. Notice that (3.9) implies that $\mathcal{M} - \lambda\mathcal{L}$ is regular and $\operatorname{rank}\begin{bmatrix} \mathcal{M} \\ -\mathcal{L} \end{bmatrix} = 2n$. Thus $[\mathcal{M}_*, \mathcal{L}_*]$ and $\left[\widetilde{\mathcal{M}}_*, \widetilde{\mathcal{L}}_*\right]$ form two different bases of the null set $\mathcal{N}(\mathcal{M}, \mathcal{L})$. Consequently, there exists a nonsingular matrix $\widetilde{W}$ such that $\left[\widetilde{\mathcal{M}}_*, \widetilde{\mathcal{L}}_*\right] = \widetilde{W}\left[\mathcal{M}_*, \mathcal{L}_*\right]$. It can be verified easily that $\widetilde{W}$ satisfies (3.10). $\square$

It is easy to verify that NME (3.7) has a symmetric nonsingular solution X if and only if X satisfies

$$\mathcal{M}\begin{bmatrix} I \\ X \end{bmatrix} = \mathcal{L}\begin{bmatrix} I \\ X \end{bmatrix} S$$

for some $S \in \mathbb{C}^{n\times n}$, where

$$\mathcal{M} \equiv \begin{bmatrix} A_{d1}^{\top} & 0 \\ -A_{d0} & -I \end{bmatrix}, \quad \mathcal{L} \equiv \begin{bmatrix} 0 & I \\ A_{d1} & 0 \end{bmatrix}.$$

Note that $\mathcal{M} - \lambda\mathcal{L}$ is in the second standard symplectic form (SSF_2) [64].

For the convergence of the SDA, we have the following adaptation of [64, Theorem 4.1]:

Theorem 3.1.3. *Let X be a symmetric invertible solution of (3.7) and let $S = X^{-1}A_{d1}$. Then the matrix sequences $\{R_k\}$, $\{Q_k\}$ and $\{P_k\}$ generated by the SDA_2 satisfy*
(a) $R_k = (X - P_k)S^{2^k}$;
(b) $Q_k - P_k = (X - P_k) + R_k^{\top}(X - P_k)^{-1}R_k$;
(c) $Q_k - X = (S^{\top})^{2^k}(X - P_k)S^{2^k}$;
provided that all the required inverses of $Q_k - P_k$ exist.

Proof. We shall apply mathematical induction. With $R_0 = A_{d1}^{\top}$, $Q_0 = -A_{d0}$ and $P_0 = 0$, denote

$$\mathcal{M}_k \equiv \begin{bmatrix} R_k & 0 \\ Q_k & -I \end{bmatrix}, \quad \mathcal{L}_k \equiv \begin{bmatrix} -P_k & I \\ R_k^{\top} & 0 \end{bmatrix}.$$

For $k = 1$, the NME in (3.7) implies the invertibility of

$$\begin{bmatrix} X & R_0 \\ R_0^{\top} & Q_0 \end{bmatrix} = \begin{bmatrix} I & 0 \\ R_0^{\top}X^{-1} & I \end{bmatrix}\begin{bmatrix} X & 0 \\ 0 & X \end{bmatrix}\begin{bmatrix} I & X^{-1}R_0 \\ 0 & I \end{bmatrix}.$$

Further computation yields

$$\begin{bmatrix} I & -R_0Q_0^{-1} \\ 0 & I \end{bmatrix}\begin{bmatrix} X & R_0 \\ R_0^{\top} & Q_0 \end{bmatrix}\begin{bmatrix} I & 0 \\ -Q_0^{-1}R_0^{\top} & I \end{bmatrix} = \begin{bmatrix} X - R_0Q_0^{-1}R_0^{\top} & 0 \\ 0 & Q_0 \end{bmatrix}.$$

Consequently,

$$X - P_1 = X - R_0Q_0^{-1}R_0^{\top} \tag{3.11}$$

is invertible, as required in (b).

From (3.7), it is easy to verify that X satisfies

$$\mathcal{M}_0\begin{bmatrix} I \\ X \end{bmatrix} = \mathcal{L}_0\begin{bmatrix} I \\ X \end{bmatrix} S$$

with $S = X^{-1}R_0$. Since $\mathcal{M}_1 - \lambda\mathcal{L}_1$ is a doubling transformation of $\mathcal{M}_0 - \lambda\mathcal{L}_0$. Part (b) in Theorem 3.1.2 implies

$$\mathcal{M}_1\begin{bmatrix} I \\ X \end{bmatrix} = \mathcal{L}_1\begin{bmatrix} I \\ X \end{bmatrix} S^2.$$

The blocks in the above equation yield

$$R_1 = (X - P_1)S^2, \quad Q_1 - X = R_1^\top S^2.$$

Together with (3.11), these imply the invertibility of

$$Q_1 - P_1 = (X - P_1) + R_1^\top (X - P_1)^{-1} R_1, \quad Q_1 - X = (S^\top)^2 (X - P_1) S^2. \quad (3.12)$$

We have proved the theorem for $k = 1$.

With the Theorem holding for all positive integers up to k, we shall prove the case for $k+1$. Since $Q_k - P_k$ is assumed to be invertible, it follows that R_{k+1}, P_{k+1} and Q_{k+1} are well defined. Similar to the proof of (3.11), the equality in (b) implies the invertibility of

$$X - P_{k+1} = (X - P_k) - R_k (Q_k - P_k)^{-1} R_k^\top.$$

On the other hand, since $\mathcal{M}_{j+1} - \lambda \mathcal{L}_{j+1}$ is a doubling transformation of $\mathcal{M}_j - \lambda \mathcal{L}_j$ for $j = 0, 1, \ldots, k$, repeat application of Part (b) in Theorem 3.1.2 implies

$$\mathcal{M}_{k+1} \begin{bmatrix} I \\ X \end{bmatrix} = \mathcal{L}_{k+1} \begin{bmatrix} I \\ X \end{bmatrix} S^{2^{k+1}}.$$

This, following the same argument leading to (3.12), implies the invertibility of

$$\begin{aligned} Q_{k+1} - P_{k+1} &= (X - P_{k+1}) + R_{k+1}^\top (X - P_{k+1})^{-1} R_{k+1}, \\ Q_{k+1} - X &= (S^\top)^{2^{k+1}} (X - P_{k+1}) S^{2^{k+1}}. \end{aligned}$$

This completes the proof for the $k+1$ case, and the induction argument. □

Convergence to the unique symmetric stable solution X_s, which the SDA seeks, is summarized in the following Corollary.

Corollary 3.1.4. *When S is stable, $R_k \to 0$ and $Q_k \to X$ quadratically as $k \to \infty$.*

3.2 SAW simulation

Quoted from [43], we consider the generalized eigenvalue problem (GEP) of the form

$$\begin{bmatrix} M_1 & G \\ F^\top & 0 \end{bmatrix} \begin{bmatrix} \psi_i \\ \psi_\ell \end{bmatrix} + \lambda \begin{bmatrix} 0 & F \\ G^\top & M_2 \end{bmatrix} \begin{bmatrix} \psi_i \\ \psi_\ell \end{bmatrix} = 0, \quad (3.13)$$

where $M_1^\top = M_1 \in \mathbb{C}^{n \times n}$, $M_2^\top = M_2 \in \mathbb{C}^{m \times m}$, F and $G \in \mathbb{C}^{n \times m}$ with $m \ll n$. If M_1 and M_2 are nonsingular, then (3.13) can be reduced as the T-palindromic quadratic eigenvalue problem (TPQEP) of the form

$$\mathcal{P}(\lambda)x \equiv (\lambda^2 A_1^\top + \lambda A_0 + A_1)x = 0, \quad (3.14)$$

where

$$\begin{aligned} x &= \psi_\ell, & \psi_i &= -M_1^{-1}(\lambda F + G)\psi_\ell, \\ A_1 &= F^\top M_1^{-1} G, & A_0 &= F^\top M_1^{-1} F + G^\top M_1^{-1} G - M_2; \end{aligned}$$

or

$$x = \psi_i, \qquad \psi_\ell = -\lambda^{-1} M_2^{-1}(F^\top + \lambda G^\top)\psi_i,$$
$$A_1 = G M_2^{-1} F^\top, \quad A_0 = F M_2^{-1} F^\top + G M_2^{-1} G^\top - M_1. \tag{3.15}$$

By taking the transpose of $\mathcal{P}(\lambda)$ in (3.14) and multiplying it by $1/\lambda^2$ it is easily seen that the eigenvalues of $\mathcal{P}(\lambda)$ appear in the reciprocal pairs $(\lambda, 1/\lambda)$ (including 0 and ∞). Since the nullity of $A_1 = GM_2^{-1}F^\top$ in (3.15) is larger or equal to $n-m$, $\mathcal{P}(\lambda)$ in (3.14) with A_0 and A_1 defined in (3.15) has $n-m$ trivial zero and infinite eigenvalues which are not interested. We are only interested in finding $2m (\ll 2n)$ nontrivial eigenpairs of $\mathcal{P}(\lambda)$.

The GEP (3.13) can be solved by traditional methods such as QZ and Arnoldi method, but there is no guarantee that the computed eigenvalues split equally inside and outside the unit circle [42]. For solving TPQEP (3.14) with small and dense matrices A_0 and A_1, some pioneering works [66, 67] have been done for preserving the reciprocity of the eigenvalues, basing on a good linearization of (3.14) which transforms (3.14) into the form $\lambda Z^\top + Z$. Some structure-preserving methods [76, 77] were proposed for solving $(\lambda Z^\top + Z)u = 0$. A structure-preserving doubling algorithm for solving (3.14) was developed in [18] via the computation of a solvent of a nonlinear matrix equation associated with (3.14). Another structure-preserving algorithm based on $(\mathcal{S} + \mathcal{S}^{-1})$-transform [63] and Patel's approach [74] was developed in [41]. For problems with large and sparse matrices A_0 and A_1, a structure-preserving algorithm using $(\mathcal{S} + \mathcal{S}^{-1})$-transform and implicity-restarted shift-and-invert Arnoldi method was also developed for searching eigenvalues in a specified region of interests [41].

An accurate and efficient eigensolver which preserves the reciprocal relationship of the associated eigenpairs is needed. In this section, we would like to compare the accuracy and computational costs of the above mentioned algorithms for computing reciprocal eigenpairs in a SAW device [86]. The SAW filter plays an important role in telecommunication filters [12, 73] and sensor technologies [3] etc. These filters are built on the physical property of piezoelectric materials, that electrical charges induce mechanical deformations and vice versa. The main component (or cell) of a SAW filter composes of a piezoelectric substrate and the input and output interdigital transducers (IDT). An input electrical signal from the input IDT produces a surface acoustic wave, traveling through periodically arranged electrodes and the output IDT picks up the output electrical signal. Depending on the material properties of the piezoelectric substrate (PZT) and the metallic electrode, and the gap length between the electrodes, frequencies in a desired range can be stopped or filtered off. In the filter design, it is important to know the stop band width and the center frequency f_c of the filter where $f_c = \frac{v_s}{\lambda_s}$ here v_s and λ_s are the wave velocity and wave length of the incident wave. The center frequency and stop band width can be determined by experiments or computation. In a computational approach, the dispersion diagram needs to be generated in which a GEP of the form (3.13) associated with each frequency in the search range has to be solved [42].

We next introduce a finite element model for a simple SAW resonator. For more finite element simulations of piezoelectric devices in two dimension (2D) and three dimension (3D), one can refer to the works done by Allik, Koshiba, Lerch, and Buchner and Mohamed [1, 11, 54, 56]. Then the four structure-preserving algorithms developed in [18, 41] can be applied to solve the TPQEP (3.14) and the GEP (3.13) resulted from our FEM model.

3.2.1 Surface Wave Propagation

To model the wave propagation in a SAW device, we assume that a large number of electrodes are placed equally-spaced along a straight line on the PZT substrate. According to the Floquet-Bloch theory, one can reduce the problem to a single cell domain with one electrode by assuming the wave ψ is quasi-periodic of the form

$$\psi(x_1, x_2) = \psi_p(x_1, x_2)e^{(\alpha+\imath\beta)x_1}, \quad \psi_p(x_1 + p, x_2) = \psi_p(x_1, x_2),$$

where x_1 is the wave propagation direction, p is the length of the unit cell (i.e. the periodic interval), α and β are the attenuation and phase shift along the wave propagation direction, respectively.

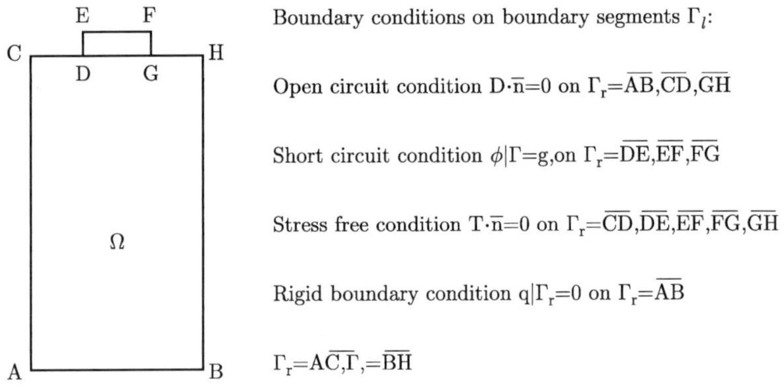

Figure 3.2 A 2D single cell domain of a LSAW resonator and boundary conditions

Let Ω denote the piezoelectric substrate with a single IDT as shown in Fig. 3.2, and Γ_ℓ and Γ_r denote the left and right boundary segments of Ω, respectively. For the general anisotropic PZT substrates, under the assumption of linear piezoelectric coupling, the elastic and electric fields interact following the general material constitution law below

$$T = c^E S - e^\top E, \quad D = eS + \varepsilon^S E, \tag{3.16}$$

where vectors T, S, D and E are the mechanical stress, strain, dielectric displacement and the electric field, respectively, and the matrices c^E, ε^S and e are the elasticity constant, dielectric constant and piezoelectric constant matrices measured at constant electric and constant strain fields at constant temperature. By applying the virtual work principle to (3.16), the equilibrium state satisfies the following equation:

$$\int_\Omega (\delta S)^\top \left[c^E S + e^\top(\nabla\phi)\right] dV + \int_\Omega (\nabla\delta\phi)^\top \left[eS - \varepsilon^S(\nabla\phi)\right] dV + \int_\Omega (\delta u)^\top \rho\ddot{u}\, dV$$
$$= \int_{\Gamma_l\cup\Gamma_r} \left[(\delta u)^\top (T\cdot\vec{n}) + (\delta\phi)^\top (D\cdot\vec{n})\right] dA, \tag{3.17}$$

where ρ is the mass density, $u = [u_1, u_2, u_3]^\top$ is the displacement vector, ϕ is the electric potential that satisfies $\nabla\phi = E$, $S = [\frac{\partial u_1}{\partial x}, \frac{\partial u_2}{\partial y}, \frac{\partial u_3}{\partial z}, \frac{\partial u_2}{\partial z} + \frac{\partial u_3}{\partial y}, \frac{\partial u_3}{\partial x} + \frac{\partial u_1}{\partial z}, \frac{\partial u_1}{\partial y} + \frac{\partial u_2}{\partial x}]^\top$, and δu, $\delta\phi$, δS are virtual displacement, potential and strain vectors, respectively. Let the notation $\psi = [u^\top, \phi]^\top$ and the subscript i, ℓ and r refer to nodal point index in the interior, the left boundary and the right boundary of the domain Ω, respectively. Using the periodic boundary conditions, proposed by Buchner [11], $T_r \cdot n_r = -\gamma T_\ell \cdot n_\ell$, $D_r \cdot n_r = -\gamma D_\ell \cdot n_\ell$ with $\gamma = e^{-(\alpha+\imath\beta)}$, the finite element discretization to (3.17) on the domain Ω [42] can be written in the following matrix form

$$C(\omega)\psi \equiv [K - \omega^2 M + \imath\omega(\kappa_1 K + \kappa_2 M)]\psi = 0, \tag{3.18}$$

where κ_1, $\kappa_2 > 0$ are the viscous damping and mass damping respectively. By ordering the nodal unknown ψ according the order of subscripts ℓ, i and r, the matrices K and M, and the vector ψ can be partitioned as following:

$$K = \begin{bmatrix} K_{\ell\ell} & K_{i\ell}^\top & 0 \\ K_{i\ell} & K_{ii} & K_{ir} \\ 0 & K_{ir}^\top & K_{rr} \end{bmatrix}, \quad M = \begin{bmatrix} M_{\ell\ell} & M_{i\ell}^\top & 0 \\ M_{i\ell} & M_{ii} & M_{ir} \\ 0 & M_{ir}^\top & M_{rr} \end{bmatrix},$$

where $K_{ii}, M_{ii} \in \mathbb{R}^{n\times n}$, $K_{\ell\ell}, K_{rr}, M_{\ell\ell}, M_{rr} \in \mathbb{R}^{m\times m}$, $K_{i\ell}, K_{ir}, M_{i\ell}, M_{ir} \in \mathbb{R}^{n\times m}$, and $\psi = [\psi_\ell^\top, \psi_i^\top, \psi_r^\top]^\top$ with $\psi_i \in \mathbb{C}^n$, $\psi_\ell, \psi_r \in \mathbb{C}^m$ $(m \ll n)$. Obviously the matrix $C(\omega)$ in (3.18) can also be partitioned into

$$C(\omega) \equiv C \equiv \begin{bmatrix} C_{\ell\ell} & C_{i\ell}^\top & 0 \\ C_{i\ell} & C_{ii} & C_{ir} \\ 0 & C_{ir}^\top & C_{rr} \end{bmatrix}$$

By setting $\psi_r = \lambda\psi_\ell$, (3.18) leads to the generalized eigenvalue problem

$$\left(\begin{bmatrix} C_{ii} & C_{i\ell} \\ C_{ir}^\top & 0 \end{bmatrix} - \lambda \begin{bmatrix} 0 & C_{ir} \\ C_{i\ell}^\top & C_{bb} \end{bmatrix} \right) \begin{bmatrix} \psi_i \\ \psi_\ell \end{bmatrix} = 0,$$

where $C_{bb} := C_{\ell\ell} + C_{rr}$.

Since the viscosity is small for PZT substrates and metals in SAW devices, the attenuation factor α of surface waves is close to zero. As a result, the propagation factor λ are generally near the unit circle thereafter denoted by $\mathbb{U}$. Furthermore, for frequency ω in the stopping band, the frequency shift parameter β shall be close to π when the periodic interval p (i.e. the domain width here) equals to half of the incident wave length λ_s. Therefore, we are interesting in finding λ close to $\mathbb{U}$, especially for those are near -1 on the complex plane.

3.3 Nano research

Quoted from [33], we shall study the nonlinear matrix equation

$$X + A^\top X^{-1} A = Q + i\eta I,$$

where $A \in \mathbb{R}^{n\times n}$, $Q = Q^\top \in \mathbb{R}^{n\times n}$ and $\eta \geq 0$. The equation arises from the non-equilibrium Green's function approach for treating quantum transport in nanodevices, where the system Hamiltonian is a semi-infinite or bi-infinite real symmetric matrix with special structures [4, 25, 51, 53, 81]. A first systematic mathematical study of the equation has already been undertaken in [35].

For the bi-infinite case, the Green's function corresponding to the scattering region $G_S \in \mathbb{C}^{n_s\times n_s}$, in which the nano scientists are interested, satisfies the relation [24, 51]

$$G_S = \left((\mathcal{E} + i0^+)I - H_S - C_{L,S}^\top G_{L,S} C_{L,S} - D_{S,R} G_{S,R} D_{S,R}^\top\right)^{-1},$$

where $\mathcal{E}$ is energy, a real number that may be negative, $H_S \in \mathbb{R}^{n_s\times n_s}$ is the Hamiltonian for the scattering region, $C_{L,S} \in \mathbb{R}^{n_\ell\times n_s}$ and $D_{S,R} \in \mathbb{R}^{n_s\times n_r}$ represent the coupling with the scattering region for the left lead and the right lead, respectively, $G_{L,S} \in \mathbb{C}^{n_\ell\times n_\ell}$ and $G_{S,R} \in \mathbb{C}^{n_r\times n_r}$ are special solutions of the matrix equations

$$G_{L,S} = \left((\mathcal{E} + i0^+)I - B_L - A_L^\top G_{L,S} A_L\right)^{-1}, \qquad (3.19)$$

$$G_{S,R} = \left((\mathcal{E} + i0^+)I - B_R - A_R G_{S,R} A_R^\top\right)^{-1}, \qquad (3.20)$$

with A_L, $B_L = B_L^\top \in \mathbb{R}^{n_\ell\times n_\ell}$ and A_R, $B_R = B_R^\top \in \mathbb{R}^{n_r\times n_r}$. Since (3.19) and (3.20) are of the same type, we only need to study (3.19), and we simplify the notation n_ℓ to n. In nano research, one is mainly interested in the values of $\mathcal{E}$ for which $G_{L,S}$ in (3.19) has a nonzero imaginary part [53].

For each fixed $\mathcal{E}$, we replace "0^+" in (3.19) by a sufficiently small positive number η and consider the matrix equation

$$X = \left((\mathcal{E} + i\eta)I - B_L - A_L^\top X A_L\right)^{-1}. \qquad (3.21)$$

It is shown in [35] that the required special solution $G_{L,S}$ of (3.19) is given by $G_{L,S} = \lim_{\eta\to 0^+} G_{L,S}(\eta)$ with $X = G_{L,S}(\eta)$ being the unique complex symmetric solution of (3.21) such that $\rho(G_{L,S}(\eta)A_L) < 1$, where $\rho(\cdot)$ denotes the spectral radius. Thus $G_{L,S}$ is a special complex symmetric solution of $X = \left(\mathcal{E}I - B_L - A_L^\top X A_L\right)^{-1}$ with $\rho(G_{L,S}A_L) \leq 1$.

The question as to when $G_{L,S}$ has a nonzero imaginary part is answered in the following result from [35], where $\mathbb{T}$ denotes the unit circle.

Theorem 3.3.1. *For $\lambda \in \mathbb{T}$, let the eigenvalues of $\psi_L(\lambda) = B_L + \lambda A_L + \lambda^{-1}A_L^\top$ be $\mu_{L,1}(\lambda) \leq \cdots \leq \mu_{L,n}(\lambda)$. Let*

$$\Delta_{L,i} = \left[\min_{|\lambda|=1} \mu_{L,i}(\lambda),\ \max_{|\lambda|=1} \mu_{L,i}(\lambda)\right],$$

and $\Delta_L = \bigcup_{i=1}^{n} \Delta_{L,i}$*. Then* $G_{L,S}$ *is a real symmetric matrix if* $\mathcal{E} \notin \Delta_L$*. When* $\mathcal{E} \in \Delta_L$*, the quadratic pencil* $\lambda^2 A_L^\top - \lambda(\mathcal{E}I - B_L) + A_L$ *has eigenvalues on* $\mathbb{T}$*. If all these eigenvalues on* $\mathbb{T}$ *are simple and nonreal, then* $G_{L,S}$ *has a nonzero imaginary part.*

By replacing X in (3.21) with X^{-1}, we get the equation

$$X + A^\top X^{-1} A = Q_\eta, \tag{3.22}$$

where $A = A_L$ and $Q_\eta = Q + i\eta I$ with $Q = \mathcal{E}I - B_L$. So Q is a real symmetric matrix dependent on the parameter $\mathcal{E}$ and is usually indefinite. For $\eta > 0$, we need the stabilizing solution X of (3.22), which is the solution with $\rho(X^{-1}A) < 1$, and then $G_{L,S}(\eta) = X^{-1}$. When $\eta = 0$ and $\mathcal{E} \in \Delta_L$, it follows from Theorem 3.3.1 that the required solution $X = G_{L,S}^{-1}$ of (3.22) is only weakly stabilizing, in the sense that $\rho(X^{-1}A) = 1$.

One way to approximate $G_{L,S}$ is to take a very small $\eta > 0$ and compute $G_{L,S}(\eta)$. It is proved in [35] that the sequence $\{X_k\}$ from the basic fixed-point iteration (FPI)

$$X_{k+1} = Q_\eta - A^\top X_k^{-1} A,$$

with $X_0 = Q_\eta$, converges to $G_{L,S}(\eta)^{-1}$. It follows that the sequence $\{Y_k\}$ from the basic FPI

$$Y_{k+1} = (Q_\eta - A^\top Y_k A)^{-1},$$

with $Y_0 = Q_\eta^{-1}$, converges to $G_{L,S}(\eta)$. However, the convergence is very slow for $\mathcal{E} \in \Delta_L$ since $\rho(G_{L,S}(\eta)A) \approx 1$ for η close to 0. It is also shown in [35] that a doubling algorithm (DA) can be used to compute the desired solution $X = G_{L,S}(\eta)^{-1}$ of (3.22) efficiently for each fixed value of $\mathcal{E}$. However, in practice the desired solution needs to be computed for many different $\mathcal{E}$ values. Since the DA is not a correction method, it cannot use the solution obtained for one $\mathcal{E}$ value as an initial approximation for the exact solution at a nearby $\mathcal{E}$ value. To compute the solutions corresponding to many $\mathcal{E}$ values, it may be more efficient to use a modified FPI together with the DA. Indeed, it is suggested in [81] that the following modified FPI be used to approximate $G_{L,S}(\eta)$:

$$Y_{k+1} = \frac{1}{2}Y_k + \frac{1}{2}(Q_\eta - A^\top Y_k A)^{-1}.$$

A variant of this FPI is given in [35] to approximate $G_{L,S}(\eta)^{-1}$:

$$X_{k+1} = \frac{1}{2}X_k + \frac{1}{2}(Q_\eta - A^\top X_k^{-1} A),$$

which requires less computational work each iteration. However, the convergence analysis of these two modified FPIs has been an open problem even for the special initial matrices $Y_0 = Q_\eta^{-1}$ and $X_0 = Q_\eta$, respectively.

Our first contribution in this section is a proof of convergence (to the desired solutions) of these two modified FPIs and their generalizations, for many choices of initial matrices. Consequently, these methods can be used as correction methods.

In this process we will show that the unique stabilizing solution $X = G_{L,S}(\eta)^{-1}$ of (3.22) is also the unique solution of (3.22) with a positive definite imaginary part. It follows that the imaginary part X_I of the matrix $G_{L,S}^{-1}$ is positive semi-definite. Our second contribution in this section is a determination of the rank of X_I in terms of the number of eigenvalues on $\mathbb{T}$ of the quadratic pencil $\lambda^2 A^\top - \lambda Q + A$. Our third contribution is a structure-preserving algorithm that is applied directly on (3.22) with $\eta = 0$. In doing so, we work with real arithmetic most of the time.

3.3.1 Rank of $\mathrm{Im}(G_{L,S})$

Equation (3.22) has a unique stabilizing solution $X_\eta = G_{L,S}(\eta)^{-1}$ for any $\eta > 0$. So

$$X_\eta + A^\top X_\eta^{-1} A = Q_\eta \tag{3.23}$$

with $\rho(X_\eta^{-1}A) < 1$. We also know that X_η is complex symmetric. Write $X_\eta = X_{\eta,R} + iX_{\eta,I}$ with $X_{\eta,R}^\top = X_{\eta,R}$, $X_{\eta,I}^\top = X_{\eta,I} \in \mathbb{R}^{n\times n}$. We know from the previous section that $\mathrm{Im}(X_\eta) = X_{\eta,I} > 0$. Let

$$\varphi_\eta(\lambda) = \lambda A^\top + \lambda^{-1} A - Q_\eta.$$

By (3.23) we have the factorization $\varphi_\eta(\lambda) = \left(\lambda^{-1} I - S_\eta^\top\right) X_\eta \left(-\lambda I + S_\eta\right)$, where $S_\eta = X_\eta^{-1} A$. Let $X = \lim_{\eta\to 0^+} X_\eta = G_{L,S}^{-1}$. Then

$$X + A^\top X^{-1} A = Q \tag{3.24}$$

with $\rho(X^{-1}A) \le 1$ and $\mathrm{Im}(X) \ge 0$. Note that $\varphi_0(\lambda) = \lambda A^\top + \lambda^{-1} A - Q$ has the factorization

$$\varphi_0(\lambda) = \left(\lambda^{-1} I - S^\top\right) X \left(-\lambda I + S\right), \tag{3.25}$$

where $S = X^{-1}A$. In particular, $\varphi_0(\lambda)$ is regular, i.e., its determinant is not identically zero. In this section we will determine the rank of $\mathrm{Im}(X)$, which is the same as the rank of $\mathrm{Im}(G_{L,S})$ since $\mathrm{Im}(G_{L,S}) = \mathrm{Im}(X^{-1}) = -X^{-1}\mathrm{Im}(X)X^{-*}$.

Let

$$\mathcal{M} = \begin{bmatrix} A & 0 \\ Q & -I \end{bmatrix}, \quad \mathcal{L} = \begin{bmatrix} 0 & I \\ A^\top & 0 \end{bmatrix}. \tag{3.26}$$

Then the pencil $\mathcal{M} - \lambda\mathcal{L}$, also denoted by $(\mathcal{M}, \mathcal{L})$, is a linearization of the quadratic matrix polynomial

$$P(\lambda) = \lambda\varphi_0(\lambda) = \lambda^2 A^\top - \lambda Q + A.$$

It is easy to check that y and z are the right and left eigenvectors, respectively, corresponding to an eigenvalue λ of $P(\lambda)$ if and only if

$$\begin{bmatrix} y \\ Qy - \lambda A^\top y \end{bmatrix}, \quad \begin{bmatrix} z \\ -\bar{\lambda} z \end{bmatrix} \tag{3.27}$$

are the right and left eigenvectors of $(\mathcal{M}, \mathcal{L})$, respectively.

Theorem 3.3.2. *Suppose that λ_0 is a semi-simple eigenvalue of $\varphi_0(\lambda)$ on the unit circle $\mathbb{T}$ with multiplicity m_0 and $Y \in \mathbb{C}^{n\times m_0}$ forms an orthonormal basis of right eigenvectors corresponding to λ_0. Then, for $\eta > 0$ sufficiently small,*

$$\lambda_{j,\eta} = \lambda_0 - \frac{\lambda_0}{d_j}\eta + O(\eta^2), \quad y_{j,\eta} = Y\xi_j + O(\eta), \quad (j = 1, \ldots, m_0) \tag{3.28}$$

are perturbed eigenvalues and the associated eigenvectors of $\varphi_\eta(\lambda)$, where d_j and ξ_j, $j = 1, \ldots, m_0$, are eigenvalues and right eigenvectors of the nonsingular Hermitian matrix $iY^*(2\lambda_0 A^\top - Q)Y$.

Proof. Since $P(\lambda_0)Y = \lambda_0\varphi_0(\lambda_0)Y = 0$ with $Y^*Y = I_{m_0}$ and $|\lambda_0| = 1$, we have

$$0^* = (P(\lambda_0)Y)^* = \frac{1}{\lambda_0^2}Y^*(\lambda_0^2 A^\top - \lambda_0 Q + A).$$

It follows that Y forms an orthonormal basis for left eigenvectors of $P(\lambda)$ corresponding to λ_0. From (3.27), we obtain that the column vectors of

$$\mathcal{Y}_R = \begin{bmatrix} Y \\ QY - \lambda_0 A^\top Y \end{bmatrix} \text{ and } \mathcal{Y}_L = \begin{bmatrix} Y \\ -\bar{\lambda}_0 Y \end{bmatrix}$$

form a basis of left and right eigenspaces of $\mathcal{M} - \lambda\mathcal{L}$ corresponding to λ_0, respectively. Since λ_0 is semi-simple, the matrix

$$[Y^*, -\lambda_0 Y^*]\mathcal{L}\begin{bmatrix} Y \\ QY - \lambda_0 A^\top Y \end{bmatrix} = -Y^*(2\lambda_0 A^\top - Q)Y = -Y^*P'(\lambda_0)Y$$

is nonsingular. Let

$$\widetilde{\mathcal{Y}}_R = -\mathcal{Y}_R(Y^*P'(\lambda_0)Y)^{-1}, \quad \widetilde{\mathcal{Y}}_L = \mathcal{Y}_L.$$

Then we have

$$\widetilde{\mathcal{Y}}_L^*\mathcal{L}\widetilde{\mathcal{Y}}_R = I_{m_0}, \quad \widetilde{\mathcal{Y}}_L^*\mathcal{M}\widetilde{\mathcal{Y}}_R = \lambda_0 I_{m_0}. \tag{3.29}$$

For $\eta > 0$, sufficiently small, we consider the perturbed equation of $P(\lambda)$ by

$$P(\lambda) - \lambda i\eta I = \lambda^2 A^\top - \lambda(Q + i\eta I) + A = \lambda\varphi_\eta(\lambda).$$

Let $\mathcal{M}_\eta = \begin{bmatrix} A & 0 \\ Q + i\eta I & -I \end{bmatrix}$. Then $\mathcal{M}_\eta - \lambda\mathcal{L}$ is a linearization of $\lambda\varphi_\eta(\lambda)$. By (3.29) and [78, Chapter VI, Theorem 2.12] there are $\widehat{\mathcal{Y}}_R$ and $\widehat{\mathcal{Y}}_L$ such that $\begin{bmatrix} \widetilde{\mathcal{Y}}_R & \widehat{\mathcal{Y}}_R \end{bmatrix}$ and $\begin{bmatrix} \widetilde{\mathcal{Y}}_L & \widehat{\mathcal{Y}}_L \end{bmatrix}$ are nonsingular and

$$\begin{bmatrix} \widetilde{\mathcal{Y}}_L^* \\ \widehat{\mathcal{Y}}_L^* \end{bmatrix}\mathcal{M}\begin{bmatrix} \widetilde{\mathcal{Y}}_R & \widehat{\mathcal{Y}}_R \end{bmatrix} = \begin{bmatrix} \lambda_0 I_{m_0} & 0 \\ 0 & \widehat{\mathcal{M}} \end{bmatrix}, \quad \begin{bmatrix} \widetilde{\mathcal{Y}}_L^* \\ \widehat{\mathcal{Y}}_L^* \end{bmatrix}\mathcal{L}\begin{bmatrix} \widetilde{\mathcal{Y}}_R & \widehat{\mathcal{Y}}_R \end{bmatrix} = \begin{bmatrix} I_{m_0} & 0 \\ 0 & \widehat{\mathcal{L}} \end{bmatrix}.$$

Then, by [78, Chapter VI, Theorem 2.15] the column vectors of $\widetilde{\mathcal{Y}}_R + O(\eta)$ span the right eigenspace of $(\mathcal{M}_\eta, \mathcal{L})$ corresponding to $(\lambda_0 I_{m_0} + E_{11} + O(\eta^2), I_{m_0})$, where

$$E_{11} = \widetilde{\mathcal{Y}}_L^* \begin{bmatrix} 0 & 0 \\ i\eta I & 0 \end{bmatrix} \widetilde{\mathcal{Y}}_R = \lambda_0 Y^*(i\eta I)Y(Y^*P'(\lambda_0)Y)^{-1} = -\lambda_0\eta(iY^*(2\lambda_0 A^\top - Q)Y)^{-1}. \tag{3.30}$$

The matrix $iY^*(2\lambda_0 A^\top - Q)Y$ in (3.30) is Hermitian since

$$\begin{aligned} iY^*(2\lambda_0 A^\top - Q)Y &= iY^*\varphi_0(\lambda_0)Y + i\lambda_0 Y^* A^\top Y - i\overline{\lambda}_0 Y^* AY \\ &= i\lambda_0 Y^* A^\top Y + (i\lambda_0 Y^* A^\top Y)^*. \end{aligned} \tag{3.31}$$

Let d_j and ξ_j, for $j = 1, \ldots, m_0$, be eigenvalues and associated eigenvectors of $iY^*(2\lambda_0 A^\top - Q)Y$. Then, for each $j \in \{1, 2, \ldots, m_0\}$, the perturbed eigenvalue $\lambda_{j,\eta}$ and the associated eigenvector $\zeta_{j,\eta}$ of $\mathcal{M}_\eta - \lambda\mathcal{L}$ with $\lambda_{j,\eta}|_{\eta=0} = \lambda_0$ can be expressed by

$$\lambda_{j,\eta} = \lambda_0 - \frac{\lambda_0}{d_j}\eta + O(\eta^2), \quad \zeta_{j,\eta} = \mathcal{Y}_R \xi_j + O(\eta). \tag{3.32}$$

The second equation in (3.28) follows from (3.32). □

Lemma 3.3.3. *Suppose that*

$$Z_\eta - R_\eta^* Z_\eta R_\eta = \eta W_\eta, \ \textit{for } \eta > 0, \tag{3.33}$$

where $W_\eta \in \mathbb{C}^{m\times m}$ is positive definite, $R_\eta = e^{i\theta} I_m + \eta E_\eta$ with $\theta \in [0, 2\pi]$ fixed, and $E_\eta \in \mathbb{C}^{m\times m}$ is uniformly bounded such that $\rho(R_\eta) < 1$. Then Z_η is positive definite. Furthermore, if Z_η converges to Z_0 and W_η converges to a positive definite matrix W_0 as $\eta \to 0^+$, then Z_0 is also positive definite.

Proof. Since $\rho(R_\eta) < 1$ and ηW_η is positive definite, it is well known that Z_η is uniquely determined by (3.33) and is positive definite.

Since E_η is bounded, we have from (3.33) that

$$\eta W_\eta = Z_\eta - \left(e^{-i\theta} I_m + \eta E_\eta^*\right) Z_\eta \left(e^{i\theta} I_m + \eta E_\eta\right) = -\eta e^{i\theta} E_\eta^* Z_\eta - \eta e^{-i\theta} Z_\eta E_\eta + O(\eta^2).$$

This implies that

$$W_\eta = -e^{i\theta} E_\eta^* Z_\eta - e^{-i\theta} Z_\eta E_\eta + O(\eta). \tag{3.34}$$

If Z_η converges to Z_0 as $\eta \to 0^+$, then Z_0 is positive semi-definite. To prove that Z_0 is positive definite, it suffices to show that Z_0 is nonsingular. Suppose that $x \in \mathbb{C}^m$ such that $Z_0 x = 0$. Then we have $Z_\eta x \to 0$ and $x^* Z_\eta \to 0$ as $\eta \to 0^+$. Multiplying (3.34) by x^* and x from the left and right, respectively, we have

$$x^* W_\eta x = -e^{i\theta} x^* E_\eta^* Z_\eta x - e^{-i\theta} x^* Z_\eta E_\eta x + O(\eta) \to 0, \quad \text{as } \eta \to 0^+.$$

Thus $x = 0$ because W_η converges to W_0 and W_0, thus Z_0, are positive definite. □

Theorem 3.3.4. *The number of eigenvalues (counting multiplicities) of* $\varphi_0(\lambda)$ *on* $\mathbb{T}$ *must be even, say* $2m$. *Let* $X = \lim_{\eta \to 0^+} X_\eta$ *be invertible and write* $X = X_R + iX_I$ *with* $X_R = X_R^\top, X_I = X_I^\top \in \mathbb{R}^{n \times n}$. *Then*

(a) $\operatorname{rank}(X_I) \leq m$;
(b) $\operatorname{rank}(X_I) = m$, *if all eigenvalues of* $\varphi_0(\lambda)$ *on* $\mathbb{T}$ *are semi-simple and* $\|S_\eta - S\|_2 = O(\eta)$ *for* $\eta > 0$ *sufficiently small;*
(c) $\operatorname{rank}(X_I) = m$, *if all eigenvalues of* $\varphi_0(\lambda)$ *on* $\mathbb{T}$ *are semi-simple and each unimodular eigenvalue of multiplicity* m_j *is preturbed to* m_j *eigenvalues (of* $\varphi_\eta(\lambda)$*) inside the unit circle or to* m_j *eigenvalues outside the unit circle.*

Proof. Consider the real quadratic pencil $P(\lambda) = \lambda\varphi_0(\lambda) = \lambda^2 A^\top - \lambda Q + A$. So $P(\lambda)$ and $\varphi_0(\lambda)$ have the same eigenvalues on $\mathbb{T}$. If $\lambda_0 \neq \pm 1$ is an eigenvalue of $P(\lambda)$ on $\mathbb{T}$ with multiplicity m_0, then so is $\overline{\lambda}_0$. Thus the total number of nonreal eigenvalues of $P(\lambda)$ on $\mathbb{T}$ must be even. Now the quadratic pencil $P_\eta(\lambda) = \lambda^2 A^\top - \lambda(Q + i\eta I) + A$ is $\top$-palindromic, and it has no eigenvalues on $\mathbb{T}$ for any $\eta \neq 0$ [35]. If 1 (or -1) is an eigenvalue of $P(\lambda)$ with multiplicity r and Q in $P(\lambda)$ is perturbed to $Q + i\eta I$, then half of these r eigenvalues are perturbed to the inside of $\mathbb{T}$ and the other half are perturbed to the outside of $\mathbb{T}$. This means that r must be even. Thus the total number of eigenvalues of $\varphi_0(\lambda)$ on $\mathbb{T}$ is also even and is denoted by $2m$.

(a) By $X_\eta + A^\top X_\eta^{-1} A = Q_\eta$ we have

$$\begin{aligned} i(Q_\eta^* - Q_\eta) &= i(X_\eta^* - X_\eta) - iA^\top (X_\eta^{-1} - X_\eta^{-*})A \\ &= i(X_\eta^* - X_\eta) - (X_\eta^{-1}A)^* i(X_\eta^* - X_\eta)(X_\eta^{-1}A). \end{aligned}$$

Thus

$$K_\eta - S_\eta^* K_\eta S_\eta = 2\eta I, \tag{3.35}$$

where $K_\eta = i\left(X_\eta^* - X_\eta\right) = 2X_{\eta,I}$. Note that the eigenvalues of $S_\eta = X_\eta^{-1}A$ are the eigenvalues of $P_\eta(\lambda)$ inside $\mathbb{T}$. Since $X = \lim_{\eta \to 0^+} X_\eta$ is invertible, we have $S = X^{-1}A = \lim_{\eta \to 0^+} S_\eta$. Let

$$S = V_0 \begin{bmatrix} R_{0,1} & 0 \\ 0 & R_{0,2} \end{bmatrix} V_0^{-1} \tag{3.36}$$

be a spectral resolution of S, where $R_{0,1} \in \mathbb{C}^{m \times m}$ and $R_{0,2} \in \mathbb{C}^{(n-m)\times(n-m)}$ are upper-triangular with $\sigma(R_{0,1}) \subseteq \mathbb{T}$ and $\sigma(R_{0,2}) \subseteq \mathbb{D} \equiv \{\lambda \in \mathbb{C} | \, |\lambda| < 1\}$, and $V_0 = [V_{0,1}, V_{0,2}]$ with $V_{0,1} \in \mathbb{C}^{n \times m}$ and $V_{0,2} \in \mathbb{C}^{n \times (n-m)}$ having unit column vectors. It follows from [78, Chapter V, Theorem 2.8] that there is a nonsingular matrix $V_\eta = [V_{\eta,1}, V_{\eta,2}]$ with $V_{\eta,1} \in \mathbb{C}^{n \times m}$ and $V_{\eta,2} \in \mathbb{C}^{n \times (n-m)}$ such that

$$S_\eta = V_\eta \begin{bmatrix} R_{\eta,1} & 0 \\ 0 & R_{\eta,2} \end{bmatrix} V_\eta^{-1}, \tag{3.37}$$

and $R_{\eta,1} \to R_{0,1}$, $R_{\eta,2} \to R_{0,2}$, and $V_\eta \to V_0$, as $\eta \to 0^+$.

From (3.35) and (3.37) we have

$$V_\eta^* K_\eta V_\eta - \begin{bmatrix} R_{\eta,1}^* & 0 \\ 0 & R_{\eta,2}^* \end{bmatrix} V_\eta^* K_\eta V_\eta \begin{bmatrix} R_{\eta,1} & 0 \\ 0 & R_{\eta,2} \end{bmatrix} = 2\eta V_\eta^* V_\eta. \tag{3.38}$$

Let

$$H_\eta = V_\eta^* K_\eta V_\eta = \begin{bmatrix} H_{\eta,1} & H_{\eta,3} \\ H_{\eta,3}^* & H_{\eta,2} \end{bmatrix}, \quad V_\eta^* V_\eta = \begin{bmatrix} W_{\eta,1} & W_{\eta,3} \\ W_{\eta,3}^* & W_{\eta,2} \end{bmatrix}. \tag{3.39}$$

Then (3.38) becomes

$$H_{\eta,1} - R_{\eta,1}^* H_{\eta,1} R_{\eta,1} = 2\eta W_{\eta,1}, \tag{3.40a}$$
$$H_{\eta,2} - R_{\eta,2}^* H_{\eta,2} R_{\eta,2} = 2\eta W_{\eta,2}, \tag{3.40b}$$
$$H_{\eta,3} - R_{\eta,1}^* H_{\eta,3} R_{\eta,2} = 2\eta W_{\eta,3}. \tag{3.40c}$$

As $\eta \to 0^+$, $R_{\eta,1} \to R_{0,1}$ with $\rho(R_{0,1}) = 1$, $R_{\eta,2} \to R_{0,2}$ with $\rho(R_{0,2}) < 1$, and $W_{\eta,2}$ and $W_{\eta,3}$ are bounded. So we have $H_{\eta,2} \to 0$ from (3.40b) and $H_{\eta,3} \to 0$ from (3.40c). It follows from (3.39) that $K_\eta = 2X_{\eta,I}$ converges to $K_0 = 2X_I$ with $\text{rank}(X_I) \le m$.

(b) Suppose that eigenvalues of $\varphi_0(\lambda)$ on $\mathbb{T}$ are semi-simple and $\|S_\eta - S\|_2 = O(\eta)$ for $\eta > 0$ sufficiently small. Then we will show that $H_{\eta,1}$ in (3.40a) converges to $H_{0,1}$ with $\text{rank}(H_{0,1}) = m$. Let $\lambda_1, \ldots, \lambda_r \in \mathbb{T}$ be the distinct semi-simple eigenvalues of S with multiplicities $m_1, \ldots, m_r$, respectively. Then (3.36) can be written as

$$S = V_0 \begin{bmatrix} D_{0,1} & 0 \\ 0 & R_{0,2} \end{bmatrix} V_0^{-1}$$

where $D_{0,1} = \text{diag}\{\lambda_1 I_{m_1}, \ldots, \lambda_r I_{m_r}\}$, $V_0 = [V_{0,\lambda_1}, \ldots, V_{0,\lambda_r}, V_{0,2}]$ and $\sum_{i=1}^r m_i = m$. Now $S_\eta = S + (S_\eta - S)$ with $\|S_\eta - S\|_2 = O(\eta)$. By repeated application of [78, Chapter V, Theorem 2.8] there is a nonsingular matrix $V_\eta = [V_{\eta,\lambda_1}, \ldots, V_{\eta,\lambda_r}, V_{\eta,2}] \in \mathbb{C}^{n\times n}$ such that

$$S_\eta = V_\eta \begin{bmatrix} D_{0,1} + \eta E_{\eta,1} & 0 \\ 0 & R_{0,2} + \eta E_{\eta,2} \end{bmatrix} V_\eta^{-1}$$

and $V_\eta \to V_0$ as $\eta \to 0^+$, where $E_{\eta,1} = \text{diag}\{E_{m_1,\eta}^1, \ldots, E_{m_r,\eta}^1\}$ with $E_{m_j,\eta}^1 \in \mathbb{C}^{m_j\times m_j}$ and $E_{\eta,2} \in \mathbb{C}^{(n-m)\times(n-m)}$ are such that $\|E_{m_j,\eta}^1\|_2 = O(1)$ for $j = 1, \ldots, r$ and $\|E_{\eta,2}\|_2 = O(1)$.

Equation (3.40a) can then be written as

$$H_{\eta,1} - (D_{0,1} + \eta E_{\eta,1})^* H_{\eta,1} (D_{0,1} + \eta E_{\eta,1}) = 2\eta W_{\eta,1}. \tag{3.41}$$

Since $D_{0,1} + \eta E_{\eta,1}$ is a block diagonal matrix and all eigenvalues of its jth diagonal block converge to λ_j, with λ_j's distinct numbers on $\mathbb{T}$, we have

$$H_{\eta,1} = \text{diag}\{H_{m_1,\eta}^1, \ldots, H_{m_r,\eta}^1\} + O(\eta),$$

where $\text{diag}\{H_{m_1,\eta}^1, \ldots, H_{m_r,\eta}^1\}$ is the block diagonal of $H_{\eta,1}$. Then (3.41) gives

$$H_{m_j,\eta}^1 - (\lambda_j I_{m_j} + \eta E_{m_j,\eta}^1)^* H_{m_j,\eta}^1 (\lambda_j I_{m_j} + \eta E_{m_j,\eta}^1) = 2\eta W_{m_j,\eta}^1 \quad (j = 1, \ldots, r)$$

where $W_{m_j,\eta}^1$ is the jth diagonal block of $W_{\eta,1}$. Since $W_{\eta,1}$ is positive definite and converges to a positive definite matrix, $W_{m_j,\eta}^1$, $j = 1, \ldots, r$, are also positive definite and converge to positive definite matrices. For $\eta > 0$, we have $\rho(\lambda_j I_{m_j} +$

$\eta E^1_{m_j,\eta}) < 1$ for $j = 1, \ldots, r$ since $\rho(S_\eta) < 1$. By the assumption that X_η converges to X, we have that $H^1_{m_j,\eta}$ converges to $H^1_{m_j,0}$ for $j = 1, \ldots, r$. From Lemma 3.3.3, we obtain that $H^1_{m_j,0}$ is positive definite for $j = 1, \ldots, r$. Hence, $H_{\eta,1}$ converges to $H_{0,1}$ with $\text{rank}(H_{0,1}) = m$. It follows from (3.39) that $K_\eta = 2X_{\eta,I}$ converges to $K_0 = 2X_I$ with $\text{rank}(X_I) = m$.

(c) It suffices to show that $\|S_\eta - S\|_2 = O(\eta)$ for $\eta > 0$ sufficiently small. Since X is a solution of $X + A^\top X^{-1} A = Q$, we have

$$\mathcal{M}\begin{bmatrix} I \\ X \end{bmatrix} = \mathcal{L}\begin{bmatrix} I \\ X \end{bmatrix} S,$$

where the pencil $(\mathcal{M}, \mathcal{L})$ is defined in (3.26) . Under the condition in (c), the column space of $\begin{bmatrix} I \\ X \end{bmatrix}$ is a simple eigenspace of $(\mathcal{M}, \mathcal{L})$, in the terminology of [78]. It follows from [78, Chapter VI, Theorems 2.12 and 2.15] that

$$\begin{bmatrix} A & 0 \\ Q + i\eta I & -I \end{bmatrix}\begin{bmatrix} I + \eta F_{\eta,1} \\ X + \eta F_{\eta,2} \end{bmatrix} = \begin{bmatrix} 0 & I \\ A^\top & 0 \end{bmatrix}\begin{bmatrix} I + \eta F_{\eta,1} \\ X + \eta F_{\eta,2} \end{bmatrix}(S + \eta E_\eta),$$

where $F_{\eta,1}, F_{\eta,2}, E_\eta \in \mathbb{C}^{n\times n}$ with $\max\{\|F_{\eta,1}\|_2, \|F_{\eta,2}\|_2, \|E_\eta\|_2\} \le c$ for $\eta > 0$ sufficiently small and $c > 0$. It is easily seen that

$$X_\eta = (X + \eta F_{\eta,2})(I + \eta F_{\eta,1})^{-1}, \quad S_\eta = X_\eta^{-1} A = (I + \eta F_{\eta,1})(S + \eta E_\eta)(I + \eta F_{\eta,1})^{-1}.$$

It follows that $\|S_\eta - S\|_2 = O(\eta)$ for $\eta > 0$ sufficiently small. □

Remark 3.3.1. *Without the additional conditions in Theorem* 3.3.4 (b) *or* (c), rank (X_I) *could be much smaller than* m. *Consider the example with* $A = I_n$ *and* $Q = 2I_n$. *Then* $\varphi_0(\lambda)$ *has all* $2n$ *eigenvalues at* 1, *with partial multiplicities* 2. *So* $m = n$, *but it is easy to see that* rank $(X_I) = 0$. *For this example, we have* $\|S_\eta - S\|_2 = O(\eta^{1/2})$ *for* $\eta > 0$ *sufficiently small. We also know that the* $2n$ *eigenvalues of* $\varphi_0(\lambda)$ *at* 1 *are perturbed to* n *eigenvalues inside the unit circle and* n *eigenvalues outside the unit circle.*

Corollary 3.3.5. *If* $\varphi_0(\lambda)$ *has no eigenvalues on* $\mathbb{T}$, *then* X *is real symmetric. Furthermore,* $\text{In}(X) = \text{In}(-\varphi_0(1))$. *Here* $\text{In}(W)$ *denotes the inertia of a matrix* W.

Proof. From Theorem 3.3.4, it is easy to see that X is a real symmetric matrix. Since X is real, $S = X^{-1}A$ is a real matrix. By setting $\lambda = 1$ in (3.25) we get $\varphi_0(1) = -(I - S^\top)X(I - S)$. Hence, $\text{In}(X) = \text{In}(-\varphi_0(1))$. □

Corollary 3.3.6. *If all eigenvalues of* $\varphi_0(\lambda)$ *are on* $\mathbb{T}$ *and simple, then* X_I *is positive definite.*

Proof. By Theorem 3.3.4 (c) immediately. □

3.3.2 A structure-preserving algorithm

As explained in [35] and also in this section, the required solution $X = G_{L,S}^{-1}$ is a particular weakly stabilizing solution of (3.24) and is given by $X = \lim_{\eta \to 0^+} X_\eta$, where X_η is the unique stabilizing solution of (3.22). We will call this particular solution *the* weakly stabilizing solution of (3.24). It can be approximated by X_η for a small η. For a fixed $\eta > 0$, X_η can be computed efficiently by the doubling algorithm studied in [35], for all energy values.

In this section we will develop a structure-preserving algorithm that can for most cases find the weakly stabilizing solution of (3.24) more efficiently and more accurately than the doubling algorithm, by working on (3.24) directly.

Consider the pencil $(\mathcal{M}, \mathcal{L})$ given by (3.26). The simple relation $\mathcal{M}\begin{bmatrix} I \\ X \end{bmatrix} = \mathcal{L}\begin{bmatrix} I \\ X \end{bmatrix} X^{-1}A$ shows that the weakly stabilizing solution of (3.24) is obtained by $X = X_2 X_1^{-1}$, where $\begin{bmatrix} X_1 \\ X_2 \end{bmatrix}$ forms (or more precisely, the columns of $\begin{bmatrix} X_1 \\ X_2 \end{bmatrix}$ form) a basis for the invariant subspace of $(\mathcal{M}, \mathcal{L})$ corresponding to its eigenvalues inside $\mathbb{T}$ and its eigenvalues on $\mathbb{T}$ that would be perturbed to the inside of $\mathbb{T}$ when Q is replaced by Q_η with $\eta > 0$.

We now assume that all unimodular eigenvalues $\lambda \neq \pm 1$ of $(\mathcal{M}, \mathcal{L})$ are semi-simple and the eigenvalues ± 1 (if exist) have partial multiplicities 2. This assumption seems to hold generically. Under this assumption, for computing the weakly stabilizing solution we need to include all linearly independent eigenvectors associated with the eigenvalues ± 1, and use Theorem 3.3.2 to determine which half of the unimodular eigenvalues $\lambda \neq \pm 1$ should be included to compute the required invariant subspace.

We may use the QZ algorithm to determine this invariant subspace, but it would be better to exploit the structure of the pencil $(\mathcal{M}, \mathcal{L})$. We will use the same approach as in [41] to develop a structure-preserving algorithm (SA) to find a basis for the desired invariant subspace of $(\mathcal{M}, \mathcal{L})$ and then compute the weakly stabilizing solution of (3.24). The algorithm is still based on the $(S + S^{-1})$-transform in [63] and Patel's algorithm in [74], but some new issues need to be addressed here.

It is well known that $(\mathcal{M}, \mathcal{L})$ is a symplectic pair, i.e., $(\mathcal{M}, \mathcal{L})$ satisfies $\mathcal{M}\mathcal{J}\mathcal{M}^\top = \mathcal{L}\mathcal{J}\mathcal{L}^\top$, where $\mathcal{J} = \begin{bmatrix} 0 & I \\ -I & 0 \end{bmatrix}$. Furthermore, the eigenvalues of $(\mathcal{M}, \mathcal{L})$ form reciprocal pairs $(\nu, 1/\nu)$, where we allow $\nu = 0, \infty$. We define the $(S + S^{-1})$-transform [63] of $(\mathcal{M}, \mathcal{L})$ by

$$\mathcal{K} := \mathcal{M}\mathcal{J}\mathcal{L}^\top\mathcal{J} + \mathcal{L}\mathcal{J}\mathcal{M}^\top\mathcal{J} = \begin{bmatrix} Q & A - A^\top \\ A^\top - A & Q \end{bmatrix},$$

$$\mathcal{N} := \mathcal{L}\mathcal{J}\mathcal{L}^\top\mathcal{J} = \begin{bmatrix} A & 0 \\ 0 & A^\top \end{bmatrix}.$$

Then $\mathcal{K}$ and $\mathcal{N}$ are both skew-Hamiltonian, i.e., $\mathcal{K}\mathcal{J} = \mathcal{J}\mathcal{K}^\top$ and $\mathcal{N}\mathcal{J} = \mathcal{J}\mathcal{N}^\top$. The relationship between eigenvalues of $(\mathcal{M}, \mathcal{L})$ and $(\mathcal{K}, \mathcal{N})$, and their Kronecker structures have been studied in [63, Theorem 3.2]. We will first extend that result to allow unimodular eigenvalues for $(\mathcal{M}, \mathcal{L})$. The following preliminary result is needed.

Lemma 3.3.7. *Let $N_r(\lambda) := \lambda I_r + N_r$, where N_r is the nilpotent matrix with $N_r(i, i+1) = 1$, $i = 1, \ldots, r-1$, and zeros elsewhere. Let $\overset{eq.}{\sim}$ denote the equivalence between two matrix pairs. Then*

(a) *For $\lambda \neq 0, \pm 1$, $(N_r(\lambda) + N_r(\lambda)^{-1}, I_r) \overset{eq.}{\sim} (N_r(\lambda + 1/\lambda), I_r)$.*
(b) *$(N_r^2 + I_r, N_r) \overset{eq.}{\sim} (I, N_r)$.*

Proof. (a) Since $\lambda \neq 0$, one can show that $N_r(\lambda)^{-1} \equiv [t_{j-i}]$ and $N_r(\lambda)+N_r(\lambda)^{-1} \equiv [s_{j-i}]$ are Toeplitz upper triangular with $t_k = (-1)^k \lambda^{-(k+1)}$, for $k = 0, 1, \ldots, r-1$, as well as $s_0 = \lambda + 1/\lambda$, $s_1 = 1 - \lambda^{-2}$ and $s_k = t_k$, for $k = 2, \ldots, r-1$. Since $\lambda \neq \pm 1$, $s_1 = 1 - \lambda^{-2}$ is nonzero. It follows that $(N_r(\lambda)+N_r(\lambda)^{-1}, I_r) \overset{eq.}{\sim} (N_r(\lambda+1/\lambda), I_r)$.
(b) $(I + N_r^2, N_r) \overset{eq.}{\sim} (I_r, N_r(I + N_r^2)^{-1}) \overset{eq.}{\sim} (I_r, N_r - N_r^3 + N_r^5 - \cdots) \overset{eq.}{\sim} (I_r, N_r)$. □

Theorem 3.3.8. *Suppose that $(\mathcal{M}, \mathcal{L})$ has eigenvalues $\{\pm 1\}$ with partial multiplicities 2. Let $\gamma = \lambda + 1/\lambda$ ($\lambda = 0, \infty$ permitted). Then λ and $1/\lambda$ are eigenvalues of $(\mathcal{M}, \mathcal{L})$ if and only if γ is a double eigenvalue of $(\mathcal{K}, \mathcal{N})$. Furthermore, for $\lambda \neq \pm 1$ (i.e., $\gamma \neq \pm 2$) γ, γ, λ and $1/\lambda$ have the same sizes of Jordan blocks, i.e., they have the same partial multiplicities; for $\lambda = \pm 1$, $\gamma = \pm 2$ are semi-simple eigenvalues of $(\mathcal{K}, \mathcal{N})$.*

Proof. By the results on Kronecker canonical form for a symplectic pencil (see [65]) and by our assumption, there are nonsingular matrices $\mathcal{X}$ and $\mathcal{Y}$ such that

$$\mathcal{Y}\mathcal{M}\mathcal{X} = \begin{bmatrix} J & D \\ 0 & I_n \end{bmatrix}, \quad \mathcal{Y}\mathcal{L}\mathcal{X} = \begin{bmatrix} I_n & 0 \\ 0 & J \end{bmatrix}, \tag{3.43}$$

where $J = J_1 \oplus J_s \oplus J_0$, $J_1 = I_p \oplus (-I_q)$, J_s is the direct sum of Jordan blocks corresponding to nonzero eigenvalues λ_j of $(\mathcal{M}, \mathcal{L})$, where $|\lambda_j| < 1$ or $\lambda_j = e^{i\theta_j}$ with $\mathrm{Im}(\lambda_j) > 0$, and J_0 is the direct sum of nilpotent blocks corresponding to zero eigenvalues, and $D = I_p \oplus I_q \oplus 0_{n-r}$ with $r = p + q$.

Let $\mathcal{X}^{-1}\mathcal{J}\mathcal{X}^{-\top} = \begin{bmatrix} X_1 & X_2 \\ -X_2^\top & X_3 \end{bmatrix}$, where $X_i \in \mathbb{C}^{n\times n}$, $i = 1, 2, 3$. So $X_1^\top = -X_1$ and $X_3^\top = -X_3$. Using (3.43) in $\mathcal{M}\mathcal{J}\mathcal{M}^\top = \mathcal{L}\mathcal{J}\mathcal{L}^\top$ we get

$$JX_1J^\top - DX_2^\top J^\top + JX_2D + DX_3D = X_1, \tag{3.44a}$$
$$JX_2 + DX_3 = X_2J^\top, \tag{3.44b}$$
$$JX_3J^\top = X_3. \tag{3.44c}$$

Let $J_{s,0} = J_s \oplus J_0$. We partition X_3 and X_2 by $X_3 = \begin{bmatrix} X_{3,1} & X_{3,2} \\ -X_{3,2}^\top & X_{3,3} \end{bmatrix}$ and $X_2 = \begin{bmatrix} X_{2,1} & X_{2,2} \\ X_{2,3} & X_{2,4} \end{bmatrix}$, respectively, where $X_{2,1}, X_{3,1} \in \mathbb{C}^{r\times r}$. Comparing the diagonal blocks in (3.44b) we get

$$J_1X_{2,1} + X_{3,1} = X_{2,1}J_1^\top, \quad J_{s,0}X_{2,4} = X_{2,4}J_{s,0}^\top. \tag{3.45}$$

From (3.45) we see that $X_{3,1}$ has the form $X_{3,1} = \begin{bmatrix} 0_p & \omega \\ -\omega^\top & 0_q \end{bmatrix}$. From (3.44c) we have $[I_p \oplus (-I_q)]X_{3,1}[I_p \oplus (-I_q)] = X_{3,1}$. It follows that $\omega = 0$ and thus $X_{3,1} = 0$. From (3.44c) we also have $J_1 X_{3,2} J_{s,0}^\top = X_{3,2}$ and $J_{s,0} X_{3,3} J_{s,0}^\top = X_{3,3}$, from which we get $X_{3,2} = 0$ and $X_{3,3} = 0$. So we have $X_3 = 0$. Then (3.44b) becomes $JX_2 = X_2 J^\top$, from which we get

$$X_{2,1} = \eta_p \oplus \eta_q, \quad X_{2,2} = X_{2,3}^\top = 0, \tag{3.46}$$

where $\eta_p \in \mathbb{C}^{p\times p}$ and $\eta_q \in \mathbb{C}^{q\times q}$. Moreover, $X_{2,1}$ and $X_{2,4}$ are nonsingular by the nonsingularity of $\mathcal{X}^{-1}\mathcal{J}\mathcal{X}^{-\top}$. Substituting (3.44b) into (3.44a) we get

$$X_1 = JX_1J^\top - DJX_2^\top + X_2 J^\top D \equiv JX_1J^\top + V, \tag{3.47}$$

where $V = X_2 J^\top D - DJX_2^\top = \begin{bmatrix} V_1 & 0 \\ 0 & 0 \end{bmatrix}$ with $V_1 = (\eta_p - \eta_p^\top) \oplus (\eta_q^\top - \eta_q)$ by (3.46). Partition $X_1 = \begin{bmatrix} X_{1,1} & X_{1,2} \\ -X_{1,2}^\top & X_{1,3} \end{bmatrix}$ with $X_{1,1} \in \mathbb{C}^{r\times r}$. From the equations for the $(1,2)$ and $(2,2)$ blocks of (3.47) we get $X_{1,2} = 0$ and $X_{1,3} = 0$, respectively. Furthermore, from the $(1,1)$ block in (3.47) we get $X_{1,1} = \xi_p \oplus \xi_q$ with $\xi_p \in \mathbb{C}^{p\times p}$ and $\xi_q \in \mathbb{C}^{q\times q}$, and we also get $\eta_p = \eta_p^\top$ and $\eta_q = \eta_q^\top$, i.e., $X_{2,1}^\top = X_{2,1}$.

From (3.43), (3.45) and Lemma 3.3.7 (b) we now have

$$
\begin{aligned}
&(\mathcal{K},\mathcal{N}) \overset{eq.}{\sim} (\mathcal{M}\mathcal{J}\mathcal{L}^\top + \mathcal{L}\mathcal{J}\mathcal{M}^\top, \mathcal{L}\mathcal{J}\mathcal{L}^\top) \overset{eq.}{\sim} \\
&\left(\begin{bmatrix} (J_1X_{1,1}+X_{1,1}J_1) \oplus 0_{n-r} & ((J_1^2+I_r)X_{2,1}) \oplus ((J_{s,0}^2+I_{n-r})X_{2,4}) \\ (X_{2,1}(J_1^2+I_r)) \oplus (X_{2,4}^\top((J_{s,0}^2)^\top+I_{n-r})) & 0_n \end{bmatrix}, \right. \\
&\left. \begin{bmatrix} X_{1,1} \oplus 0_{n-r} & J_1X_{2,1} \oplus J_{s,0}X_{2,4} \\ X_{2,1}J_1 \oplus X_{2,4}^\top J_{s,0}^\top & 0_n \end{bmatrix} \right) \\
&\overset{eq.}{\sim} \left(\begin{bmatrix} 2\xi_p \oplus (-2\xi_q) & 2I_r \\ 2I_r & 0_r \end{bmatrix} \oplus (J_s + J_s^{-1}) \oplus I \oplus (J_s + J_s^{-1}) \oplus I, \right. \\
&\left. \begin{bmatrix} \xi_p \oplus \xi_q & I_p \oplus (-I_q) \\ I_p \oplus (-I_q) & 0_r \end{bmatrix} \oplus I \oplus J_0 \oplus I \oplus J_0 \right) \\
&\overset{eq.}{\sim} \left(2I_{2p} \oplus (-2I_{2q}) \oplus (J_s+J_s^{-1}) \oplus I \oplus (J_s+J_s^{-1}) \oplus I, I_{2p} \oplus I_{2q} \oplus I \oplus J_0 \oplus I \oplus J_0\right).
\end{aligned}
$$

The proof is completed by using Lemma 3.3.7 (a). □

It is helpful to keep in mind that the transform $\lambda \to \gamma$ achieves the following:

$$\{0,\infty\} \to \infty,\ \mathbb{T} \to [-2,2],\ \mathbb{R}\setminus\{\pm 1, 0\} \to \mathbb{R}\setminus[-2,2],\ \mathbb{C}\setminus(\mathbb{R}\cup\mathbb{T}) \to \mathbb{C}\setminus\mathbb{R}.$$

By Theorem 3.3.8 and our assumption on the unimodular eigenvalues of $(\mathcal{M},\mathcal{L})$, all eigenvalues of $(\mathcal{K},\mathcal{N})$ in $[-2,2]$ are semi-simple.

Based on Patel approach [74] we first reduce $(\mathcal{K},\mathcal{N})$ to a block triangular matrix pair

$$\mathcal{U}^\top\mathcal{K}\mathcal{Z} = \begin{bmatrix} K_{11} & K_{12} \\ 0 & K_{11}^\top \end{bmatrix}, \quad \mathcal{U}^\top\mathcal{N}\mathcal{Z} = \begin{bmatrix} N_{11} & N_{12} \\ 0 & N_{11}^\top \end{bmatrix}, \tag{3.48}$$

where K_{11} and $N_{11} \in \mathbb{R}^{n\times n}$ are quasi-upper and upper triangular, respectively, K_{12} and N_{12} are skew symmetric, $\mathcal{U}$ and $\mathcal{Z} \in \mathbb{R}^{2n\times 2n}$ are orthogonal satisfying $\mathcal{U}^\top \mathcal{J}\mathcal{Z} = \mathcal{J}$.

From (3.48) we see that the pair (K_{11}, N_{11}) contains half of eigenvalues of $(\mathcal{K}, \mathcal{N})$. We then reduce (K_{11}, N_{11}) to the quasi-upper and upper block diagonal matrix pair

$$\widehat{U}^\top K_{11}\widehat{Z} = \mathrm{diag}\{I_{m_0}, \Gamma_1, \Gamma_2, \ldots, \Gamma_r\}, \quad \widehat{U}^\top N_{11}\widehat{Z} = \mathrm{diag}\{\Gamma_0, I_{m_1}, I_{m_2}, \ldots, I_{m_r}\} \tag{3.49}$$

by solving some suitable Sylvester equations, where $m_0 + m_1 + \cdots + m_r = n$, Γ_0 is nilpotent, $\Gamma_1 = \mathrm{diag}\{g_1, \ldots, g_{m_1}\}$ with $g_i \in [-2, 2]$, and $\sigma(\Gamma_j) = \{\gamma_j\} \subseteq \mathbb{R}\backslash[-2, 2]$ or $\sigma(\Gamma_j) = \{\gamma_j, \overline{\gamma}_j\} \subseteq \mathbb{C}\backslash\mathbb{R}$ with $\sigma(\Gamma_j)\cap\sigma(\Gamma_i) = \emptyset$, $i \neq j$, $i, j = 2, \ldots, r$. Partition $\widehat{Z} = [\widehat{Z}_0, \widehat{Z}_1, \ldots, \widehat{Z}_r]$ with $\widehat{Z}_i \in \mathbb{R}^{n\times m_i}$ according to the block sizes of (3.49). It holds that

$$K_{11}\widehat{Z}_0\Gamma_0 = N_{11}\widehat{Z}_0, \quad K_{11}\widehat{Z}_j = N_{11}\widehat{Z}_j\Gamma_j, \quad j = 1, \ldots, r. \tag{3.50}$$

It follows that $\mathcal{Z}\begin{bmatrix}\widehat{Z}_j\\ 0\end{bmatrix}$ forms a basis for an invariant subspace of $(\mathcal{K}, \mathcal{N})$ for $j = 0, 1, \ldots, r$. In particular, the columns of $\mathcal{Z}\begin{bmatrix}\widehat{Z}_1\\ 0\end{bmatrix}$ are real eigenvectors of $(\mathcal{K}, \mathcal{N})$ corresponding to real eigenvalues in $[-2, 2]$. We then need to get a suitable invariant subspace of $(\mathcal{M}, \mathcal{L})$ from each of these invariant subspaces for $(\mathcal{K}, \mathcal{N})$.

We start with two lemmas about solving the quadratic equation $\gamma = \lambda + 1/\lambda$ in the matrix form.

Lemma 3.3.9. *Given a real quasi-upper triangular matrix*

$$\Gamma_s = \begin{bmatrix}\gamma_{11} & \cdots & \gamma_{1m}\\ \vdots & & \vdots\\ 0 & \cdots & \gamma_{mm}\end{bmatrix}, \tag{3.51}$$

where γ_{ii} is 1×1 or 2×2 block with $\sigma(\gamma_{ii}) \subseteq \mathbb{C}\setminus[-2, 2]$, $i = 1, \ldots, m$. Then the quadratic matrix equation

$$\Lambda_s^2 - \Gamma_s\Lambda_s + I = 0 \tag{3.52}$$

of Λ_s is uniquely solvable with Λ_s being real quasi-upper triangular with the same block form as Γ_s in (3.51) and $\sigma(\Lambda_s) \subseteq \mathbb{D} \equiv \{\lambda \in \mathbb{C} | \ |\lambda| < 1\}$.

Proof. Let

$$\Lambda_s = \begin{bmatrix}\lambda_{11} & \cdots & \lambda_{1m}\\ \vdots & & \vdots\\ 0 & \cdots & \lambda_{mm}\end{bmatrix}$$

have the same block form as Γ_s. We first solve the diagonal blocks $\{\lambda_{ii}\}_{i=1}^m$ of Λ_s from the quadratic equation $\lambda^2 - \gamma_{ii}\lambda + I_{[i]} = 0$, where $[i]$ denotes the size of γ_{ii}. Note that the scalar equation

$$\lambda^2 - \gamma\lambda + 1 - 0 \tag{3.53}$$

has no solutions on $\mathbb{T}$ for $\gamma \in \mathbb{C} \setminus [-2, 2]$. So it always has one solution inside $\mathbb{T}$ and the other outside $\mathbb{T}$.

For $i = 1, \ldots, m$, if $\gamma_{ii} \in \mathbb{R} \setminus [-2, 2]$, then $\lambda_{ii} \in (-1, 1)$ is uniquely solved from (3.53) with $\gamma = \gamma_{ii}$. If $\gamma_{ii} \in \mathbb{R}^{2\times 2}$ with $\gamma_{ii} z = \gamma z$ for $z \neq 0$ and $\gamma \in \mathbb{C} \setminus \mathbb{R}$, then $\gamma_{ii} = [z, \overline{z}]\, \mathrm{diag}\,\{\gamma, \overline{\gamma}\}\, [z, \overline{z}]^{-1}$ and the required solution is $\lambda_{ii} = [z, \overline{z}]\, \mathrm{diag}\,\{\lambda, \overline{\lambda}\}\, [z, \overline{z}]^{-1} \in \mathbb{R}^{2\times 2}$, where $\lambda \in \mathbb{D}$ is uniquely solved from (3.53).

For $j > i$, comparing the (i, j) block on both sides of (3.52) and using $\lambda_{ii} - \gamma_{ii} = -\lambda_{ii}^{-1}$, we get

$$\lambda_{ij}\lambda_{jj} - \lambda_{ii}^{-1}\lambda_{ij} = \gamma_{ij}\lambda_{jj} + \sum_{\ell=i+1}^{j-1} (\gamma_{i\ell} - \lambda_{i\ell})\lambda_{\ell j}.$$

Since $\sigma(\lambda_{ii}^{-1}) \cap \sigma(\lambda_{jj}) = \emptyset$, $i, j = 1, \ldots, m$, the strictly upper triangular part of Λ_s can be determined by the following recursive formula.

For $d = 1, \ldots, m-1$,
    For $i = 1, \ldots, m-d$, $j = i+d$,
        $A := \lambda_{jj}^{\top} \otimes I_{[i]} - I_{[j]} \otimes \lambda_{ii}^{-1}$,
        $b := \gamma_{ij}\lambda_{jj} + \sum_{\ell=i+1}^{j-1} (\gamma_{i\ell} - \lambda_{i\ell})\lambda_{\ell j}$,
        $\lambda_{ij} = \mathrm{vec}^{-1}(A^{-1}\mathrm{vec}(b))$,
    end i,
end d.

Here $\otimes$ denotes the Kronecker product, vec is the operation of stacking the columns of a matrix into a vector, and vec^{-1} is its inverse operation. □

Lemma 3.3.10. *Given a nilpotent matrix $\Gamma_0 = [\gamma_{ij}] \in \mathbb{R}^{e\times e}$. The quadratic matrix equation*

$$\Gamma_0\Lambda_0^2 - \Lambda_0 + \Gamma_0 = 0 \text{ in } \Lambda_0 = [\lambda_{ij}] \in \mathbb{R}^{e\times e} \tag{3.54}$$

with Λ_0 being nilpotent is uniquely solvable.

Proof. From (3.54) the matrix Λ_0 is uniquely determined by $\lambda_{i,i+j} = \gamma_{i,i+j}$, $i = 1, \ldots, e-2$, $j = 1, 2$, $\lambda_{e-1,e} = \gamma_{e-1,e}$ and

For $j = 3, \ldots, e$,
    For $i = 1, \ldots, e-j+1$,
        $\widehat{\lambda}_{i,i+j-1} = \sum_{\ell=i+1}^{i+j-2} \lambda_{i,\ell}\lambda_{\ell,i+j-1}$,
    end i,
    For $i = 1, \ldots, e-j$,
        $\lambda_{i,i+j} = \gamma_{i,i+j} + \sum_{\ell=i+1}^{i+j-2} \gamma_{i,\ell}\widehat{\lambda}_{\ell,i+j}$,
    end i,
end j.

□

Theorem 3.3.11. *Let Z_s form a basis for an invariant subspace of $(\mathcal{K}, \mathcal{N})$ corresponding to Γ_s with $\sigma(\Gamma_s) \subseteq \mathbb{C} \setminus [-2, 2]$, i.e., $\mathcal{K}Z_s = \mathcal{N}Z_s\Gamma_s$. Suppose that Λ_s*

solves $\Gamma_s = \Lambda_s + \Lambda_s^{-1}$ *as in Lemma* 3.3.9 *with* $\sigma(\Lambda_s) \subseteq \mathbb{D} \setminus \{0\}$. *If the columns of* $\mathcal{J}(\mathcal{L}^\top \mathcal{J} Z_s \Lambda_s - \mathcal{M}^\top \mathcal{J} Z_s)$ *are linearly independent, then they form a basis for a stable invariant subspace of* $(\mathcal{M}, \mathcal{L})$ *corresponding to* Λ_s.

Proof. Since

$$\mathcal{K} Z_s - \mathcal{N} Z_s \Gamma_s = (\mathcal{M}\mathcal{J}\mathcal{L}^\top + \mathcal{L}\mathcal{J}\mathcal{M}^\top)\mathcal{J} Z_s - \mathcal{L}\mathcal{J}\mathcal{L}^\top \mathcal{J} Z_s(\Lambda_s + \Lambda_s^{-1}) = 0,$$

we have $\mathcal{M}\mathcal{J}\left(\mathcal{L}^\top \mathcal{J} Z_s \Lambda_s - \mathcal{M}^\top \mathcal{J} Z_s\right) = \mathcal{L}\mathcal{J}\left(\mathcal{L}^\top \mathcal{J} Z_s \Lambda_s - \mathcal{M}^\top \mathcal{J} Z_s\right)\Lambda_s$. □

Remark 3.3.2. *Let*

$$Z_s = \begin{bmatrix} Z_1 \\ Z_2 \end{bmatrix}, \quad \begin{bmatrix} X_1 \\ X_2 \end{bmatrix} = \mathcal{J}(\mathcal{L}^\top \mathcal{J} Z_s \Lambda_s - \mathcal{M}^\top \mathcal{J} Z_s),$$

where each of X_1, X_2, Z_1, *and* Z_2 *has* n *rows. Then direct computation gives* $X_1 = Z_2\Lambda_s - Z_1$. *But it is more convenient to get* $X_2 = QX_1 - A^\top X_1 \Lambda_s$ *from* $\mathcal{M}\begin{bmatrix} X_1 \\ X_2 \end{bmatrix} = \mathcal{L}\begin{bmatrix} X_1 \\ X_2 \end{bmatrix}\Lambda_s$.

We now explain how we can get eigenvectors of $(\mathcal{M}, \mathcal{L})$ from eigenvectors of $(\mathcal{K}, \mathcal{N})$ corresponding to eigenvalues in $[-2, 2]$.

Theorem 3.3.12. *Let* $v = \begin{bmatrix} v_1 \\ v_2 \end{bmatrix}$ *be a real eigenvector of* $(\mathcal{K}, \mathcal{N})$ *corresponding to eigenvalue* $\gamma \in [-2, 2]$. *Let* λ *be a solution of* (3.53) *and let* $u_1 = \lambda v_2 - v_1$, $u_2 = Qv_1 - \lambda A^\top v_1$. *Then* $u = \begin{bmatrix} u_1 \\ u_2 \end{bmatrix}$ *is an eigenvector of* $(\mathcal{M}, \mathcal{L})$ *corresponding to eigenvalue* λ *if* $u \neq 0$. *Moreover, we indeed have* $u \neq 0$ *for each* $\gamma \in (-2, 2)$.

Proof. The first part is proved by direct computation as in the proof of Theorem 3.3.11. For the second part, we simply note that λ is not real when $\gamma \in (-2, 2)$, and then $u_1 \neq 0$ since v is a nonzero real vector. □

Remark 3.3.3. *When* $\gamma \in (-2, 2)$ *is an eigenvalue of* $(\mathcal{K}, \mathcal{N})$ *with multiplicity* $2k$ *(or an eigenvalue of* (K_{11}, N_{11}) *with multiplicity* k*), we can use Theorem* 3.3.12 *to get* k *eigenvectors of* $(\mathcal{M}, \mathcal{L})$ *corresponding to eigenvalue* λ, *from the* k *linearly independent eigenvectors of* $(\mathcal{K}, \mathcal{N})$ *we have already obtained. However, there is no guarantee that the* k *eigenvectors of* $(\mathcal{K}, \mathcal{N})$ *so obtained are also linearly independent.*

When A is singular, $(\mathcal{M}, \mathcal{L})$ has eigenvalues at 0 and ∞. The following result will then be needed.

Theorem 3.3.13. *Let* $Z_\infty \in \mathbb{R}^{2n \times m}$ *span an infinite invariant subspace of* $(\mathcal{K}, \mathcal{N})$ *corresponding to* (I, Γ_0), *where* Γ_0 *is nilpotent, i.e.,* $\mathcal{N} Z_\infty = \mathcal{K} Z_\infty \Gamma_0$. *Suppose that* Λ_0 *solves* $\Gamma_0 \Lambda_0^2 - \Lambda_0 + \Gamma_0 = 0$ *as in Lemma 3.3.10 with* Λ_0 *being nilpotent. If the columns of* $\mathcal{J}(\mathcal{L}^\top \mathcal{J} Z_\infty \Lambda_0 - \mathcal{M}^\top \mathcal{J} Z_\infty)$ *are linearly independent, then they form a basis for a zero invariant subspace of* $(\mathcal{M}, \mathcal{L})$ *corresponding to* (Λ_0, I).

Proof. Since $\Gamma_0 = \Lambda_0(I + \Lambda_0^2)^{-1}$, we have

$$\begin{aligned}\mathcal{N} Z_\infty(I + \Lambda_0^2) &= \mathcal{M}\mathcal{J}\mathcal{M}^\top \mathcal{J} Z_\infty + \mathcal{L}\mathcal{J}\mathcal{L}^\top \mathcal{J} Z_\infty \Lambda_0^2 = \mathcal{K} Z_\infty \Lambda_0 \\ &= \mathcal{L}\mathcal{J}\mathcal{M}^\top \mathcal{J} Z_\infty \Lambda_0 + \mathcal{M}\mathcal{J}\mathcal{L}^\top \mathcal{J} Z_\infty \Lambda_0,\end{aligned}$$

and then

$$\mathcal{M}\mathcal{J}(\mathcal{L}^\top \mathcal{J} Z_\infty \Lambda_0 - \mathcal{M}^\top \mathcal{J} Z_\infty) = \mathcal{L}\mathcal{J}(\mathcal{L}^\top \mathcal{J} Z_\infty \Lambda_0 - \mathcal{M}^\top \mathcal{J} Z_\infty)\Lambda_0.$$

□

We can now present a structure-preserving algorithm (SA) for the computation of the weakly stabilizing solution of (3.24). Here e_k denotes the kth column of I.

Algorithm 3.3.1 (SA). ***Input:*** $A \in \mathbb{R}^{n\times n}$, $Q = Q^\top \in \mathbb{R}^{n\times n}$.
Output: *The weakly stabilizing solution* X *of* $X + A^\top X^{-1} A = Q$.

Step 1: *Form the matrix pair* $(\mathcal{K}, \mathcal{N})$ *as in* (3.42);

Step 2: *Reduce* $(\mathcal{K}, \mathcal{N})$ *as in* (3.48): $\mathcal{K} \leftarrow \mathcal{U}^\top \mathcal{K} \mathcal{Z} = \begin{bmatrix} K_{11} & K_{12} \\ 0 & K_{11}^\top \end{bmatrix}$, $\mathcal{N} \leftarrow \mathcal{U}^\top \mathcal{N} \mathcal{Z} = \begin{bmatrix} N_{11} & N_{12} \\ 0 & N_{11}^\top \end{bmatrix}$, *where* K_{11} *and* $N_{11} \in \mathbb{R}^{n\times n}$ *are quasi-upper and upper triangular, respectively,* $\mathcal{U}$ *and* $\mathcal{Z}$ *are orthogonal satisfying* $\mathcal{U}^\top \mathcal{J} \mathcal{Z} = \mathcal{J}$ *(see a pseudo code in* [41, *Appendix*]);

Step 3: *Compute eigenmatrix pairs* $K_{11}\widehat{Z}_0\Gamma_0 = N_{11}\widehat{Z}_0$, $K_{11}\widehat{Z}_j = N_{11}\widehat{Z}_j\Gamma_j$ *of* (K_{11}, N_{11}), $j = 1, \ldots, r$, *as in* (3.50) *by solving suitable Sylvester equations to get* (3.49);

Step 4: $Z_\infty = \mathcal{Z}\begin{bmatrix} \widehat{Z}_0 \\ 0 \end{bmatrix} \equiv \begin{bmatrix} Z_{\infty,1} \\ Z_{\infty,2} \end{bmatrix}$, $Z_j = \mathcal{Z}\begin{bmatrix} \widehat{Z}_j \\ 0 \end{bmatrix} \equiv \begin{bmatrix} Z_{j,1} \\ Z_{j,2} \end{bmatrix}$, $j = 1, \ldots, r$;

Step 5-1: *Use Lemma* 3.3.10 *to solve the nilpotent* Λ_0 *for* $\Gamma_0\Lambda_0^2 - \Lambda_0 + \Gamma_0 = 0$;
Compute $X_{\infty,1} = Z_{\infty,2}\Lambda_0 - Z_{\infty,1}$, $X_{\infty,2} = QX_{\infty,1} - A^\top X_{\infty,1}\Lambda_0$;
Set $X_1 \leftarrow X_{\infty,1}$, $X_2 \leftarrow X_{\infty,2}$ *(by Theorem* 3.3.13);

Step 5-2: *For* $k = 1, \ldots, m_1$,
Solve $\lambda = e^{i\theta}$ *with* $Im(\lambda) \geq 0$ *from* (3.53) *with* $\gamma = g_k$;
Compute $x_1 = z_2 e^{i\theta} - z_1$, $x_2 = Qx_1 - e^{i\theta}A^\top x_1$, *where* $z_1 = Z_{1,1}e_k$, $z_2 = Z_{1,2}e_k$;
If $Im(e^{i\theta}x_1^* A^\top x_1) > 0$, *then* $x_1 \leftarrow \overline{x}_1$, $x_2 \leftarrow \overline{x}_2$;
Set $X_1 \leftarrow [X_1|x_1]$, $X_2 \leftarrow [X_2|x_2]$ *(by Theorems* 3.3.12 *and* 3.3.2);
end k;

Step 5-3: *For* $j = 2, \ldots, r$,
Use Lemma 3.3.9 *to solve* $\Lambda_j = \Lambda_s$ *for* (3.52) *with* $\Gamma_s = \Gamma_j$;
Compute $X_{j,1} = Z_{j,2}\Lambda_j - Z_{j,1}$, $X_{j,2} = QX_{j,1} - A^\top X_{j,1}\Lambda_j$;
Set $X_1 \leftarrow [X_1|X_{j,1}]$, $X_2 \leftarrow [X_2|X_{j,2}]$ *(by Theorem* 3.3.11);
end j;

Step 6: *Compute* $X = X_2 X_1^{-1}$.

To find the weakly stabilizing solution of (3.24), the total number of flops is roughly $120n^3$ for SA, and $440n^3$ for QZ. Note that a flop denotes a multiplication or an addition in real arithmetic. In Step 5-2 of SA we have assumed that the nonsingular Hermitian matrix $iY^*(2\lambda_0 A^\top - Q)Y$ in Theorem 3.3.2 (which is the matrix in (??)) is definite. This assumption is the same as the assumption in Theorem 3.3.4 (iii). We could have used Theorem 3.3.2 to choose the right eigenvectors when the Hermitian matrix is not definite, but this will increase computational work. We have thus chosen to use the simpler Step 5-2. We can always check in the end whether the imaginary part of the computed X is (numerically) positive semidefinite. If not, we can use a more complicated Step 5-2 according to Theorem 3.3.2 to re-compute X.

3.4 Sensitivity of time-delayed systems

Quoted from [59], we consider the numerical solution of the quadratic eigenvalue problem (QEP)

$$Q(\lambda)x \equiv (\lambda^2 B + \lambda C + A)x = 0, \tag{3.55}$$

where $A,\ B,\ C \in \mathbb{C}^{n\times n}$ satisfy $P\overline{B}P = \varepsilon A$, $P\overline{A}P = \varepsilon B$, $P\overline{C}P = \varepsilon C$ with $\varepsilon = \pm 1$, $P \in \mathbb{C}^{n\times n}$ being an idempotent matrix (i.e., $P^2 = I_n$), and $\overline{B}$ denoting the complex conjugate of B. In our application for the time-delay system (3.57), the matrix P is a real involuntary matrix (more details later). The scalar $\lambda \in \mathbb{C}$ and the vector $x \in \mathbb{C}^n \setminus \{0\}$ are the eigenvalue and the associated eigenvector of the quadratic pencil $Q(\lambda)$, and the pair (λ, x) is called an *eigenpair* of $Q(\lambda)$.

We shall propose a structure-preserving doubling algorithm (SDA; [16, 17, 18, 19, 64]) for the solution of (3.55). We define the `conjugate` and the `reverse` of a quadratic pencil, respectively, by

$$\overline{Q}(\lambda) \equiv \lambda^2\overline{B} + \lambda\overline{C} + \overline{A}, \quad \texttt{rev}(Q(\lambda)) \equiv \lambda^2 A + \lambda C + B.$$

It is easy to see that QEP in (3.55) satisfies

$$Q(\lambda) = \varepsilon\, P\,\texttt{rev}(\overline{Q}(\lambda))\, P. \tag{3.56}$$

From (3.56), we see that (λ, x) is an eigenpair of $Q(\lambda)$ if and only if $(1/\overline{\lambda}, P\overline{x})$ is also an eigenpair of $Q(\lambda)$ [28].

As in [67], a quadratic pencil $Q(\lambda)$ in (3.55) is said to be palindromic if $Q(\lambda) = \texttt{rev}(\overline{Q}(\lambda))$. Similar to the terminology of $(\star, \varepsilon)$-palindromic QEPs ($\star$= H or $\top$) in [67], the QEP (3.55) is referred to as

$(\varepsilon = +1)$ P-Conjugate-P Palindromic QEP (PCP_PQEP), or
$(\varepsilon = -1)$ anti-P-Conjugate-P Palindromic QEP ($-$PCP_PQEP).

PCP_PQEPs as in (3.55) were first proposed in [47] from the stability analysis of *retarded* time-delayed systems (TDS), and generalized in [28] the more general *neutral* TDS (see [5, 10, 27, 37, 38, 71, 72, 87] and the references therein). Consider a neutral linear time-delayed system with m constant delays $h_0 = 0 < h_1 < \cdots <$

h_m:

$$\sum_{k=0}^{m} D_k \dot{x}(t-h_k) = \sum_{k=0}^{m} A_k x(t-h_k) \quad (t>0); \quad x(t)=\varphi(t) \quad (t \in [-h_m, 0]) \tag{3.57}$$

with $x : [-h_m, \infty) \to \mathbb{R}^N$, $\varphi \in \mathbb{C}^1([-h_m, 0])$, and $A_k, D_k \in \mathbb{R}^{N \times N}$ $(k = 1, \ldots, m)$. When $D_0 = I_N$ and $D_k = 0$ $(k > 0)$, we have the retarded time-delayed systems.

The stability of the TDS (3.57) can be determined by its characteristic equation

$$\left(s \sum_{k=0}^{m} D_k e^{-h_k s} - \sum_{k=0}^{m} A_k e^{-h_k s} \right) v = 0 \quad (v \neq 0) \tag{3.58}$$

with eigenpairs (s, v) from the nonlinear eigenvalue problem.

A TDS is said to be *critical* if and only if some eigenvalues s is purely imaginary. The set of all points $(h_1, \ldots, h_m)$ in the delay-parameter space for which the TDS (3.57) is critical are called critical curves $(m = 2)$ or surfaces $(m > 2)$. Under certain continuity assumptions, the boundary of the stability domain of a TDS is a subset of the critical curves/surfaces. Consequently, purely imaginary eigenvalues of (3.58) are of great interest. See [47] and the references therein for approaches to compute critical surfaces. In [28], the following parameterization of critical surfaces gives rise to an associated PCP_PQEP. Detailed discussion on palindromic linearizations, a Schur-like canonical form and other useful results can also be found in [28].

For a given eigenpair (s, v) of (3.58) with $\|v\|_2 = 1$, a point $(h_1, \ldots, h_m)$ in the delay-parameter space is critical if and only if there exist $\varphi_k \in [-\pi, \pi]$ $(k = 1, \ldots, m-1)$ and $\omega \in \mathbb{R}$ such that

$$h_k = \frac{\varphi_k + 2p_k\pi}{\omega}, \quad p_k \in \mathbb{Z} \ (k = 1, \ldots, m-1); \quad h_m = \frac{-\mathrm{Arg}\, z + 2p_m\pi}{\omega}, \quad p_m \in \mathbb{Z}$$

giving rise to the QEP

$$(z^2 E + zF + G)u = 0, \tag{3.59}$$

where the unimodular eigenvalue $z = e^{-i\omega h_m}$, and the corresponding eigenvector $u = \mathrm{vec}\, vv^* = v \otimes \bar{v}$,

$$\omega = \left(\frac{-i}{w^* w} \right) w^* \left(A_m z + \sum_{k=0}^{m-1} A_k e^{-i\varphi_k} \right) v, \quad w = \left(D_m z + \sum_{k=0}^{m-1} D_k e^{-i\varphi_k} \right) v,$$

and

$$E = \left(\sum_{k=0}^{m-1} D_k e^{i\varphi_k} \right) \otimes A_m + \left(\sum_{k=0}^{m-1} A_k e^{i\varphi_k} \right) \otimes D_m \in \mathbb{C}^{N^2 \times N^2},$$

$$F = \left(\sum_{k=0}^{m-1} D_k e^{i\varphi_k} \right) \otimes \left(\sum_{k=0}^{m-1} A_k e^{-i\varphi_k} \right) + \left(\sum_{k=0}^{m-1} A_k e^{i\varphi_k} \right) \otimes \left(\sum_{k=0}^{m-1} D_k e^{-i\varphi_k} \right)$$
$$+ D_m \otimes A_m + A_m \otimes D_m \in \mathbb{C}^{N^2 \times N^2},$$

$$G = D_m \otimes \left(\sum_{k=0}^{m-1} A_k e^{-i\varphi_k} \right) + A_m \otimes \left(\sum_{k=0}^{m-1} D_k e^{-i\varphi_k} \right) \in \mathbb{C}^{N^2 \times N^2}.$$

Here $\otimes$ denotes the usual Kronecker product. As $M_1 \otimes M_2$ and $M_2 \otimes M_1$ both contain products of elements of M_1 and M_2 at different positions, we can find an involuntary matrix $P \in \mathbb{R}^{N^2 \times N^2}$ $(P^{-1} = P^\top = P)$ such that $M_1 \otimes M_2 = P(M_2 \otimes M_1)P$ [40, Corollary 4.3.10]. Here, $P = \sum_{i,j=1}^N E_{ij} \otimes E_{ij}^\top = [E_{ij}^\top]_{i,j=1}^N$, $E_{ij} = e_i \otimes e_j^\top \in \mathbb{R}^{N \times N}$ and e_j is the jth column of I_N. Consequently from the structures in E, F and G, we can easily show that (3.59) is a PCP_PQEP because $E = P\overline{G}P$ and $F = P\overline{F}P$.

Remark 3.4.1. *In [14], it has been proved that unimodular eigenvalues occur quite often for PCP_PQEPs (and other palindromic eigenvalue problems with eigenvalue pairs $(\lambda, \varepsilon/\overline{\lambda})$). These eigenvalues can stay unimodular under perturbation, thus are numerically stable to compute. Furthermore, the probability of having too many of them is low and multiple unimodular eigenvalues are rare. This makes our problem of computing the unimodular eigenvalues z of (3.59) well-posed, unlike for complex-T palindromic eigenvalue problems.*

Before proceeding further, we would like to point out some recent developments in the numerical solution of palindromic eigenvalue problems. The QEP (3.55) is said to be $\star$-palindromic if $B^\star = A$ and $C^\star = C$, with $\star = \top$ or H. The train vibration problem and the associated $\top$-palindromic QEP were discussed in [46, 66] and structure-preserving palindromic linearizations for (3.55) were suggested in [67]. An SDA algorithm [19] and a backward-stable generalized (Arnoldi-)Patel algorithm [41] were proposed for the $\top$-palindromic QEP, and can be extended to H-palindromic QEPs. For other approaches for $\star$-palindromic QEPs, see [68, 76, 77]. In [16], the SDA algorithm was generalized for the g-palindromic QEPs, which do not include the PCP_PQEP. For further information on the numerical solution of palindromic EVPs, see [18]. For general survey of matrix polynomials, the associated eigenvalue problems and their applications, see [31, 80].

3.4.1 SDA Algorithm for εPCP_PQEP

For a given PCP_PQEP (3.55), we define

$$\mathcal{M} = \begin{bmatrix} A & 0 \\ -C & -I \end{bmatrix}, \quad \mathcal{L} = \begin{bmatrix} D & I \\ B & 0 \end{bmatrix}. \tag{3.60}$$

With $D = 0$ in (3.60), we have

$$\begin{bmatrix} A & 0 \\ -C & -I \end{bmatrix} \begin{bmatrix} x_1 \\ x_2 \end{bmatrix} = \lambda \begin{bmatrix} 0 & I \\ B & 0 \end{bmatrix} \begin{bmatrix} x_1 \\ x_2 \end{bmatrix}$$

leading to

$$Ax_1 = \lambda x_2, \quad -Cx_1 - x_2 = \lambda B x_1. \tag{3.61}$$

Multiplying the second equation in (3.61) by λ and substituting the first, we obtain $(\lambda^2 B + \lambda C + A)x_1 = 0$.

We have shown that the pencil $\mathcal{M} - \lambda\mathcal{L}$ is a linearization of PCP_PQEP (3.55) with $D = 0$. Based on the SDA algorithm proposed in [64], we develop a new SDA for solving the PCP_PQEP.

For the pencil $\mathcal{M} - \lambda\mathcal{L}$ defined in (3.60), we compute

$$\mathcal{M}_* = \begin{bmatrix} -AK^{-1} & 0 \\ BK^{-1} & I \end{bmatrix}, \quad \mathcal{L}_* = \begin{bmatrix} I & AK^{-1} \\ 0 & -BK^{-1} \end{bmatrix},$$

where $K \equiv C - D$ is assumed to be invertible. It is easy to check that $\mathcal{M}_*\mathcal{L} = \mathcal{L}_*\mathcal{M}$. Direct calculation gives rise to

$$\widehat{\mathcal{M}} = \mathcal{M}_*\mathcal{M} = \begin{bmatrix} \widehat{A} & 0 \\ -\widehat{C} & I \end{bmatrix}, \quad \widehat{\mathcal{L}} = \mathcal{L}_*\mathcal{L} = \begin{bmatrix} \widehat{D} & I \\ \widehat{B} & 0 \end{bmatrix}, \tag{3.62}$$

where

$$\widehat{A} \equiv -AK^{-1}A, \quad \widehat{B} \equiv -BK^{-1}B, \tag{3.63a}$$
$$\widehat{C} \equiv C - BK^{-1}A, \quad \widehat{K} \equiv K - (BK^{-1}A + AK^{-1}B), \tag{3.63b}$$
$$\widehat{D} = \widehat{C} - \widehat{K}.$$

Theorem 3.4.1. *The pencil $\widehat{\mathcal{M}} - \lambda\widehat{\mathcal{L}}$ has the doubling property, i.e., if*

$$\mathcal{M}\begin{bmatrix} X_1 \\ X_2 \end{bmatrix} = \mathcal{L}\begin{bmatrix} X_1 \\ X_2 \end{bmatrix} S,$$

where $X_1, X_2 \in \mathbb{C}^{n\times m}$ and $S \in \mathbb{C}^{m\times m}$, then

$$\widehat{\mathcal{M}}\begin{bmatrix} X_1 \\ X_2 \end{bmatrix} = \widehat{\mathcal{L}}\begin{bmatrix} X_1 \\ X_2 \end{bmatrix} S^2.$$

Proof. From (3.62) and the relation $\mathcal{M}_*\mathcal{L} = \mathcal{L}_*\mathcal{M}$, we have

$$\begin{aligned} \widehat{\mathcal{M}}\begin{bmatrix} X_1 \\ X_2 \end{bmatrix} &= \mathcal{M}_*\mathcal{M}\begin{bmatrix} X_1 \\ X_2 \end{bmatrix} = \mathcal{M}_*\mathcal{L}\begin{bmatrix} X_1 \\ X_2 \end{bmatrix} S = \mathcal{L}_*\mathcal{M}\begin{bmatrix} X_1 \\ X_2 \end{bmatrix} S \\ &= \mathcal{L}_*\mathcal{L}\begin{bmatrix} X_1 \\ X_2 \end{bmatrix} S^2 = \widehat{\mathcal{L}}\begin{bmatrix} X_1 \\ X_2 \end{bmatrix} S^2. \end{aligned}$$

□

The iteration in (3.63) is structure-preserving for εPCP_PQEP, as shown in the following theorem:

Theorem 3.4.2. *For a pencil $\mathcal{M} - \lambda\mathcal{L}$ given in (3.60), suppose that*

$$P\overline{A}P = \varepsilon B, \; P\overline{B}P = \varepsilon A, \; P\overline{K}P = \varepsilon K, \tag{3.64}$$

where $K = C - D$. Then it holds that

$$P\overline{\widehat{A}}P = \varepsilon\widehat{B}, \; P\overline{\widehat{B}}P = \varepsilon\widehat{A}, \; P\overline{\widehat{K}}P = \varepsilon\widehat{K},$$

where $\widehat{A}$, $\widehat{B}$, $\widehat{K}$ are defined in (3.63).

Proof. From (3.63b) and (3.64), we have

$$\begin{aligned} P\overline{\widehat{K}}P &= P\overline{K}P - \left[P\overline{B}P(P\overline{K}P)^{-1}P\overline{A}P + P\overline{A}P(P\overline{K}P)^{-1}P\overline{B}P\right] \\ &= \varepsilon K - \varepsilon^3(AK^{-1}B + BK^{-1}A) = \varepsilon\widehat{K}. \end{aligned}$$

Similarly, from (3.63a) and (3.64), we have

$$P\overline{\widehat{A}}P = -P\overline{A}P(P\overline{K}P)^{-1}P\overline{A}P = -\varepsilon^3 BK^{-1}B = \varepsilon\widehat{B}$$

and then $P\overline{\widehat{B}}P = \varepsilon\widehat{A}$. □

The SDA for a basis for the stable invariant subspace of $(\mathcal{M}, \mathcal{L})$ can then be constructed, as summarized in Theorems 3.4.1 and 3.4.2.

3.4.2 Convergence of SDA

Consider the matrix pair $(\mathcal{M}, \mathcal{L})$ as in (3.60). In order to ensure that the SDA converges to a basis of the stable invariant subspace of $(\mathcal{M}, \mathcal{L})$, we suppose that all eigenvalues of $(\mathcal{M}, \mathcal{L})$ on the unit circle are semisimple (generically, multiple unimodular eigenvalues are rare; see [14]).

From the Kronecker Canonical form [28, 29], there exist nonsingular matrices Q and Z such that

$$Q\mathcal{M}Z = \begin{bmatrix} J_1 & 0 \\ 0 & I_n \end{bmatrix} \equiv J_{\mathcal{M}}, \quad Q\mathcal{L}Z = \begin{bmatrix} I & 0 \\ 0 & J_2 \end{bmatrix} \equiv J_{\mathcal{L}}, \tag{3.65}$$

where

$$\begin{aligned} &J_1 = \Omega_1 \oplus J_s, \quad J_2 = \Omega_2 \oplus \varepsilon\overline{J}_s, \\ &\Omega_1 = \operatorname{diag}\{e^{i\omega_1}, \ldots, e^{i\omega_\ell}\}, \quad \Omega_2 = \operatorname{diag}\{e^{i\omega_{\ell+1}}, \ldots, e^{i\omega_{2\ell}}\}, \end{aligned}$$

and J_s is the stable Jordan block of size m (i.e., $\rho(J_s) < 1$) with $m = n - \ell$. Here ρ denotes the spectral radius and $\oplus$ the direct sum of matrices. Since $J_{\mathcal{M}}$ and $J_{\mathcal{L}}$ commute with each other, it follows from (3.65) that

$$\mathcal{M}ZJ_{\mathcal{L}} = Q^{-1}J_{\mathcal{L}}J_{\mathcal{M}} = \mathcal{L}ZJ_{\mathcal{M}}. \tag{3.66}$$

Let $\{(\mathcal{M}_k, \mathcal{L}_k)\}_{k=0}^{\infty}$ be the sequence with

$$\mathcal{M}_k = \begin{bmatrix} A_k & 0 \\ -C_k & -I_n \end{bmatrix}, \quad \mathcal{L}_k = \begin{bmatrix} D_k & I_n \\ B_k & 0 \end{bmatrix}, \tag{3.67}$$

where A_k, B_k, C_k and D_k are generated by the SDA. With $\mathcal{M}_0 = \mathcal{M}$ and $\mathcal{L}_0 = \mathcal{L}$, it follows from (3.66) and Theorem 3.4.1 that

$$\mathcal{M}_k Z J_{\mathcal{L}}^{2^k} = \mathcal{L}_k Z J_{\mathcal{M}}^{2^k}. \tag{3.68}$$

Theorem 3.4.3. *Let $(\mathcal{M}, \mathcal{L})$ be given in (3.60) with all its unimodular eigenvalues being semisimple. Write Z in (3.65) in the form*

$$Z = \begin{bmatrix} Z_1 & Z_3 \\ Z_2 & Z_4 \end{bmatrix}, \quad Z_i \in \mathbb{C}^{n \times n}, \; i = 1, \ldots, 4. \tag{3.69}$$

Suppose that the sequence $\{A_k, B_k, C_k, D_k\}_{k=1}^{\infty}$ generated by the SDA is well defined and $\{A_k\}_{k=1}^{\infty}$ is uniformed bounded on k. If Z_1 and Z_3 are invertible, we have
(a) $A_k Z_{12} = O(\rho(J_s^{2^k})) \to 0$ *as* $k \to \infty$*;*
(b) $C_k Z_{12} = -Z_{22} + O(\rho(J_s^{2^k})) \to -Z_{22}$ *as* $k \to \infty$*;*
(c) $D_k Z_{12} = -Z_{42} + O(\rho(\overline{J}_s^{2^k})) \to -Z_{42}$ *as* $k \to \infty$*,*
where $Z_{12} = Z_1(:, \ell+1, \ldots, n)$, $Z_{22} = Z_2(:, \ell+1, \ldots, n)$, $Z_{32} = Z_3(:, \ell+1, \ldots, n)$, $Z_{42} = Z_4(:, \ell+1, \ldots, n)$ *and* $\rho(\cdot)$ *is the spectrum radius.*

Proof. Substituting $(\mathcal{M}_k, \mathcal{L}_k)$ from (3.67), $J_{\mathcal{M}}$ and $J_{\mathcal{L}}$ in (3.65) as well as Z in (3.69) into (3.68), we have

$$A_k Z_1 = (D_k Z_1 + Z_2)(\Omega_1^{2^k} \oplus J_s^{2^k}), \tag{3.70a}$$

$$A_k Z_3(\Omega_2^{2^k} \oplus \overline{J}_s^{2^k}) = D_k Z_3 + Z_4, \tag{3.70b}$$

$$-(C_k Z_1 + Z_2) = B_k Z_1(\Omega_1^{2^k} \oplus J_s^{2^k}), \tag{3.70c}$$

$$-(C_k Z_3 + Z_4)(\Omega_2^{2^k} \oplus \overline{J}_s^{2^k}) = B_k Z_3.$$

From (3.70b), it follows that

$$D_k = -Z_4 Z_3^{-1} + A_k Z_3(\Omega_2^{2^k} \oplus \overline{J}_s^{2^k}) Z_3^{-1}. \tag{3.71}$$

Substituting (3.71) into (3.70a), we get

$$\begin{aligned} A_k & \left[I - Z_3 \left(\Omega_2^{2^k} \oplus \overline{J}_s^{2^k} \right) Z_3^{-1} Z_1 \left(\Omega_1^{2^k} \oplus J_s^{2^k} \right) Z_1^{-1} \right] \\ & = \left(-Z_4 Z_3^{-1} Z_1 + Z_2 \right) \left(\Omega_1^{2^k} \oplus J_s^{2^k} \right) Z_1^{-1}. \end{aligned}$$

Since $\Omega_1^{2^k}$ and $\Omega_2^{2^k}$ are uniformly bounded (independent of k), and $\rho(J_s) < 1$, we then have

$$\begin{aligned} A_k & \left[I - \left(z_3 \Omega_2^{2^k} \omega_3^\top \right) \left(z_1 \Omega_1^{2^k} \omega_1^\top \right) + O\left(\rho(J_s^{2^k}) \right) \right] \\ & = \left(-Z_4 Z_3^{-1} z_1 + z_2 \right) \Omega_1^{2^k} \omega_1^\top + O\left(\rho(J_s^{2^k}) \right), \end{aligned}$$

where

$$z_1 = Z_1(:, 1:\ell), \; z_2 = Z_2(:, 1:\ell), \; z_3 = Z_3(:, 1:\ell),$$
$$\omega_1^\top = Z_1^{-1}(1:\ell, :), \; \omega_3^\top = Z_3^{-1}(1:\ell, :).$$

By the assumption that A_k is uniformly bounded on k, we have

$$A_k = a_k \omega_1^\top + O(\rho(J_s^{2^k}))$$

for some suitable $a_k \in \mathbb{C}^{n\times\ell}$ with $\|a_k\|$ being uniformly bounded on k. Thus we have

$$A_k Z_{12} = O(\rho(J_s^{2^k})) \to 0, \text{ as } k \to \infty,$$

where $Z_{12} = Z_1(:, \ell+1, \ldots, n)$ being orthogonal to ω_1. This proves (i).

Since $B_k = P\overline{A}_k P$ by Theorem 3.4.2, from (3.70c) we obtain

$$C_k = -Z_2 Z_1^{-1} + c_k \omega_1^\top + O(\rho(J_s^{2^k}))$$

for some suitable $c_k \in \mathbb{C}^{n\times\ell}$ with $\|c_k\|$ being uniformly bounded on k. Thus, we have

$$C_k Z_{12} = -Z_2 Z_1^{-1} Z_{12} + c_k \omega_1^\top Z_{12} + O(\rho(J_s^{2^k})) = -Z_2 \begin{bmatrix} 0 \\ I_{n-\ell} \end{bmatrix} + O(\rho(J_s^{2^k})) \to -Z_{22}$$

as $k \to \infty$, where $Z_{22} = Z_2(:, \ell+1, \ldots, n)$. This proves (ii).

Similarly from (3.71), we see, as $k \to \infty$, that

$$D_k Z_{32} = -Z_4 Z_3^{-1} Z_{32} + d_k \omega_3^\top Z_{32} + O(\rho(\overline{J}_s^{2^k})) \to -Z_{42} \to 0,$$

where $Z_{32} \equiv Z_3(:, \ell+1, \ldots, n)$, $Z_{42} \equiv Z_4(:, \ell+1, \ldots, n)$ and $d_k \in \mathbb{C}^{n\times\ell}$ is some suitable matrix which is uniformly bounded on k. □

3.5 PDA

Quoted from [57], we develop the palindromic doubling algorithm (PDA) for the numerical solution of the palindromic generalized eigenvalue problem (PGEP)

$$A^* x = \lambda A x, \tag{3.72}$$

where A is a real or complex $N \times N$ matrix, $\lambda \in \mathbb{C}$ and $x \in \mathbb{C}^N \backslash \{0\}$ are eigenvalue and the corresponding eigenvector of (3.72), respectively. Here, the symbol "$*$" $= \top$ (Transpose) or H (Hermitian). The pencil $A^* - \lambda A$ and the pair (A^*, A) are usually called a palindromic linear pencil and a palindromic matrix pair, respectively. It is easily seen that the eigenvalues of (3.72) satisfy the reciprocal property, i.e., they appear in pairs as in $\{\lambda, 1/\lambda^*\}$.

The PGEPs with complex coefficient matrices were firstly suggested as "good" linearizations [66, 67] of palindromic polynomial/quadratic matrix pencils, arising from the study of vibration analysis [46, 86]. A PGEP with real coefficient matrices can also be shown to be equivalent to the generalized continuous/discrete-time algebraic Riccati equations, associated with the continuous/discrete-time, linear-quadratic optimal control problems (see [85] for details).

The standard approach for solving the PGEP is to compute its generalized Schur form (e.g., by `qz` in MATLAB), ignoring its symmetric or palindromic structure in (A^*, A). However, the reciprocal property of eigenvalues of (3.72) is not preserved by computation generally, producing large numerical errors [68]. Recently, a QR-like algorithm [76] and a hybrid method [68] (which combines Jacobi-type method with the Laub's trick) were proposed for the PGEP. The QR-like algorithm generally requires $O(N^4)$ flops and the hybrid method requires $O(N^3 \log N)$ flops. Alternatively, for methods of cubic complexity, a URV-decomposition based structured method [77] and a structure-preserving algorithm [41] for PGEPs were proposed, producing eigenvalues which are paired to working precision. Unfortunately for PGEPs, these more efficient (and equivalent) methods require the transformation of the PGEP to the quadratic form $(\mu^2 A^* + \mu \cdot 0 + A)x = 0$, leading to operations in larger $2N \times 2N$ matrices. The PDA is a unique and more direct, thus more efficient, algorithm for the PGEP.

The purpose of this section is to develop the PDA for solving the PGEP structurally. We establish quadratic convergence and linear convergence with rate 1/2 of the PDA, respectively, when (A^*, A) has no unimodular eigenvalues and has unimodular eigenvalues with partial multiplicities two. In application to discrete-time optimal control problems, we especially develop a new algorithm combined with the PDA (as in Algorithm 3.3.1) for solving the optimal control of *singular* descriptor linear systems. To our knowledge, the associated generalized discrete-time algebraic Riccati equation (GDARE) has not been solved successfully in a structure-preserving manner.

We shall denote the open left-half plane and the open unit disk by $\mathbb{C}_-$ and $\mathbb{D}_1$, respectively.

3.5.1 Palindromic doubling algorithm

For a given palindromic matrix pair (A^*, A), we shall develop a doubling algorithm for solving the associated PGEP which preserves the palindromic structure at each iterative step.

Suppose $-1 \notin \sigma(A^*, A)$ (the assumption can be removed later in Remark 3.5.1). We then have

$$\begin{aligned} A^*(A^* + A)^{-1}A &= ((A^* + A) - A)(A^* + A)^{-1}(A + A^* - A^*) \\ &= (I - A(A^* + A)^{-1})((A^* + A) - A^*) \\ &= A(A^* + A)^{-1}A^*. \end{aligned} \tag{3.73}$$

From (3.73), it is easily seen that

$$\left[A(A^* + A)^{-1},\ A^*(A^* + A)^{-1}\right] \begin{bmatrix} A^* \\ -A \end{bmatrix} = 0.$$

We now define the doubling transform $A \to \hat{A}$ by

$$\hat{A} = A(A^* + A)^{-1}A. \tag{3.74}$$

Theorem 3.5.1. *The matrix pair $(\hat{A}^*, \hat{A})$ has the doubling property; i.e., if*

$$A^*U = AUS, \tag{3.75}$$

where $U \in \mathbb{C}^{N \times \ell}$ and $S \in \mathbb{C}^{\ell \times \ell}$, then

$$\hat{A}^*U = \hat{A}US^2. \tag{3.76}$$

Proof. Multiplying the both sides of (3.75) by $A^*(A^*+A)^{-1}$, and (3.73) and (3.75) imply (3.76). □

From Theorem 3.5.1, we see that the doubling transform (3.74) preserves the palindromic structure. So, for a palindromic matrix pair (A_0^*, A_0) with $A_0 \in \mathbb{C}^{N\times N}$, we can develop the PDA to generate the sequence $\{(A_k^*, A_k)\}$ if no breakdown occurs in the iterative process.

Algorithm 3.5.1 (PDA). *Given $A_0 \in \mathbb{C}^{N\times N}$, τ (a small tolerance),*
for** $k = 0, 1, 2, \ldots$, **compute

$$A_{k+1} = A_k(A_k^* + A_k)^{-1}A_k, \tag{3.77}$$

***if** dist(Null(A_{k+1}), Null(A_k)) $< \tau$, then stop,*
end for
Here, "Null($\cdot$)" denotes the null space of the given matrix and "dist($\cdot,\cdot$)"denotes the distance between two subspaces.

To develop the PDA further, denote

$$A_k = H_k + K_k, \tag{3.78}$$

where

$$H_k = \frac{1}{2}(A_k^* + A_k) = H_k^*, \quad K_k = \frac{1}{2}(A_k - A_k^*) = -K_k^*$$

are the $*$-symmetric and $*$-anti-symmetric parts of A_k, respectively. Then the iteration (3.77) can be rewritten as

$$A_{k+1} = H_{k+1} + K_{k+1} = \frac{1}{2}(H_k + K_k)H_k^{-1}(H_k + K_k) = \frac{1}{2}(H_k + K_kH_k^{-1}K_k) + K_k.$$

The iteration (3.77) in the PDA can be simplified to

$$H_{k+1} = \frac{1}{2}(H_k + K_0H_k^{-1}K_0), \quad K_{k+1} = K_k = \cdots = K_0.$$

3.5.2 Convergence of PDA

Let $A_0 \in \mathbb{C}^{N\times N}$. Suppose the eigenvalue "1" of (A_0^*, A_0) (if exists) has partial multiplicity one or two, and the other unimodular eigenvalues of (A_0^*, A_0) (if exist) have exactly partial multiplicities two. By the theorem of Kronecker canonical form there are nonsingular matrices Q and Z such that

$$QA_0^*Z = \begin{bmatrix} J_0 \oplus \Omega_0 & 0_{\ell,\tilde{\ell}} \oplus I_r \\ 0_{\tilde{n},n} & I_{\tilde{\ell}} \oplus \Omega_0 \end{bmatrix} \equiv C_0, \quad QA_0Z = \begin{bmatrix} I_n & 0_{n,\tilde{n}} \\ 0_{\tilde{n},n} & \tilde{J}_0^* \oplus I_r \end{bmatrix} \equiv D_0, \tag{3.79}$$

where $\Omega_0 = \mathrm{diag}(e^{i\omega_1}, \ldots, e^{i\omega_r})$, $J_0 \in \mathbb{C}^{\ell\times\ell}$ consists of stable Jordan blocks (i.e., $\rho(J_0) < 1$, where $\rho(\cdot)$ is the radius of the spectrum) and $\tilde{J}_0^* - J_0 \oplus I_m$ with

$n = \ell + r$, $\tilde{\ell} = \ell + m$, $\tilde{n} = n + m = \ell + r + m$ and $N = 2n + m$. Here "$\oplus$" denotes the direct sum of matrices.

Since $C_0 D_0 = D_0 C_0$, from (3.79) we have that $A_0^* Z D_0 = A_0 Z C_0$. From Theorem 3.5.1 and steps in the PDA, it follows that

$$A_k^* Z D_0^{2^k} = A_k Z C_0^{2^k}. \tag{3.80}$$

Substituting (3.79) into (3.80), we get

$$A_k^* Z \begin{bmatrix} I_n & 0_{n,\tilde{n}} \\ 0_{\tilde{n},n} & (\tilde{J}_0^*)^{2^k} \oplus I_r \end{bmatrix} = A_k Z \begin{bmatrix} J_0^{2^k} \oplus \Omega_0^{2^k} & 0_{\ell,\tilde{\ell}} \oplus \Gamma_k \\ 0_{\tilde{n},n} & I_\ell \oplus \Omega_0^{2^k} \end{bmatrix}, \tag{3.81}$$

where $\Gamma_k = 2^k \Omega_0^{2^k - 1}$. On the other hand, we can interchange the role of (A_0^*, A_0) by considering the pair (A_0, A_0^*) which has the same Kronecker structure as (A_0^*, A_0). Therefore, there are nonsingular P and Y such that

$$PA_0 Y = \begin{bmatrix} J_0^* \oplus \Omega_0^* & 0_{\ell,\tilde{\ell}} \oplus I_r \\ 0_{\tilde{n},n} & I_{\tilde{\ell}} \oplus \Omega_0^* \end{bmatrix} \equiv E_0, \quad PA_0^* Y = \begin{bmatrix} I_n & 0_{n,\tilde{n}} \\ 0_{\tilde{n},n} & \tilde{J}_0 \oplus I_r \end{bmatrix} \equiv F_0, \tag{3.82}$$

Since $E_0 F_0 = F_0 E_0$, we deduce that $A_0 Y F_0 = A_0^* Y E_0$. Using the similar arguments as in (3.80)–(3.81), we obtain

$$A_k Y \begin{bmatrix} I_n & 0_n \\ 0_n & \tilde{J}_0^{2^k} \oplus I_r \end{bmatrix} = A_k^* Y \begin{bmatrix} (J_0^*)^{2^k} \oplus (\Omega_0^*)^{2^k} & 0_{\ell,\tilde{\ell}} \oplus \Gamma_k^* \\ 0_{\tilde{n},n} & I_{\tilde{\ell}} \oplus (\Omega_0^*)^{2^k} \end{bmatrix}. \tag{3.83}$$

We partition A_k, H_k and K_0 in (3.78) into four sub-blocks as in

$$A_k = \begin{bmatrix} A_{k1} & A_{k3} \\ A_{k2} & A_{k4} \end{bmatrix}, \; H_k = \begin{bmatrix} H_{k1} & H_{k2}^* \\ H_{k2} & H_{k4} \end{bmatrix}, \; K_0 = \begin{bmatrix} K_{01} & -K_{02}^* \\ K_{02} & K_{04} \end{bmatrix}, \tag{3.84}$$

where A_{k1}, H_{k1}, $K_{01} \in \mathbb{C}^{n \times n}$, A_{k2}^*, A_{k3}, H_{k2}^*, $K_{02}^* \in \mathbb{C}^{n \times \tilde{n}}$ and A_{k4}, H_{k4}, $K_{04} \in \mathbb{C}^{\tilde{n} \times \tilde{n}}$. From (3.78) and (3.84), we also have

$$\begin{aligned} A_{k1} &= H_{k1} + K_{01}, \quad & A_{k2} &= H_{k2} + K_{02}, \\ A_{k3} &= H_{k2}^* - K_{02}^*, \quad & A_{k4} &= H_{k4} + K_{04}. \end{aligned}$$

Furthermore, we partition Z in (3.81) and Y in (3.83) as in

$$Z = \begin{bmatrix} Z_1 & Z_3 \\ Z_2 & Z_4 \end{bmatrix}, \quad Y = \begin{bmatrix} Y_1 & Y_3 \\ Y_2 & Y_4 \end{bmatrix}, \tag{3.85}$$

where Z_1, $Y_1 \in \mathbb{C}^{n \times n}$; Z_2^*, Z_3, Y_2^*, $Y_3 \in \mathbb{C}^{n \times \tilde{n}}$ and Z_4, $Y_4 \in \mathbb{C}^{\tilde{n} \times \tilde{n}}$. For convenience, we denote

$$\begin{aligned} Z_{i,a} &\equiv Z_i(:, 1 : \ell), \quad & Y_{i,a} &\equiv Y_i(:, 1 : \ell), \quad & i &= 3, 4; \\ Z_{i,b} &\equiv Z_i(:, \ell + 1 : n), \quad & Y_{i,b} &\equiv Y_i(:, \ell + 1 : n), \quad & i &= 1, 2. \end{aligned}$$

Theorem 3.5.2. *Let $A_0 \in \mathbb{C}^{N\times N}$. Suppose that the eigenvalue "1" of (A_0^*, A_0) (if exists) has partial multiplicity one or two, the other unimodular eigenvalues of $(A_0^*,\ A_0)$ (if exist) have exactly partial multiplicities two, and (3.79) and (3.82) hold with $\tilde{n} \le 2\ell$ (i.e., $r+m \le \ell$). Suppose that Z_1 and Y_1 in (3.85) are invertible, and $W \equiv [\Phi Z_{3,a} - Z_{4,a} | \Psi Y_{3,a} - Y_{4,a}] \in \mathbb{C}^{\tilde{n}\times 2\ell}$ is of full row rank, where $\Phi \equiv Z_2 Z_1^{-1}$, $\Psi \equiv Y_2 Y_1^{-1}$. If $-1 \notin \bigcup_{j=1}^{r}\{e^{2^k i\omega_j}, k \ge 0\}$, then the sequence $\{(A_k^*, A_k)\}$ generated by the PDA is well defined and satisfies*

$$A_k^* \begin{bmatrix} Z_1 \\ Z_2 \end{bmatrix},\ A_k \begin{bmatrix} Y_1 \\ Y_2 \end{bmatrix} \to 0,$$

linearly as $k \to \infty$ with convergence rate $1/2$, where span$\left\{\begin{bmatrix} Z_1 \\ Z_2 \end{bmatrix}\right\}$ and span$\left\{\begin{bmatrix} Y_3 \\ Y_4 \end{bmatrix}\right\}$ form the weakly stable and the unstable invariant subspaces of (A_0^*, A_0) corresponding to $(J_0 \oplus \Omega_0, I_n)$ and $(I_n, J_0^* \oplus \Omega_0^*)$, respectively.

Proof. Since $-1 \notin \bigcup_{j=1}^{r}\{e^{2^k i\omega_j}, k \ge 0\}$, from (3.77) we see that $-1 \notin \sigma(A_k^*, A_k)$, thus, $A_k^* + A_k$ is invertible for all k.

From (3.84), (3.81) and (3.85), we have

$$A_{k1}^* Z_1 + A_{k2}^* Z_2 = A_{k1} Z_1 (J_0^{2^k} \oplus \Omega_0^{2^k}) + A_{k3} Z_2 (J_0^{2^k} \oplus \Omega_0^{2^k}), \tag{3.86}$$

$$A_{k3}^* Z_1 + A_{k4}^* Z_2 = A_{k2} Z_1 (J_0^{2^k} \oplus \Omega_0^{2^k}) + A_{k4} Z_2 (J_0^{2^k} \oplus \Omega_0^{2^k}), \tag{3.87}$$

$$\begin{aligned} & A_{k1}^* Z_3 ((\tilde{J}_0^*)^{2^k} \oplus I_r) + A_{k2}^* Z_4 ((\tilde{J}_0^*)^{2^k} \oplus I_r) \\ =& A_{k1}[Z_1(0_{\ell,\tilde{\ell}} \oplus \Gamma_k) + Z_3(I_{\tilde{\ell}} \oplus \Omega_0^{2^k})] \\ &+ A_{k3}[Z_2(0_{\ell,\tilde{\ell}} \oplus \Gamma_k) + Z_4(I_{\tilde{\ell}} \oplus \Omega_0^{2^k})], \end{aligned} \tag{3.88}$$

$$\begin{aligned} & A_{k3}^* Z_3 ((\tilde{J}_0^*)^{2^k} \oplus I_r) + A_{k4}^* Z_4 ((\tilde{J}_0^*)^{2^k} \oplus I_r) \\ =& A_{k2}[Z_1(0_{\ell,\tilde{\ell}} \oplus \Gamma_k) + Z_3(I_{\tilde{\ell}} \oplus \Omega_0^{2^k})] \\ &+ A_{k4}[Z_2(0_{\ell,\tilde{\ell}} \oplus \Gamma_k) + Z_4(I_{\tilde{\ell}} \oplus \Omega_0^{2^k})]. \end{aligned} \tag{3.89}$$

Post-multiplying (3.88) by $0_{\tilde{\ell},\ell} \oplus \Gamma_k^{-1}\Omega_0^{2^k}$, we get

$$\begin{aligned} & A_{k1}^* Z_3 (0_{\tilde{\ell},\ell} \oplus \Gamma_k^{-1}\Omega_0^{2^k}) + A_{k2}^* Z_4 (0_{\tilde{\ell},\ell} \oplus \Gamma_k^{-1}\Omega_0^{2^k}) \\ &= (A_{k1} Z_1 + A_{k3} Z_2)(0_\ell \oplus \Omega_0^{2^k}) \\ &\quad + (A_{k1} Z_3 + A_{k3} Z_4)(0_{\tilde{\ell},\ell} \oplus \Omega_0^{2^k}\Gamma_k^{-1}\Omega_0^{2^k}). \end{aligned} \tag{3.90}$$

Substituting (3.90) into (3.86) and using $\Omega_0^{2^k}\Gamma_k^{-1}\Omega_0^{2^k} = 2^{-k}\Omega_0^{2^k+1}$, we have

$$\begin{aligned} & A_{k1}^* Z_1 + A_{k2}^* Z_2 \\ =& (A_{k1} Z_1 + A_{k3} Z_2)(J_0^{2^k} \oplus 0_r) + (A_{k1} Z_1 + A_{k3} Z_2)(0_\ell \oplus \Omega_0^{2^k}) \\ =& (A_{k1} Z_1 + A_{k3} Z_2)(J_0^{2^k} \oplus 0_r) + (A_{k1}^* Z_3 + A_{k2}^* Z_4)(0_{\tilde{\ell},\ell} \oplus \Gamma_k^{-1}\Omega_0^{2^k}) \\ & - (A_{k1} Z_3 + A_{k3} Z_4)(0_{\tilde{\ell},\ell} \oplus 2^{-k}\Omega_0^{2^k+1}). \end{aligned} \tag{3.91}$$

Using (3.84) and re-arranging (3.91), we get

$$\begin{aligned} H_{k1} &\left\{ Z_1[I_n - (J_0^{2^k} \oplus 0_r)] - Z_3[0_{\tilde{\ell},\ell} \oplus 2^{-k}\Omega_0(I_r - \Omega_0^{2^k})] \right\} \\ &+ H_{k2}^* \left\{ Z_2[I_n - (J_0^{2^k} \oplus 0_r)] - Z_4[0_{\tilde{\ell},\ell} \oplus 2^{-k}\Omega_0(I_r - \Omega_0^{2^k})] \right\} \\ =& K_{01} \left\{ Z_1[I_n + (J_0^{2^k} \oplus 0_r)] - Z_3[0_{\tilde{\ell},\ell} \oplus 2^{-k}\Omega_0(I_r + \Omega_0^{2^k})] \right\} \\ &- K_{02}^* \left\{ Z_2[I_n + (J_0^{2^k} \oplus 0_r)] - Z_4[0_{\tilde{\ell},\ell} \oplus 2^{-k}\Omega_0(I_r + \Omega_0^{2^k})] \right\}. \end{aligned} \tag{3.92}$$

Denote

$$\epsilon_k \equiv \max\{\rho(J_0)^{2^k}, 2^{-k}\} \to 0, \quad \text{as } k \to \infty. \tag{3.93}$$

Since $\|\Omega_0^{2^k}\|$ is bounded and Z_1 is invertible, by letting $\Phi \equiv Z_2 Z_1^{-1}$, (3.92) can be simplified to

$$H_{k1} = -H_{k2}^*(\Phi + O(\epsilon_k)) + K_{01} - K_{02}^*\Phi + O(\epsilon_k). \tag{3.94}$$

Post-multiplying (3.88) by $I_{\tilde{\ell}} \oplus \Gamma_k^{-1}$, we have

$$\begin{aligned} A_{k1}^* Z_3((\tilde{J}_0^*)^{2^k} \oplus \Gamma_k^{-1}) &+ A_{k2}^* Z_4((\tilde{J}_0^*)^{2^k} \oplus \Gamma_k^{-1}) \\ &= A_{k1}[Z_1(0_{\ell,\tilde{\ell}} \oplus I_r) + Z_3(I_{\tilde{\ell}} \oplus \Omega_0^{2^k}\Gamma_k^{-1})] \\ &\quad + A_{k3}[Z_2(0_{\ell,\tilde{\ell}} \oplus I_r) + Z_4(I_{\tilde{\ell}} \oplus \Omega_0^{2^k}\Gamma_k^{-1})]. \end{aligned} \tag{3.95}$$

From (3.84) and (3.93), (3.95) becomes

$$\begin{aligned} H_{k1}&[Z_3(I_\ell \oplus 0_{m+r}) + Z_1(0_{\ell,\tilde{\ell}} \oplus I_r) + O(\epsilon_k)] \\ &+ H_{k2}^*[Z_4(I_\ell \oplus 0_{m+r}) + Z_2(0_{\ell,\tilde{\ell}} \oplus I_r) + O(\epsilon_k)] \\ =& - K_{01}[Z_3(I_\ell \oplus 2I_m \oplus 0_r) + Z_1(0_{\ell,\tilde{\ell}} \oplus I_r) + O(\epsilon_k)] \\ &+ K_{02}^*[Z_4(I_\ell \oplus 2I_m \oplus 0_r) + Z_2(0_{\ell,\tilde{\ell}} \oplus I_r) + O(\epsilon_k)]. \end{aligned} \tag{3.96}$$

Substituting (3.94) into (3.96) we get

$$\begin{aligned} H_{k2}^* &\left\{ (\Phi + O(\epsilon_k))[Z_3(I_\ell \oplus 0_{m+r}) + Z_1(0_{\ell,\tilde{\ell}} \oplus I_r) + O(\epsilon_k)] \right. \\ &\left. - Z_4(I_\ell \oplus 0_{m+r}) - Z_2(0_{\ell,\tilde{\ell}} \oplus I_r) + O(\epsilon_k) \right\} = O(1). \end{aligned}$$

Since $\Phi Z_{1,b} = Z_{2,b}$, it holds that

$$H_{k2}^*([\Phi Z_{3,a} - Z_{4,a}] + O(\epsilon_k)) = O(1) \in \mathbb{C}^{\tilde{n} \times \ell}. \tag{3.97}$$

On the other hand, from (3.84), (3.83) and (3.85), we have

$$A_{k1}Y_1 + A_{k3}Y_2 = A_{k1}^* Y_1((J_0^*)^{2^k} \oplus (\Omega_0^*)^{2^k}) + A_{k2}^* Y_2((J_0^*)^{2^k} \oplus (\Omega_0^*)^{2^k}), \tag{3.98}$$

$$A_{k2}Y_1 + A_{k4}Y_2 = A_{k3}^* Y_1((J_0^*)^{2^k} \oplus (\Omega_0^*)^{2^k}) + A_{k4}^* Y_2((J_0^*)^{2^k} \oplus (\Omega_0^*)^{2^k}), \tag{3.99}$$

$$A_{k1}Y_3(\tilde{J}_0^{2^k} \oplus I_r) + A_{k3}Y_4(\tilde{J}_0^{2^k} \oplus I_r) = (A_{k1}^*Y_3 + A_{k2}^*Y_4)(I_{\tilde{\ell}} \oplus (\Omega_0^*)^{2^k}) + (A_{k1}^*Y_1 + A_{k2}^*Y_2)(0_{\ell,\tilde{\ell}} \oplus \Gamma_k^*), \quad (3.100)$$

$$A_{k2}Y_3(\tilde{J}_0^{2^k} \oplus I_r) + A_{k4}Y_4(\tilde{J}_0^{2^k} \oplus I_r) = (A_{k3}^*Y_3 + A_{k4}^*Y_4)(I_{\tilde{\ell}} \oplus (\Omega_0^*)^{2^k}) + (A_{k3}^*Y_1 + A_{k4}^*Y_2)(0_{\ell,\tilde{\ell}} \oplus \Gamma_k^*). \quad (3.101)$$

As in (3.90)–(3.92), post-multiplying (3.100) by $0_{\tilde{\ell},\ell} \oplus (\Gamma_k^*)^{-1}(\Omega_0^*)^{2^k}$ and substituting it into (3.98), we have

$$A_{k1}Y_1 + A_{k3}Y_2 = (A_{k1}^*Y_1 + A_{k2}^*Y_2)((J_0^*)^{2^k} \oplus 0_r) + (A_{k1}^*Y_3 + A_{k3}^*Y_4)(0_{\tilde{\ell},\ell} \oplus 2^{-k}\Omega_0^*) - (A_{k1}^*Y_3 + A_{k2}^*Y_4)(0_{\tilde{\ell},\ell} \oplus 2^{-k}(\Omega_0^*)^{2^k+1}). \quad (3.102)$$

From (3.84) and (3.93), (3.102) becomes

$$\begin{aligned} H_{k1} &\left\{Y_1[I_n - ((J_0^*)^{2^k} \oplus 0_r)] - Y_3[0_{\tilde{\ell},\ell} \oplus 2^{-k}\Omega_0^*(I_r - (\Omega_0^*)^{2^k})]\right\} \\ &+ H_{k2}^*\left\{Y_2[I_n - ((J_0^*)^{2^k} \oplus 0_r)] - Y_4[0_{\tilde{\ell},\ell} \oplus 2^{-k}\Omega_0^*(I_r - (\Omega_0^*)^{2^k})]\right\} \\ = &- K_{01}\left\{Y_1[I_n + ((J_0^*)^{2^k} \oplus 0_r)] - Y_3[0_{\tilde{\ell},\ell} \oplus 2^{-k}\Omega_0^*(I_r + (\Omega_0^*)^{2^k})]\right\} \\ &+ K_{02}^*\left\{Y_2[I_n + ((J_0^*)^{2^k} \oplus 0_r)] - Y_4[0_{\tilde{\ell},\ell} \oplus 2^{-k}\Omega_0^*(I_r + (\Omega_0^*)^{2^k})]\right\}, \end{aligned} \quad (3.103)$$

and then

$$H_{k1} = -H_{k2}^*(\Psi + O(\epsilon_k)) - K_{01} + K_{02}^*\Psi + O(\epsilon_k), \quad \Psi = Y_2Y_1^{-1}.$$

Post-multiplying (3.100) by $I_{\tilde{\ell}} \oplus (\Gamma_k^*)^{-1}$ and substituting (3.103) into it, we have

$$H_{k2}^*\left\{Y_4(I_\ell \oplus 0_{m+r}) + Y_2(0_{\ell,\tilde{\ell}} \oplus I_r) + O(\epsilon_k) - (\Psi + O(\epsilon_k))[Y_3(I_\ell \oplus 0_{m+r}) + Y_1(0_{\ell,\tilde{\ell}} \oplus I_r) + O(\epsilon_k)]\right\} = O(1).$$

Since $\Psi Y_{1,b} = Y_{2,b}$, it holds that

$$H_{k2}^*([\Psi Y_{3,a} - Y_{4,a}] + O(\epsilon_k)) = O(1) \in \mathbb{C}^{\tilde{n}\times\ell}. \quad (3.104)$$

Combining (3.97) and (3.104) we get

$$H_{k2}^*([\Phi Z_{3,a} - Z_{4,a}|\Psi Y_{3,a} - Y_{4,a}] + O(\epsilon_k)) = O(1) \in \mathbb{C}^{\tilde{n}\times 2\ell}. \quad (3.105)$$

By the assumption that $W \equiv [\Phi Z_{3,a} - Z_{4,a}|\Psi Y_{3,a} - Y_{4,a}] \in \mathbb{C}^{\tilde{n}\times 2\ell}$ is of full row rank, it follows that $\|H_{k2}^*\|$ is uniformly bounded on k. Consequently, (3.94) implies that $\|H_{k1}\|$, and in turn $\|A_{k1}\|$ and $\|A_{k2}^*\|$, are uniformly bounded on k. From (3.91), it follows that

$$A_{k1}^*Z_1 + A_{k2}^*Z_2 = O(\epsilon_k) \to 0, \quad \text{as } k \to \infty. \quad (3.106)$$

Applying the similar argument as in (3.90)–(3.92) to (3.99) and (3.101), we deduce that

$$H_{k4} = -H_{k2}(Y_2Y_1^{-1} + O(\epsilon_k)) + K_{04} - K_{02}Y_2Y_1^{-1} + O(\epsilon_k).$$

Thus, (3.105) implies that $\|H_{k4}\|$, and in turn $\|A_{k4}\|$, are uniformly bounded on k.

To show $A_{k3}^* Z_1 + A_{k4}^* Z_2 = O(\epsilon_k)$, we post-multiply (3.89) by $0_{\tilde{\ell},\ell} \oplus \Gamma_k^{-1}\Omega_0^{2^k}$ and obtain

$$\begin{aligned} &A_{k3}^* Z_3(0_{\tilde{\ell},\ell} \oplus \Gamma_k^{-1}\Omega_0^{2^k}) + A_{k4}^* Z_4(0_{\tilde{\ell},\ell} \oplus \Gamma_k^{-1}\Omega_0^{2^k}) \\ &\quad = (A_{k2}Z_1 + A_{k4}Z_2)(0_\ell \oplus \Omega_0^{2^k}) + (A_{k2}Z_3 + A_{k4}Z_4)(0_{\tilde{\ell},\ell} \oplus 2^{-k}\Omega_0^{2^k+1}) \end{aligned} \quad (3.107)$$

Substituting (3.107) into (3.87), as in (3.91) we have

$$\begin{aligned} A_{k3}^* Z_1 + A_{k4}^* Z_2 =&(A_{k2}Z_1 + A_{k4}Z_2)(J_0^{2^k} \oplus 0_r) + (A_{k3}^* Z_3 + A_{k4}^* Z_4)(0_{\tilde{\ell},\ell} \oplus 2^{-k}\Omega_0) \\ &- (A_{k2}Z_3 + A_{k4}Z_4)(0_{\tilde{\ell},\ell} \oplus 2^{-k}\Omega_0^{2^k+1}) = O(\epsilon_k) \to 0, \quad \text{as} \to \infty. \end{aligned} \quad (3.108)$$

Combining (3.106) and (3.108), we have shown that

$$\begin{bmatrix} A_{k1}^* & A_{k2}^* \\ A_{k3}^* & A_{k4}^* \end{bmatrix} \begin{bmatrix} Z_1 \\ Z_2 \end{bmatrix} = O(\epsilon_k) \to 0, \quad \text{as } k \to \infty.$$

Similarly, as in (3.90)-(3.91), from (3.99) and (3.101) we have

$$\begin{aligned} A_{k2}Y_1 + A_{k4}Y_2 =&(A_{k3}^* Y_1 + A_{k4}^* Y_2)((J_0^*)^{2^k} \oplus 0_r) + (A_{k2}Y_3 + A_{k4}Y_4)(0_{\tilde{\ell},\ell} \oplus 2^{-k}\Omega_0^*) \\ &- (A_{k3}^* Y_3 + A_{k4}^* Y_4)(0_{\tilde{\ell},\ell} \oplus 2^{-k}(\Omega_0^*)^{2^k+1}) = O(\epsilon_k). \end{aligned} \quad (3.109)$$

Using the boundedness of $\|A_{ki}\|$, $i = 1, \ldots, 4$, and combining (3.102) and (3.109), we have shown that

$$\begin{bmatrix} A_{k1} & A_{k3} \\ A_{k2} & A_{k4} \end{bmatrix} \begin{bmatrix} Y_1 \\ Y_2 \end{bmatrix} = O(\epsilon_k) \to 0, \quad \text{as } k \to \infty,$$

because $\frac{1}{2^k}$ dominates $\rho(J_0)^{2^k}$ in (3.93) for sufficiently large values of k. □

Remark 3.5.1. *Consider the assumption* $-1 \notin U \equiv \bigcup_{j=1}^r \{e^{2^k i\omega_j}, k \geq 0\}$ *in Theorem 3.5.2. Since* U *is a countable set (possibly dense on the unit circle only when* $r \to \infty$*), there exist an* $-e^{i\theta_0} \notin U$. *With* $A_{\text{new}}^* \equiv e^{-i\theta_0/2}A_0^*$, *we have* $A_{\text{new}}^* + A_{\text{new}} = e^{i\theta_0/2}A_0 + e^{-i\theta_0/2}A_0 = e^{-i\theta_0/2}(A_0^* + e^{i\theta_0}A_0)$ *being invertible. It is unclear how the "optimal"* θ_0 *can be found.*

Theorem 3.5.3. *Suppose* (A_0^*, A_0) *has no unimodular eigenvalues. The sequence* $\{(A_k^*, A_k)\}$ *generated by the PDA satisfies*

$$A_k^* \begin{bmatrix} Z_1 \\ Z_2 \end{bmatrix}, \; A_k \begin{bmatrix} Y_1 \\ Y_2 \end{bmatrix} \to 0,$$

quadratically, as $k \to \infty$, *with convergence rate* $\rho(J_0)$.

Proof. Since (A_0^*, A_0) has no unimodular eigenvalues, Theorem 3.5.2 implies (A_k^*, A_k) has no unimodular eigenvalues and $(A_k^* + A_k)$ is invertible. So, the PDA is well-defined.

From (3.84), (3.81) and (3.85), we have

$$A_{k1}^* Z_1 + A_{k2}^* Z_2 = A_{k1} Z_1 J_0^{2^k} + A_{k3} Z_2 J_0^{2^k}, \tag{3.110}$$

$$A_{k3}^* Z_1 + A_{k4}^* Z_2 = A_{k2} Z_1 J_0^{2^k} + A_{k4} Z_2 J_0^{2^k}. \tag{3.111}$$

From (3.84), it holds that

$$H_{k1} Z_1 + H_{k2}^* Z_2 = (K_{01} Z_1 - K_{01}^* Z_2)(I + J_0^{2^k})(I - J_0^{2^k})^{-1}.$$

Therefore, $\|H_{k1} Z_1 + H_{k2}^* Z_2\| \leq O(1)$, and this implies

$$\|(H_{k1} + K_{01}) Z_1 + (H_{k2}^* - K_{02}^*) Z_2\| = \|A_{k1} Z_1 + A_{k3} Z_2\| \leq O(1). \tag{3.112}$$

From (3.110) and (3.112), we have

$$A_{k1}^* Z_1 + A_{k2}^* Z_2 = O(\rho(J_0)^{2^k}) \to 0, \quad \text{as } k \to \infty.$$

Similarly, from (3.111), we obtain

$$H_{k2} Z_1 + H_{k4}^* Z_2 = (K_{02} Z_1 + K_{04}^* Z_2)(I + J_0^{2^k})(I - J_0^{2^k})^{-1},$$

which is uniformly bounded on k. This implies

$$A_{k3}^* Z_1 + A_{k4}^* Z_2 = O(\rho(J_0)^{2^k}) \to 0, \quad \text{as } k \to \infty.$$

This shows that $A_k^* \begin{bmatrix} Z_1 \\ Z_2 \end{bmatrix} \to 0$, quadratically, with convergence rate $\rho(J_0)$. Similarly, from (3.84), (3.83) and (3.85), we can also show that $A_k \begin{bmatrix} Y_1 \\ Y_2 \end{bmatrix} \to 0$ quadratically, with rate $\rho(J_0)$. □

3.5.3 Applications of PDA

In this section, we want to apply the PDA to find all the eigenpairs of a general PGEP, and solve the c-/d-stabilizing solutions of generalized continuous/discrete-time algebraic Riccati equations (GCARE/GDARE). We especially apply the Algorithm 3.5.2 for the computation of the d-semi-stabilizing solution of GDAREs arising in the optimal control of singular descriptor linear systems. To our knowledge, Algorithm 3.5.2 is the first structure-preserving algorithm for solving GDAREs associated with singular descriptor systems.

For operation counts or complexity, it depends on the details in the individual applications and whether efficiency can be squeezed from these fine structures. From the PDA, it suffices to say that the algorithm is of $O(N^3)$ complexity per iteration. In addition, for problems without unimodular eigenvalues, the convergence is quadratic and typically less than ten iterations are required for convergence to machine accuracy.

PGEP

We can apply the PDA to solve the PGEP $A_0^* x = \lambda A_0 x$, where $A_0 \in \mathbb{C}^{2n\times 2n}$. First, we apply the PDA to A_0 until convergence to A_k. Then we compute the bases Z_s,$Y_s \in \mathbb{C}^{2n\times n}$ for the right and left null spaces of A_k^*, respectively, satisfying

$$A_k^* Z_s = 0, \quad Y_s^* A_k^* = 0.$$

This implies that there are S and $T \in \mathbb{C}^{n\times n}$ with $\rho(S) \leq 1$ and $\rho(T) \leq 1$ such that

$$A_0^* Z_s = A_0 Z_s S, \quad A_0 Y_s = A_0^* Y_s T. \tag{3.113}$$

From (3.113), S and T can be computed by

$$S = (Y_s^* A_0 Z_s)^{-1}(Y_s^* A_0^* Z_s) \equiv S_1^{-1} S_2, \quad T = (Z_s^* A_0^* Y_s)^{-1}(Z_s^* A_0 Y_s) \equiv S_1^{-*} S_2^*.$$

Rewrite the second equation of (3.113) as

$$A_0(Y_s S_1^{-*}) = A_0^*(Y_s S_1^{-*}) S_2^* S_1^{-*} = A_0^*(Y_s S_1^{-*}) S^*.$$

Compute $S g_j = \lambda_j g_j$ and $S^* h_j = \lambda_j^* h_j$, as well as $z_j = Z_s g_j$ and $y_j = (Y_s S_1^{-*}) h_j$, for $j = 1, \ldots, n$. It holds that

$$A_0^* z_j = \lambda_j A_0 z_j, \quad \lambda_j^* A_0^* y_j = A_0 y_j, \quad j = 1, \ldots, n.$$

GCARE

Here we are interested in finding the c-stabilizing solution of the generalized continuous-time algebraic Riccati equation (GCARE)

$$A_c^\top X_c E_c + E_c^\top X_c A_c - (N_c + E_c^\top X_c B_c) R_c^{-1} (N_c + E_c^\top X_c B_c)^\top + M_c = 0, \tag{3.114}$$

which solves the continuous-time linear-quadratic control problem

$$\min_u \frac{1}{2} \int_0^\infty \begin{bmatrix} x \\ u \end{bmatrix}^\top \begin{bmatrix} M_c & N_c \\ N_c^\top & R_c \end{bmatrix} \begin{bmatrix} x \\ u \end{bmatrix} dt$$

subject to the descriptor linear system

$$E_c \dot{x} = A_c x + B_c u, \quad x(0) = x^0,$$

where $E_c, A_c, M_c = M_c^\top, X_c = X_c^\top \in \mathbb{R}^{n\times n}$, $B_c, N_c \in \mathbb{R}^{n\times m}$ and $R_c = R_c^\top \in \mathbb{R}^{m\times m}$ with E_c and R_c being nonsingular. Furthermore, the c-stabilizing closed-loop matrix pencil of (3.115) is given by $A_c + B_c K_c - \lambda E_c$ with the $\sigma(A_c + B_c K_c, E_c) \subseteq \mathbb{C}_-$, where $K_c \equiv -R_c^{-1}(B_c^\top X_c E_c + N_c^\top)$.

Let

$$\mathcal{M}_c - \lambda\mathcal{L}_c \equiv \begin{bmatrix} 0 & A_c & B_c \\ A_c^\top & M_c & N_c \\ B_c^\top & N_c^\top & R_c \end{bmatrix} - \lambda \begin{bmatrix} 0 & E_c & 0 \\ -E_c^\top & 0 & 0 \\ 0 & 0 & 0 \end{bmatrix}. \quad (3.116)$$

One common approach to solve (3.114) is to compute the n-dimensional, c-stable invariant subspace $\mathcal{U}_c$ of the symmetric/skew-symmetric pencil $\mathcal{M}_c - \lambda\mathcal{L}_c$ corresponding to the eigenvalue matrix pair (S_c, E_c) with $\sigma(S_c, E_c) \subseteq \mathbb{C}_-$, where $\mathcal{U}_c$ is the column space of $U_c \in \mathbb{R}^{(2n+m)\times n}$ which satisfies $\mathcal{M}_c U_c E_c = \mathcal{L}_c U_c S_c$.

We assume that the matrix pencil $\mathcal{M}_c - \lambda\mathcal{L}_c$ has no eigenvalues on the imaginary axis. The generalized eigenvalues of $(\mathcal{M}_c, \mathcal{L}_c)$ can be arranged by

$$\lambda_1, \ldots, \lambda_{2n}; \bar{\lambda}_1, \ldots, \bar{\lambda}_{2n}; \underbrace{\infty, \ldots, \infty}_{m},$$

where $\lambda_i \in \mathbb{C}_-$, for $1 \leq i \leq 2n$. The m trivial infinity eigenvalues are from the nonsingularity of R_c.

With

$$U_c = \begin{bmatrix} X_c E_c \\ I_n \\ K_c \end{bmatrix} \begin{matrix} \}\, n \\ \}\, n \\ \}\, m \end{matrix},$$

X_c is the c-stabilizing solution of GCARE (3.114) and K_c is the optimal controller for (3.115)[85].

In order to utilize the PDA to compute an orthogonal basis $V = \left[V_1^\top, V_2^\top, V_3^\top\right]^\top$ for $\mathcal{U}_c$ with $V_1, V_2 \in \mathbb{R}^{n\times n}$, we consider the Cayley transformation

$$\mathcal{A}_0^\top - \lambda\mathcal{A}_0 = (\mathcal{M}_c + \mathcal{L}_c) - \lambda(\mathcal{M}_c - \mathcal{L}_c),$$

where

$$\mathcal{A}_0 = \mathcal{M}_c - \mathcal{L}_c = \begin{bmatrix} 0 & A_c - E_c & B_c \\ A_c^\top + E_c^\top & M_c & N_c \\ B_c^\top & N_c^\top & R_c \end{bmatrix}.$$

Then the c-stabilizing solution X_c for GCARE (3.114) can be obtained by $X_c = V_1 V_2^{-1} E_c^{-1}$.

GDARE

Here we are interested in finding the d-semi-stabilizing solution of the generalized discrete-time algebraic Riccati equation (GDARE)

$$A_d^\top X_d A_d - E_d^\top X_d E_d - (N_d + A_d^\top X_d B_d)(R_d + B_d^\top X_d B_d)^{-1}(N_d + A_d^\top X_d B_d)^\top + M_d = 0 \quad (3.117)$$

which solves the discrete-time linear-quadratic control problem

$$\min_{u_k} \frac{1}{2}\sum_{k=0}^{\infty} \begin{bmatrix} x_k \\ u_k \end{bmatrix}^\top \begin{bmatrix} M_d & N_d \\ N_d^\top & R_d \end{bmatrix} \begin{bmatrix} x_k \\ u_k \end{bmatrix}$$

subject to the singular descriptor linear system

$$E_d x_{k+1} = A_d x_k + B_d u_k, \quad x_0 = x^0,$$

where $E_d, A_d, M_d = M_d^\top, X_d = X_d^\top \in \mathbb{R}^{n\times n}$, $B_d, N_d \in \mathbb{R}^{n\times m}$ and $R_d = R_d^\top \in \mathbb{R}^{m\times m}$ with E_d being singular. Furthermore, the d-semi-stabilizing closed-loop matrix pencil of (3.118) is given by $A_d + B_d K_d - \lambda E_d$ with the $\sigma(A_d + B_d K_d, E_d) \subseteq \mathbb{D}_1 \cup \{\infty\}$, where

$$K_d \equiv -(R_d + B_d^\top X_d B_d)^{-1}(B_d^\top X_d A_d + N_d^\top).$$

Let

$$\mathcal{M}_d - \lambda \mathcal{L}_d \equiv \begin{bmatrix} 0 & A_d & B_d \\ E_d^\top & M_d & N_d \\ 0 & N_d^\top & R_d \end{bmatrix} - \lambda \begin{bmatrix} 0 & E_d & 0 \\ A_d^\top & 0 & 0 \\ B_d^\top & 0 & 0 \end{bmatrix}.$$

One common approach to solve (3.117) is to compute the n-dimensional, d-semi-stable invariant subspace $\mathfrak{U}_d$ of the matrix pencil $\mathcal{M}_d - \lambda \mathcal{L}_d$ corresponding to the eigenvalue matrix pair (S_d, E_d) with $\sigma(S_d, E_d) \subseteq \mathbb{D}_1 \cup \{\infty\}$, where $\mathfrak{U}_d$ is the column space of $U_d \in \mathbb{R}^{(2n+m)\times n}$ which satisfies $\mathcal{M}_d U_d E_d = \mathcal{L}_d U_d S_d$.

With

$$U_d = \begin{bmatrix} X_d E_d \\ I_n \\ K_d \end{bmatrix} \begin{matrix} \}\, n \\ \}\, n \\ \}\, m \end{matrix},$$

X_d is the d-semi-stabilizing solution of GDARE (3.117) and K_d is the optimal controller for (3.118)[85].

Assume that the matrix pencil $\mathcal{M}_d - \lambda \mathcal{L}_d$ has no eigenvalues on the unit circle, r_d =nullity(E_d) and $\mathrm{ind}_\infty(A_d, E_d) \le 1$. From [85] we see that

$$\sigma(\mathcal{M}_d, \mathcal{L}_d) = \sigma(A_d + B_d K_d, E_d) \cup \sigma(E_d^\top, (A_d + B_d K_d)^\top) \cup \sigma(0_m, I_m). \quad (3.119)$$

So, the generalized eigenvalues of $(\mathcal{M}_d, \mathcal{L}_d)$ corresponding to (3.119) can be arranged by

$$\{\lambda_1, \ldots, \lambda_{n-r_d}, \underbrace{\infty, \ldots, \infty}_{r_d}\} \cup \{\underbrace{0, \ldots, 0}_{r_d}, \lambda_1^{-1}, \ldots, \lambda_{n-r_d}^{-1}\} \cup \{\underbrace{\infty, \ldots, \infty}_{m}\}, \quad (3.120)$$

where $\lambda_i \in \mathbb{D}_1$ (can possibly be zero) , $i = 1, \ldots, n - r_d$. For convenience, we apply the convention that 0 and ∞ are mutually reciprocal. The r_d infinity and r_d zero eigenvalues in (3.120) are from the assumption r_d = nullity(E_d). The last trivial m infinity eigenvalues are from the last m columns of $\mathcal{L}_d$. In fact, $(A_d + B_d K_d, E_d)$ is an eigenvalue matrix pair associated with the d-semi-stable invariant subspace $\mathfrak{U}_d$.

We now introduce an elegant transformation between the coefficient matrices of the GDARE (3.117) and GCARE (3.114) proposed by [85]. We define

$$f_W : (E_d, A_d, B_d, M_d, N_d, R_d) \to (E_c, A_c, B_c, M_c, N_c, R_c),$$

where the matrices $E_c, A_c, B_c, M_c, N_c, R_c$ satisfy

$$\begin{bmatrix} E_c & 0 \\ A_c & B_c \end{bmatrix} = \chi \begin{bmatrix} E_d & 0 \\ A_d & B_d \end{bmatrix} W_d^\top = \frac{1}{\sqrt{2}} \begin{bmatrix} A_d + E_d & B_d \\ A_d - E_d & B_d \end{bmatrix} W_d^\top,$$
$$\begin{bmatrix} M_c & N_c \\ N_c^\top & R_c \end{bmatrix} = W_d \begin{bmatrix} M_d & N_d \\ N_d^\top & R_d \end{bmatrix} W_d^\top,$$

in which $\chi = \frac{1}{\sqrt{2}} \begin{bmatrix} I_n & I_n \\ -I_n & I_n \end{bmatrix}$, and $\begin{bmatrix} A_d + E_d & B_d \end{bmatrix} = \begin{bmatrix} H & 0 \end{bmatrix} W_d$ is the QR-factorization with W_d being orthogonal and H being lower triangular.

By the important property of f_W in [85], it is assumed that $\text{rank}(\begin{bmatrix} A_d + E_d & B_d \end{bmatrix}) = n$ and $(\mathcal{M}_d, \mathcal{L}_d)$ has no eigenvalue "-1". Thus, the coefficient matrix tuple $(E_c, A_c, B_c, M_c, N_c, R_c)$ corresponds to a GCARE (3.114) with E_c and R_c being nonsingular. Furthermore, the GDARE (3.117) and the GCARE (3.114) share the same stabilizing solutions, i.e., $X_d = X_c$.

We construct $(\mathcal{M}_c, \mathcal{L}_c)$ by $(E_c, A_c, B_c, M_c, N_c, R_c)$ as in (3.116) which satisfies

$$\mathcal{M}_c + \mathcal{L}_c = W^{-1} \mathcal{M}_d W, \tag{3.122}$$

where $W \equiv \text{diag}(\sqrt{2} I_n, W_d^\top)$. Let

$$(\mathcal{A}_0^\top, \mathcal{A}_0) = (\mathcal{M}_c + \mathcal{L}_c, \mathcal{M}_c - \mathcal{L}_c) \tag{3.123}$$

be the Cayley transformation of $(\mathcal{M}_c, \mathcal{L}_c)$. From (3.121) and (3.122), we see that the eigenvalues $\lambda_d \in \sigma(\mathcal{M}_d, \mathcal{L}_d)$, $\mu \in \sigma(\mathcal{M}_c, \mathcal{L}_c)$ and $\lambda \in \sigma(\mathcal{A}_0^\top, \mathcal{A}_0)$ satisfy the relationship in Table 1, in which $\mu = (\lambda - 1)(\lambda + 1)^{-1}$. From Table 1, we see that the key property of the transformation $\lambda_d \to \lambda$ is to transform m trivial infinity eigenvalues to m trivial -1 while preserving other eigenvalues (including nontrivial ∞) unchanged.

Table 1. Correspondence among λ_d, μ and λ

λ_d	$0 < \lvert\lambda_d\rvert < 1$	$\lvert\lambda_d\rvert > 1$	0	∞	m trivial ∞
μ	$\text{Re}(\mu) < 0$	$\text{Re}(\mu) > 0$	-1	1	m trivial ∞
λ	$\lambda = \lambda_d$	$\lambda = \lambda_d$	0	∞	m trivial -1

In the following, we use the PDA and the special structure of $(\mathcal{M}_d, \mathcal{L}_d)$ for the computation of the d-semi-stabilizing solution X_d of GDARE (3.117).

Firstly, we apply the PDA to the matrix $\mathcal{A}_0$ until convergence to $\mathcal{A}_k$. Then we compute the orthogonal bases $\mathcal{N}_r$ and $\mathcal{N}_\ell \in \mathbb{R}^{(2n+m)\times n}$ for the right and left null spaces of $\mathcal{A}_k^\top$; i.e.,

$$\mathcal{A}_k^\top \mathcal{N}_r = 0, \quad \mathcal{A}_k \mathcal{N}_\ell = 0, \tag{3.124}$$

which form orthogonal bases for the d-stable invariant subspaces of $(\mathcal{A}_0^\top, \mathcal{A}_0)$ and $(\mathcal{A}_0, \mathcal{A}_0^\top)$, respectively.

We then compute the QR-factorization $\mathcal{A}_0 \mathcal{N}_r = Q_1 R_1$, where Q_1 is orthogonal and R_1 is upper triangular. Next compute

$$S = Q_s^\top \mathcal{A}_0^\top \mathcal{N}_r, \quad T = Q_s^\top \mathcal{A}_0 \mathcal{N}_r,$$

where $Q_s = Q_1(:, 1:n)$. We see that (S, T) forms the d-stable eigenvalue matrix pair of $(\mathcal{A}_0^\top, \mathcal{A}_0)$ associated with $\mathcal{N}_r$, and T is clearly nonsingular.

We would like to separate the invariant subspaces of $(\mathcal{A}_0^\top, \mathcal{A}_0)$ corresponding to the zero and nonzero d-stable eigenvalues. Let $G = T^{-1}S$. By Van Dooren's algorithm [82], there is an orthogonal matrix $\Phi \in \mathbb{R}^{n\times n}$ such that

$$\Phi G \Phi^\top = \begin{bmatrix} G_{11} & G_{12} \\ 0 & G_{22} \end{bmatrix},$$

where $G_{11} \in \mathbb{R}^{s\times s}$ with $\sigma(G_{11}) = \{\lambda \in \sigma(\mathcal{A}_0^\top, \mathcal{A}_0) | 0 < |\lambda| < 1\}$ and $G_{22} \in \mathbb{R}^{(n-s)\times(n-s)}$ with $\sigma(G_{22}) = \{0\}$. Since $\sigma(G_{11}) \bigcap \sigma(G_{22}) = \phi$, there is a $\Psi = \begin{bmatrix} I_s & \Psi_{12} \\ 0 & I_{n-s} \end{bmatrix}$ such that

$$\Psi^{-1}\Phi^\top G \Phi \Psi = \begin{bmatrix} G_{11} & 0 \\ 0 & G_{22} \end{bmatrix},$$

where Ψ_{12} solves the Sylvester equation $G_{11}\Psi_{12} - \Psi_{12}G_{22} = G_{12}$ uniquely. Then

$$V_0 = \mathcal{N}_r \Phi \Psi(:, s+1:n), \quad \hat{V}_s = \mathcal{N}_r \Phi \Psi(:, 1:s) \tag{3.125}$$

span the invariant subspaces of $(\mathcal{A}_0^\top, \mathcal{A}_0)$ corresponding to the zero and nonzero d-stable eigenvalues, respectively.

Let $\hat{\zeta}_\ell$ spans the left null space of E_d. Then $\zeta_\ell = \begin{bmatrix} \hat{\zeta}_\ell^\top, & 0 \end{bmatrix}^\top \in \mathbb{R}^{(2n+m)\times r_d}$ contains the r_d eigenvectors of $(\mathcal{M}_d, \mathcal{L}_d)$ corresponding to the trivial zeros. From the transformation (3.122), we see that $W^{-1}\zeta_\ell$ contains the r_d eigenvectors of $(\mathcal{A}_0^\top, \mathcal{A}_0)$ corresponding to trivial zeros. Now, we want to extract $W^{-1}\zeta_\ell$ from span$\{V_0\}$.

Compute the QR-factorization $\begin{bmatrix} W^{-1}\zeta_\ell & V_0 \end{bmatrix} = Q_0 R_0$, where Q_0 is orthogonal and R_0 is upper triangular. Let

$$\hat{V}_0 = Q_0(:, r_d + 1 : n - s) \tag{3.126}$$

which forms the eigenvectors of $(\mathcal{A}_0^\top, \mathcal{A}_0)$ corresponding to zero eigenvalues of (S_d, E_d).

We will next find the invariant space U_∞ of $(\mathcal{A}_0^\top, \mathcal{A}_0)$ corresponding to the infinity eigenvalues. Compute the QR-factorization $\mathcal{A}_0 \mathcal{N}_\ell = Q_\infty R_\infty$, where Q_∞ is orthogonal and R_∞ is upper triangular. Let

$$\mathcal{N}_\infty = \mathcal{N}_\ell Q_\infty(:, s+1:n) \equiv \begin{bmatrix} \mathcal{N}_{\infty,1} \\ \mathcal{N}_{\infty,2} \\ \mathcal{N}_{\infty,3} \end{bmatrix} \begin{matrix} \}\, n \\ \}\, n \\ \}\, m \end{matrix}, \quad V_s = \begin{bmatrix} \hat{V}_s & \hat{V}_0 \end{bmatrix} \equiv \begin{bmatrix} V_{s1} \\ V_{s2} \\ V_{s3} \end{bmatrix} \begin{matrix} \}\, n \\ \}\, n \\ \}\, m \end{matrix}.$$

From the Cayley transform, there is a full rank matrix $Z \in \mathbb{R}^{(n-s)\times r_d}$ so that

$$V = \begin{bmatrix} V_{s1} & \mathcal{N}_{\infty,1} Z \\ V_{s2} & \mathcal{N}_{\infty,2} Z \\ V_{s3} & \mathcal{N}_{\infty,3} Z \end{bmatrix} \equiv \begin{bmatrix} V_1 \\ V_2 \\ V_3 \end{bmatrix} \tag{3.127}$$

is a basis of an invariant subspace of $(\mathcal{A}_0^\top, \mathcal{A}_0)$, satisfying $\mathrm{Span}\{V\} = \mathrm{Span}\left\{\begin{bmatrix} X_c E_c \\ I_n \\ K_c \end{bmatrix}\right\}$.

To determine Z, (3.127) and the fact $X_c = X_c^\top$ imply

$$\begin{bmatrix} V_{s2}^\top \\ Z^\top \mathcal{N}_{\infty,2}^\top \end{bmatrix} E_c^\top \begin{bmatrix} V_{s1} \ \mathcal{N}_{\infty,1} Z \end{bmatrix} = \begin{bmatrix} V_{s1}^\top \\ Z^\top \mathcal{N}_{\infty,1}^\top \end{bmatrix} E_c \begin{bmatrix} V_{s2} \ \mathcal{N}_{\infty,2} Z \end{bmatrix}.$$

That is,

$$V_{s2}^\top E_c^\top V_{s1} = V_{s1}^\top E_c V_{s2}, \tag{3.128a}$$

$$Z^\top \mathcal{N}_{\infty,2}^\top E_c^\top V_{s1} = Z^\top \mathcal{N}_{\infty,1}^\top E_c V_{s2}, \tag{3.128b}$$

$$V_{s2}^\top E_c^\top \mathcal{N}_{\infty,1} Z = V_{s1}^\top E_c \mathcal{N}_{\infty,2} Z, \tag{3.128c}$$

$$Z^\top \mathcal{N}_{\infty,2}^\top E_c^\top \mathcal{N}_{\infty,1} Z = Z^\top \mathcal{N}_{\infty,1}^\top E_c \mathcal{N}_{\infty,2} Z. \tag{3.128d}$$

Since E_c is nonsingular, from the isotropic property of $\begin{bmatrix} V_{s1} \\ V_{s2} \end{bmatrix}$ and $\begin{bmatrix} N_{\infty 1} \\ N_{\infty 2} \end{bmatrix}$, (3.128a) and (3.128d) hold automatically. Since (3.128b) is the transpose of (3.128c), the matrix Z is solved by finding the basis of $\mathrm{Null}(V_{s2}^\top E_c^\top \mathcal{N}_{\infty,1} - V_{s1}^\top E_c \mathcal{N}_{\infty,2})$.

Finally, we have the d-semi-stabilizing solution X_d for GDARE (3.117) can be obtained by

$$X_d = X_c = V_1 V_2^{-1} E_c^{-1}. \tag{3.129}$$

We summarize the computational steps (3.123)–(3.129) for X_d in Algorithm 3.5.2.

Algorithm 3.5.2. *(for GDARE* (3.117)*)*

Input:	$E_d, A_d, B_d, M_d, N_d, R_d$; τ *(a small tolerance);*
Output:	*The d-semi-stabilizing solution* X_d *of* (3.117).
1.	*Construct* $\mathcal{A}_0$ *via* (3.123).
2.	*Apply PDA to* $(\mathcal{A}_0^\top, \mathcal{A}_0)$ *until* $dist(Null(\mathcal{A}_k), Null(\mathcal{A}_{k-1})) < \tau$.
3.	*Compute bases* $\mathcal{N}_r$, $\mathcal{N}_\ell$ *for the right and left null spaces of* $\mathcal{A}_k^\top$ *as in* (3.124).
4.	*Compute bases* V_0, $\hat{V}_s$ *for d-stable invariant subspaces of* $(\mathcal{A}_0^\top, \mathcal{A}_0)$ *as in* (3.125).
5.	*Compute eigenvectors* $\hat{V}_0$ *of* $(\mathcal{A}_0^\top, \mathcal{A}_0)$ *corresponding to zeros as in* (3.126) .
6.	*Determine* Z *by* (3.128c).
7.	*Compute* $X_d = V_1 V_2^{-1} E_c^{-1}$ *as in* (3.129).

3.5.4 Additional remarks

We have developed the palindromic doubling algorithm (PDA) for solving the palindromic generalized eigenvalue problem (PGEP) $A^* x = \lambda A x$ structurally. We prove quadratic convergence and linear convergence with rate 1/2 of the PDA, when (A^*, A) has no unimodular eigenvalues and has unimodular eigenvalues with partial multiplicities two (one or two for eigenvalue 1), respectively. Algorithm 3.5.2 is specially developed for the computation of the d-semi-stabilizing solution of the generalized discrete-time algebraic Riccati equation (GDARE) for the singular descriptor linear system. It is the first structure-preserving algorithm for *singular* descriptor systems.

Our numerical experience indicates that the PDA is not necessarily better than other specialist algorithms (if exist) for solving the original problem, without linearizing the associated palindromic matrix polynomials. Such specialist algorithms may be able to better utilize the finer structures of the original problems. Our numerical examples showed selected applications for which the PDA was better or when no specialist structure-preserving algorithms exist. For future work, research will be conducted on how the finer structures can be fully utilized for individual applications. For a general PGEP without finer structures, the PDA is the only structure-preserving algorithm which performs reasonably efficiently. Consequently, the "good" vibrations from "good" linearizations [66, 67] can always be computed using the PDA, in the absence of better methods. Of course, numerical solutions from the PDA or other methods may be refined using the finer structures in the original problems, if feasible.

3.6 Large-scale problems

As an example, we shall only sketch the result from [60] on large-scale DAREs. For large-scale CAREs, NAREs, NMEs and Stein/Lyapunov equations, please consult [21, 22, 61, 62].

Let the system matrix A be large and sparse, possibly with band structures. The discrete-time algebraic Riccati equation (DARE) in (1.1):

$$\mathcal{D}(X) \equiv -X + A^\top X(I + GX)^{-1}A + H = 0,$$

with the low-ranked

$$G = BR^{-1}B^\top, \quad H = CT^{-1}C^\top, \tag{3.130}$$

where $B \in \mathbb{R}^{n\times m}$, $C \in \mathbb{R}^{n\times l}$ and $m, l \ll n$, arises often in linear-quadratic optimal control problems.

The solution of DAREs and CAREs has been an extremely active area of research; see, e.g., [23, 55, 70]. The usual solution methods such as the Schur vector method, symplectic SR methods, the matrix sign function, the matrix disk function or the doubling method have not made (full) use of the sparsity and structure in A, G and H. Requiring in general $O(n^3)$ flops and workspace of size $O(n^2)$, these methods are obviously inappropriate for the large-scale problems we are interested in here.

For control problems for parabolic PDEs and the balancing based model order reduction of large linear systems, large-scale CAREs, DAREs, Lyapunov and Stein equations have to be solved [6, 7], [48]–[50]. As stated in [7], "the basic observation on which all methods for solving such kinds of matrix equations are based, is that often the (numerical) rank of the solution is very small compared to its actual dimension and therefore it allows for a good approximation via low rank solution factors". Importantly, without solving the corresponding algebraic Riccati equations, alternative solutions to the optimal control problem require the deflating subspace of the corresponding Hamiltonian matrices or (generalized) symplectic pencils which are prohibitively expensive to compute.

Benner has done much on large-scale algebraic Riccati equations; see [6, 75] and the references therein. They built their methods on (inexact) Newton's methods with inner iterations for the associated Lyapunov and Stein equations. We shall adapt the structure-preserving doubling algorithm (SDA) [15, 17, 64], making use of the sparsity in A and the (numerically) low-ranked structures in G, H and X. For other applications of the SDA, see [19].

We shall concentrate on the numerical solution of large-scale DAREs in this section. The results for large-scale CAREs and Stein/Lyapunov equations will be presented elsewhere [61, 62].

3.6.1 Structure-preserving doubling algorithm for DAREs

The structure-preserving doubling algorithm (SDA) [17], assuming $(I + GH)^{-1}$ exists, has the form in (1.2). We can apply the Sherman-Morrison-Woodbury formula (SMWF) to $(I + GH)^{-1}$ and make use of the low-ranked forms of G and H in (3.130). It is then obvious that $(I + GH)^{-1}$ is a low-rank update of the identity matrix, and from (1.2) that $\widehat{G}$, $\widehat{H}$ and $\widehat{A}$ are the same of the corresponding previous iterates. So we can organize the SDA in the low-rank update form

$$A_k = A_{k-1}^2 - D_k^{(1)} S_k^{-1} \left[D_k^{(2)}\right]^\top, \quad G_k = B_k R_k^{-1} B_k^\top, \quad H_k = C_k T_k^{-1} C_k^\top.$$

The previous attempts in generalizing the SDA for large-scale problems were abandoned mainly because of the difficulties presented by the computation of A_k, when it fills up as the iteration proceeds. There are different ways of handing this problem, dependent on the structure of A, H and G. Under favourable assumptions on the band-width of A, we may just compute A_k explicitly, with the growth in its exponentially growing band-widths out run by the fast convergence of the SDA. Alternatively, we may assume that the multiplication of A and $A^\top$ to arbitrary vectors to be of $O(n)$ computational complexity. For given vectors u and v, we can then compute $A_k u$ and $A_k^\top v$ recursively, in terms of A_{k-1} and eventually $A_0 = A$. Assuming that n is large and the convergence of the SDA is fast, we obtain the important result that the SDA is of $O(n)$ computational complexity and memory requirement.

Note that the rank of X has been observed to be numerically low-ranked. Under suitable assumptions, the convergence of the SDA [17] implies the convergence of $A_k = O(|\lambda|^{2^k}) \to 0$, for some $|\lambda| < 1$. We then see that B_{k+1} and C_{k+1} equal, respectively, the sums of B_k and C_k and the diminishing components $A_k B_k$ and $A_k^\top C_k$. Thus the observation about the numerical ranks of X is shown to be true.

Obviously, as the SDA converges, increasingly smaller but fatter low-ranked components are added to B_k and C_k. Apparently, the growth in the sizes and ranks of these iterates is potentially exponential. To reduce the dimensions of B_k,

C_k, $D_k^{(1)}$ and $D_k^{(2)}$, we may compress their columns by the QR decomposition with column pivoting.

4 Conclusions

We have told the story about the (structure-preserving) doubling algorithms, how it was "folk-lore", presumably once popular until the seventies of the last century, then forgotten, and rediscovered and redeveloped in the past decade. Since the paper [17] by Chu, Fan, Lin and Wang in 2004, there have been approximately fifty published papers and reports by our extended group of authors based in Australia, Monash, Beijing, and Hsinchw. Other authors are starting to realize the importance, power and usefulness of the doubling algorithms. Our papers are widely cited. In addition, others are starting their research in the area and there have been a handful of papers from other authors since 2010, on the method's theoretical aspects as well as applications. It has been an interesting and rewarding journey, and it is continuing.

Acknowledgements

This research work is partially supported by the National Science Council and the National Centre for Theoretical Sciences. Part of the work was completed when the first author visited the National Centre for Theoretical Sciences at the Chiao Tung University, Hsinchu. This manuscript was written when the first authors visited Southeast University, Nanjing, China. We would like to acknowledge the support from these institutions.

References

[1] H. ALLIK AND T. HUGHES. Finite element method for piezoelectric vibration, *Int. J. Numer. Methods Eng.*, **2** (1970) 151–157.

[2] B.D.O. ANDERSON AND J.B. MOORE. *Optimal Filtering*, Prentice-Hall, Englewood Cliffs, NJ, 1979.

[3] M.B. ANGEL, M.I. ROCHA-GASO, M.I. CARMEN, AND A.V. ANTONIO. Surface generated acoustic wave biosensors for detection of pathogens: a review, *Sensors*, **9** (2009) 5740–5769.

[4] I. APPELBAUM, T. WANG, J.D. JOANNOPOULOS, AND V. NARAYANAMURTI. Ballistic hot-electron transport in nanoscale semiconductor heterostructures: Exact self-energy of a three-dimensional periodic tight-binding Hamiltonian, *Physical Review B*, **69** (2004), Article Number 165301, 6 pp.

[5] R. BELLMAN AND K.L. COOKE. *Differential Difference Equations.* Academic Press, New York, 1963.

[6] P. BENNER. Editorial of special issue on "Large-Scale Matrix Equations of Special Type", *Numer. Lin. Alg. Appl.*, **15** (2008) 747–754.

[7] P. BENNER AND J. SAAK. A Galerkin-Newton-ADI method for solving large-scale algebraic Riccati equations, *DFG Priority Programme 1253 "Optimization with Partial Differential Equations"*, Preprint SPP1253-090, January 2010.

[8] D.S. BERNSTEIN AND W.M. HADDAD. LQG control with an H_∞ performance bound: a Riccati equation approach, *IEEE Trans. Automat. Control*, **34** (1989) 293–305.

[9] D.A. BINI, B. IAMMAZZO, AND F. POLONI. A fast Newton's method for a nonsymmetric algebraic Riccati equation, *SIAM J. Matrix Anal. Appl.*, **30** (2008) 276–290.

[10] E.K. BOUKAS AND Z.K. LIU. *Deterministic and Stochastic Time Delay Systems*, Birkhäuser, Boston, 2002.

[11] M. BUCHNER, W. RUILE, A. DIETZ, AND R. DILL. FEM analysis of the reflection coefficient of SAWS in an infinite periodic array, In *Proc. IEEE Ultrason. Symp.*, 371–375, 1991.

[12] C.K. CAMPBELL. *Surface Acoustic Wave Devices for Mobile and Wireless Communications*, Academic Press, New York, 1998.

[13] C.-Y. CHIANG, E.K.-W. CHU, C.-H. GUO, T.-M. HUANG, W.-W. LIN, AND S.-F. XU. Convergence analysis of the doubling algorithm for several nonlinear matrix equations in the critical case, *SIAM J. Matrix Anal. Applic.*, **31** (2009) 227–247.

[14] C.Y. CHIANG, E.K.-W. CHU, AND T.-M. HUANG. Unimodular eigenvalues for palindromic eigenvalue problems. Technical report, *NCTS Preprints in Mathematics*, Tsing Hua University, Hsinchu, 2008.

[15] E.K.-W. CHU, H.-Y. FAN, AND W.-W. LIN. A structure-preserving doubling algorithm for continuous-time algebraic Riccati equations, *Lin. Alg. Applic.*, **396** (2005) 55–80.

[16] E.K.-W. CHU, T.-M. HUANG, AND W.-W. LIN. Structured doubling algorithms for solving g-palindromic quadratic eigenvalue problems, Technical report, *NCTS Preprints in Mathematics*, Tsing Hua University, Hsinchu, 2008.

[17] E.K.-W. CHU, H.-Y. FAN, W.-W. LIN, AND C.-S. WANG. Structure-preserving algorithms for periodic discrete-time algebraic Riccati equations, *Int. J. Control*, **77** (2004) 767–788.

[18] E.K.-W.CHU, T.-M. HWANG, W.-W. LIN, AND C.-T. WU. Vibration of fast trains, palindromic eigenvalue problems and structure-preserving doubling algorithms, *J. Comput. Appl. Math.*, **219** (2008) 237–252.

[19] E.K.-W. CHU, T.M. HUANG, W.-W. LIN, AND C.-T. WU. Palindromic eigenvalue problems: a brief survey, *Taiwanese J. Math.*, **14** (2010) 743–779.

[20] E.K.-W. CHU, T. LI, J. JUANG, AND W.-W. LIN. Solution of a nonsymmetric algebraic Riccati equation from a one-dimensional multi-state transport model, *IMA J. Numer. Anal.*, **31** (2011) 1453–1467.

[21] E.K.-W. CHU, C.-Y. WENG, Y.-C. KUO, AND W.-W. LIN. Solving large-scale nonlinear matrix equations by doubling, *Technical Report*, NCTS, Chiao Tung University, Taiwan, 2011.

[22] E.K.-W. CHU, C.-Y. WENG, T. LI, AND W.-W. LIN. Solving large-scale nonsymmetric algebraic Riccati equations by doubling, *Technical Report*, NCTS,

Chiao Tung University, Hsinchu, 2011.

[23] B. DATTA. *Numerical Methods for Linear Control Systems*, Elsevier Academic Press, Boston, 2004.

[24] S. DATTA, *Electronic Transport in Mesoscopic Systems*, Cambridge University Press, Camb ridge, 1995.

[25] S. DATTA, Nanoscale device modeling: the Green's function method, *Superlattices and Microstructures*, **28** (2000) 253–278.

[26] B. DE MOOR AND J. DAVID. Total linear least squares and the algebraic Riccati equations, *Systems Control Lett.*, **18** (1992) 329–337.

[27] O. DIEKMANN, S.A. VAN GILS, S.M. VERDUYN LUNEL, AND H. O. WALTHER. *Delay Equations: Functional-, Complex-, and Nonlinear Analysis*, Applied Mathematical Sciences **110**, Springer-Verlag, New York, 1995.

[28] H. FASSBENDER, N. MACKEY, D.S. MACKEY, AND C. SCHRÖDER. A structured polynomial eigenproblem arising in the analysis of time delay systems and related polynomial eigenproblems, Technical report, TU Braunschweig, 2007.

[29] F.R. GANTMACHER. *The Theory of Matrices*, Vol. II. Chelsea Publishing Company, New York, 1977.

[30] I. GOHBERG AND M.A. KAASHOEK. An inverse spectral problem for rational matrix functions and minimal divisibility, *Integral Equations and Operator Theory*, **10** (1987) 437–465.

[31] I. GOHBERG, P. LANCASTER, AND L. RODMAN. *Matrix Polynomials*. Academic Press, New York, 1982.

[32] C.-H. GUO. A new class of nonsymmetric algebraic Riccati equations, *Lin. Alg. Applic.*, **426** (2007) 636–649.

[33] C.-H. GUO, Y.-C. KUO, AND W.-W. LIN. On a nonlinear matrix equation arising in nano research, *SIAM J. Matrix Anal. Applic.*, **33** (2012) 235–262.

[34] C.-H. GUO AND W.-W. LIN. Solving a structured quadratic eigenvalue problem by a structure-preserving doubling algorithm, *SIAM J. Matrix Anal. Applic.*, **31** (2010) 2784–2801.

[35] C.-H. GUO AND W.-W. LIN. The matrix equation $X + A^{\top} X^{-1} A = Q$ and its application in nano research, *SIAM J. Sci. Comput.*, **32** (2010) 3020–3038.

[36] X.-X. GUO, W.-W. LIN, AND S.-F. XU. A structure-preserving doubling algorithm for nonsymmetric algebraic Riccati equations, *Numer. Math.*, **103** (2006) 393–412.

[37] K. GU, V. KHARITONOV, AND J. CHEN. *Stability of Time-Delay Systems*. Control Engineering, Birkhäuser, Boston, 2003.

[38] J. HALE AND S.M. VERDUYN LUNEL. *Introduction to Functional Differential Equations*. Springer-Verlag, New York, 1993.

[39] D. HINRICHSEN, B. KELB, AND A. LINNEMANN. An algorithm for the computation of the structured complex stability radius, *Automatica*, **25** (1989) 771–775.

[40] R.A. HORN AND C.R. JOIHNSON. *Topics in Matrix Analysis*, Cambridge University Press, Cambridge, 1995.

[41] T.-M. HUANG, W.-W. LIN, AND J. QIAN. Structure-preserving algorithms for palindromic quadratic eigenvalue problems arising from vibration on fast

trains, *SIAM J. Matrix Anal. Applic.*, **30** (2009) 1566–1592.

[42] T.-M. HUANG, W.-W. LIN, AND C.-T. WU. Structure-preserving Arnoldi-type algorithms for solving palindromic quadratic eigenvalue problems in leaky surface wave propagation. *NCTS Preprints in Mathematics*, No. **2011-2-001**, Tsing Hua University, 2011.

[43] T.-M. HUANG, C.-T. WU, AND T. LI. Numerical study of structure-preserving algorithms for surface acoustic wave simulation, *Technical Report*, NCTS, Chiao Tung University, Hsinchu, 2011.

[44] T.-M. HWANG, E.K.-W. CHU, AND W.-W. LIN. A generalized structure-preserving doubling algorithm for generalized discrete-time algebraic Riccati equations, *Int. J. Control*, **78** (2005) 1063–1075.

[45] T.-M. HWANG AND W.-W. LIN. Structured doubling algorithms for weak Hermitian solutions of algebraic Riccati equations, Technical report, *NCTS Preprints in Mathematics*, Tsing Hua University, Hsinchu, **2006-7-009**, 2006.

[46] C.F. IPSEN. Accurate eigenvalues for fast trains, *SIAM News*, **37**, 2004.

[47] E. JARLEBRING. *The Spectrum of Delay-Differential Equations: Computational Analysis*, PhD thesis, Institute Computational Mathematics, TU Braunschweig, Germany, 2008.

[48] K. JBILOU. Block Krylov subspace methods for large continuous-time algebraic Riccati equations, *Numer. Algorithms*, **34** (2003) 339–353.

[49] K. JBILOU. An Arnoldi based algorithm for large algebraic Riccati equations, *Appl. Math. Lett.*, **19** (2006) 437–444.

[50] K. JBILOU AND A. RIQUET. Projection methods for large Lyapunov matrix equations, *Lin. Alg. Applic.*, **415** (2006) 344–358.

[51] D.L. JOHN AND D.L. PULFREY, Green's function calculations for semi-infinite carbon nanotubes, *Physica Stutus Solidi B – Basic Solid State Physics*, **243** (2006) 442–448.

[52] J. JUANG. Existence of algebraic matrix Riccati equations arising in transport theory, *Lin. Alg. Applic.*, **230** (1995) 89–100.

[53] A. KLETSOV, Y. DAHNOVSKY, AND J.V. ORTIZ. Surface Green's function calculations: A nonrecursive scheme with an infinite number of principal layers, *J. Chemical Physics*, **126** (2007), Article Number 134105.

[54] M. KOSHIBA, S. MITOBE, AND M. SUZUKI. Finite-element solution of periodic waveguides for acoustic waves, *IEEE Trans. Ultrason. Ferroelectr. Freq. Control*, **34** (1987) 472–477.

[55] P. LANCASTER AND L. RODMAN, *Algebraic Riccati Equations*, Clarendon Press, Oxford, 1995.

[56] R. LERCH. Simulation of piezoelectric devices by two- and three-dimensional finite elements, *IEEE Trans. Ultrason. Ferroelectr. Freq. Control*, **37** (1990) 233–247.

[57] T. LI, C.-Y. CHIANG, E.K.-W. CHU, AND W.-W. LIN. The palindromic generalized eigenvalue problem $A^*x = \lambda Ax$: numerical solution and applications, *Lin. Alg. Applic.*, **434** (2011) 2269–2284.

[58] T. LI, E.K.-W. CHU, J. JUANG, AND W.-W. LIN. Solution of a nonsymmetric algebraic Riccati equation from a two-dimensional transport model, *Lin.*

Alg. Applic., **434** (2011) 201–214.

[59] T. LI, E.K.-W. CHU, AND W.-W. LIN. A structure-preserving doubling algorithm for quadratic eigenvalue problems arising from time-delay systems, *J. Comp. Appl. Math.*, **223** (2010) 1799–1745.

[60] T. LI, E.K.-W. CHU, AND W.-W. LIN. Solving large-scale discrete-time algebraic Riccati equations by doubling, *Technical Report*, NCTS, Chiao Tung University, Hsinchu, 2011.

[61] T. LI , E.K.-W. CHU, W.-W. LIN, AND C.-Y. WENG. Solving large-scale continuous-time algebraic Riccati equations by doubling, *Technical Report*, NCTS, Chiao Tung University, Hsinchu, 2011.

[62] T. LI , C.-Y. WENG, E.K.-W. CHU, AND W.-W. LIN. Solving large-scale Stein and Lyapunov equations by doubling, *Technical Report*, NCTS, Chiao Tung University, Hsinchu, 2011.

[63] W.-W. LIN. A new method for computing the closed-loop eigenvalues of a discrete-time algebraic Riccati equation, *Lin. Alg. Applic.*, **96** (1987)157–180.

[64] W.-W. LIN AND S.-F. XU, Convergence analysis of structure-preserving doubling algorithms for Riccati-type matrix equations, *SIAM J. Matrix Anal. Applic.*, **28** (2006) 26–39.

[65] W.-W. LIN, V. MEHRMANN, AND H. XU. Canonical forms for Hamiltonian and symplectic matrices and pencils, *Lin. Alg. Appl.*, **302/303** (1999) 469–533.

[66] D. S. MACKEY, N. MACKEY, C. MEHL, AND V. MEHRMANN. Vector spaces of linearizations for matrix polynomials. *SIAM J. Matrix Anal. Applic.*, **28** (2006) 971–1004.

[67] D. S. MACKEY, N. MACKEY, C. MEHL, AND V. MEHRMANN. Structured polynomial eigenvalue problems: Good vibrations from good linearizations, *SIAM J. Matrix Anal. Applic.*, **28** (2006) 1029–1051.

[68] D. S. MACKEY, N. MACKEY, C. MEHL, AND V. MEHRMANN. Numerical methods for palindromic eigenvalue problems, *Technical Report*, TU Berlin, MATHEON, Germany, 2007.

[69] V.L. MARKINE, A.P. DE MAN, S JOVANOVIC, AND C. ESVELD. Optimal design of embedded rail structure for high-speed railway lines, In *Railway Engineering 2000, 3rd Int. Conf.*, London, 2000.

[70] V.L. MEHRMANN, *The Autonomous Linear Quadratic Control Problem*, Lecture Notes in Control and Information Sciences, Vol. **163**, Springer Verlag, Berlin, 1991.

[71] W. MICHIELS AND S.-I. MICULESCU. *Stability and Stabilization of Time-Delay Systems: An Eigenvalue-Based Approach*, Advances in Design and Control 12, SIAM Publications, Philadelphia, 2007.

[72] W. MICHIELS AND T. VYHLÍDAL. An eigenvalue based approach for the stabilization of linear time-delay systems of neutral type, *Automatica*, **41** (2005) 991–998.

[73] M. MOHAMED, M.M. EL GOWINI, AND W.A. MOUSSA. A finite element model of a MEMS-based surface acoustic wave hydrogen sensor, *Sensors*, **10** (2010) 1232–1250.

[74] R.V. PATEL. On computing the eigenvalues of a symplectic pencil, *Lin. Alg.*

Applic., **188** (1993) 591–611.

[75] J. SAAK, H. MENA, AND P. BENNER. *Matrix Equation Sparse Solvers (MESS): a Matlab Toolbox for the Solution of Sparse Large-Scale Matrix Equations*, Chemnitz University of Technology, Germany, 2010.

[76] C. SCHRÖDER. A QR-like algorithm for the palindromic eigenvalue problem, *Technical Report*, Preprint **388**, TU Berlin, Matheon, Germany, 2007.

[77] C. SCHRÖDER. URV decomposition based structured methods for palindromic and even eigenvalue problems, *Technical Report*, Preprint **375**, TU Berlin, MATHEON, Germany, New York, 2007.

[78] G.W. STEWART AND J.G. SUN. *Matrix Perturbation Theory*, Academic Press, 1990.

[79] J.W. STRUTT (LORD RAYLEIGH). *Theory of Sound*, Dover, New York, 1945.

[80] F. TISSEUR AND K. MEERBERGEN. A survey of the quadratic eigenvalue problem, *SIAM Rev.*, **43** (2001) 234–286.

[81] J. TOMFOHR AND Q.F. SANKEY. Theoretical analysis of electron transport through organic molecules, *J. Chemical Physics*, **120** (2004) 1542–1554.

[82] P. VAN DOOREN. The computation of Kronecker's canonical form of a singular pencil, *Lin. Alg. Applic.*, **27** (1979) 103–141.

[83] D. WILLIAMS. A potential-theoretical note on the quadratic Wiener-Hopf equation for Q-matrices, in *Seminar on Probability XVI*, Lecture Notes in Mathematics **920**, pp. 91–94, Springer-Verlag, Berlin, 1982.

[84] Y.S. WU, Y.B. YANG, AND E.D. YAU. Three-dimensional analysis of train-rail-bridge interaction problems, *Vehicle Syst. Dyn.*, **36** (2001) 1–35.

[85] H. XU. Transformations between discrete-time and continuous-time algebraic Riccati equations, *Lin. Alg. Applic.*, **425** (2007) 77–101.

[86] S. ZAGLMAYR. *Eigenvalue Problems in SAW-Filter Simulations*, Diplomarbeit, Institute of Computational Mathematics, Johannes Kepler University Linz, Linz, Austria, 2002.

[87] Q.C. ZHONG. *Robust Control of Time-Delay Systems*, Springer Verlag, New York, 2006.

Open Problems and Surveys of Contemporary Mathematics
SMM 6, pp. 195–235

Platonic Solids, Binary Polyhedral Groups Kleinian Singularities and Lie Algebras of Type ADE*

Joris van Hoboken†

Contents

*Master's Thesis under supervision of Prof. Dr. E. M. Opdam.

†University of Amsterdam, Faculty of Mathematics, Plantage Muidergracht 24, 1018 TV Amsterdam, The Netherlands. Email: vanhoboken@1v1r.nl

1 Introduction

This thesis is an attempt to show an astonishing relation between basic objects from different fields in mathematics. Most peculiarly it turns out that their classification is "the same": the ADE classification. Altogether these objects and the connections between them form a coherent web.

The connections are accomplished by direct mathematical constructions leading to bijections between these classes of objects. These constructions however do not always explain or give satisfactory intuition why these classifications are, or to say it better, why they should be related in this way. Therefore the deeper reason remains mysterious and when discovered will have to be of great depth. This gives a high motivation to look for new concepts and it shows that simple and since long understood mathematics can still raise very interesting questions, show paths for new research and give a glance at the mystery of mathematics. In my opinion to be aware of a certain truth without having its reason is fundamental to the practise of mathematics.

In this thesis I will demonstrate the following correspondences, in this order:

Platonic solids	$\leftrightarrow$	binary polyhedral groups
binary polyhedral groups	$\leftrightarrow$	Kleinian singularities
Kleinian singularities	$\leftrightarrow$	Dynkin diagrams
representations of bin. pol. groups	$\leftrightarrow$	affine Dynkin diagrams
representations of bin. pol. groups	$\leftrightarrow$	Kleinian singularities
Lie algebras of type A, D, E	$\leftrightarrow$	Kleinian singularities

The first connection was found in the end of the nineteenth century. It was Felix Klein who studied the theory of the regular polyhedra and their symmetries, in his book on the icosahedron, see [18]. He found the relation between the possible finite subgroups of the special unitary group $SU_2(\mathbb{C})$ and the platonic solids. For these groups he computed a fundamental set of invariant polynomials which he found to be subject to one essential relation. He used these results, which we will discuss in the first sections, for his theory of equations of fifth degree.

The results of Klein on the invariant theory of the binary polyhedral groups, $\Gamma \subset SU_2(\mathbb{C})$, were a starting point for later developments. In the 1920's du Val obtained a resolution of the Kleinian singularities $\mathbb{C}^2/\Gamma$; the quotient singularities associated with the binary polyhedral groups. He was able to give a characterization that made them a particular class of singularities: "Isolated singularities of surfaces which do not affect the conditions of adjunction". After this other characterizations of these singularities have been found, for instance by Arnold. We will show that with the use of a minimal resolution $\tilde{S} \to \mathbb{C}^2/\Gamma$ with exceptional set E they can be classified by the Dynkin diagrams Δ of type A_r, D_r, E_6, E_7 and E_8.

Let us summarize this in a picture:

$$\begin{array}{rcl} & & \mathbb{C}^2 \\ & & \downarrow \pi \\ E = \cup E_i \subset \tilde{S} & \to & \mathbb{C}^2/\Gamma \end{array}$$

These Dynkin diagrams classify another special class of objects, the simple Lie algebras of Lis groups which have root systems with all roots the same length.

In this field of Lie theory and algebraic group theory great progress had been made; the classification of semisimple Lie algebras by Dynkin diagrams, the theory of reflection groups and Weyl groups and their representations, the study of conjugacy classes in algebraic groups. With this theory various other surprising connections with binary polyhedral groups and Kleinian singularities were found.

As stated above the Dynkin diagrams of type A, D and E arise naturally in the resolution of the quotient singularities of binary polyhedral groups. Then in 1980 McKay found there is a direct connection between their affine versions and the irreducible representations of these groups and soon after this it was shown that also the geometry of the resolution of the quotient singularity and the irreducible representations are directly related.

After his work on singularity theory and the work of Kostant, Steinberg, and Springer on the adjoint quotients $\chi : \mathbf{g} \to \mathbf{h}/\mathcal{W}$, $\chi : G \to T/\mathcal{W}$ and the resolution of the nilpotent, unipotent variety, Brieskorn was able to proof the conjecture, stated by Grothendieck, that the quotient singularities emerge as the intersection of the nilpotent variety and a transversal slice to the orbit of a subregular nilpotent element in the simple Lie algebra $\mathbf{g}$ of type A, D or E, or analogously in the algebraic group G of this type with unipotent instead of nilpotent.

In view of this theorem, which we will illustrate in the final section of this thesis, it is interesting to see what Klein once said about his work and its relation to the work of Lie:

> Meine Verpflichtungen gegen Hrn. Lie gehen in die Jahre 1879-70 zurück, wo wir in engem Verkehre mit einander unsere Studienzeit in Berlin und Paris abschlossen. Wir fassten damals gemeinsam den Gedanken, überhaupt solche geometrische oder analytische Gebilde in Betracht zu zieben, welche durch Gruppen von Aenderungen in sich selbst transformirt werden. Dieser Gedanke is für unsere beiderseitigen späteren Arbeiten, soweit dieselben auch auseinander zu liegen scheinen, bestimmt geblieben. Während ich selbst in erster Linie Gruppen discreter Operationen in Auge fasste und also insbesondere zur Untersuchung der regulären Körper und ihrer Beziehung zur Gleichungstheorie geführt wurde, hat Hrn.Lie von vorneherein die schwierigere Theorie der continuirlichen Transformationsgruppen und somit die Differentialgleichungen in Angiff genommen.

These two viewpoints turned out to have more connections than Klein could foresee.

P. Slodowy has written various articles and gave lectures about this topic the last twenty years, see [32–36]. There is a book of Lamotke, see [23], that exhibits the correspondences between the platonic solids, their symmetries and the quotient singularities and resolutions. On the connection of the ADE classification with quivers and the representation theory of algebras, there is a recent survey article by Idun Reiten, see [30]. We do not discuss this connection in this thesis.

There are many recent developments in the subject discussed here. Its widespread presence and significance in mathematics and physics, i.e. string theory, caused many to study it, both mathematicians and physicists. It is of great interest to see if, and if so, in what way the theory can be generalized. We will conclude this thesis by making some remarks about this. However in this thesis my goal was to give a clear picture of the list of connections stated above. On various occasions the large amount of theory that is covered by these different topics has made it necessary to omit details and proofs, many of which are standard. They can be found in the references. I have nevertheless tried to illustrate the discussed theory with examples.

2 Platonic solids and symmetry

The platonic solids are a class of objects that have been present in mathematics since ancient times. They were already described and classified by the Greek mathematicians.

Euclid's work culminates in the description and classification of the regular polyhedra. These results are contained in Book 13 of Euclid's elements; see [12]. Plato (427—347 B.C.) used the regular polyhedra to describe matter in terms of a mathematical model. In his dialogue "Timaios", he associates the four solids tetrahedron, cube, octahedron and icosahedron to the basic four elements of matter: fire, earth, air and water. The dodecahedron represented the universe. Roughly 2000 years later, in 1596, Kepler used the platonic solids in his description of the solar system. He associated a platonic solid to each of the six planets known at that time, in such a way that when inscribed in each other this gave an explanation of the configuration of the planets. For a modern account on the regular polyhedra' see [7] or [44].

We recall briefly, that a polyhedron P is an intersection of finitely many closed half-spaces H_i in 3-dimensional Euclidean space E. Let $P = \cap_{i=1}^{m} H_i$ be a bounded polyhedron. P is a convex compact subset of E. Let the V_i be the defining planes of the H_i. Then a finite intersection of these planes intersected with P gives us, if not empty, a polygon, a line segment, of a point. These are called the faces, edges and vertices of P respectively. Such a P obtained is called a regular polyhedron, or a platonic solid, if its faces are congruent regular polygons, and if the number of these faces meeting in a vertex is the same for all vertices.

For every $q \geqslant 3$ there exists a unique (up to size) regular convex polygon

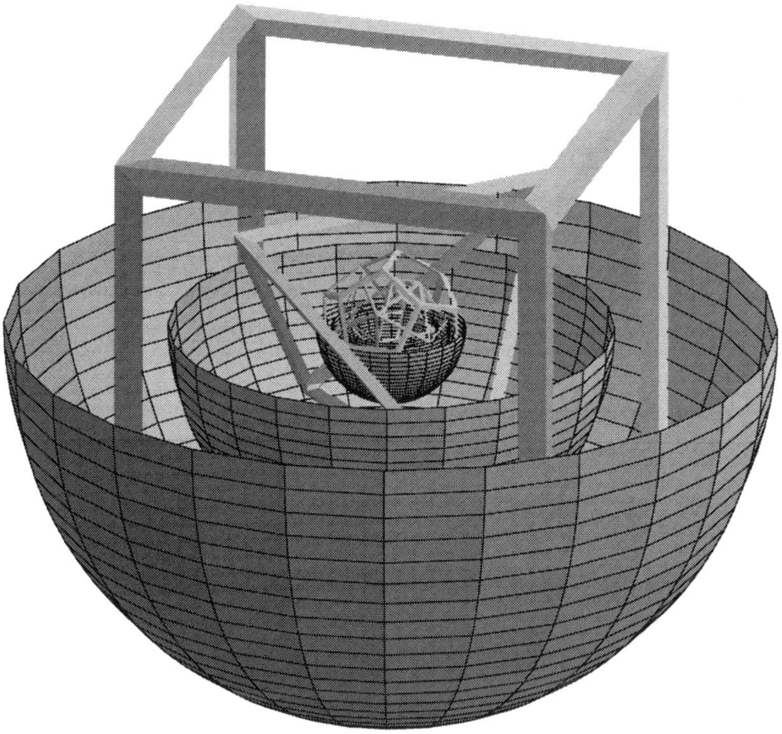

Figure 1 Kepler's proof of the configuration of orbits of the planets.

with q edges and q vertices. It's group of rotation symmetries is the cyclic group of order q:

$$\mathbb{Z}/q\mathbb{Z} \hookrightarrow SO(2),$$

and its full group of symmetries (rotations and reflections), is the dihedral group of order $2q$:

$$\mathbb{D}_q = \mathbb{Z}/q\mathbb{Z} \rtimes \mathbb{Z}/2\mathbb{Z} \hookrightarrow O(2).$$

So we can define a regular polyhedron P by a pair of integers $\{q, p\}$, the Schläfli symbol, each face of P being a regular q-gon and p of them meeting in each vertex of P. A regular q-gon has interior angle $\alpha_q = \pi(1 - \frac{2}{q})$, whence

$$2\pi > p\alpha_q = \pi\left(p - \frac{2p}{q}\right).$$

This gives us:

$$\frac{1}{p} + \frac{1}{q} + \frac{1}{2} > 1. \tag{2.1}$$

We see that the only possibilities for $\{q, p\}$, if we admit the degenerate cases in which one of them is 2,are the following:

Table 1 Schläfli symbols. f faces, e edges and v vertices

$\{q,p\}$	P	f	e	v
$\{q,2\}$	dihedron	2	q	q
$\{2,p\}$	hosohedron	p	p	2
$\{3,3\}$	tetrahedron	4	6	4
$\{4,3\}$	cube	6	12	8
$\{3,4\}$	octahedron	8	12	6
$\{5,3\}$	dodecahedron	12	30	20
$\{3,5\}$	icosahedron	20	30	12

The dihedron consists of 2 regular q-gons glued together at the edges. The hosohedron can be considered the dual of this; a sphere divided into q equal parts by great circles on the sphere intersecting in the north and south pole.

Table 2 The five platonic solids

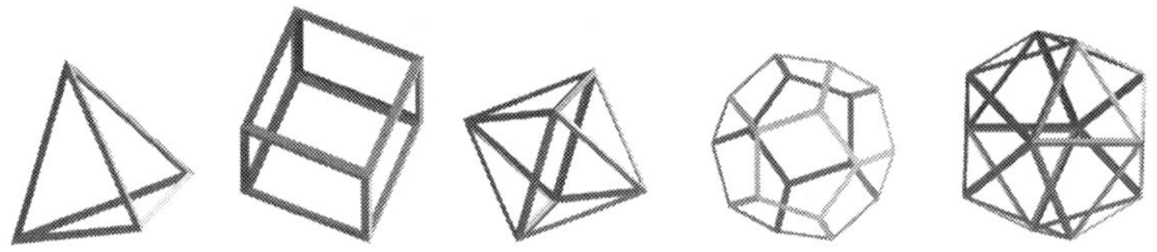

tetrahedron cube octahedron dodecahedron icosahedron

The reason of the interest in these objects is their high degree of regularity which is represented by the rich symmetry groups they possess. We will therefore for each object in the list above determine its group of rotation symmetries G.

Clearly the axis of a rotation symmetry should always intersect P in a vertex, the midpoint of a face, or in the midpoint of an edge. We will denote the order of the respective symmetries with n_1, n_2, n_3. This gives us the following symmetry groups G, with orders N:

Table 3 Symmetry groups of the regular polyhedra

P	n_1	n_2	n_3	N	G
dihedron	2	p	2	$2p$	D_p
hosohedron	q	2	2	$2q$	D_q
tetrahedron	3	3	2	12	$T \cong A_4$
cube	3	4	2	24	$O \cong S_4$
octahedron	4	3	2	24	$O \cong S_4$
dodecahedron	3	5	2	60	$I \cong A_5$
icosahedron	5	3	2	60	$I \cong A_5$

Let us mention the well known duality; If we take a regular polyhedron P we get a dual P'; it has its vertices in the midpoints of faces of P. This gives the duality (dihedron, hosohedron), (cube, octahedron), (dodecahedron, icosahedron). The tetrahedron is self-dual. The symmetry groups of dual solids are isomorphic.

We get the descriptions as permutation groups by looking at the action of G on a well chosen subset of P, namely the set of 4 vertices of the tetrahedron, the set of 4 body-diagonals of the cube, and the set of 5 so called "Dreibeine" in the icosahedron, each consisting of 3 pairwise orthogonal axes through the midpoints of opposite edges.

If we put the center of gravity of the above objects in the origin of E, we see that the groups G are naturally embedded in $SO_3(\mathbb{R})$. In fact we will see that up to the case of the cyclic group of order N they exhaust all the possibilities of finite subgroups of $SO_3(\mathbb{R})$.

3 Finite subgroups of $SU_2(\mathbb{C})$

Let $SU_2(\mathbb{C}) \subset SL_2(\mathbb{C})$ be the special unitary group, the subgroup of unitary transformations in $GL_2(\mathbb{C})$ with determinant 1. It has a group action on the complex projective line $\mathbb{P} := \mathbb{P}^1(\mathbb{C})$:

$$SU_2(\mathbb{C}) \times \mathbb{P} \longrightarrow \mathbb{P}$$
$$\left(\begin{pmatrix} a & b \\ c & d \end{pmatrix}, (z:1)\right) \longmapsto \left(\frac{az+b}{cz+d} : 1\right),$$

for all points $(z:1)$. The point at infinity $(1:0) = \infty$ goes to $(ac^{-1}:1)$ if $c \neq 0$ and else to ∞. The element $-I$ acts trivially so the action factors through an action of $PSU_2(\mathbb{C})$.

Take a finite subgroup $\Gamma \subset SU_2(\mathbb{C})$ and let G be its image in $PSU_2(\mathbb{C})$. We want to classify the possible groups G. To do so assume that G is not trivial and consider its action on $\mathbb{P}$. Let $F \subset \mathbb{P}$ be the set of points which are fixed by some non-identity element of G. Each non-identity element has exactly 2 fixed points, hence F is a finite set and G permutates it. Choose representatives $p_1, \ldots, p_d \in \mathbb{P}$ of the distinct orbits of G in F. Let G_i be the corresponding stabilizer subgroups of these points in G;

$$G_i = G_{p_i} = \{g \in G | g \cdot p_i = p_i\}$$

and let n, n_i, be the orders of the groups G, G_i, respectively. We see now that

$$\begin{aligned} 2(n-1) &= \sum_{p \in \mathbb{P}} (|G_p| - 1) = \sum_{p \in F} (|G_p| - 1) \\ &= \sum_{p \in F} (|G_p|) - |F| = \sum_{i=1}^{d} \frac{n}{n_i} n_i - \sum_{i=1}^{d} \frac{n}{n_i}. \end{aligned}$$

Rewriting this[1] gives us

$$d + \frac{2}{n} = 2 + \sum_{i=1}^{d} \frac{1}{n_i}. \tag{3.2}$$

Clearly all $n_i \geqslant 2$, so our equation implies that we must have $d + \frac{2}{n} \leqslant 2 + \frac{d}{2}$, which gives us $d = 1$, 2 or 3.

[1] This equality is also known as the theorem of Luroth.

$d = 1$ Then $2 = n + \frac{n}{n_i}$, so $n = 1$, G consists of the identity element alone.

$d = 2$ Then $\frac{2}{n} = \frac{1}{n_1} + \frac{1}{n_2}$, whence $n = n_1 = n_2$ and $G = G_1 = G_2$ cyclic groups of order n.

$d = 3$ Then the equation becomes

$$1 + \frac{2}{n} = \frac{1}{n_1} + \frac{1}{n_2} + \frac{1}{n_3}. \tag{3.3}$$

so what we see is that $\sum_{i=1}^{3} \frac{1}{n_i} > 1$, so if we take $2 \leqslant n_3 \leqslant n_2 \leqslant n_1$, we see that $n_3 = 2$ and $n_2 \leqslant 3$. Using this we find the following possibilities for (n, n_1, n_2, n_3):

$$(2p, p, 2, 2), (12, 3, 3, 2), (24, 4, 3, 2), (60, 5, 3, 2). \tag{3.4}$$

These are the so-called polar configurations of our subgroups of $PSU_2(\mathbb{C})$.

$\mathbb{P}$ is isomorphic to the sphere $S^2 \subset E$: We identify $\mathbb{C} \subset \mathbb{P}$ with the plain in E through the equator and extend the stereographic projection by sending the point at infinity of the north pole of the sphere. This isomorphism gives us an isomorphism of groups:

$$PSU_2(\mathbb{C}) \cong SO_3(\mathbb{R})$$

so there is a bijection of the finite subgroups of theses groups. We see that the rotation-symmetry groups of the objects of the last paragraph give realizations of the possible polar configurations, except for the (trivial) cyclic case. A realization of the cyclic group of order n as a rotation symmetry group, is given by the pyramid with n sides. We can therefore conclude:

Theorem 3.1. *If G is a finite subgroup in $SO_3(\mathbb{R})$, then it is isomorphic to a cyclic group of order n, C_n, the dihedral group of order $2n$, D_n, the alternating group on 4 elements, A_4, the symmetric group of 4 elements, S_4, or the alternating group on 5 elements, A_5.*

Proof. The cyclic case is clear. The rest follows from comparing the results in Table 3 and the listed polar configurations in (4). □

We now use the double covering:

$$\pi : SU_2(\mathbb{C}) \longrightarrow SO_3(\mathbb{R}).$$

to get a classification of all possible finite subgroups of $SU_2(\mathbb{C})$. If G is a finite subgroup of $SO_3(\mathbb{R})$ we see that $|\pi^{-1}(G)| = 2|G|$.

Proposition 3.2. *Let $\Gamma \subset SU_2(\mathbb{C})$ be a finite subgroup. Then $\Gamma = \pi^{-1}(G)$, for some $G \subset SO_3(\mathbb{R})$ if and only if Γ is not cyclic of odd order.*

Proof. $\Gamma = \pi^{-1}(G)$ if and only if $-I \in \Gamma$, which is true if and only if $|\Gamma|$ is even. In the case that $|\Gamma|$ is odd we see that $\Gamma \cong \pi(\Gamma)$, which makes Γ a subgroup of $SO_3(\mathbb{R})$ of odd order, hence cyclic by the last theorem. □

The subgroups Γ that are inverse images of subgroups in $SO_3(\mathbb{R})$ are called binary. There is the binary cyclic group $\mathcal{C}_{2n} = C_n^*$ of order $2n$, the binary dihedral groups $\mathcal{D}_n = D_n^*$, of order $4n$, the binary tetrahedral group $\mathcal{T} = T^*$ of order 24, the binary octahedral group $\mathcal{O} = T^*$ of order 48, and the binary icosahedral group $\mathcal{I} = I^*$ of order 120. And there is the cyclic group $\mathcal{C}_{2n-1}$ of odd order.

We have a classification of the finite subgroups of $SU_2(\mathbb{C})$. Since every finite subgroup of $SL_2(\mathbb{C})$ is conjugated to a subgroup in $SU_2(\mathbb{C})$ we also have a classification of the finite subgroups of $SL_2(\mathbb{C})$.

Let us finally mention here that the three special orbits Gp_i in $PSU_2(\mathbb{C})$ are corresponding to the three special sets of points in the regular polyhedra, namely the set of vertices, the set of midpoints of edges, and the set of midpoints of faces.

4 Invariant theory of binary polyhedral groups

Let us first collect some general remarks and results about the theory of invariants. Take V a finite dimensional complex vector space, let $(e_i)_{i=1}^n$ be a basis of V and let $f_i \in V^*$ be the corresponding linear functions on V defined by $f_i(e_j) = \delta_{ij}$. Then the f_i generate the symmetric algebra $S = S(V^*)$ of V, which is isomorphic to the polynomial algebra on V, $\mathbb{C}[Z_1, \ldots, Z_n]$. S is a graded algebra, its components being the homogeneous polynomials of degree d;

$$S = \bigoplus_{d \geqslant 0} S_d.$$

Let $GL(V)$ denote the group of all invertible linear transformations of V. Every subgroup G of $GL(V)$ has an action on the algebra S, coming from the contragradient action:

$$g.f(v) = f(g^{-1}.v), \quad g \in G, f \in S, v \in V.$$

This action preserves the grading.

We denote the subalgebra of G-invariant polynomial functions by S^G;

$$S^G := \{f \in S | g.f = f \quad \forall g \in G\} = \bigoplus_{d \geqslant 0} (S^G \cap S_d).$$

Theorem 4.1. *For a reductive linear algebraic group the algebra of invariants S^G is finitely generated over $\mathbb{C}$. In particular for finite G the algebra S^G is finitely generated over $\mathbb{C}$.*

Proof. This result in the case of G finite was already proven by E. Noether. For a proof of the more general statement see [38]. □

The invariant theory for finite subgroups of $SU_2(\mathbb{C})$ was solved by Klein in his study concerning quintic equations, see [18]. We will follow mainly the same approach to establish the results he obtained. Other sources are [2], [23], [44] and Springer [38]. Springer uses some invariant theory of complex reflection groups. Finally we mention here the article of Pinkham [28]. He uses more advanced methods from algebraic geometry. We will first state the results and then (Partly) give a proof of them.

What Klein found was that the ring of invariants S^Γ, Γ a finite subgroup of $SU_2(\mathbb{C})$, is generated by 3 fundamental invariants X, Y, Z, subject to one essential relation $R(X, Y, Z) = 0$, which is weighted homogeneous;

$$S^\Gamma = \mathbb{C}[Z_1, Z_2]^\Gamma = \mathbb{C}[X, Y, Z]/(R) =: V.$$

We see that the map $\Phi : \mathbb{C}^2 \to \mathbb{C}^3$,

$$\Phi(z_1, z_2) = (X(z_1, z_2), Y(z_1, z_2), Z(z_1, z_2)),$$

factorizes over the quotient space $\mathbb{C}^2/\Gamma$. The image of Φ is the zero set of R, $V(R)$, so Φ induces an isomorphism of varieties $\mathbb{C}^2/\Gamma \cong V$.

$$\begin{array}{ccc} \mathbb{C}^2 & \xrightarrow{\Phi} & V(R) \subset \mathbb{C}^3 \\ \downarrow & & \uparrow id \\ \mathbb{C}/\Gamma & \xrightarrow{\sim} & V(R) \end{array}$$

For later use we make the following note here. Let the multiplicative group of the complex numbers $\mathbb{C}^*$ act on l-dimensional complex space $\mathbb{C}^l$ with integer weights $w_1, \ldots, w_l \in \mathbb{Z}$; for $x = (x_1, \ldots, x_l) \in \mathbb{C}^l$, and $t \in \mathbb{C}^*$,

$$t.x = t.(x_1, \ldots, x_l) = (t^{w_1} x_1, \ldots, t^{w_l} x_l).$$

Let us recall that each linear $\mathbb{C}^*$-action is of this type. If there is another such action on p-dimensional space with weights $d_1, \ldots, d_p$, we say that a $\mathbb{C}^*$-equivariant map $f : \mathbb{C}^l \to \mathbb{C}^p$ is weighted homogeneous of type $(d_1, \ldots, d_p; w_1, \ldots, w_l)$, with weights $w_1, \ldots, w_l$ and degrees $d_1, \ldots, d_p$. The monomials $a_{i_1,\ldots,i_l} x_1^{i_1} \cdots x_l^{i_l}$ that appear in a component f_i, a weighted homogeneous polynomial of type $(d_1; w_1, \ldots, w_l)$, satisfy $\sum_{j=1}^{l} i_j w_j = d_i$.

We have listed the results for the fundamental invariants X, Y and Z, conveniently chosen, in table 4 and we have added the weighted homogeneous type of R.

Table 4 Essential relations R for the fundamental invariants of G.

Γ	$R(X, Y, Z)$	$(d; w_1, w_2, w_3)$
$\mathcal{C}_n$	$X^n + YZ$	$(2n; 2, n+1, n+1)$
$\mathcal{D}_n$	$X^{n+1} + XY^2 + Z^2$	$(2n+2; 2, n, n+1)$
$\mathcal{T}$	$X^4 + Y^3 + Z^2$	$(12; , 3, 4, 6)$
$\mathcal{O}$	$X^3Y + Y^3 + Z^2$	$(18; 4, 6, 9)$
$\mathcal{I}$	$X^5 + Y^3 + Z^2$	$(30; 6, 10, 15)$

Proposition 4.2. *Let $f : \mathbb{C}^3 \to \mathbb{C}$ be weighted homogeneous of one of the types in table 4 and R the corresponding polynomial. Then there is a $\mathbb{C}^*$-equivariant automorphism α of $\mathbb{C}^3$ such that $f \circ \alpha = R$.*

Proof. For a proof of this statement see [32]. □

Now let us see how to obtain these results. To start with we have a finite subgroup Γ of $SU_2(\mathbb{C})$, we take $V = \mathbb{C}^2$ and we want to determine invariant polynomials. We know that the action preserves the grading of $\mathbb{C}[Z_1, Z_2]$, so we can restrict our inquiry to the analysis of the action on homogeneous polynomials.

The cyclic and dihedral case are easy and can be dealt with directly. We compute invariants for the generators of Γ. They give us the invariants for Γ.

Let Γ be cyclic of order N. A generator for Γ will be $\begin{pmatrix} \zeta & 0 \\ 0 & \zeta^{-1} \end{pmatrix}$, with ζ a primitive N-th root of unity; it acts on $\mathbb{C}[Z_1, Z_2]$ by $Z_1 \mapsto \zeta^{-1} Z_1$, $Z_2 \mapsto \zeta Z_2$. We get the invariants $X = Z_1 Z_2$, $Y = Z_1^N$ and $Z = Z_2^N$, which span the whole of S^Γ and satisfy the relation $X^N = YZ$.

Let Γ now be binary dihedral of order $4N$. Generators of Γ are $\begin{pmatrix} \zeta & 0 \\ 0 & \zeta^{-1} \end{pmatrix}$ and $\begin{pmatrix} 0 & i \\ i & 0 \end{pmatrix}$. The first has invariants listed above, the second has the invariants $Z_1^2 - Z_2^2$, $Z_1 Z_2 (Z_1^2 + Z_2^2)$ and $Z_1^2 Z_2^2$. We get invariants of Γ, $X = Z_1^2 Z_2^2$, $Y = Z_1^{2N} + (-1)^N Z_2^{2N}$, $Z = Z_1 Z_2 (Z_1^{2N} - (-1)^N Z_2^{2N})$. They span the whole of S^Γ and satisfy the relation

$$Z^2 - XY^2 = 4(-X)^{N+1}.$$

So if we substitute $-X$ for X, $2Y$ for Y, and $2Z$ for Z, they satisfy the relation $R(X, Y, Z) = 0$ listed above.

In the other cases this method is not a good choice. The generators of Γ get more complicated and the invariants of them have high degrees. It is easier to calculate the semi-invariants of Γ.

Definition 4.3. *f is a semi-invariant of Γ if*

$$\exists_\chi \in HOM(\Gamma, \mathbb{C}^*) \text{ such that } \quad \gamma.f = \chi(\gamma) f \quad \forall \gamma \in \Gamma.$$

Such characters χ of Γ factor through the abelianized group, $\mathcal{A}\Gamma$.

We have an action of Γ on $\mathbb{P}$ which factors through the action of $G = \pi(\Gamma) \subset SO_3(\mathbb{R})$ and we take as before representatives $p_1, \ldots, p_d$ of the distinct orbits of the action of G on F, the set of points fixed by some non-identity element in G. So we can define an action on the Abelian group of divisors on $\mathbb{P}$, $\mathfrak{D}$:

$$\gamma.D(x) = D(\gamma^{-1}.x), \text{ with } D \in \mathfrak{D} \text{ and } x \in \mathbb{P}.$$

Every homogeneous polynomial f of degree d defines a divisor $D_f \in \mathfrak{D}$ of degree d. Namely for $x \in \mathbb{P}$ we let $D(x)$ be the order of vanishing of f on the line in $\mathbb{C}^2$ defined by x, and a divisor likewise defines a homogeneous polynomial, uniquely up to a scalar factor.

If D_f is Γ-invariant we see that f is Γ-semi-invariant. The Γ-invariant divisors clearly form a subgroup of $\mathfrak{D}$. An orbit $\Sigma \subset \mathbb{P}$ defines a Γ-invariant divisor D;

$$D(x) = 1 \text{ if } x \in \Sigma, \quad D(x) = 0 \text{ else}.$$

Such D are called simple and are denoted by the same letter Σ. In our case we have 2 (cyclic case) or 3 exceptional orbits $\Sigma_i = \Gamma p_i$. They define exceptional divisors Σ_i, and the following semi-invariants f_i, of degree $|Gp_i|$:

$$f_i(Z_1, Z_2) = \prod_{(a_1:a_2)\in G(p_i)} (a_2 Z_1 - a_1 Z_2)$$

With every divisor D we get a well-defined character χ_D, since we can associate a semi-invariant f with D.

Proposition 4.4. *Let Γ be a finite subgroup of $SU_2(\mathbb{C})$ with its canonical action on polynomials in two variables. Let $\mathcal{D}$ be the abelian group of divisors on $\mathbb{P}$, with an action of Γ defined as above. Then the following holds:*

1. *Every positive invariant divisor $D \in \mathcal{D}$ can uniquely (up to order) be written as a finite sum:*

$$D = \sum_{i=1}^{k} r_i \Sigma_i$$

   of different simple divisors Σ_i, with $r_i \in \mathbb{Z}_{>0}$.
2. *There is a bijection between the group of positive Γ-invariant divisors and the ring of Γ-semi-invariant homogeneous polynomials; Take $D \in \mathcal{D}$ positive and invariant. It defines a homogeneous semi-invariant f by*

$$f(Z_1, Z_2) = \prod_{a=(a_1:a_2)\in\mathbb{P}} (a_2 Z_1 - a_1 Z_2)^{D(a)}.$$

   Take f a semi-invariant. It defines a divisor $D \in \mathcal{D}$ by

$$D([a_1 : a_2]) = ord_f((a_1, a_2)).$$

3. *If f is a Γ-semi-invariant homogeneous polynomial, then f is $[\Gamma, \Gamma,]$-invariant.*

Proof. Take a positive non zero divisor D. $D \neq 0$ in a finite number of points. D is constant on orbits and there are only finitely many orbits, say $\Sigma_1, \ldots, \Sigma_k$ on which D is strictly positive, of value $r_1, \ldots, r_k$. This proofs the first part of the proposition. The second part follows from remarks made above. For the third part, let f be Γ-semi-invariant. The character χ_f factorizes through $\mathcal{A}\Gamma = \Gamma/[\Gamma, \Gamma]$. □

In the icosahedral case $\mathcal{A}\Gamma$ is trivial and we get the 3 fundamental invariants directly from the three exceptional orbits, of order 12, 20, and 30 respectively. It can be shown that the space of invariant homogeneous polynomials of degree 60 is 2-dimensional. This gives us that these three invariants span the whole of S^Γ and by scaling them they satisfy the relation in the list.

In the other 2 cases, the tetrahedral and the octahedral case, we get three semi-invariants, coming from the exceptional orbits. Here $\mathcal{A}\Gamma$ is not trivial so we need some explicit calculation of the characters to get the results listed above. See the references for details.

5 Kleinian singularities

5.1 Characterizations of Kleinian singularities

We have shown that the quotient variety $\mathbb{C}/\Gamma$ is isomorphic to the affine orbit variety V. We see that in all cases we have a unique isolated singularity at 0. These particular singularities are called Kleinian singularities, but they are also denoted by rational double points, quotient singularities or simple singularities. They form a special class of singularities, having various equivalent characterizations.

Proposition 5.1. *The following characterizations of singular germ $(V, 0)$ of a surface are equivalent:*

1. *Let the finite group $G \subset SL_2(\mathbb{C})$ act naturally on complex 2-dimensional space. Then the orbit space $\mathbb{C}^2/G$ is a complex space and $(V, 0)$ is isomorphic to the germ of the set $(\mathbb{C}^2/G, 0)$. Germs of this type are called quotient singularities.*
2. *$(V, 0)$ is isomorphic to the germ $(V', 0)$ of the affine variety $V' = \mathbb{C}[X, Y, Z]/ < R >$, with R one of the polynomials from table 4.*
3. *Let $\rho : W \to V$ be a minimal resolution of the singularity $(V, 0)$. Then the exceptional set $\rho^{-1}(0)$ is a finite union of projective lines, which have intersection matrix the negative of a Cartan matrix of a Dynkin diagram of type A, D or E.*
4. *The singularity $(V, 0)$ is normal, has embedding dimension $\leqslant 3$ and is rational: Let $\rho : V^* \to V$ be a minimal resolution of $0 \in V$. Then the higher direct images of the structure sheaf are zero;*

$$(R^i\rho_*\mathcal{O}_{V^*})_0 = 0, \quad \forall i > 0,$$

Proof. $1 \Longleftrightarrow 2$, follows from section 4. To proof the equivalence $1 \Leftrightarrow 3$, we will give resolutions of the quotient singularities and show that they are of the type stated here. Moreover they will turn out to be characterizing; the minimal resolution is unique. For the equivalence with 4 and other characterizations of Kleinian singularities we refer to [8]. □

A point p on an analytic variety is called regular if a neighborhood of p is isomorphic to a complex manifold. A non-regular point is called singular. For a general account of singularities see [2]. Furthermore the survey article of Brieskorn [5] is recommended. He takes the hyper-geometric differential equation as a starting point. It is closely related to our discussion and has a long and rich history in mathematics.

5.2 Resolutions of Kleinian singularities

The goal of this section is to give a minimal resolution of the quotient singularity, which as stated gives a characterization of the singularity. Let us first give some definitions.

Definition 5.2. *A modification of a variety V at a point v is a morphism*

$$\mu : \overline{V} \longrightarrow V$$

with the following properties:

1. *μ is proper (In the case of a complex variety this means that the inverse image of compact subsets are compact).*
2. *μ restricted to $\mu^{-1}(V\backslash\{v\})$ is an isomorphism to its image $V\backslash\{v\}$ and $\mu^{-1}(V\backslash\{v\})$ is dense in $\overline{V}$.*

The subvariety $E := \mu^{-1}(v)$ is called the exceptional set.

Definition 5.3. *And if (V, v) is a singularity it is called a resolution of the singularity if in addition*

3. *every point of $\overline{V}$ is regular.*

For curve and surface singularities it is since long known that resolutions exist. In characteristic zero there is a theorem by Hironaka that states existence of resolutions of singularities in the general case. For the theory of surface singularities see Laufer [24]. The method used to obtain a resolution is the blow-up of singular points. The singular point is replaced by all the lines through it. A resolution ρ is minimal if all other resolutions factor through ρ. Minimal resolutions exist and are unique up to isomorphism.

We now give a construction for the minimal resolutions of the quotient singularities coming from the binary polyhedral groups. We will follow mainly the approach of Lamotke, see [23].

As usual the cyclic case is the easiest one. Let $\Gamma \subset SU_2(\mathbb{C})$ be the cyclic subgroup of order N in $SU_2(\mathbb{C})$,

$$\Gamma =< \begin{pmatrix} \zeta & 0 \\ 0 & \zeta^{-1} \end{pmatrix} > .$$

We will give a resolution of the corresponding quotient singularity $\mathbb{C}^2/\Gamma$. The singularity is isomorphic to the hyperplane singularity $(V, 0)$ defined by the relation $X^N = YZ$.

We consider the line bundle $\Sigma(b)$ (This is $\mathcal{O}(-b)$),

$$\Sigma(b) := \{((z_1, z_2), (w_1 : w_2)) \in \mathbb{C}^2 \times \mathbb{P} | z_1 w_2^b = z_2 w_1^b\},$$

and we give holomorphic charts U_1 and U_2 for $\Sigma(b)$, induced from the 2 charts covering $\mathbb{P}$:

$$U_i := \{(z, w) \in \Sigma(b) | w_i \neq 0\},$$

$$\Sigma(b) \supset U_1 \to \mathbb{C}^2, (z, w) \mapsto \left(z_1, \frac{w_2}{w_1}\right), w_1 \neq 0$$

and

$$\Sigma(b) \supset U_2 \to \mathbb{C}^2, (z, w) \mapsto \left(z_2, \frac{w_1}{w_2}\right), w_2 \neq 0.$$

We see that the U_i are isomorphic to $\mathbb{C}^2$ and the transition is given by ϕ : $(u,v) \mapsto (\frac{1}{v}, u^b v)$. The z_1 and v_2 are called linear fiber coordinates and $\frac{w_1}{w_2}$, $\frac{w_1}{w_2}$ affine base coordinates.

We use this to form a manifold M by pasting more of these copies together. We take

$$M = \bigcup_{i=1}^{N} U_i, \text{ with } U_i \cong \mathbb{C}^2, \text{ and } U_i \cup U_{i+1} =: L_i \cong \Sigma(2) \cong T^*\mathbb{P},$$

the U_i having coordinates (u_i, v_i) such that u_i is an affine base coordinate of L_{i-1} and a linear fiber coordinate of L_i and v_i is a linear fiber coordinate of L_{i-1} and an affine base coordinate of L_i.

The transition, pasting of the U_i, is done by

$$\begin{aligned} \phi_i : U_i \backslash \{v_i = 0\} &\to U_{i+1} \backslash \{u_{i+1} = 0\} \\ (u_i, v_i) &\mapsto \left(\frac{1}{v_i}, u_i v_i^2\right) = (u_{i+1}, v_{i+1}) \end{aligned}$$

for $i = 1, \ldots, N-1$.

$$M = \bigcup_{i=1}^{N-1} L_i, L_i \cap L_{i+1} = U_{i+1}.$$

We have the following result:

Proposition 5.4. *The minimal resolution for $\mathbb{C}^2/\Gamma$, Γ cyclic of order N, is given by*

$$\begin{aligned} &\rho : M \to V \\ &(u_i, v_i) \mapsto (x, y, z) = (u_i v_i, u_i^i v_i^{i-1}, u_i^{N-i} v_i^{N-i+1}). \end{aligned}$$

The exceptional set of ρ, E, is the union of the zero sections of the line bundles L_i,

$$E = \bigcup_{i=1}^{N-1} E_i, \text{ with } E_i := \{u_i = v_{i+1} = 0\}.$$

Here E_i and E_{i+1} intersect transversely in one point and $E_i \cap E_j = \emptyset$ for $|i-j| > 1$. The self-intersection number of the components E_i is -2.

Proof. We will only give a sketch. For details see [23]. □

The key ingredient of the proof is the following result about cyclic quotients:

Let H be a finite cyclic subgroup in $GL_n(\mathbb{C})$. Up to linear coordinate transformations we can assume that H is generated by $diag(\zeta_r, \zeta_s)$, with ζ_i a primitive i-th root of unity. Take $p \in \mathbb{Z}$ such that rp is the least common multiple of r and s and let H^* be the cyclic subgroup of order r generated by $diag(\zeta_r, \zeta_s^p)$.

Lemma 5.5. *The mapping $f(z_1, z_2) = (z_1, z_2^p)$ induces and isomorphism of complex spaces*

$$\mathbb{C}^2/H \cong \mathbb{C}^2/H^*.$$

Proof. The map $\tilde{f}$ is induced by the following commutative diagram:

$$\begin{array}{ccc} \mathbb{C}^2 & \xrightarrow{f} & \mathbb{C}^2 \\ \downarrow \pi & & \downarrow \pi^* \\ \mathbb{C}^2/H & \xrightarrow{\tilde{f}} & \mathbb{C}^2/H^* \end{array}$$

The non-trivial part of the proof is the injectivity of $\tilde{f}$. If $\tilde{f}([u_1, u_2]) = \tilde{f}([v_1, v_2])$ then $\pi^* \circ f(u_1, u_2) = \pi^* \circ f(v_1, v_2)$. This gives us $u_1 = \zeta_r^q v_1$ and $u_2^p = \zeta_s^{pq} v_2^p$, for some integer q. We get $u_2 = \zeta_p \zeta_s^q v_2$, and since ζ_s^r is a primitive p-th root of unity there is an integer l, such that $\zeta_p = \zeta_s^{rl}$. Take $t = rl + q$. Then clearly $u_1 = \zeta_r^t v_1$ and $u_2 = \zeta_s^t v_2$, whence $[u_1, u_2] = [v_1, v_2]$. □

Using the lemma, we see that $\mathbb{C}^2/\Gamma \cong \mathbb{C}^2/\Gamma^*$, with

$$\Gamma^* =< \begin{pmatrix} \zeta_N & 0 \\ 0 & \zeta_{N-1}\zeta_N \end{pmatrix} > .$$

Then we blow-up of the unique singular point $0 \in \mathbb{C}^2/\Gamma^*$ by the canonical map $\Sigma(2) \to \mathbb{C}^2$ and extending the action of Γ^* to an action of $\Sigma(2)$, making the following diagram commutative:

$$\begin{array}{ccc} \Sigma(2) & \longrightarrow & \mathbb{C}^2 \\ \downarrow & & \downarrow \\ \Sigma(2)/\Gamma^* & \longrightarrow & \mathbb{C}^2/\Gamma^* \end{array}$$

We look new for singularities in $\Sigma(2)/\Gamma^*$. The local charts U_i are Γ^*-invariant and any singularity must be in the exceptional set which is given by the zero section of $\Sigma(2)$, $\{(z, w) \in \Sigma(2) | z = 0\}$. This is a projective line.

The generator of Γ^* acts on $\Sigma(2)$;

$$((z_1, z_2), (w_1 : w_2)) \mapsto ((\zeta_N z_1, \zeta_{N-1}\zeta_N z_2)(w_1 : \zeta_{N-1} w_2)).$$

Now we look in the charts U_i. On U_i we can fix $w_i = 1$ and we see that we have the following induced actions:

$$\begin{aligned} &\text{on } U_1 : (z_1, w_2) \mapsto (\zeta_N z_1, \zeta_{N-1} w_2), \\ &\text{on } U_2 : (z_2, w_1) \mapsto (\zeta_{N-1}\zeta_N z_2, \zeta_{N-1}^{-1}). \end{aligned}$$

We get that

$$U_1/\Gamma^* \cong U_1/ < \begin{pmatrix} \zeta_N & 0 \\ 0 & \zeta_{N-1} \end{pmatrix} >,$$

but the last is isomorphic to $U_1/\{Id\}$ using our lemma twice. We get that

$$U_2/\Gamma^* \cong U_2/ < \begin{pmatrix} \zeta_{N-1} & 0 \\ 0 & \zeta_{N-1}^{-1} \end{pmatrix} > .$$

We have a cyclic quotient singularity of order one less in one of the charts, whereas the other chart has no singular points.

So we can repeat this construction leading to cyclic quotient singularities of smaller order and we get the construction of our manifold M, with all the properties mentioned above.

Minimality of a resolution can be characterized by the fact that it doesn't have exceptional curves with self-intersection -1. Hence this follows from the fact that the self intersection number for the components of E is -2. In fact the Euler number of $\mathcal{O}(-2)$ is -2. This follows from the fact that it admits a meromorphic section s without zeros and with 2 simple poles:

$$s : \mathbb{P} \to \Sigma(2),\ [w_1 : w_2] \mapsto \left(\frac{w_1^2}{w_1^2 - w_2^2}, \frac{w_2^2}{w_1^2 - w_2^2}, [w_1 : w_2] \right).$$

Definition 5.6. *Let $\rho : M \to V$ be a resolution of the singularity $(V, 0)$, with exceptional set E a union of projective lines; $E = \bigcup_{i \in I} E_i$, $E_i \cong \mathbb{P}$, where the intersections between the components, if not empty, are transversal.*

Then the graph Δ of the resolution has a vertex for each component E_i and an edge between E_i and E_j if and only if $E_i \cap E_j \neq \emptyset$ for $i, j \in I$, $i \neq j$.

We have demonstrated that the quotient singularity for the cyclic group of order N has the Dynkin graph of type A_{N-1} as a resolution graph.

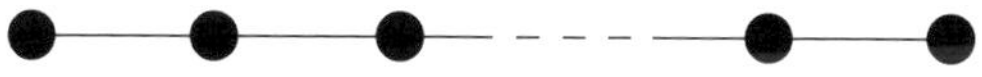

Figure 2 Dynkin diagram A_{N-1}.

Even in the easiest case possible, namely in the case of Γ cyclic of order 2, the real picture of the minimal resolution, see Figure 3, gives a good feeling for what is going on. This singularity is defined by the equation $X^2 + Y^2 - Z^2 = 0$.

Having dealt with the cyclic case we can proceed with the other. We have a quotient singularity $\mathbb{C}^2/\Gamma$, Γ a non-cyclic binary polyhedral group.

We first blow-up the point $0 \in \mathbb{C}^2$,

$$\begin{aligned} \mu : \Sigma(1) &\to \mathbb{C}^2 \\ (z, w) &\mapsto z \end{aligned}$$

and lift the action of Γ to $\Sigma(1)$, obtaining a commutative diagram

$$\begin{array}{ccc} \Sigma(1) & \xrightarrow{\mu} & \mathbb{C}^2 \\ \downarrow & & \downarrow \\ \Sigma(1)/\Gamma & \xrightarrow{\mu^2} & \mathbb{C}^2/\Gamma \end{array}$$

The map μ^* is a modification of the singularity. We will find a resolution for $\Sigma(1)/\Gamma$, thus resolving $\mathbb{C}^2/\Gamma$.

We look at the action of Γ on $\Sigma(1)$, given by $\gamma(z, w) = (\gamma.z, \gamma.w)$. The exceptional set $E = \mu^{-1}(0)$ is isomorphic to the projective line (hence isomorphic to the sphere) and contains all the points which are a fix-point of some non-identity operation in Γ.

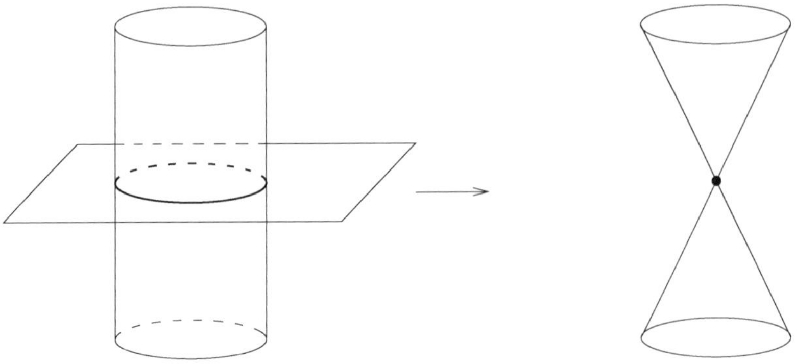

Figure 3 Resolution of cyclic quotient singularity of order 2.

Table 5 Resolution graphs for the quotient singularities.

Γ	type of Δ	Δ
$\mathcal{C}_N$	A_{N-1}	
$\mathcal{D}_N$	D_{N+2}	
$\mathcal{T}$	E_6	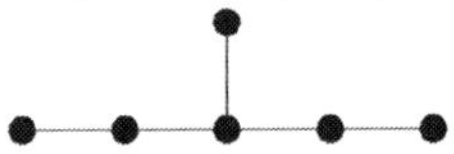
$\mathcal{O}$	E_7	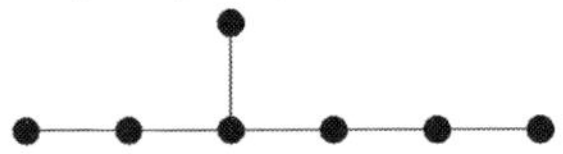
$\mathcal{I}$	E_8	

It is easy to see that the action of Γ on E factors through the action of G on the sphere, where G is the image of Γ in $SO_3(\mathbb{R})$. (Note that $\pm I$ acts trivially.) The projection $\Sigma(1) \to \Sigma(2)$, $(z, w) \mapsto (z^2, w)$ induces an action of Γ on $\Sigma(2)$ and we obtain an isomorphism $\Sigma(2)/G \cong \Sigma(1)/\Gamma$. We have obtained the following result:

Proposition 5.7. *There is a modification*

$$\psi : \Sigma(2)/G \to \mathbb{C}^2/\Gamma.$$

The only singular points come from the exceptional orbits of Γ on E. We know there are 3 such orbits corresponding to stabilizer subgroups in G which are cyclic subgroups of order p, q and 2, the polar configurations of G being $(\mathrm{p}, \mathrm{q}, 2)$. Hence $\Sigma(2)/G$ is a smooth complex surface, except from 3 isolated cyclic quotient singularities of order p, q and 2, which all three lie on the exceptional set of $\psi^{-1}(0) \cong \mathbb{P}$. We know how to solve these singularities. If we give a resolution for these three singularities we obtain a resolution of the quotient singularity $\mathbb{C}^2/\Gamma$. To proof minimality the criterion of the self intersection numbers is used.

The resolution graph has one special node from where three branches of length $p-1$, $q-1$ and $r-1$ start. This gives the resolution graphs as listed in Table 5.

We have sketched the real picture for the resolution of $\mathbb{C}^2/\mathcal{D}_2$ in Figure 4.

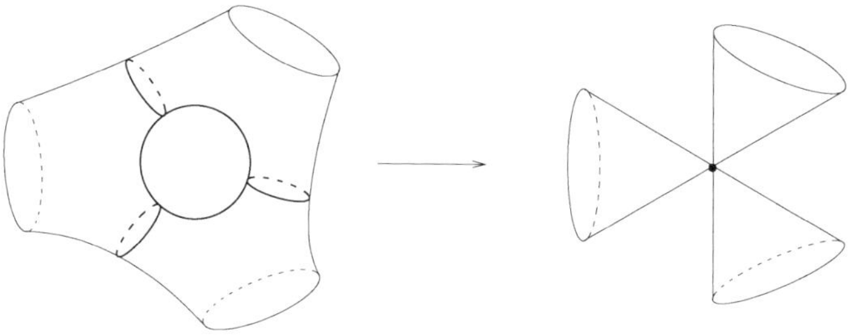

Figure 4 Minimal resolution for $\mathbb{C}^2/\mathcal{D}_2$.

5.3 Semi-universal deformations

For later use we briefly discuss the notion of deformations. For details see [17] and [27].

Definition 5.8. *Let V be a variety. A deformation of V is a pair (p,i), where p is a flat morphism of varieties, $p: X \to (U,u)$ from the total space X to a pointed base space (U,u), and $i: V \to p^{-1}(u)$ is an isomorphism.*

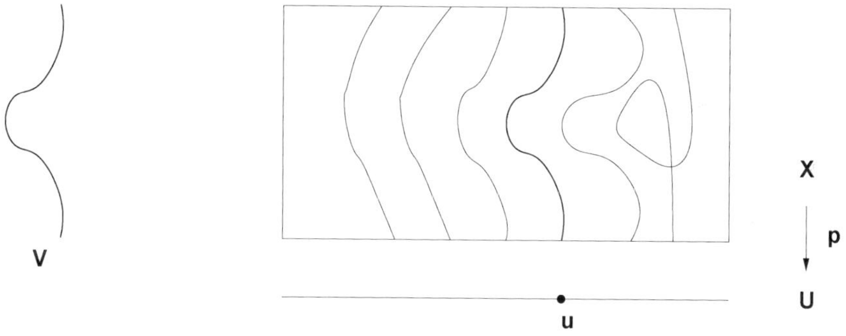

Figure 5 A deformation of a variety.

If we have a deformation (p,i), and a morphism of pointed varieties, ϕ : $(U',u') \to (U,u)$, we get an induced deformation (by the base change ϕ), $(\phi * (p), \phi * (i))$, with total space $X \times_U U'$ and base space U'.

Two deformations are isomorphic if they have the same pointed base space and there is an isomorphism of the total spaces that is an isomorphism of the special fiber.

The theory of deformations comes from the study of moduli spaces. In our situation we will consider Kleinian singularities, singular germs $(V,0)$ and we want to find deformations of them. As in the case of resolutions we would also like to have some "minimal" deformation, if possible unique.

Definition 5.9. *A deformation (p, i) of a variety V, with total space X and base space U, is called semi-universal if and only if both the following hold:*

1. *Up to isomorphism any deformation of V can be induced from (p, i) by some base change $\phi(U', u') \to (U, u)$.*
2. *The differential $D_{u'}\phi : T_{u'}U' \to T_u U$ is uniquely determined.*

In the present case it turns out to be too much to ask for the base change to be uniquely determined in the special point. The condition of the differential map of the tangent space in the special point to be uniquely determined guaranties that up to isomorphism the semi-universal deformation is unique. See the references for details.

Let $(V, 0)$ now be the Kleinian singularity defined as $\mathbb{C}[X, Y, Z]/R$, with R from Table 4. Then

$$\mathbb{C}[[Z_1, Z_2, Z_3]]/\left(R, \frac{\partial R}{\partial Z_1}, \frac{\partial R}{\partial Z_2}, \frac{\partial R}{\partial Z_3}\right)$$

in finite dimensional so we can choose a basis for it, say $b_1, \ldots, b_r$.

Define the spaces $X := \{(z, u) \in \mathbb{C}^3 \times \mathbb{C}^r | R(z) + \sum_{i=1}^r u_i b_i(z) = 0\}$ and $(U, u) := (\mathbb{C}^r, 0)$. Let the morphisms $p : X \to (U, u)$ and $i : V \hookrightarrow X$ be the projection on the second coordinate, and the embedding of $(V, 0)$ into X respectively. Then we have the following result.

Theorem 5.10. *The pair (p, i) is a semi-universal deformation of the singularity $(V, 0)$.*

Consider the case that we have a cyclic quotient singularity of order n. Then we have the variety V defined by $R = X^n - Y^2 - Z^2 = 0$ hence the quotient algebra

$$\mathbb{C}[X, Y, Z]/\left(R, \frac{\partial R}{\partial X}, \frac{\partial R}{\partial Y}, \frac{\partial R}{\partial Z}\right) = \mathbb{C}[X, Y, Z]/(X^{n-1}, Y, Z) = \mathbb{C}[X]/(X^{n-1}).$$

A semi-universal deformation is then given by the second projection from X to $(U, u) = (\mathbb{C}^{n-1}, 0)$, where

$$X := \{(x, y, z, u_1, \ldots, u_{n-1}) \in \mathbb{C}^{3+n-1} | x^n + \sum_{i=2}^{n} u_i x^{n-i} - y^2 - z^2 = 0\}.$$

$n = 2$ X is a family of surfaces depending on one parameter $u \in \mathbb{C}$. The only singular fiber is the fiber above 0. Let us sketch the possible fibers

$n = 4$ The singularity is of type A_3 and we obtain a 3-parameter family of surfaces:

$$x^4 - y^2 - z^2 + u_1 x + u_2 x^2 + u_3 x^3$$

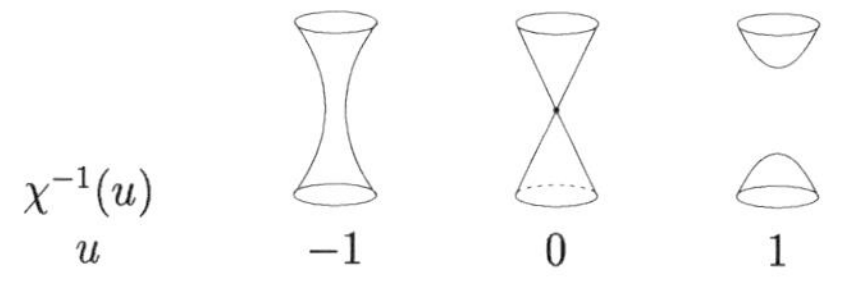

The singular fibers occur exactly for values of u_1, u_2, u_3 for which $x^4 + u_1x+u_2x^2+u_3x^3$ has multiple roots. This defines the discriminant locus, which in this case is a so-called swallow-tail. The singularities of type A_1 A_2 and A_3 also occur as fibers of χ. This phenomenon makes it possible to define a partial ordering on the singularity types, which corresponds with the partial ordering on diagrams. See for instance [34] for details. For the present example we have sketched the singular locus with the occurring singularity types in Figure 6.

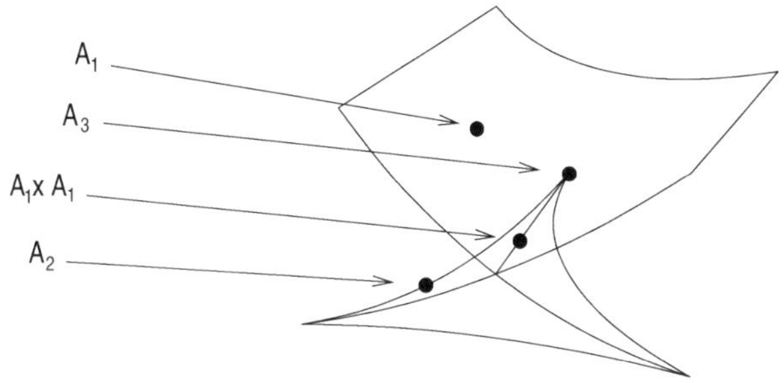

Figure 6 The swallow-tail, parameterizing the singular fibers.

6 The McKay correspondence

In the last section we have shown that to every finite $\Gamma \subset SU_2(\mathbb{C})$ we can associate a Dynkin diagram, namely the resolution graph of the quotient singularity. Quite recently a direct group theoretical proof of this correspondence has been found by McKay.

Let us consider the finite subgroups of $SU_2(\mathbb{C})$ and their representations. We take such a subgroup Σ and we let $\widehat{\Gamma} = \{\rho_0, \dots, \rho_r\}$ be the set of equivalence classes of irreducible finite dimensional complex representations of Γ, with ρ_0 the trivial representation.

Now let $\rho : \Gamma \to SU_2(\mathbb{C})$ be the canonical 2-dimensional representation of Γ. Then we can define for each group Γ a matrix $A(\Gamma)$, by decomposing the tensor products:

$$\rho_i \otimes \rho = \sum_{i=0}^{r} A(\Gamma)_{ij}\rho_j.$$

We see that A is a $(l+1) \times (l+1)$-matrix with all $a_{ij} \in \mathbb{Z}_{\geqslant 0}$.

It was McKay who made the following remarkable observation, see [25], [26]:

Theorem 6.1. *With each* Γ *we can associate a complex simple simply laced Lie-algebra* $\mathbf{g}$ *of rank* r *such that*

$$A(\Gamma) = 2I - C(\widetilde{\mathbf{g}}),$$

where $\widetilde{\mathbf{g}}$ *is the affine Kac-Moody Lie-algebra associated to* $\mathbf{g}$ *and* $C(\widetilde{\mathbf{g}})$ *its Cartan matrix. The Dynkin diagram that is associated to* Γ *in this way is the extended Dynkin diagram of the diagram associated to* Γ *by the resolution of its quotient singularity.*

Proof. This theorem can be proved (observed) by checking, since on both sides of the correspondence the structure is well known. There is a unified proof, not depending on this, by Steinberg, see [43]. □

Let us first collect some results to put this result in perspective. We take a complex simple Lie algebra $\mathbf{g}$ of rank r, and in it we take a Cartan subalgebra $\mathbf{h}$. We obtain a root-system in the dual $\mathbf{h}'$ of $\mathbf{h}$ and in it we make a choice for a positive root-system Φ. We get a set of simple roots and put an ordering on it;

$$\Delta = \{\alpha_1, \ldots, \alpha_r\} \subset \mathbf{h}'.$$

Then the Cartan-matrix of $\mathbf{g}$ is defined by:

$$(C(\mathbf{g}))_{ij} = \frac{2(\alpha_i, \alpha_j)}{(\alpha_i, \alpha_i)},$$

the symmetric non-degenerate bilinear form (,) on $\mathbf{h}'$ induced by the Killing form of $\mathbf{g}$.

With $\mathbf{g}$ we can associate the affine Kac-Moody Lie algebra mentioned above, which we take modulo its central extension. For details see [16]. The Cartan subalgebra $\widetilde{\mathbf{h}}$ of $\widetilde{\mathbf{g}}$ has dimension $r+1$ and we can take an ordered set of simple roots of $\widetilde{\mathbf{h}}'$ such that $\widetilde{\Delta} = \{\alpha_0, \ldots, \alpha_r\}$. This is a basis for $\widetilde{\mathbf{h}}'$ and we can extend (,) to a symmetric, but degenerate bilinear form on $\widetilde{\mathbf{h}}'$ by putting:

$$(\alpha_0, \alpha_i) = (-\psi, \alpha_i), \text{ and } (\alpha_0, \alpha_0) = (\psi, \psi),$$

where ψ is the unique highest root in Φ.

This bilinear form defines the Cartan matrix, of $\widetilde{\mathbf{g}}$, $C(\widetilde{\mathbf{g}})$, which contains the Cartan matrix of $\mathbf{g}$ as a principal $r \times r$ minor. We have the affine Weyl group $\widetilde{W}$ and to the Cartan matrix we can associate the extended Dynkin diagram; to each simple root we associate a node and there is an edge between two nodes if $C(\widetilde{\mathbf{g}})_{ij} = -1$, this being the case when not equal to zero. The theorem above states that the matrix $A(\Gamma)$, that is obtained for every finite subgroup $\Gamma \subset SU_2(\mathbb{C})$, minus twice the identity matrix is negative the Cartan matrix of $\widetilde{\mathbf{g}}$. We have listed the extended Dynkin diagrams in Table 6.

By the correspondence each node of the extended Dynkin diagram corresponds to a unique simple root in Δ and a representation in $\widehat{\Gamma}$. There is a bijection between the set of irreducible representations and the set of simple roots,

Table 6 The McKay correspondence

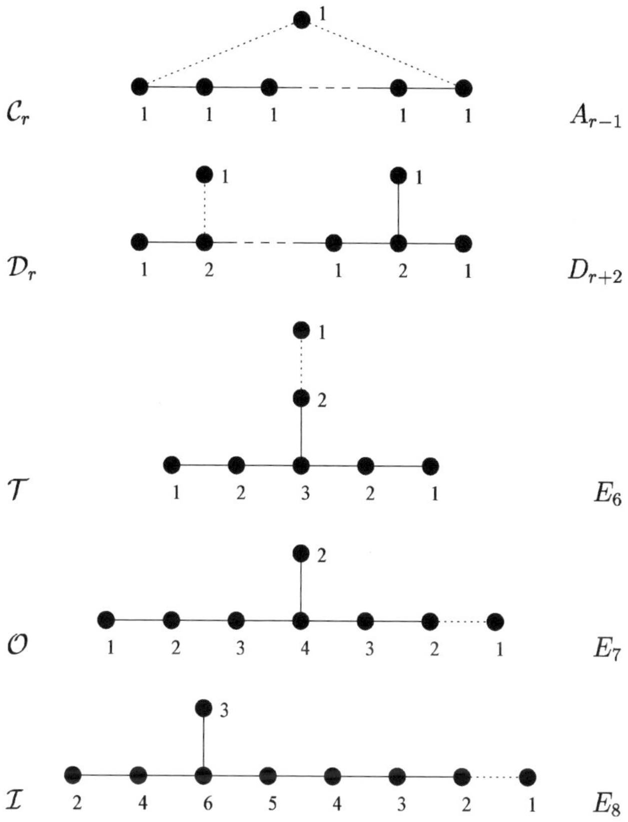

$\rho_i \mapsto \alpha_i, i = 0, \ldots, r$. The trivial representation corresponds to the extra node of the extended Dynkin diagram.

Let us take a look at one other way of viewing the correspondence. With each representation ρ_i, we get a character χ_i, and we can form the character table. Since the number of irreducible representations of a finite group Γ equals the number of conjugacy classes in it, we can choose representatives $x_0 = 1, x_1, \ldots, x_r$ for the latter and we then form a $(r+1) \times (r+1)$-matrix, the character table, by in the i-th row and j-th column evaluating χ_i in x_j. We now have the following result:

Proposition 6.2. *The i-th column of the character table of Γ, $\mathbf{v} := (\chi_j(x_i))_j$, is an eigenvector of the Cartan matrix of* $\mathbf{g}$ *with eigenvalue $2 - \chi(x_i)$.*

Proof. Take $x \in \Gamma$. We have $\chi(x)\chi_i(x) = \sum_j A(\Gamma)_{ij}\chi_j(x)$. This gives us $(2 - \chi(x_i))\mathbf{v} = 2\mathbf{v} - \sum_{k,j} A(\Gamma)_{kj}\chi_j(x_i) = (2 - A(\Gamma))\mathbf{v} = C\mathbf{v}$. □

So we can translate the theorem in terms of the diagrams; if we tensor a representation ρ_i with our fixed 2-dimensional representation ρ we see that this decomposes into the representations ρ_j neighboring ρ_i in the diagram.

Corollary 6.3.

$$\rho_j \otimes \rho = \sum_{k \in N(j)} \rho_k, \tag{6.5}$$

where $N(j)$ is the set of nodes in the diagram that has an edge with the j-th node.

We can decompose the vector $\alpha_0 + \psi \in \widetilde{\mathbf{h}}$, which is orthogonal to every other vector in $\widetilde{\mathbf{h}}$;

$$\alpha_0 + \psi = \sum_{i=0}^{r} d_i \alpha_i$$

In the diagrams we have added to each node the corresponding coefficient d_i. There is the following result relating these coefficients to the dimensions of the irreducible representations:

Proposition 6.4. *Let the d_i and ρ_i be as above and let h be the Coxeter number of the Lie algebra $\mathbf{g}$. Then the following holds for all $j = 0, \ldots, r$:*

$$dim(\gamma_i) = d_i. \quad \text{and } h = \sum_{i=0}^{r} dim(\rho_i)$$

Proof. Let $\mathbf{d} = \{d_i\}$ and $\mathbf{e} = \{dim(\rho_i)\}$ be vectors in $\mathbb{R}^{r+1}$. We want to show that they are equal. We will first show they are both eigenvectors of eigenvalue 2 for the matrix $A(\Gamma)$. This is the case for $\mathbf{e}$, since $\mathbf{e} = \{\chi_j(id)\}$, which has eigenvalue 0 for $C = 2I - A(\Gamma)$, by Proposition 6.2. The vector $\mathbf{d}$ is orthogonal to every vector in $\widetilde{\mathbf{h}}'$, hence $\widetilde{\mathcal{W}}$-invariant. This gives us

$$2d_i = \sum_{j \in N(i)} d_j \tag{6.6}$$

so also $\mathbf{d}$ is eigenvector of eigenvalue 2. The space spanned by eigenvectors of $A(\Gamma)$ of eigenvalue 2, is one dimensional since they are eigenvectors of the extended Cartan matrix, of eigenvalue 0; this $r+1 \times r+1$-matrix contains the non singular $r \times r$-matrix $C(\mathbf{g})$ as a principal minor. The fact that $d_0 = 1 = dim(\rho_0) = e_0$ concludes the proof of the first statement.

Now the second follows from the well-known fact that h is the sum of the coefficients of the highest roots plus one. □

If we consider the numbers $d_i = dim(\rho_i)$ to be the image of a function f of the modes of the graph to the strictly positive integers $\mathbb{Z}_{>0}$, the equality in (6.6) is the condition of f to be an additive function of the graph. It can be shown that the existence of such an additive function on a connected graph with no loops is actually defining for it to be an extended Dynkin diagram of type A, D, or E. See [30].

We would like to mention some results obtained by using the McKay correspondence. Agood example of the use of it is illustrated in the article of Kostant, see [22]. In this article he attacks the following question: Let $S = \bigoplus_{n=0}^{\infty} S^n(\mathbb{C}^2)$ be the symmetric algebra over $\mathbb{C}^2$ and let π_n be the representation of $SU_(\mathbb{C})$ on the n-th component. These representations are irreducible for all n and set up a basis for the unitary dual of $SU_2(\mathbb{C})$. Given a finite subgroup $\Gamma \subset SU_2(\mathbb{C})$, we want to see how the restriction of π_n to Γ decomposes.

Now Kostant shows that the root-structure of the Lie algebra that is associated to Γ contains all the information about the multiplicities of the decomposition of the representations mentioned above. These multiplicities come in a beautiful way from the orbit structure of the Coxeter element on the roots of $\mathbf{g}$.

The results obtained by Kostant in this paper were proven later in a slightly different way, however with the use of the McKay correspondence, by Springer, see [39].

The observation made by McKay and the ideas discussed above have been a starting point for many others, either using the McKay correspondence to get new results, as for instance Kostant in the article mentioned above, or trying to give an explanation or generalization of it. One of these directions is discussed in the next section.

7 Geometric McKay correspondence

We have seen that to each finite Γ in $SU_2(\mathbb{C})$ we can associate a Dynkin graph, the resolution graph of the quotient singularity. On the other hand the McKay correspondence gives a correspondence between the affine version of these graphs and the irreducible representations of Γ. The question now rises whether these representations and the resolution of the singularity are more directly related.

This is indeed the case. Gonzalez-Springberg and Verdier have proven a theorem which states such a connection, see [10]. There is another proof of this by H. Knörrer [19], which is more accessible. This connection is sometimes referred to as the 'geometric' McKay correspondence, where the McKay correspondence discussed in the last section is called 'algebraic'.

In short the theorem states there is a bijection between the irreducible representations of a binary polyhedral group and irreducible divisors of the minimal resolution of the corresponding quotient singularity.

Let Γ be binary polyhedral, $V := \mathbb{C}^2$, $S := V/\Gamma$ and $\phi : \tilde{S} \to S$ the minimal resolution with exceptional set D. Let $\rho : \Gamma \to GL(E)$ be an irreducible representation of Γ. There is a natural action of Γ on V, that makes it possible to define the Γ-equivariant vector bundle $\mathcal{E} := V \times E \to V$, on which Γ acts by $\gamma.(\mathbf{z},\mathbf{v}) = (\gamma(\mathbf{z}), \gamma(\mathbf{v}))$.

Now the associated locally free sheaf on V is equal to $\mathcal{O}_V \otimes_{\mathbb{C}} E$. It has a canonical Γ-action. We denote it, by abuse of notation, with $\mathcal{E}$.

The group Γ acts freely on $V - \{0\}$ and $\mathcal{E}$ is a Γ-vector bundle, hence $\mathcal{E}$ defines a vector bundle $\mathcal{E}'$ on the quotient $S - \{0\} = (V - \{0\})/\Gamma$. Let $\tilde{\mathcal{E}} := \phi^*(\mathcal{E}')$ be the pullback of this on $\tilde{S} - D \cong S - \{0\}$. Let $i : \tilde{S} - D \to \tilde{S}$ be the inclusion map. If s is an invariant global section of $\mathcal{E}$, $s \in \Gamma(V, \mathcal{E})^\Gamma$, then s induces a global section on $\mathcal{E}'$ and $\tilde{\mathcal{E}}$. So s defines a section $\pi(s)$ of the sheaf $i_*(\tilde{\mathcal{E}})$ on $\tilde{S}$. Denote the subsheaf of $i_*(\tilde{\mathcal{E}})$ generated by the sections $\pi(s)$, $s \in \Gamma(V, \mathcal{E})^\Gamma$, by $\pi(\rho)$. By $c_1(\pi(\rho))$ is meant the first Chern class of the sheaf $\pi(\rho)$. Let $\mathrm{Irr}(D)$ be the set of irreducible components of D and let $\widehat{\Gamma}$ be the set of non-trivial irreducible representations. By a_{ij} we denote the coefficient of ρ_j in the decomposition of $\rho \otimes \rho_i$, this ρ being the canonical two-dimensional representation of Γ.

Theorem 7.1. *For each $\rho \in \widehat{\Gamma}$ the sheaf $\pi(\rho)$ on $\tilde{S}$ is locally free of $deg(\rho)$. There is a bijection $\psi : \widehat{\Gamma} \to Irr(D)$ such that*

$$c_1(\pi(\rho)) \cdot d = \begin{cases} 0 & if\ d \neq \psi(\rho) \\ 1 & if\ d = \psi(\rho) \end{cases}$$

and

$$\psi(\rho_i) \cdot \psi(\rho_j) = a_{ij} \ for\ all\ i \neq j.$$

So for each non trivial irreducible representation ρ_i, a locally free sheaf $\pi(\rho_i)$ on $\tilde{S}$ is constructed, which by its first Chern class is represented by a divisor. This divisor meets the exceptional set D transversally in one point, hence in a unique component $D_i \in D$, which is taken to be the image of ρ_i under ψ.

8 Brieskorn's theorem

8.1 Introduction

In this section we discuss a theorem by Brieskorn, which establishes a direct connection between the binary polyhedral groups with their corresponding singularities and the Lie algebras of the type of the Dynkin diagrams obtained in the resolution of the singularities. We will see that the nilpotent variety of such a Lie algebra intersected with a slice transversal to the subregular nilpotent orbit is a singularity with the Dynkin diagram of this Lie algebra as its resolution graph.

At many points there are generalizations possible, but we do not want to explore in this direction. The goal here is to demonstrate the statement above in the simplest case.

The results presented in this section have a history in the studies of the connection between quotient singularities and Dynkin diagrams, presented in Sect. 5, and the work by Steinberg, Kostant and Springer on conjugacy classes and the unipotent variety of an algebraic group. Brieskorn studied further connections between the resolutions of singularities and structures related to their Dynkin diagrams, such as the Weyl group and Coxeter transformations. That led Grothendieck to conjecture the above mentioned connection between the simple singularities and Lie algebras or groups of type A, D, E.

We want to mention here the fundamental papers by Kostant [20, 21], Steinberg [41], and Springer [37].

To start with we take a complex simple algebraic group G. We let G act on itself by the conjugation action:

$$\begin{aligned} G \times G &\longrightarrow G \\ (g, h) &\mapsto ghg^{-1}. \end{aligned}$$

This action is called the adjoint action. We have the linearization of this action, the adjoint representation of G on the Lie algebra of G, $\mathbf{g}$:

$$\begin{aligned} G \times \mathbf{g} &\longrightarrow \mathbf{g} \\ (g, h) &\mapsto ad(g)(h). \end{aligned}$$

Let T be a maximal torus of G with character group $X^*(T) \cong \mathbb{Z}^r$, so $\text{rank}(G) = \text{rank}(\mathbf{g}) = r$. Let $N = N_G(T)$ be the normalizer of T in G, leading us to the Weyl group of G, $\mathcal{W} = N/T$. If we restrict the adjoint representation to T we have the decomposition of $\mathbf{g}$ into a direct sum of eigenspaces:

$$\mathbf{g} = \bigoplus_{\alpha \in X^*(T)} \mathbf{g}_\alpha.$$

This defines the set of roots of T in $\mathbf{g}$. By choosing a Borel subgroup of G, containing T, we make a choice for a set of positive roots and a corresponding basis of simple roots, $\Delta := \{\alpha_1, \ldots, \alpha_r\}$, and we have the fundamental dominant weights λ_i, $i = 1, \ldots, r$, defined by

$$\frac{2(\lambda_i, \alpha_j)}{(\alpha_j, \alpha_j)} = \delta_{ij},$$

$(\, , \,)$ being the $\mathcal{W}$-invariant scalar product on the r-dimensional Euclidean space, spanned by the roots.

We recall that for each of these fundamental dominant weights we have an irreducible highest weight representation of G, ρ_λ, with character $\chi_\lambda \in \mathbb{C}[G]$, which is constant on conjugacy classes of G.

We have a Jordan decompositions in G and $\mathbf{g}$. In G it realizes a decomposition of each $x \in G$ in its semisimple and unipotent part:

$$x = x_s \cdot x_u, \; x_s \text{ semisimple}, \; x_u \text{ unipotent, such that } x_s \text{ and } x_u \text{ commute.}$$

In the Lie algebra it decomposes each $x \in \mathbf{g}$ in its semisimple and nilpotent part:

$$x = x_s + x_n, \; x_s \text{ semisimple}, \; x_n \text{ nilpotent, with } [x_s, x_n] = 0.$$

These decompositions are unique and representations respect them.

8.2 The adjoint quotient

Let us consider the adjoint representation of G on the Lie algebra of G, $\mathbf{g}$. We want to study the orbit structure, i.e., look for a quotient map. This was done extensively by Kostant in his papers mentioned above.

Let $\mathbf{h}$ be the Lie algebra of the maximal torus T. On it we have a natural action of $\mathcal{W}$. We want to study the ring of G-invariant polynomials in $\mathbf{g}$, $\mathbb{C}[\mathbf{g}]^G$. We have the following result.

Theorem 8.1. *The algebras $\mathbb{C}[\mathbf{g}]^G$ and $\mathbb{C}[\mathbf{h}]^{\mathcal{W}}$ are isomorphic. There exist r homogeneous G-invariant polynomials γ_i of degrees $d_i = m_i + 1$, that generate $\mathbb{C}[\mathbf{g}]^G$ freely as a $\mathbb{C}$-algebra. The m_i are the exponents of the Weyl group of $\mathbf{g}$.*

Proof. Consider the mapping

$$\mathbb{C}[\mathbf{g}] \to \mathbb{C}[\mathbf{h}],$$

realized by restricting polynomials to the subalgebra $\mathbf{h}$. Its restriction to the algebra of invariants $\mathbb{C}[\mathbf{g}]^G$ gives an isomorphism with $\mathbb{C}[\mathbf{h}]^{\mathcal{W}}$. It is a result due to Chevalley that the last is freely generated as a $\mathbb{C}$-algebra by r homogeneous polynomials p_i of degrees $d_i = m_i + 1$, the m_i being the exponents of the Weyl group of $\mathbf{g}$. For a complete proof see [4], paragraph 8, no. 3, Thm. 1 and Cor. 1. □

The semisimple conjugacy classes in $\mathbf{g}$, under the adjoint action of G, are in natural correspondence with elements of $\mathbf{h}_{/\mathcal{W}}$; $s, t \in \mathbf{h}$ are conjugated under G if and only if they are in the same $\mathcal{W}$-orbit. Furthermore the γ_i's separate semisimple orbits: Let $s, t \in \mathbf{g}$ be semisimple. Then

$$\begin{aligned} &s \text{ and } t \text{ conjugate under } G \iff \\ &s \text{ and } t \text{ conjugate under } \mathcal{W} \iff \\ &\gamma_i(s) = \gamma_i(t) \quad \forall i \in \{1, \ldots, r\}. \end{aligned}$$

The embedding of $\mathbb{C}[\mathbf{h}]^{\mathcal{W}}$ into $\mathbb{C}[\mathbf{g}]$, induces a morphism $\mathbf{g} \to \mathbf{h}/\mathcal{W}$, called the adjoint quotient of $\mathbf{g}$;

$$\gamma : \mathbf{g} \to \mathbf{h}/\mathcal{W}.$$

It can be realized by generators γ_i of the algebra of invariants, thus taking a concrete form and making $\mathbf{h}/\mathcal{W}$ isomorphic to affine r-dimensional complex space.

$$\begin{aligned} \gamma : \mathbf{g} &\to \mathbb{C}^r \\ x &\mapsto (\gamma_1(x), \ldots, \gamma_r(x)). \end{aligned}$$

We now want to study the fibers of the adjoint quotient. These fibers must be unions of G-orbits in $\mathbf{g}$. Let us collect some results on the adjoint quotient and its fibers.

Proposition 8.2. *Let $\gamma : \mathbf{g} \to \mathbf{h}/\mathcal{W}$ be the adjoint quotient defined above. Then we have the following:*

1. *γ is a flat morphism: all fibers of γ have dimension $dim(\mathbf{g}) - r$.*
2. *Each fiber is a union of finitely many G-orbits in $\mathbf{g}$.*
3. *The orbit dimensions have to be even.*

Proof. For a proof see [21]. □

8.3 Regular and subregular elements, and the nilpotent variety

The elements for which the orbit dimension is maximal are called regular. The orbit dimension is maximal if the dimension of the stabilizer subgroup, $C_G(x)$, is minimal, $C_G(x) := \{g \in G | g.x = x\}$ for $x \in \mathbf{g}$. The minimal dimension of $C_G(x)$ is equal to the rank of $\mathbf{g}$, r. It can be shown that the Lie algebra of $C_G(x)$, $\mathbf{c}_G(x)$, contains an r-dimensional commutative subalgebra.

This gives us that if O is an orbit that

$$dim(O) \leqslant dim(G) - r.$$

So if an orbit is not regular, it has to have dimension at least 2 less then $\dim(G)-r$, the dimension of the centralizer $C_G(x)$ is $r+2i$, $i = 1, 2, \ldots$. Irregular elements for which the centralizer has maximal dimension, that is dimension $r + 2$, are called subregular.

Proposition 8.3. *Each fiber contains a unique regular orbit, which is dense in it.*

Proof. See [21]. □

Proposition 8.4. *An element $x \in \mathbf{g}$ is nilpotent $\Longleftrightarrow \gamma(x) = \gamma(0)$.*

Proof. This is a special case of the theorem that for $x \in \mathbf{g}$, and $f \in \mathbb{C}[\mathbf{g}]^G$, $f(x) = f(x_s)$. For a proof of this more general statement see [42]. □

This allows us to define the nilpotent variety,

$$Nil(\mathbf{g}) := \{x \in \mathbf{g} | \gamma(x) = \gamma(0)\} = \gamma^{-1}(\gamma(0)).$$

Proposition 8.5. *There is a unique subregular nilpotent G-orbit in $\mathbf{g}$.*

Proof. See [42], 3.10 Thm. 1. □

8.4 Example: $\mathbf{sl}_n$

To illustrate the theory of the last sections we look at the case $G = SL_{r+1}$. Then the Lie algebra $\mathbf{sl}_{r+1}$ is given by the $r + 1 \times r + 1$-matrices of trace zero. The adjoint representation of G is given by conjugation:

$$\begin{aligned} G \times \mathbf{g} &\to \mathbf{g} \\ (g, x) &\mapsto gxg^{-1} \end{aligned}$$

If we choose T the set of diagonal matrices in G, we get the corresponding Cartan subalgebra $\mathbf{h}$ the diagonal matrices in $\mathbf{g}$. It has dimension r. The Weyl group $\mathcal{W}$ is isomorphic to S_{r+1}, the symmetric group on $r+1$ elements. It acts on T and $\mathbf{h}$ by permuting the entries.

The adjoint quotient γ can be realized by the map

$$\begin{aligned} \gamma : \mathbf{sl}_{r+1} &\to \mathbb{C}^r \\ x &\mapsto (\gamma_1(x), \ldots, \gamma_r(x)), \end{aligned}$$

such that

$$det(\lambda \cdot Id - x) = \lambda^{r+1} + \gamma_1(x)\lambda^{r-1} + \cdots + \gamma_r(x),$$

is the characteristic polynomial of x.

Let us take a look at the fibers of γ. Take $c = (c_1, \ldots, c_r) \in \mathbb{C}^r$. It defines a polynomial $p(c) = \lambda^r + c_1\lambda^{r-1} + \cdots + c_r$, which in turn defines a partition

$m(c) = (m_1, \ldots, m_k)$ of $r+1$ by taking the multiplicities of the roots, i.e. the eigenvalues, of p, say $t_1, \ldots, t_k$. We take all partitions decreasing. Each m_i is the total size of the blocks with one eigenvalue, t_i; there are partitions s_i of the m_i, $i = 1, \ldots, k$, $s_i = (s_{i1}, \ldots, s_{il_1})$. In the Jordan normal form of x, s_{ij} is the size of the j-th block for the i-th eigenvalue t_i. Each $x \in \chi^{-1}(c)$ is up to conjugacy determined by such a sequence $s_1, \ldots, s_k$ of partitions of the m_i. Let us illustrate this by putting x into Jordan normal form:

$$\begin{pmatrix} M_1 & & & \\ & M_2 & & \\ & & \ddots & \\ & & & M_k \end{pmatrix}$$

with each M_i a $m_i \times m_i$-matrix with all eigenvalues t_i and Jordan form

$$M_i = \begin{pmatrix} S_{i1} & & & \\ & S_{i2} & & \\ & & \ddots & \\ & & & S_{il_i} \end{pmatrix}, \quad S_{ij} = \begin{pmatrix} t_i & 1 & & \\ & t_i & \ddots & \\ & & \ddots & 1 \\ & & & t_i \end{pmatrix},$$

an $s_{ij} \times s_{ij}$-matrix.

So clearly there are finitely many orbits in each fiber of γ. We see that

- x is semisimple exactly when all the S_{ij} are one-dimensional, the partitions s_i are $(1, 1, \ldots, 1)$.
- x is regular exactly when $s_i = (m_i) \forall i = 1, \ldots, k$
- x is subregular exactly when $s_i = (m_i) \forall i = 1, \ldots, k$ but one, say j, for which $s_j = (m_j - 1, 1)$.

If we restrict to x nilpotent, so $k = 1$, we see that for x to be semisimple, regular, subregular, it means that x is conjugated to

$$\begin{pmatrix} 0 \end{pmatrix}, \begin{pmatrix} 0 & 1 & & & \\ & 0 & 1 & & \\ & & 0 & \ddots & \\ & & & \ddots & 1 \\ & & & & 0 \end{pmatrix}, \begin{pmatrix} 0 & 1 & & & & \\ & 0 & 1 & & & \\ & & 0 & \ddots & & \\ & & & \ddots & 1 & \\ & & & & 0 & 0 \\ & & & & & 0 \end{pmatrix},$$

respectively. It is clear that the regular and the subregular nilpotent orbits are unique.

An interesting question is how the nilpotent variety decomposes into G-orbits. We know there is one unique dense orbit, the regular orbit, and that every other orbit has smaller dimension. The semisimple orbit, which is a single point lies in the closure of every other orbit in $Nil(\mathbf{g})$. We can define the closure relation between arbitrary orbits in $\mathbf{g}$. Let O and O' be two nilpotent orbits, defined by

```
partition    dim
(5)          20
 |            |
(4,1)        18
 |            |
(3,2)        16
 |            |
(3,1,1)      14
 |            |
(2,2,1)      12
 |            |
(2,1,1,1)     8
 |            |
(1,1,1,1,1)   0
```

```
       partition            dim
          (6)               30
           |                 |
         (5,1)              28
           |                 |
         (4,2)              26
         /    \             / \
   (4,1,1)  (3,3)        24   24
         \    /             \ /
        (3,2,1)             22
         /    \             / \
 (3,1,1,1)  (2,2,2)      18   18
         \    /             \ /
       (2,2,1,1)            16
           |                 |
      (2,1,1,1,1)           10
           |                 |
     (1,1,1,1,1,1)           0
```

partitions $\pi = (n_1, \ldots, n_{l'})$ and $\pi' = (n'_1, \ldots, n'_{l'})$. We define the partial ordering on the set of orbits by

$$O \leqslant O' \Longleftrightarrow O \text{ is contained in } \overline{O'}.$$

This is the partial order on partitions

$$O \leqslant O' \Longleftrightarrow \pi \leqslant \pi' \Longleftrightarrow n_1 + n_2 + \cdots + n_i \leqslant n'_1 + n'_2 + \cdots + n'_i \quad \forall i.$$

We have illustrated this, for r=5 and r=6.

8.5 Brieskorn's theorem

The subregular nilpotent orbit has a particular property. Let us first make the following definition:

Definition 8.6. *A transversal slice to the G-orbit of an element $x \in \mathbf{g}$ in the point x is a smooth locally closed subvariety $S \subset \mathbf{g}$, such that $x \in S$, $dim(S) = codim(Gx)$, and the map $G \times S \to \mathbf{g}$ is a submersion.*

In our case we are interested in a slice transversal to the orbit of a subregular nilpotent element $x \in \mathbf{g}$. This is a subspace $S \subset \mathbf{g}$, such that $x \in S$ and S is complementary to the affine tangent space at x of the orbit Gx.

Now we can state the main theorem we want to illustrate in this section, conjectured by Grothendieck, and later on proved by Brieskorn:

Theorem 8.7. *Let x be a subregular nilpotent element of a simple complex Lie algebra $\mathbf{g}$ of type $\Delta = A_r$, D_r, or E_r, and let S be a transversal slice to the G-orbit of x in the point x. Then*

1. *$S \cap Nil(\mathbf{g})$ is a surface with a Kleinian singularity of type Δ.*
2. *The restriction $\gamma_{|S} : S \to \mathbf{h}/W$ is a semi-universal deformation of the singularity $(S \cap Nil(\mathbf{g}), x)$.*

Proof. The possible outlines of a proof, for the first part, were given by Brieskorn, see [6]. For a complete proof of this theorem see Slodowy [32]. It gives a thorough account of this result, making generalizations where possible. For a proof of the second part of the proof the reader is referred to this book. In what follows we will discuss the first part. □

8.6 Identification of the subregular singularity

There are two different strategies to get the identification of the singularities. The first is by giving a resolution of the singularity. The second is by giving a $\mathbb{C}^*$-action on S and on $\mathbf{h}/\mathcal{W}$, such that the restriction of γ to S becomes $\mathbb{C}^*$-equivariant, that is weighted homogeneous of the same type as the polynomial R in Table 4. Then the identification follows from Proposition 4.2.

We will sketch the outlines of the second method, and conclude with some remarks on the first. It is of great interest and historically the first viewpoint.

Let us recall the following result:

Lemma 8.8. *(Jacobson-Morozov) Let* $\mathbf{g}$ *be a reductive Lie algebra and let* $x \in \mathbf{g}$ *be nilpotent. Then there exist* $y, h \in \mathbf{g}$ *and an injective homomorphism of Lie algebras,*

$$\rho : \mathbf{sl}_2 \to \mathbf{g},$$

with $\rho = \left(\begin{pmatrix} 0 & 1 \\ 0 & 0 \end{pmatrix}\right) = x$, $\rho = \left(\begin{pmatrix} 1 & 0 \\ 0 & -1 \end{pmatrix}\right) = h$, *and* $\rho = \left(\begin{pmatrix} 0 & 0 \\ 1 & 0 \end{pmatrix}\right) = y$. *This gives us the relations:*

$$[h, x] = 2x, [h, y] = -2y, and\ [x, y] = h.$$

Let us take $x \in \mathbf{g}$ nilpotent. The tangent space to the orbit of x is given by

$$T_x(Gx) = x + [\mathbf{g}, x].$$

We want to take a transversal slice S to the G-orbit of x in x. This has to be of the form

$$S := \{s \in \mathbf{g} | s = x + e,\ e \text{ in the linear complement of } [\mathbf{g}, x]\}$$

To determine $[\mathbf{g}, x]$ and a linear complement we now use the Jacobson-Morozov lemma to get $y, h \in \mathbf{g}$, such that $A := < x, h, y >$ is a $\mathbf{sl}_2$-subalgebra of $\mathbf{g}$. $\mathbf{g}$ becomes an A-module by the adjoint representation of $\mathbf{g}$. We have a complete description for such $\mathbf{sl}_2$-modules. They are the direct sum of irreducible $n + 1$-dimensional $\mathbf{sl}_2$-modules V_{n+1}, which with respect to $h \in A$, are the direct sum of one-dimensional subspaces $V_{n+1}(\lambda)$, $\lambda = -n, -n+2, \ldots, n-2, n$. Here λ is the eigenvalue of h acting on V_{n+1}.

So we get a string $\{n_i\} \in \mathbb{Z}^k_{\geqslant 0}$, such that $\mathbf{g}$ splits

$$\mathbf{g} = \bigoplus_{i=1}^{k} V_{n_i+1}$$

into irreducible submodules, as an A-module. And we see that

$$[\mathbf{g}, x] = ad(x)(\mathbf{g}) = \bigoplus_{i=1}^{k} ad(x)(V_{n_i+1}).$$

Since in an A-module V_{n_i+1}, $ad(x)$ maps $V_{n_i+1}(k)$ onto $V_{n_i+1}(k+2)$, for $k \geqslant -n$, we see that in V_{n_i+1} a complement to $ad(x)(V_{n_i+1})$ is spanned by the vector of weight $-n_i$, i.e. the space $S_i := V_{n_i+1}(-n_i)$. We see that a linear complement to $[\mathbf{g}, x]$ is given by $\bigoplus_{i=1}^{k} S_i$. This is exactly the centralizer of the element $y \in \mathbf{g}$. It has dimension equal to the dimension of the centralizer of x; we see that for x regular nilpotent, $k = r$, and for x subregular nilpotent $k = r+2$. We have $S = x + \bigoplus_{i=1}^{k} S_i$ is a transversal slice to the G-orbit of x in the point x.

Now that we have a transversal slice to the orbit of nilpotent $x \in \mathbf{g}$ we introduce an action of $\mathbb{C}^*$ on it. This action will have to stabilize S, so that the restriction of γ to S can be made equivariant with respect to this action.

Firstly we have the action of scalar multiplication on $\mathbf{g}$, which we denote by σ_1:

$$\begin{aligned} \mathbb{C}^* \times \mathbf{g} &\to \mathbf{g} \\ (t, x) &\mapsto \sigma_1(t)x = tx. \end{aligned}$$

We have another action of $\mathbb{C}^*$, with the use of our Jacobson-Morozov homomorphism constructed above. The subalgebra $< h > \subset \mathbf{sl}_2$ corresponds to a multiplicative group $\mathbb{C}^*$ in $T \subset G$, so to action of h we can associate an action of $\mathbb{C}^*$. Let us take vectors $s_i \in S_i$. Now an element of $s \in S$ can be written as $s = x + \sum_{i=1}^{k} c_i s_i$. The action mentioned above on such a vector is as follows:

$$(t, s) \mapsto \sigma_2(t)(x + \sum_{i=1}^{k} c_i s_i) = t^2 x + \sum_{i=1}^{k} t^{-n_i} c_i s_i.$$

Neither of the actions σ_1 and σ_2 stabilizes the subspace S. An easy computation shows that these actions on S commute. This allows us to consider a certain product of these actions, namely $\sigma(t) = \sigma_1(t^2)\sigma_2(t^{-1})$, acting on a vector $s \in S$ by

$$\sigma(t)(x + \sum_{i=1}^{k} c_i s_i) = x + \sum_{i=1}^{k} t^{n_i+2} c_i s_i.$$

The action σ does stabilize the subspace S, and with respect to the coordinate system $\{s_i\}$ it is linear with weights $n_i + 2$.

As we have stated above the adjoint quotient $\mathbf{g} \to \mathbf{h}/\mathcal{W} \cong \mathbb{C}^r$ can be realized by the γ_i, these being G-invariant homogeneous of degree $m_i + 1$. So we have

$$\gamma_i(\sigma_1(t)v) = t^{m_i+1}\gamma_i(v) \quad \forall i \in \{1, \ldots, r\},$$

by homogeneity and

$$\gamma_i(\sigma_2(t)v) = \gamma_i(v) \quad \forall i \in \{1, \ldots, r\},$$

by G-invariance.

We want to specify a $\mathbb{C}^*$-action on $\mathbb{C}^r$ that makes the restriction of $\gamma = (\gamma_1, \ldots, \gamma_r)$ to S $\mathbb{C}^*$-equivariant, having the σ-action on S. On a vector $s = x + \sum_{i=1}^{k} c_i s_i$ we have

$$\gamma_j(\sigma(t)s) = \gamma_j(\sigma_1(t^2)\sigma_2(t^{-1})s) = \gamma_j(\sigma_1(t^2)s) = t^{m_j+2}\gamma_j(x).$$

So we get the desired result if we let $\mathbb{C}^*$ act on $\mathbf{h}/\mathcal{W} \cong \mathbb{C}^r$ by the weights $d_j := 2m_j + 2$. Then γ becomes weighted homogeneous of type $(d_1, \ldots, d_r; n_1 + 2, \ldots, n_k + 2)$.

Table 7 Weights and degrees for the adjoint quotient.

	d_1	d_2	d_3	$\ldots$	$\ldots$	d_{r-2}	d_{r-1}	d_r	w_r	w_{r+1}	w_{r+2}
A_r	4	6	8	$\ldots$	$\ldots$	$2r-2$	$2r$	$2r+2$	2	$r+1$	$r+1$
D_r	4	8	12	$\ldots$	$\ldots$	$4r-8$	$2r$	$4r-4$	4	$2r-4$	$2r-2$
E_6	4	10	12			16	18	24	6	8	12
E_7	4	12	16		20	24	28	36	8	12	18
E_8	4	16	24	28	36	40	48	60	12	20	30

As stated before in the case that x is a regular element, the number of irreducible A-modules must be equal to the rank of $\mathbf{g}$, $k = r$. If we consider the restriction of γ to S it becomes weighted homogeneous of type $(2m_1+2, \ldots, 2m_r+2; n_1+2, \ldots, n_r+2)$. It is a well-known fact that in this case the degrees and the weights are equal, hence $2m_i = n_i$, for all $i = 1, \ldots, r$. In fact this equality of weights and degrees can be used to compute the exponents m_i, from the weights n_i.

Varadarajan [45] showed that this result together with the linear independence of the differentials $d\gamma_i$ in all regular elements of $\mathbf{g}$, follows from a more general formula by Harish-Chandra [11] on differential operators on $\mathbf{g}$.

For subregular nilpotent x we have listed the weights $w_1, \ldots, w_{r+2}$ and the degrees $d_1, \ldots, d_r$ in table 7. The computation of the values d_i and w_i shows that the first $r-1$ values of the weights and the degrees are equal; $d_i = w_i$ for $i = 1, \ldots, r-1$, so we have not listed the first $r-1$ values of w_i. We get the d_i from the exponents of $\mathbf{g}$. They can be found in the tables in Bourbaki, see [3], Planches. The n_i are the highest weights of $\mathbf{g}$ as the particular $\mathbf{sl}_2$-module constructed above. The values $\alpha(h)$, for the simple roots α can be taken from the valuated Dynkin diagram of the subregular nilpotent x. Then all the eigenvalues of $ad(h)$ on $\mathbf{g}$ can be calculated. This gives the values n_i. See [9] for tables.

As an example we consider the subregular class in the simple Lie algebra of type A_3. The valuated Dynkin diagram then has the form where the nodes correspond from left to right to the simple roots α_1, α_2 and α_3. The valuation is realized by the eigenvalue of $ad(h)$ on the root space $\mathbf{g}_{\alpha_i}$. The set of positive roots,

Π, is given by:

$$\Pi = \{\alpha_1, \alpha_2, \alpha_3, \alpha_1 + \alpha_2, \alpha_2 + \alpha_3, \alpha_1 + \alpha_2 + \alpha_3\}.$$

On $\mathbf{n}^+ := \bigoplus_{\alpha \in \Pi} \mathbf{g}_\alpha$ $ad(h)$ acts with eigenvalues 2, 0, 2, 2, 2, 4 respectively. On h it has three times eigenvalue 0, and on $\mathbf{n}^-$ it acts with eigenvalues -2, 0, -2, -2, -2, -4. So $\mathbf{g}$ splits into five irreducible $\mathbf{sl}_2$-modules, one of highest weight 0, three of highest weight 2 and one of highest weight 4. This gives the values 2, 4, 4, 4, 6. for the weights of the restriction $\gamma_{|S}$.

We are almost ready to identify the surface singularity $Nil(\mathbf{g}) \cap S$ as a Kleinian singularity. There is one more lemma we need.

Lemma 8.9. *The restriction of the adjoint quotient to the transversal slice S, $\chi_{|S} \to \mathbf{h}/\mathcal{W}$, has rank $r-1$ in x.*

Proof. This is Lemma 8.3.1 in [32]. □

Using the lemma and the fact that $w_i = d_i$ for $i = 1, \ldots, r-1$, we may suppose that the adjoint quotient $\chi_{|S} : S \to \mathbf{h}/\mathcal{W}$ is of the form

$$S = M \oplus B \to E \oplus B$$
$$(a, b) \mapsto (\delta(a, b), p(b))$$

where $\dim(B) = r - 1$, $\dim(M) = 3$, $\dim(E) = 1$, p is a polynomial isomorphism and δ the unique singular component weighted homogeneous of type $(d_r; w_r, w_{r+1}, w_{r+2})$. Hence the singularity $S \cap Nil(\mathbf{g})$ can be defined as a hypersurface in M by a polynomial f, weighted homogeneous of type $(d_r; w_r, w_{r+1}, w_{r+2})$. Comparing these values to the values in Table 4 and using Prop. 4.2 this f can be taken to have the following normal form:

A_r	$X_{r+1} + YZ$
D_r	$X_{r-1} + XY^2 + Z^2$
E_6	$X^4 + Y^3 + Z^2$
E_7	$X^3Y + Y^3 + Z^2$
E_8	$X^5 + Y^3 + Z^2$

As we have seen these polynomials are defining for the Kleinian singularities. This concludes the proof of the first part of the theorem.

Let us now illustrate Brieskorn's theorem with an example. We take $\mathbf{g}$ of type A_3, that is $\mathbf{g} = \mathbf{sl}_4$. For the subregular nilpotent element x we take

$$x = \begin{pmatrix} 0 & 1 & 0 & 0 \\ 0 & 0 & 1 & 0 \\ 0 & 0 & 0 & 0 \\ 0 & 0 & 0 & 0 \end{pmatrix}$$

This element lies in the $\mathbf{sl}_2$-triple $\{x, h, y\} \subset \mathbf{g}$, with

$$h = \begin{pmatrix} -2 & 0 & 0 & 0 \\ 0 & -4 & 0 & 0 \\ 0 & 0 & -6 & 0 \\ 0 & 0 & 0 & 12 \end{pmatrix} \quad y = \begin{pmatrix} 0 & 0 & 0 & 0 \\ -2 & 0 & 0 & 0 \\ 0 & -6 & 0 & 0 \\ 0 & 0 & 0 & 0 \end{pmatrix}.$$

An element in a transversal slice S to the orbit of x is of the form $s = x + c$, with c in the centralizer of y. This leads to the following:

$$S = \{\begin{pmatrix} x & 1 & 0 & 0 \\ v & x & 1 & 0 \\ w & 3v & x & y \\ z & 0 & 0 & -3x \end{pmatrix} |(v, w, x, y, z) \in \mathbb{C}^5\}.$$

The characteristic polynomial of an element $s \in S$ is

$$\begin{aligned} p(v, w, x, y, z, \lambda) = \lambda^4 + (-6x^2 - 4v)\lambda^2 + (-8vx - 8x^3 - w)\lambda \\ +(-3x^4 - zy + 12vx^2 - 3wx), \end{aligned}$$

so the restriction of the adjoint quotient to S is the map

$$\begin{aligned} \chi : S \to \mathbb{C}^3 \\ (v, w, x, y, z) \mapsto (-6x^2 - 4v, -8vx - 8x^3 - w, -3x^4 - zy + 12vx^2 - 3wx) \end{aligned}$$

after substituting $v = (-u_1 + 6x^2)/4$ and $w = 20x^3 + 2u_1x - u_2$ we get

$$\chi(u_1, u_2, x, y, z) = (u_1, u_2, -81x^4 - yz - 9u_1x^2 + 3u_2x),$$

which is in fact the semiuniversal deformation of a singularity of type A_3. That is the second statement of Brieskorn's theorem. The nilpotent variety intersected with S is equal to $\chi^{-1}(0, 0, 0)$, the surface singularity defined by $X^4 - YZ$.

8.7 A resolution of the adjoint quotient

To conclude the discussion of Brieskorn's theorem let us now quickly look at the other strategy available to identify the singularity. It uses the resolution of the unipotent variety by Springer. This gives a minimal resolution of the singularity and by showing that this resolution has the properties it should have, see Proposition 8.3, the identification follows. For details see [32]. There is a recent review of the theory of conjugacy classes in semisimple algebraic groups by Humphreys [14].

Let G, T, B, $\mathbf{g}$, $\mathbf{h}$ be as before. Analogously to the adjoint quotient for the Lie algebra there is an adjoint quotient for the algebraic group G. There is an isomorphism $\mathbb{C}[G]^G \to \mathbb{C}[T]^W$. The algebras of invariants $\mathbb{C}[G]^G$ and $\mathbb{C}[T]^W$, are freely generated as $\mathbb{C}$-algebras by the fundamental characters $\chi_i := \chi\lambda_i$, and their restrictions $\chi_{i|T}$ to T respectively.

The adjoint quotient for G is then the mapping

$$\begin{aligned} \chi : G \to T/W \\ x \mapsto (\chi_1(x), \ldots, \chi_r(x)), \end{aligned}$$

realized by the fundamental characters of G.

For all $f \in \mathbb{C}[G]^G$ we have $f(x) = f(x_s)$, so the unipotent variety U, is the fiber above e, the semisimple part of a unipotent element.

Now let $\mathcal{B}$ be the projective variety of Borel groups, which we can identify with the quotient G/B by $G/B \to \mathcal{B}$, $gB \mapsto gBg^{-1}$. Take the subvariety of $\mathcal{B} \times G$,

$$\mathcal{C} := \{(B, x) \in \mathcal{B} \times G | x \in B\}.$$

We have natural morphisms $\tau : \mathcal{C} \to T$, $\pi : \mathcal{C} \to G$, such that we get the following commutative diagram:

$$\begin{array}{ccc} \mathcal{C} & \to & G \\ \downarrow & & \downarrow \\ T & \to & T/W \end{array}$$

Theorem 8.10. *The diagram above is a resolution of the adjoint quotient. In particular for each $t \in T$ the induced map*

$$\pi : \tau^{-1}(t) \to \chi^{-1}(\bar{t})$$

is a resolution of singularities of the fiber of t.

A special case of this theorem, conjectured by Grothendieck, is Springer's resolution of the unipotent variety U:

$$\pi : \mathcal{C}_U \to U.$$

Here we take $\mathcal{C}_U$ the subvariety of $\mathcal{B} \times G$,

$$\mathcal{C}_U := \{(B, x) \in \mathcal{B} \times G | x \in B, x \in U\}.$$

We can identify $\mathcal{C}_U$ with the associated bundle $G \times^B U$, a smooth irreducible variety of dimension $\dim(G) - \text{rank}(G)$, by the morphism $\tau : G \times^B U \to \mathcal{C}_U$, $g * u \mapsto (gB, gug^{-1})$. Now the morphism π factors through the second projection $p : \mathcal{B} \times U \to U$, so $\pi = p \circ \tau$ is proper.

We have the following result on regular unipotent elements in G.

Proposition 8.11. *Let $x \in G$ be unipotent. Then the following are equivalent:*

- *x is regular,* $\dim(C_G(x)) = r$.
- *x is contained in a unique Borel subgroups of G.*
- *The adjoint quotient $\chi : G \to T/W$ is regular, i.e. smooth at x.*

Proof. For a proof see [14]. □

So onto the regular points of U, π is one to one. The variety U is the closure of the regular orbit, the smooth points of U. This concludes the proof of the claim that π is a resolution of the singularities of U.

The way to proceed now is to proof that this resolution is of the desired form. It can be shown that this is in fact the case, namely the exceptional set of a subregular nilpotent element is a union of projective lines which intersect transversally and have self-intersection number -2. They have the Dynkin diagram corresponding to G as a resolution graph.

9 Conclusion

Let us try to summarize the connections we have discussed. We have seen three theories that fall into the ADE classification pattern. We can make the following table:

Table 8 The ADE correspondence

	Theory	Nodes	Matrices
I	Finite subgroup $SU_2(\mathbb{C})$	Irreducible representations	decomposition coefficients a_{ij}
II	Lie algebra of type ADE	Simple roots	Cartan matrix
III	Minimal resolution of $\mathbb{C}^2/\Gamma$	Components E_i of the exceptional set	Intersection coefficients

First of all there is a structural correspondence between these theories. They have a classification of the same type. We have shown that by computations within these theories, it is possible to relate the objects and matrices listed above. See the characterization of Kleinian singularities in Prop. 5.1, the McKay correspondence in Theorem 6.1, and the Geometric McKay correspondence in Theorem 7.1.

Secondly there are direct links between the theories, one of the most remarkable the fact that the subregular singularity in the Lie algebra is exactly the quotient singularity corresponding to this Lie algebra.

There is a generalization of the McKay correspondence to higher dimensions. For a recent review see [29]. The picture we sketched in the introduction is then generalized to arbitrary dimension n:

$$\begin{array}{c} \mathbb{C}^n \\ \downarrow \pi \\ \tilde{S} \rightarrow \mathbb{C}^n/\Gamma. \end{array}$$

Here Γ is a finite subgroup of $SL_n(\mathbb{C})$, such that the Gorenstein singularity $\mathbb{C}^n/\Gamma$ allows a crepant resolution. For $n = 3$ it is proven that there is a McKay correspondence. The first indication for this came in fact from the work of string theorists, who proved that the Euler number of $\tilde{S}$ equals the number of conjugacy classes in Γ.

For recent articles on generalizations of the McKay correspondence and related topics see for instance: http://homepages.warwick.ac.uk/~masda/McKay.

I want to finish here with a 'second-hand' citation of Proust, that I took from [13], and which presents a very nice way to look at the theory discussed in my thesis:

"...si les choses se répètent, c'est avec de grandes variations."
Proust, La Prisonnièrre

References

[1] V. I. Arnold, *On matrices depending on parameters*, Russian Math. Surveys 26, no.2, 29–43, 1971.

[2] D. Bättig, H. Knörrer, *Singularitäten*, Birkhäuser Verlag, Basel, 1991.

[3] N. Bourbaki, *Groupes et algèbres de Lie*, Chapitre 4, 5 et 6, Éléments de mathématique, Hermann, Paris, 1968.

[4] N. Bourbaki, *Groupes et algèbres de Lie*, Chapitre 7 et 8, Éléments de mathématique, Hermann, Paris, 1975.

[5] E. Brieskorn, *Singularitäten*, Jber. Deutsch. Math.-Verein 78, 93–112, 1976.

[6] E. Brieskorn, *Singular elements of semisimple algebraic groups*, Actes, Congrès internat. Math. Nice , Tome 2, 279–284, 1970.

[7] H. S. M. Coxeter, *Regular complex polytopes*, Cambridge University Press, Cambridge, 1974.

[8] A. H. Durfee, *Fifteen characterizations of rational double points and simple critical points*, Ens. Math. 25, 131–163, 1979.

[9] E. B. Dynkin, *Semisimple subalgebras of semisimple Lie algebras*, A. M. S. translations, Ser. 2, 6, 111–245, 1957.

[10] G. Gonzales-Springberg and J.-L. Verdier, *Construction géométrique de la correspondence de McKay*, Ann. Sci. Ecole Norm. Sup., t. 16, no. 3, 410–449, 1983.

[11] Harish-Chandra, *Invariant distributions on Lie algebras*, Amer. J. Math. 86, 271–309, 1964.

[12] T. L. Health, *The Thirteen Books of Euclid's Elements*, Dover, New York, 1956.

[13] J. E. Humphreys, *Reflection Groups and Coxeter Groups*, Cambridge University Press, Cambridge, 1990.

[14] J. E. Humphreys, *Conjugation Classes in Semisimple Algebraic Groups*, Mathematical Surveys and Monographs 43, AMS, 1995.

[15] J. E. Humphreys, *Linear Algebraic Groups*, Graduate texts in Mathematics, Springer-Verlag, New York, 1975.

[16] V. Kac, *Infinite Dimensional Lie algebras*, Birkhauser, Boston, 1983.

[17] A. Kas, M. Schlessinger, *On the versal deformation of a complex space with an isolated singularity*, Math. Ann. 196, 23–29, 1972,

[18] F. Klein, *Vorlesungen über das Ikosaeder und die Auflösung der Gleichungen vom fünften Grade*, Teubner, Leibniz, 1884.

[19] H. Knörrer, *Group Representations and the Resolution of Rational Double points*, Contemporary Mathematics 45, 175–222, AMS, 1985.

[20] B. Kostant, *The principal three-dimensional subgroup and the Betty numbers of a complex simple Lie group*, Amer. J. Math. 81, 973–1032, 1959.

[21] B. Kostant, *Lie group representations on polynomial rings*, Amer. J. Math. 85, 327–404, 1963.

[22] B. Kostant, *The McKay Correspondence, the Coxeter Element and Representation Theory*, Société de Mathématique de France, Astérisque, hors série, 209–255, 1985.

[23] K. Lamotke, *Regular Solids and Isolated Singularities*, Vieweg, Braun-

schweig, 1986.

[24] H. B. Laufer, *Normal Two-dimensional Singularities*, Ann. of Math. Studies, Princeton Univ. Press, Princeton, 1971.

[25] J. McKay, *Graphs, singularities and finite groups*, Proc. Symp. Pure math., t. 37, 183–186, 1980.

[26] J. McKay, *Cartan matrices, finite groups of quaternions, and Kleinian singularities*, Proc. A.M.S., 1981, 153–154.

[27] V. P. Palamodov, *Deformations of complex spaces*, Russian Math. Surveys 31, no. 3, 129–197, 1976,

[28] H. Pinkham, *Singularites de Klein I, II*, Sém. sing. des surfaces, Lecture notes in Mathematics 777, Springer, 1980.

[29] M. Reid, *La correspondence de McKay*, Séminaire Bourbaki, 52ème année, novembre 1999, no. 867, to appear in Astérisque 2000, preprint math.AG/9911165.

[30] I. Reiten, *Dynkin diagrams and the representation theory of algebras*, Notices of the AMS 44, no. 5, 546–556, 1997.

[31] I. R. Shafarevich, *Basic Algebraic Geometry*, Springer-Verlag, Berlin-Heidelberg-New York, 1974.

[32] P. Slodowy, *Simple singularities and simple algebraic groups*, Lecture Notes in mathematics 815, Berlin and New York, Springer-Verlag, 1981.

[33] P. Slodowy, *Platonic solids, Kleinian singularities and Lie groups*, in "Algebraic Geometry", Proc., Ann Arbor 1981, Ed. I. Dolgachev, Springer Lecture Notes in Mathematics 1008, 1983.

[34] P. Slodowy, *Groups and special singularities*, in "Singularity Theory", Ed. D. T. Lê, K. Saito, B. Teissier, World Scientific, Singapore, 731–799, 1995.

[35] P. Slodowy, *Algebraic Groups and Resolutions of Kleinian Singularities*, Report Kyoto University, 1996.

[36] P. Slodowy, *Four lectures on simple groups and singularities*, Communications of the Math. Inst. Rijksun. Utrecht, 11, 1980.

[37] T. A. Springer, *The unipotent variety of a semisimple group*, Proc. of the Bombay Colloquium in Algebraic Geometry, 373–391, ed. S. Abhyankar, Oxford Univ. Press, London, 1969.

[38] T. A. Springer, *Invariant Theory*, Lecture Notes in Mathematics 585, Springer-Verlag, Berlin-Heidelberg-New York, 1977.

[39] T. A. Springer, *Poincaré series of binary polyhedral groups and Mckay's correspondence*, Math. Ann. 278, 99–116, Springer-Verlag, 1987.

[40] T. A. Springer and R. Steinberg, *Conjugacy classes*, Seminar on algebraic groups and related finite groups, Lecture Notes in Mathematics 131, Springer, Berlin-Heidelberg-New York, 1970.

[41] R. Steinberg, *Regular elements of semisimple algebraic groups*, Publ. Math. I.H.E.S. 25, 49–80, 1965.

[42] R. Steinberg, *Conjugacy Classes in Algebraic Groups*, Lecture notes in mathematics 366, Springer, Berlin-Heidelberg-New York, 1974.

[43] R. Steinberg, *Finite subgroups of SU_2, Dynkin diagrams and affine Coxeter elements*, Pacific Journal of Mathematics, 118, no.2, 587–598, 1985.

[44] P. du Val, *Homographies, Quaternions and Rotations*, Clarendon Press, Ox-

ford, 1964.

[45] V. S. Varadarajan, *On the ring of invariant polynomials of a semisimple Lie algebra*, Amer. J. Math. 90, 1968, 308–317.

Open Problems and Surveys of Contemporary Mathematics
SMM 6, pp. 237–324

Open Problems on Discrete Subgroups of Lie Groups

Lizhen Ji*

Abstract

In this expository paper, we discuss some open problems on discrete subgroups, mainly lattices and arithmetic subgroups, of Lie groups from the perspectives of geometry, topology, analysis, and combinatorial group theory. They are closely related to questions on locally homogeneous spaces, in particular, locally symmetric spaces. There are several types of open problems, and together they might give a reasonable glimpse of a rich landscape of discrete subgroups. To make it easier for the reader, we also include some definitions and preliminary results on arithmetic groups, and discussion of some known results to motivate the open problems later. In some sense, this paper can also be considered as a problem-oriented introduction to some aspects of discrete subgroups of Lie groups and locally symmetric spaces.

Contents

*Partially supported by an NSF grant DMS 1104696, Department of Mathematics, University of Michigan, Ann Arbor, MI 48109, USA. Email: `lji@umich.edu`

1 Introduction and general motivations

A subgroup Γ of a Lie group G is called *a discrete subgroup* if the induced subspace topology on Γ is discrete. Note that every group can be given the discrete topology and hence can be viewed as a discrete group. But not every discrete group can be realized as a discrete subgroup of a Lie group (recall that a Lie group is usually assumed to have only finitely many connected components) or more generally as a discrete subgroup of a non-discrete locally compact topological group. For example, a realization of a group Γ as a linear subgroup over $\mathbb{C}$, i.e., a subgroup of $GL(n, \mathbb{C})$, amounts to a faithful finite dimensional representation over $\mathbb{C}$. Realizing Γ as a *discrete* subgroup of $GL(n, \mathbb{C})$ imposes a *stronger* condition. There are several long outstanding problems about linearity of some naturally occurring groups such as the mapping class groups, i.e., embedding them into $GL(n, F)$ for some field F.

Before we start to describe more specific open problems, we discuss several general questions on open problems.[1]

Question 1.1. *Why do we discuss open problems?*

Abundance of problems in a subject shows that the subject is full of life. Formulating and solving problems is certainly an important part of mathematical life. It forces us to think and helps us to understand known theories and results at a deeper level. As many people might agree, finding a good problem is a challenging process, and viewing a problem from the right perspective is often helpful, or even crucial, to a solution of the problem. Similarly, finding a research direction is also important and difficult.[2] In this sense, putting problems together might provide a helpful global point of view. With a steady supply of problems, there are new things to be learnt and understood, and the subject is still growing. Probably one cannot do better than quote from the great Hilbert:

"Who of us would not be glad to lift the veil behind which the future lies hidden; to cast a glance at the next advances of our science and at the secrets of its development during future centuries? What particular goals will there be toward which the leading mathematical spirits of coming generations will strive? ... The deep significance of certain problems for the advance of mathematical science in general and the important role which they play in the work of the individual investigator are not to be denied. As long as a branch of science offers an abundance of problems, so long is it alive: a lack of problems foreshadow extinction or the cessation of independent development. Just as every human

[1] These questions are intriguing to the author. Hope that the random thoughts below might be helpful, or at least entertaining to the reader.

[2] Many people might still remember the difficult time and struggle in deciding a research direction and finding workable problems right after getting Ph.D. degree or at other stages of one's career. In some sense, picking a research direction might even be more important than picking a concrete mathematical problem.

undertaking pursues certain objectives, so also mathematical research requires its problems. It is by the solution of problems that the strength of the investigator is hardened; he finds new methods and new outlooks, and gains a wider and freer horizon."

Question 1.2. *What problems to pick?*

This is a difficult question to answer. Different people will have different tastes and answers. Of course, there are always some famous and difficult problems in each subject, which have been attacked by experts. Chances are probably small to solve them, but one never knows (think of a long list of open problems solved by naive people outside of the subject or by brave students who do not known many failed attempts.) A more realistic choice would be a problem that seems *reasonable, feasible and interesting.* Of course, all these quantifiers are subjective. According to my colleague Ralf Spatzier, one criterion for selecting good problems is as follows: *"If you had a last wish to solve one more problem before you die, which one would you pick?"* This might be one concrete manifestation of the golden rule. If one is ambitious, one should try to solve or formulate such a problem. Probably most of the open problems listed in this paper are not up to this standard, and one can only do what one can. Since one of the purposes in writing this paper is to help the author to gain a perspective on some important aspects of discrete subgroups of Lie groups, some problems are known ones, and no much originality is claimed. Our idea is that putting them together might shed a different light on them. We also include some vaguely stated problems which are not the usual open problems but rather can be interpreted as possible research directions. This is often an important initial step towards formulating and finding good concrete problems.

Question 1.3. *Why do we discuss open problems on discrete subgroups of Lie groups?*

Briefly, they are *beautiful, useful* and *important.* More specifically, they have rich structures: both internal algebraic properties and external properties such as actions on spaces which arise from their embedding into the ambient Lie groups. These structures and properties can often distinguish them from the general discrete groups (note that the existence of a suitable action of a space with some desirable properties often imposes strong restrictions on the group). They often have unexpected connections with many topics, basically with all the major areas of mathematics. For example, finite subgroups of Lie groups are obvious discrete subgroups. Finite subgroups of the orthogonal group $SO(3)$ of the Euclidean space $\mathbb{R}^3$ are related to regular polytopes and also to algebraic geometry, singularity theory and string theory. These properties and connections make them beautiful.

Discrete subgroups of Lie groups have many different applications and are effective tools to attack various problems, and hence are useful. The interaction between discrete subgroups and Lie groups plays a crucial role in such applications. For example, once many counting problems (or functions) are expressed in terms of coefficients of automorphic functions with respect to special discrete subgroups acting on symmetric spaces, they can be solved relatively easily and effectively. Discrete subgroups and their associated locally symmetric spaces also motivated important theories and have been actively pursued by some of the most important mathematicians. For example, the celebrated Langlands program resulted from the spectral theory of automorphic forms, i.e., spectral theory of the Laplace operator of locally symmetric spaces. Another example consists of Shimura varieties, which are special Hermitian locally symmetric spaces.

Discrete subgroups of Lie groups arise naturally from many different contexts of the modern mathematics. For example, they arise from arithmetic subgroups of linear algebraic groups in number theory and automorphic representations. The simplest example is $\mathbb{Z} \subset \mathbb{R}$, and the most basic example of non-abelian discrete groups is $SL(2, \mathbb{Z}) \subset SL(2, \mathbb{R})$. In fact, as we will show below, arithmetic subgroups are the special and most important discrete subgroups of Lie groups.

Arithmetic subgroups also occur as symmetries of various structures. For example, symmetry groups of crystals (or lattices) in $\mathbb{R}^n$ give rise to crystallographic groups, and many tilings of the Poincaré upper half-plane are symmetric (or invariant) with respect to $SL(2, \mathbb{Z})$ or its subgroups finite index. Many modularity properties such as the Monstrous moonshine of the largest sporadic finite simple group are also controlled by the modular group $SL(2, \mathbb{Z})$.

Actions of discrete subgroups of Lie groups on the ambient Lie groups and associated homogeneous spaces provide some of the most interesting actions (or transformation groups) and spaces. For example, locally symmetric spaces associated with discrete subgroups of semisimple Lie groups provide special manifolds in differential geometry with remarkable rigidity properties, moduli spaces of special varieties such as abelian varieties and $K3$-surfaces in algebraic geometry, and classifying spaces in geometric topology and algebraic topology.

Discrete subgroups of Lie groups have been extensively studied since the pioneering work of H. Poincaré and F. Klein on discrete subgroups of $SL(2, \mathbb{R})$ and $PSL(2, \mathbb{C})$ (called Fuchsian and Kleinian groups respectively). A systematic, well-known but little read classic is the two huge volumes by Fricke and Klein on these groups [63]. There are still many fascinating problems waiting to be studied and understood for Fuchsian groups and especially for Kleinian groups. Of course, much less is understood for more general discrete subgroups of general Lie groups, or even for arithmetic subgroups of semisimple Lie groups. This is basically the point of this paper, which will be explained in more detail below.

Besides their own applications, discrete subgroups of Lie groups, in particular arithmetic subgroups are important examples of infinite groups in geometric group theory, combinatorial group theory and homological algebra, and they often motivate problems and results for general groups in these theories.

Results and problems on actions of discrete subgroups on natural homogeneous spaces of Lie groups provide guiding examples for various discrete transformation groups arising from topology and geometry. For example, a lot of recent work on mapping class groups of surfaces and Teichmüller spaces, and the outer automorphism groups $Out(F_n)$ of the free group F_n and the outer spaces are motivated by arithmetic subgroups of semisimple Lie groups. In fact, there is a long dictionary between similar properties of these three important classes of discrete groups which is keeping on growing. In many instances, it is more than an analogy. There are natural homomorphisms from mapping class groups of surfaces and $Out(F_n)$ to arithmetic subgroups which can be used for various purposes.

Since problems of a subject give snapshots of active parts of a subject, a systematic discussion of open problems might give a helpful introduction to students and others who would like to get a glimpse of the subject. Since there are so many topics related to discrete subgroups of Lie groups,[3] no single paper can do justice to them and give a reasonable comprehensive coverage, and this task on open problems on discrete subgroups is clearly beyond the capability of the author. On the other hand, I would like to learn what are some interesting problems on discrete subgroups of Lie groups, what are some important directions of research, and also to learn relevant theories, and I have tried to read some papers and asked some experts for problems which are interesting to them. In spite of all these efforts, the scope of this paper is still very limited. Many problems in this paper are well-known to experts on the topics or some natural extensions. We also put in some problems which are vaguely stated on purpose, since it is not clear what kind of results are possible, but we feel that it is fruitful to work in the general directions indicated. In any case, finding and formulating problems precisely are part of solution of the general problems.

In many cases, research is carried out by starting with a general direction and some vaguely stated programs, and *one often does not solve the exact problem which one sets out to solve.*

To help organize the materials and also force myself to learn some relevant materials, I decided to give several lectures on some open problems on discrete subgroups and their actions on homogeneous spaces associated with their ambient Lie groups, and then to expand the lecture notes into the present expository paper. We have tried to keep the informal style of the lectures. Writing this paper is a struggling but fruitful process and we hope that you might enjoy this paper too and find it helpful in some ways.

Acknowledgments. I would like to thank Martin Bridson, Marc Burger, Tom Far-

[3]For example, it includes the vast and very active subject of hyperbolic manifolds. See the survey [93] and the book [94] for some results and references. The current paper concentrates on problems related to discrete subgroups of general Lie groups and semisimple Lie groups of higher rank, though discrete subgroups of semisimple Lie groups of rank 1 and Kleinian groups naturally provide important motivations and examples.

rell, Bill Harvey, Xiaojun Huang, Eduard Looijenga, Andrei Rapinchuk, Mark Sapir, and Scott Wolpert for helpful conversations, suggestions and references, and especially Ralf Spatzier for many concrete suggestions and his enthusiasm for this project. I would also like thank the Mathematical Sciences Center at Tsinghua University for an opportunity to lecture on some of the open problems mentioned in this paper.

2 A bit of history on discrete subgroups of Lie groups

Discrete groups of the motion group (i.e., the isometry group) of the Euclidean spaces $\mathbb{R}^2$ and $\mathbb{R}^3$ have been used and studied since the ancient time, probably not systematically, before the concept of group was introduced. For example, regular solids (Platonic solids) were studied and classified in Chapter 13 of Euclid's classical book *Elements*. They have most symmetries, i.e., the largest isometry groups among polytopes.

After the concept of groups was introduced by Galois, Camille Jordan wrote a book *Traité des substitutions et des équations algébriques* in 1870 to discuss the concept of groups and applications to algebraic equations systematically. In a famous paper in 1868, *Mémoire sur les groupes de mouvements*, Jordan also discussed the symmetry group of crystals, i.e., discrete subgroups of the isometry group of $\mathbb{R}^3$ that preserve the crystals. These are infinite discrete subgroups, and they are called crystallographic groups for the obvious reason.

This paper of Jordan is one of the founding papers on geometric transformation groups. It might be helpful to point out that both Klein and Lie studied with Jordan in Paris when Jordan was mainly concerned with group theories and emphasized the importance of group theories in geometry. Jordan's points of view had great influence on Klein's and Lie's works on Lie groups and discrete subgroups.

Though Klein is famous for the Erlangen program and putting Lie groups into the heart of geometry, and also into a large part of modern mathematics (or rather a large part of modern sciences), it was said that Klein and Lie split discrete groups and continuous groups (i.e., Lie groups) between them: discrete groups for Klein, and continuous groups for Lie. There is some truth in this statement. Probably one of the most important contributions of Klein is concerned with Kleinian groups, i.e., discrete subgroups of $SL(2,\mathbb{C})$ or the discrete isometric groups of the real hyperbolic space of dimension 3. (One point of this paper is to show that the embedding of discrete groups into Lie groups really makes both of them interesting. So Klein and Lie's contributions should be viewed together).

It is true that the foundation of the theory of discrete subgroups of Lie groups was partially built on the early work of Klein on Fuchsian groups, subgroups of $SL(2,\mathbb{Z})$ and in particular congruence subgroups of $SL(2,\mathbb{Z})$. The famous two volumes [63] by Fricke and Klein on automorphic functions set directions of research on discrete subgroups of Lie groups for many years, and these works had also motivated a lot of later works. Of course, Poincaré's works on more geometrical description of Fuchsian groups and automorphic forms (see [141] for a selection) and his competition with Klein on automorphic forms and Fuchsian groups were also crucial and equally important to the foundation of discrete subgroups of Lie groups. These works of Klein and Poincaré were carried out around 1890s—1900s.

Before Klein and Poincaré, there was earlier works by Lagrange and Gauss on the reduction theory of binary quadratic forms, i.e., fundamental domains for $SL(2,\mathbb{Z})$ acting on the upper half plane, and later generalizations by Hermite and Minkowski which were developed into the important and beautiful subject of geometry of numbers. They were not interested in general discrete subgroups. Probably these works of Lagrange, Gauss and Hermite might not have too much impact on Klein and Poincaré. On the other hand, the works of Riemann and Schwartz on tilings of the upper half plane by triangles and groups generated by reflections with respect to the sides of the triangles might have a more direct influence on Klein and Poincaré, and especially the work of Fuchs on the monodromy of ordinary differential equations with regular singularities had a crucial impact on Poincaré.

Indeed, in the theory of functions of one complex variable, integration of algebraic functions and solution of differential equations of second order with algebraic coefficients led to monodromy groups which can be realized as discrete subgroups of $SL(2,\mathbb{R})$. Some particular examples of such groups were studied by Hermite, Dedekind and Fuchs around 1850 and later by Picard.

Poincaré named discrete subgroups of $SL(2,\mathbb{R})$ Fuchsian groups because of his inspiration from Fuchs' work. There is also rumor saying that Poincaré thought that these discrete subgroups of $SL(2,\mathbb{C})$ are not so important and interesting, and therefore named them after Klein. One possible explanation for this is that one of the final goals of the competition between Klein and Poincaré is the uniformization of Riemann surfaces of finite type. They correspond to Kleinian groups whose limit sets are circles, and hence are conjugated to Fuchsian groups. Later it turned out that Kleinian groups are important, especially in 3-dimensional geometry and topology for uniformization and description of hyperbolic 3-manifolds. It is also fair to point out that Fuchsian groups are important too, since they describe hyperbolic surfaces and the important Shimura curves are obtained from special Fuchsian groups.

At the beginning of the 20th century, automorphic functions in several variables and their associated arithmetic groups were introduced and studied. Contributors include Hilbert (The Hilbert modular groups were studied by him in work from 1893-4 and continued by his student Blumenthal in 1901).

For a history of the early work on discrete subgroups of $SL(2,\mathbb{R})$ and other related Lie groups from the point of view of combinatorial group theory, see [38,

I.6].

Around the same time, Bieberbach made fundamental contributions to crystallographic groups, i.e., three basic theorems on the structure and classification of crystallographic groups in the general dimension, after classification of the three dimensional crystallographic groups by E. S. Fedorov and A. Schoenflies in 1890s.

Then Siegel studied more examples of arithmetic subgroups of classical Lie groups, their fundamental domains in connection with quadratic forms, and also started a general theory of discrete subgroups of Lie groups. His work motivated a lot of work on lattices of semisimple Lie groups. There were also some works by H. Weyl and B. A. Venkov on some other examples of arithmetic subgroups of classical Lie groups and fundamental domains for them.

There were also a lot of work on lattices of non-semisimple Lie groups, for example, the beautiful work of Malcev on lattices of nilpotent Lie groups and later work of Mostow, L. Auslander, H. C. Wang on lattices of solvable Lie groups. We will discuss these results in more detail later.

In some sense, Borel, and Harish-Chandra continued the work of Siegel on the reduction theory of arithmetic subgroups of semisimple Lie groups. The newly developed theory of linear algebraic groups by Chevalley, Borel and others played a foundational role by providing the general framework of constructing and understanding arithmetic groups.

Siegel also started compactifications of arithmetic locally symmetric spaces. This theory was continued by Satake, Baily-Borel, Borel-Serre, and also Mumford and his collaborators (see [27] for references).

Selberg and later Piateskii-Shapiro studied lattices and arithmetic subgroups from a different perspective, in particular, arithmeticity of lattices and rigidity of lattices (see [85] for a summary and related references).

The Selberg trace formula and the theory of Eisenstein series studied lattices and arithmetic subgroups from yet another totally different point of view, i.e., the spectral theory of locally symmetric spaces and automorphic forms. (This is a vast subject. See the reference book [81] for a list of selected topics and references.)

Weil also contributed to foundational works on both infinitesimal rigidity of lattices of semisimple Lie groups and classification of arithmetic subgroups. For example, the classification of all arithmetic subgroups of $SL(2, \mathbb{R})$ was due to him.

Then we have the important work of Mostow on rigidity of lattices of semisimple Lie groups [132] (following his earlier work on rigidity of solvmanifolds), and the work of Margulis on arithmeticity and super-rigidity of irreducible lattices of semisimple Lie groups of higher rank (see [119] for an authoritative account). These results inspired a lot of work on rigidity and arithmeticity of more general classes of groups.

There are also works of many other people, for example, Kazhdan, Langlands, Gelfand, Tits, and Serre on different aspects of lattices and arithmetic subgroups of semisimple Lie groups.

There is another direction on discrete subgroups initiated by Poincaré, Dehn and Nielsen on Fuchsian groups and geometrical and topological approaches, which have had a huge impact on geometric group theory.

As mentioned before, arithmetic subgroups and lattices of Lie groups are useful and important due to their many connections to and applications in subjects such as automorphic forms and representations and arithmetic of algebraic groups. They also serve as special examples in many theories, for example, in geometric group theory and topology.

Mentioning all leading figures on many more recent topics related to arithmetic subgroups is difficult and will make a long list. It is also difficult to identify boundaries between topics on arithmetic groups proper and other neighboring subjects. In some sense, these difficulties and connections make arithmetic subgroups and lattices of Lie groups interesting: they are natural infinite groups which unify many things.

Perhaps we should conclude this section by emphasizing that the above brief discussion shows that discrete groups manifest their power through actions, and their embedding in the ambient Lie groups and resulting actions on homogeneous spaces are crucial for many applications.

3 Definition, examples and basic results

Before discussing open problems, we will need to define more carefully the objects we are discussing.

As mentioned before, the most basic example of arithmetic subgroups and also of discrete subgroups of a Lie group is $\mathbb{Z} \subset \mathbb{R}$. Though the structure of $\mathbb{Z}$ is simple, the embedding $\mathbb{Z} \subset \mathbb{R}$ and the action of $\mathbb{Z}$ on $\mathbb{R}$ have motivated and introduced many general results. It also includes many deep results in mathematics. For example, the Poisson summation formula, and its application to meromorphic continuation to the whole complex plane of the Riemann zeta function, can be understood from studying relations between the geometry and analysis of the compact Riemannian manifold (or rather homogeneous space) $\mathbb{Z}\backslash\mathbb{R}$.

When we talk about a discrete subgroup Γ, it comes with an embedding into an ambient locally compact topological group G. For simplicity, we will concentrate on pairs of Lie groups G and their discrete subgroups Γ. We often call any abstract group *an arithmetic group* if it is isomorphic to an arithmetic subgroup of a Lie group, which will be defined later. In some cases, we concentrate on internal properties of the discrete groups without emphasizing embedding into Lie groups and related locally homogeneous spaces coming from the embedding.

An immediate generalization of the infinite cyclic group $\mathbb{Z}$ is the discrete group $\mathbb{Z}^n$ for $n \geq 1$. It can be realized as the subgroup $\mathbb{Z}^n \subset \mathbb{R}^n$ and also as any lattice $\Lambda = \mathbb{Z}v_1 + \cdots + \mathbb{Z}v_n \subset \mathbb{R}^n$, where $v_1, \ldots, v_n$ are linearly independent vectors. If $\mathbb{R}^n$ is given the Euclidean metric, then the geometry of the flat manifold $\Lambda\backslash\mathbb{R}^n$ depends on the lattice Λ, or the embedding of $\mathbb{Z}^n$ into $\mathbb{R}^n$. For example, each embedding of $i : \mathbb{Z} \to \mathbb{R}$ determines and is determined up to orientation by the length of the shortest closed geodesic of the compact manifold $i(\mathbb{Z})\backslash\mathbb{R}$.

The space of isometric equivalence classes of embedding of $\mathbb{Z}^n$ into $\mathbb{R}^n$ is given by the locally symmetric space $GL(n,\mathbb{Z})\backslash GL(n,\mathbb{R})/O(n)$. An important point here is that *understanding all these different realizations of $\mathbb{Z}^n$ as discrete subgroups of $\mathbb{R}^n$, or equivalently the study of this locally symmetric space $GL(n,\mathbb{Z})\backslash GL(n,\mathbb{R})/O(n)$, leads to the subject of geometry of numbers, a subject which was initiated by Minkowski and has many unexpected applications.* For example, the problem of finding the densest lattice sphere packings is related to finding maximum points of a packing density function on the locally symmetric space $GL(n,\mathbb{Z})\backslash GL(n,\mathbb{R})/O(n)$.

Therefore, it is important to consider a discrete group together with an embedding into an ambient Lie group G. It is perhaps helpful to point out that the naive question how many realizations of a given group as arithmetic subgroups is closely related to the important rigidity theory of Margulis to be discussed below.

The groups $\mathbb{Z}$ and $\mathbb{Z}^n$ are abelian. Probably the best known non-abelian discrete group is the modular group $SL(2,\mathbb{Z}) \subset SL(2,\mathbb{R})$. (The reason why it is called the modular group is probably related to the fact that the quotient of the upper half plane by $SL(2,\mathbb{Z})$ is the moduli space of elliptic curves, i.e., Riemann surfaces of genus 1.) Since $\mathbb{Z}$ is discrete in $\mathbb{R}$, $SL(2,\mathbb{Z})$ is a discrete subgroup of the Lie group $SL(2,\mathbb{R})$. Since it consists of integral matrices, it is *an arithmetic subgroup.* In some sense, this is the most basic or important example of arithmetic subgroups of semisimple Lie groups. (According to MathWorld, *Arithmetic is the branch of mathematics dealing with integers or, more generally, numerical computation.*)

Definition of linear algebraic groups

The basic objects of this paper are discrete subgroups Γ of Lie groups G. A natural problem is how to produce them, in particular, with desirable properties. For example, pick a nontrivial element $\gamma \in G$. It is not true in general that the cyclic subgroup generated by γ is a discrete subgroup of G. Counterexamples can be constructed by using compact Lie groups, for example, products of compact and noncompact Lie groups. The situation is more complicated for subgroups generated by finitely many elements. One effective and realistic source of discrete subgroups comes from arithmetic subgroups. For this purpose, we need to introduce linear algebraic groups and relate them to Lie groups.

Recall that a *linear algebraic group* (or more concretely *a matrix group*) $\mathbf{G}$ is a subgroup of $GL(n,\mathbb{C})$ such that (1) $\mathbf{G}$ is a subvariety, i.e., defined by a collection of polynomial equations in entries of the matrix, (2) and the group operations are morphisms between varieties.

If the defining polynomial equations have coefficients in $\mathbb{Q}$, then the algebraic group $\mathbf{G}$ is said to be defined over $\mathbb{Q}$. Similarly for any field F containing $\mathbb{Q}$, if all coefficients of the defining polynomials are contained in F, $\mathbf{G}$ is said to be defined over F.

Example 3.1. (1) One standard example is the special linear group

$$SL(n,\mathbb{C}) = \{g \in GL(n,\mathbb{C}) \mid \det g = 1\}.$$

It should be pointed out that the embedding $SL(n,\mathbb{C}) \to GL(n,\mathbb{C})$ is important for identifying $SL(n,\mathbb{C})$ as a linear algebraic group.

(2) Another example is the orthogonal group

$$O(n,\mathbb{C}) = \{g \in GL(n,\mathbb{C}) \mid {}^t gg = I_n\},$$

and the special orthogonal group

$$SO(n,\mathbb{C}) = \{g \in SL(n,\mathbb{C}) \mid {}^t gg = I_n\}.$$

(3) More generally, for any quadratic form of n variables, $Q(x,y)$, $x,y \in \mathbb{C}^n$, define its orthogonal group

$$O(Q,\mathbb{C}) = \{g \in GL(n,\mathbb{C}) \mid Q(gx,gy) = Q(x,y), \text{ for all } x,y \in \mathbb{C}^n\}.$$

It is a linear algebraic group. If the coefficients of Q are contained in $\mathbb{Q}$, then $O(Q,\mathbb{C})$ is defined over $\mathbb{Q}$. To see this, write $Q(x,y) = {}^t xAy$, where A is a symmetric $n \times n$ matrix. Then $O(Q,\mathbb{C}) = \{g \in GL(n,\mathbb{C}) \mid {}^t gAg = A\}$. Clearly this provides a large class of linear algebraic groups which are defined over $\mathbb{Q}$ (and also many examples which are not defined over $\mathbb{Q}$).

(4) We can also consider a skew-symmetric bilinear form J, for example,

$$J = \begin{pmatrix} 0 & I_n \\ -I_n & 0 \end{pmatrix},$$

and the associated symplectic group

$$Sp_J(2n,\mathbb{C}) = \{g \in GL(2n,\mathbb{C}) \mid {}^t gJg = J\}.$$

The above examples in (1) — (4) provide some of the typical classical linear algebraic groups. Though they exhaust all classical algebraic groups defined over $\mathbb{C}$ up to isomorphism, there are finitely many more algebraic groups defined over $\mathbb{R}$, and infinitely many more algebraic groups defined over $\mathbb{Q}$. There are also exceptional linear algebraic groups.

Remark 3.2. We can also define algebraic groups as groups with a compatible structure of algebraic varieties. In some sense, this might be more natural and is similar to the definition of topological groups and Lie groups. One difference with

Lie groups is that any projective (in particular compact) algebraic group defined over $\mathbb{C}$ is commutative and hence an abelian variety. Any algebraic subvariety of $GL(n,\mathbb{C})$ is noncompact. This is one reason for considering or rather concentrating on linear algebraic groups. Of course, a more important reason is that important noncompact algebraic groups are given by the classical linear algebraic groups given above, which can be defined and studied explicitly.

Definition of arithmetic subgroups

Let $\mathbf{G} \subset GL(n,\mathbb{C})$ be a linear algebraic group defined over $\mathbb{Q}$. Let

$$G = \mathbf{G}(\mathbb{R}) = \mathbf{G} \cap GL(n,\mathbb{R})$$

be its real locus, which is a real Lie group with finitely many connected components by a general result of Whitney on the real locus of algebraic varieties defined over $\mathbb{Q}$. (Note that when we talk about a linear algebraic group, we are always referring to its complex locus $\mathbf{G} = \mathbf{G}(\mathbb{C})$). Define the rational locus

$$\mathbf{G}(\mathbb{Q}) = \mathbf{G} \cap GL(n,\mathbb{Q}),$$

and the integral locus $\mathbf{G}(\mathbb{Z})$ with respect to the embedding $\mathbf{G} \subset GL(n,\mathbb{C})$ by

$$\mathbf{G}(\mathbb{Z}) = \mathbf{G} \cap GL(n,\mathbb{Z}).$$

Definition 3.3. *A subgroup* $\Gamma \subset \mathbf{G}(\mathbb{Q})$ *is called* **an arithmetic subgroup** *if it is commensurable with* $\mathbf{G}(\mathbb{Z})$, *i.e.,* $\Gamma \cap \mathbf{G}(\mathbb{Z})$ *has finite index in both* Γ *and* $\mathbf{G}(\mathbb{Z})$.

Remark 3.4. Note that the $\mathbb{R}$-locus G and the $\mathbb{Q}$-locus $\mathbf{G}(\mathbb{Q})$ do not depend on a particular embedding of $\mathbf{G}$ into $GL(n,\mathbb{C})$ as long as it is defined over $\mathbb{Q}$ and compatible with the $\mathbb{Q}$-structure of $\mathbf{G}$, but $\mathbf{G}(\mathbb{Z})$ will change with the embedding. On the other hand, the resulting groups $\mathbf{G}(\mathbb{Z})$ are commensurable and hence the class of arithmetic subgroups does not change.

Example 3.5. Any subgroup of finite index of $SL(n,\mathbb{Z})$ is an arithmetic subgroup of the linear algebraic group $SL(n,\mathbb{C}) \subset GL(n,\mathbb{C})$ embedded in the standard way. A special subclass of them consists of congruence subgroups of $SL(n,\mathbb{Z})$.

Remark 3.6. According to this definition, an arithmetic Γ is a subgroup of the Lie group G. More general, any subgroup $\Gamma \subset G$ is called *an arithmetic subgroup* of the Lie group G with respect to the linear algebraic group defined over $\mathbb{Q}$ if it is commensurable with $\mathbf{G}(\mathbb{Z})$. The point is that now Γ does not have to be contained in $\mathbf{G}(\mathbb{Q})$. For some purpose, we also call any subgroup of $\mathbf{G} = \mathbf{G}(\mathbb{C})$ which is commensurable with $\mathbf{G}(\mathbb{Z})$ an arithmetic subgroup of the linear algebraic group.

Proposition 3.7. *Arithmetic subgroups* Γ *of* $\mathbf{G}(\mathbb{Q})$ *are discrete subgroups of the real Lie group* G.

Proof. Since the discreteness of induced topology is preserved by commensurability, it is clear from the definition and the fact that $\mathbb{Z} \subset \mathbb{R}$ is a discrete subgroup.

Remark 3.8. There are important examples of algebraic groups $\mathbf{G}$ and subgroups Γ of G which are not discrete subgroups in G but are isomorphic to arithmetic subgroups of other Lie groups (or rather the real locus of algebraic groups). For example, let $\mathbb{Q}(\sqrt{d})$ be a real quadratic number field, where d is a positive square-free integer, and $\mathcal{O} = O_{\mathbb{Q}(\sqrt{d})}$ its ring of integers. Then the associated Hilbert modular group is $SL(2, \mathcal{O})$. Under each embedding $\mathbb{Q}(\sqrt{d}) \hookrightarrow \mathbb{R}$, $SL(2, \mathcal{O})$ is *not a discrete* subgroup of $SL(2, \mathbb{R})$. But it is an arithmetic subgroup of the Lie group $SL(2, \mathbb{R}) \times SL(2, \mathbb{R})$ when the latter is realized as the real locus of a suitable linear algebraic group defined over $\mathbb{Q}$. The reason is that under the two real embeddings of $\mathbb{Q}(\sqrt{d})$, $\mathcal{O}$ can be embedded diagonally into $\mathbb{R} \times \mathbb{R}$ as a discrete subgroup. Similarly, there are also examples of subgroups Γ of $\mathbf{G} = \mathbf{G}(\mathbb{C})$ that are not contained in $G = \mathbf{G}(\mathbb{R})$ but are arithmetic subgroups of $\mathbf{G}(\mathbb{C})$ when it is viewed as the real locus of another linear algebraic group. For example, let d be a positive square-free integer as above. Let $\mathcal{O} = \mathcal{O}_{\mathbb{Q}(\sqrt{-d})}$ be the ring of integers of the imaginary number field $\mathbb{Q}(\sqrt{-d})$. Then $SL(2, \mathcal{O})$ is a discrete subgroup of $SL(2, \mathbb{C})$ of finite covolume, called a *Bianchi group*. (This is an important example of arithmetic Kleinian group.)

The above remark raises the two questions:

1. *What is an arithmetic subgroup of a Lie group?*
2. *Does every Lie group always admit arithmetic subgroups?*

Given a Lie group G, if there exists a linear algebraic group $\mathbf{G}$ defined over $\mathbb{Q}$ such that its real locus $\mathbf{G}(\mathbb{R})$ is equal to G, then we have a class of arithmetic subgroups of G which are commensurable with $\mathbf{G}(\mathbb{Z})$. One important point to note is that in this case, there could be more than one linear algebraic groups whose real locus is equal to G, and they will lead to different (i.e., noncommensurable) arithmetic subgroups.

Clearly, this class of arithmetic subgroups of G depends on existence of the linear algebraic group $\mathbf{G}$ and the fact that it is defined over $\mathbb{Q}$. As mentioned before, it can happen that there are *different, i.e., non-isomorphic over* $\mathbb{Q}$, linear algebraic groups defined over $\mathbb{Q}$ whose real loci are the same and equal to G. For example, given two quadratic forms $Q_1(x, y) = x^2 - y^2$, and $Q_2(x, y) = x^2 - 2y^2$. They are isomorphic under the linear change of variable $(x, y) \to (x, \sqrt{2}y)$, but they are not isomorphic to each other under linear change of variables with $\mathbb{Q}$-coefficients. Their associated orthogonal groups $\mathbf{G}_1 = O(Q_1, \mathbb{C})$, $\mathbf{G}_2 = O(Q_2, \mathbb{C})$ are not isomorphic as linear algebraic groups over $\mathbb{Q}$, but their real locus $G_1 = \mathbf{G}_1(\mathbb{R})$, $G_2 = \mathbf{G}_2(\mathbb{R})$ are isomorphic Lie groups. (In fact, they are isomorphic as linear algebraic groups defined over $\mathbb{R}$.) Arithmetic subgroups of $G = G_1 =$

G_2 with respect to the linear algebraic groups $\mathbf{G}_1, \mathbf{G}_2$ are not commensurable. Though $G_1 = G_2$ is not a semi-simple Lie group, we can produce other examples of semisimple Lie groups using quadratic forms in more than 3 variables in the same way.

One problem with the above approach in defining arithmetic subgroups is that not every Lie group G is equal to the real locus of a linear algebraic group. For example, any Lie group that is not linear, i.e., does not admit a finite dimensional faithful representation, gives an example satisfying this property.

Remark 3.9. Though it is well-known that there are non-linear Lie groups, it is not so easy to find explicit references on examples and proofs that they are non-linear. One general construction goes as follows. Let $\mathbf{G}$ be a simply connected linear algebraic group, i.e., $\mathbf{G}(\mathbb{C})$ is simply connected as a topological space. Assume that the real locus $G = \mathbf{G}(\mathbb{R})$ is not simply connected. Then every nontrivial cover of G is a nonlinear Lie group. Examples include $\mathbf{G} = SL(2), Sp(2n)$, and double covers of $SL(2, \mathbb{R})$ and $Sp(2n, \mathbb{R})$ are metaplectic groups.

Even if a Lie group can be realized as the real locus of a linear algebraic group, the linear algebraic group may not be defined over $\mathbb{Q}$. Many simply connected nilpotent Lie groups provide such examples.

In order to define arithmetic subgroups of general Lie groups, we introduce the following definition.

Definition 3.10. *A $\mathbb{Q}$-structure of a Lie group G is a linear algebraic group $\mathbf{H}$ defined over $\mathbb{Q}$ together with a Lie group homomorphism $\varphi : \mathbf{H}(\mathbb{R}) \to G$ with a compact kernel and compact cokernel.*

Remark 3.11. One could also consider homomorphisms $G \to \mathbf{H}(\mathbb{R})$, when G is a metaplectic group mentioned above. For the most general case, we should define a $\mathbb{Q}$-structure on G as a Lie group homomorphism $\varphi : \mathbf{H}(\mathbb{R}) \to G/F$ with a compact kernel and compact cokernel, where F is a finite normal subgroup of G.

In many practical applications, for example, in constructing cocompact arithmetic groups of $SL(2, \mathbb{R})$, the real locus of the linear algebraic groups that appear is mapped surjectively to $SL(2, \mathbb{R})$ with compact kernel, i.e., by forgetting some compact factors. One reason is that the identity component of the isometry group of a symmetric space of noncompact type is a noncompact semisimple Lie group without compact factors (and also with trivial center), and it might be more convenient to consider homomorphisms in this direction when we view a group homomorphism $\varphi : \mathbf{H}(\mathbb{R}) \to G$ essentially as an action on the symmetric space G/K.

Example 3.12. To understand the notion of $\mathbb{Q}$-structures, it is helpful to consider the example of $G = \mathbb{R}^n$. A $\mathbb{Q}$-structure of $\mathbb{R}^n$ is a $\mathbb{Q}$-vector subspace $V_{\mathbb{Q}}$ of dimension n such that $V_{\mathbb{Q}} \otimes \mathbb{R} = \mathbb{R}^n$, which is equivalent to the condition that $V_{\mathbb{Q}}$

is spanned by n vectors $v_1, \ldots, v_n$ which are linearly independent over $\mathbb{R}$ (not only over $\mathbb{Q}$). (Note that the condition that the dimension of $V_{\mathbb{Q}}$ is n is important.) Then it can be shown that every lattice $\Lambda \subset \mathbb{R}^n$ defines a $\mathbb{Q}$-structure of $\mathbb{R}^n$, and every $\mathbb{Q}$-structure of $\mathbb{R}^n$ arises this way; furthermore, two lattices induce the same $\mathbb{Q}$-structure if and only if they are commensurable. This implies that $\mathbb{R}^n$ admits *uncountably infinitely many* $\mathbb{Q}$*-structures.*

Once a Lie group G is given a $\mathbb{Q}$-structure, we can define its arithmetic subgroups.

Definition 3.13. *For any* $\mathbb{Q}$*-structure of a Lie group* G *as above,* $\varphi : \mathbf{H}(\mathbb{R}) \to G$*, a discrete subgroup* $\Gamma \subset G$ *is called an arithmetic subgroup with respect to this* $\mathbb{Q}$*-structure if there exists an arithmetic subgroup* $\Gamma' \subset \mathbf{H}(\mathbb{Q})$ *such that* $\varphi(\Gamma')$ *is commensurable with* Γ.

It is clear that the above definition does not depend on the choice of arithmetic subgroup $\Gamma' \subset \mathbf{H}(\mathbb{Q})$ since they are all commensurable. But it does depend on the $\mathbb{Q}$-structure of $\mathbf{H}$ and the homomorphism $\varphi : \mathbf{H}(\mathbb{R}) \to G$. This is quite clear in the case of $G = \mathbb{R}^n$.

Remark 3.14. The class of arithmetic subgroups of G depends on the choice of $\mathbb{Q}$-structure. When we talk about an arithmetic subgroup of a Lie group G, we have imposed a $\mathbb{Q}$-structure on it implicitly or explicitly. *If* G *does not admit any* $\mathbb{Q}$*-structure, then* G *does not admit any arithmetic subgroup.*

Since the existence of arithmetic subgroups of a Lie group is equivalent to existence of $\mathbb{Q}$-structures on G, a natural question is whether every Lie group admits a $\mathbb{Q}$-structure.

Proposition 3.15. *If* G *is a semisimple Lie group with finitely many connected components, then it admits* $\mathbb{Q}$*-structures.*

Proof. The basic idea of the proof is as follows. The existence of $\mathbb{Q}$-structure on G is equivalent to the existence of $\mathbb{Q}$-structure on its Lie algebra $\mathfrak{g}$, which is equivalent to the existence of a basis $v_1, \ldots, v_n$ of $\mathfrak{g}$ such that the structure constants of this basis are rational numbers, i.e., the Lie bracket $[v_i, v_j]$ is a $\mathbb{Q}$-linear combination of $v_1, \ldots, v_n$ for all pairs i, j. The existence of such basis follows from the celebrated Chevalley theorem on existence of *integral* basis. See [26] for example.

On the other hand, there are nilpotent and solvable Lie groups which do not admit any $\mathbb{Q}$-structure. We will mention this below in connection with lattices of nilpotent Lie groups.

Example 3.16. The basic Lie group $SL(2, \mathbb{R})$ admits infinitely many nonisomorphic $\mathbb{Q}$-structures and they give rise to infinitely many non-commensurable classes of arithmetic subgroups. For example, with the standard $\mathbb{Q}$-structure from the embedding $SL(2,\mathbb{C}) \subset GL(2,\mathbb{C})$, we obtain the arithmetic subgroup $SL(2,\mathbb{Z})$ and subgroups of $SL(2,\mathbb{R})$ commensurable with it. It is a classical fact that $SL(2,\mathbb{Z})$ is not a cocompact subgroup, i.e., $SL(2,\mathbb{Z})\backslash SL(2,\mathbb{R})$ is noncompact. On the other hand, arithmetic subgroups from other $\mathbb{Q}$-structures on $SL(2,\mathbb{R})$ are cocompact subgroups. They come from quaternion algebras over number fields that split only at one real place but do not split over the number fields. See [95] and [115].

A basic fact due to Borel and Harish-Chandra is the following result. See [81] for references on more details of reduction theories of arithmetic subgroups.

Proposition 3.17. *If G is a semisimple Lie group, then every arithmetic subgroup Γ of G is a cofinite discrete subgroup, i.e., the volume of $\Gamma\backslash G$ with respect to any invariant Riemannian metric on G is finite.*

This is an important conclusion of reduction theory of arithmetic groups and has many applications. It is also the starting point of many questions. For example, one important question, a now solved conjecture of Weil on the Tamagawa number, concerns the value of the volume of $\Gamma\backslash G$ with respect to a suitably normalized measure or volume form on G.

Motivated by the notion of lattices in $\mathbb{R}^n$, the following definition is standard.

Definition 3.18. *A discrete subgroup Γ of a locally compact subgroup G is called a* **lattice** *if with respect to a measure induced from a left invariant measure on G, the total measure of the quotient $\Gamma\backslash G$ is finite.*

From the point of view of measure theory (hence ergodic theory and dynamics) and analysis, lattices are coarse approximations to the ambient Lie groups and hence form a natural class of discrete subgroups. As it will become clear later, they also turn out to have amazing rich structures and properties from the perspectives of geometry, topology, geometric group theory and algebras etc.

Remark 3.19. Not every locally compact topological group G, in particular, a Lie group, admits a lattice Γ. One necessary condition is that G is unimodular, i.e., a left invariant Haar measure is also right invariant. It can be shown by example that this necessary condition is not sufficient. See [162] for such examples.

The above Proposition 3.17 says that *arithmetic subgroups of semisimple Lie groups are lattices.* In fact, they provide the most important and interesting lattices. The question when a lattice of a Lie group is an arithmetic subgroup of the Lie group is natural and important and has been extensively studied. The most famous result is the arithmeticity of irreducible lattices of semisimple Lie groups of rank at least 2, which will be discussed later.

Arithmetic subgroups and lattices of Lie groups are natural and important classes of discrete subgroups. This does not mean that they are the only interesting discrete subgroups. Many Kleinian groups provide important discrete subgroups which are not lattices. One natural problem discussed below is to find other interesting discrete subgroups.

We state several basic results lattices here.

Proposition 3.20. *If G is a noncompact semisimple Lie group, then it admits both cocompact and nocompact arithmetic subgroups, which are both lattices.*

It is easier to see the existence of non-uniform arithmetic subgroups. For example, the $\mathbb{Q}$-structure corresponding to the Chevalley basis of the Lie algebra gives non-uniform arithmetic subgroups (actually they are maximally non-cocompact in some sense. For example, they have the maximal $\mathbb{Q}$-rank). The existence of uniform arithmetic subgroups is more subtle and is due to Borel [26]. These lattices correspond to different $\mathbb{Q}$-structures on the Lie group, and a seemingly more general definition of arithmetic subgroups of linear algebraic groups defined over number fields is crucial for this purpose, which is equivalent to the one defined above via the functor of restriction of scalar.

Corollary 3.21. *Every noncompact semisimple Lie group G admits lattices.*

On the other hand, nonsemisimple Lie groups may not always admit lattices. This reflects the rich and rigid structures of semisimple Lie groups.

Proposition 3.22. *If N is a simply connected nilpotent Lie group, then N admits a lattice Γ if and only if there exists a linear algebraic group $\mathbf{N}$ defined over $\mathbb{Q}$ such that N is equal to the real locus $\mathbf{N}(\mathbb{R})$ and Γ is an arithmetic subgroup with respect to the $\mathbb{Q}$-structure on N defined by $\mathbf{N}$.*

This is clear in the case when $N = \mathbb{R}^n$, since each lattice defines a $\mathbb{Q}$-structure as pointed out earlier.

The same result is not known for solvable Lie groups and might not be true. Indeed, structures of solvable Lie groups and their lattices are still not well-understood. (They are too far away from both simply connected abelian Lie groups $\mathbb{R}^n$ and semisimple Lie groups. The former is sort of soft and the structure is simple and clear, and the latter has rich structures, which impose strong restriction on the groups. Nilpotent groups are closer to $\mathbb{R}^n$ and share many properties with $\mathbb{R}^n$.)

The following result is true [148, Theorem 4.34].

Proposition 3.23. *If S is a simply connected solvable Lie group, then every lattice Γ is cocompact, i.e., $\Gamma\backslash S$ is compact, and there exists a faithful integral representation $\rho : S \to GL(n, \mathbb{R})$ such that $\rho(\Gamma) \subset GL(n, \mathbb{Z})$.*

On the other hand, there are lattices in simply connected solvable Lie groups which are not arithmetic [148, p. 77].

Remark 3.24. It is not clear how big $\rho(\Gamma)$ is in $GL(n,\mathbb{Z})$. Certainly it is not of finite index, since $SL(n,\mathbb{Z})$ is of finite index in $GL(n,\mathbb{Z})$ and $SL(n,\mathbb{R})$ is a semisimple Lie group, but S is a solvable Lie group. Though it is known that any simply connected solvable Lie group can be realized as a subgroup of the group U of the upper triangular matrices in $GL(m,\mathbb{R})$, its image and the intersection with the arithmetic subgroup $U(\mathbb{Z})$ is not clear and complicated.

Remark 3.25. It is clear from the definition that one key point in defining arithmetic subgroups is the ring $\mathbb{Z}$ of integers. Of course, this is a basic and important ring in number theory. On the other hand, there is a class of closely related rings of S-integers, which are important for various applications in number theory. Briefly, for any finite set S of primes, $S = \{p_1, \dots, p_k\}$, define the S-ring of integers $\mathbb{Z}_S = \mathbb{Z}[\frac{1}{p_1}, \dots, \frac{1}{p_k}]$. Then a generalization of arithmetic subgroup consists of S-arithmetic subgroups. Specifically, for any linear algebraic group $\mathbf{G}$ defined over $\mathbb{Q}$, any subgroup $\Gamma \subset \mathbf{G}(\mathbb{Q})$ is called *an S-arithmetic subgroup* if it is commensurable with $\mathbf{G}(\mathbb{Z}_S)$. They are not discrete subgroups of the real Lie group $\mathbf{G}(\mathbb{R})$ in general, but are discrete subgroups of the product group $\mathbf{G}(\mathbb{R}) \times \mathbf{G}(\mathbb{Q}_{p_1}) \times \cdots \times \mathbf{G}(\mathbb{Q}_{p_k})$, which is a locally compact topological group, where $\mathbb{Q}_p$ is the locally compact field of p-adic numbers.

Besides providing a large class of discrete groups, S-arithmetic subgroups are also natural in the following sense. For any linear algebraic group $\mathbf{G}$ defined over $\mathbb{Q}$, pick any finitely many elements $\gamma_1, \dots, \gamma_k \in \mathbf{G}(\mathbb{Q})$, then the subgroup $\langle \gamma_1, \dots, \gamma_k \rangle$ generated by them is in general not contained in an arithmetic subgroup of $\mathbf{G}(\mathbb{Q})$. Certainly they are contained in an S-arithmetic subgroup by taking S to contain all the primes which appear in the denominators of the matrix entries of these elements. As discussed below, a natural, but difficult, problem is to decide when $\langle \gamma_1, \dots, \gamma_k \rangle$ is an arithmetic subgroup or an S-arithmetic subgroup.

After the above discussions, a natural question is *why we study arithmetic subgroups*. Certainly, everyone will agree that the subgroup $\mathbb{Z} \subset \mathbb{R}$ is important and useful, and many people will also agree that the modular group $SL(2,\mathbb{Z})$ is also natural and interesting. Probably the real reason for studying arithmetic subgroups of Lie groups comes from the fact they are often the only systematic way to produce lattices of Lie groups and interesting discrete subgroups of Lie groups. Another reason is their actions on homogeneous spaces associated with the ambient Lie groups. Indeed, the quotient spaces obtained from these actions provide the most interesting and important examples of spaces in differential geometry, algebraic geometry, algebraic and geometric topology, number theory, and analysis, and these actions provide the most interesting and important examples of group actions, or transformation groups, and have served as guiding lights for geometric group theory.[4] These will be discussed in Section 5.

[4]According to the Erlangen program proposed by Klein, symmetries are crucial in many

4 Intrinsic properties of lattices and arithmetic subgroups of Lie groups

Arithmetic subgroups and lattices of Lie groups enjoy many properties. For more comprehensive listing of these properties and references on them, see a book of references [81]. The expository papers [21, 158, 161, 162] contain good summaries and references on arithmetic groups. The paper [80] also contains an extensive list of properties of arithmetic subgroups of semisimple Lie groups.

A basic fact on lattices of Lie groups is the following property.

Proposition 4.1. *Suppose that G is a Lie group with finitely many connected components. Then any lattice Γ of G is finitely generated.*

This is a basic property of a discrete group and is well expected and known. But its proof is not entirely trivial. This can be proved easily once the reduction theory for lattices and arithmetic subgroups (i.e., Proposition 5.3) is established. Without the reduction theory, Kazhdan introduced the notion of property T and used it to prove finite generation of irreducible lattices of semisimple Lie groups of higher rank.

Remark 4.2. For a while, it seemed that the only way to prove finite generation of lattices of semisimple Lie groups is to use the property T for irreducible lattices of semisimple Lie groups of higher rank and the reduction theory for lattices of semisimple Lie groups of rank 1 by Garland and Raghunathan. For a recent uniform proof of the finite generation property, see [66].

This might seem surprising due to the following observation: Suppose that Γ is an irreducible lattice of Lie group of rank at least 2. Then Margulis arithmeticity theorem implies that Γ is an arithmetic subgroup with respect to a suitable $\mathbb{Q}$-structure on G, and the reduction theory of arithmetic subgroups can be applied to Γ to prove its finite generation. But right at the beginning of the proof of the Margulis arithmeticity theorem, the finite generation of Γ is used. See [20] for the definition of property T and its applications.

One important application of finite generation of Γ is that we can endow Γ with word metrics. Briefly, giving a finite symmetric generating set S of Γ (the symmetry means that $\gamma \in S$ if and only if $\gamma^{-1} \in S$), we can define a *word length* on elements of γ: for any $\gamma \in \Gamma$,

$$|\gamma|_S = \inf\{k \mid \gamma = \gamma_1 \cdots \gamma_k, \gamma_1, \ldots, \gamma_k \in S\},$$

and for any elements $\gamma_1, \gamma_2 \in \Gamma$, the *word metric* (*or distance*)

$$d_S(\gamma_1, \gamma_2) = |\gamma_1^{-1}\gamma_2|_S.$$

different contexts. In geometry and physics, Lie groups provide the most important classes of symmetry groups. On the other hand, as far as discrete symmetry groups are concerned, $SL(2, \mathbb{Z})$ and its variants are the most important examples.

The word metric turns Γ into a metric space (Γ, d_S). If we prefer a connected geodesic space, then we can use the Cayley graph of Γ associated with the generating set S, and (Γ, d_S) is isometrically embedded into the Cayley graph. (Recall that the Cayley graph is the graph whose vertices are elements of Γ and two vertices γ_1, γ_2 are connected by an edge if and only if $\gamma_1\gamma_2^{-1}$ belongs to S.)

A slightly more difficult result to prove is finite presentation.

Proposition 4.3. *Every lattice Γ of a Lie group with finitely many connected components is finitely presented.*

To prove this, the reduction theory for arithmetic subgroups of semisimple Lie groups and for lattices of semisimple Lie groups of real rank 1 is needed. This confirms the point of this paper that actions of groups on suitable spaces are needed to understand groups.

Once we have shown that lattices are finitely generated and finitely presented, we can study them from the points of view of combinatorial group theory and geometric group theory. For some open problems on combinatorial group theory, see [16].

There are many other properties of lattices of Lie groups, and lattices of semisimple Lie groups in particular. We will mention more later when we discuss some open problems.

A basic point is that structures and properties of lattices depend on the types of Lie groups we consider. We will consider Lie groups G and hence their lattices Γ in four cases.

Case (1). *G is isomorphic to a vector space $\mathbb{R}^n$.*

Proposition 4.4. *Every discrete subgroup Γ of $\mathbb{R}^n$ is finite generated and finitely presented and is of the form $\mathbb{Z}v_1 + \cdots + \mathbb{Z}v_k$, where $v_1, \ldots, v_k$ are linearly independent vectors over $\mathbb{R}$.*

Such a result on finite generation and finite presentation might not be true for other Lie groups, at least it is not known in general.

By definition, a discrete subgroup Γ of $\mathbb{R}^n$ is called *a lattice* if and only if it is a free abelian group of rank n, i.e., in the above presentation, $k = n$, and $\Lambda = \mathbb{Z}v_1 + \cdots + \mathbb{Z}v_n$. The best known example is the standard integral (or square) lattice $\mathbb{Z}^n$.[5]

Proposition 4.5. *The following conditions on a discrete subgroup Γ of $\mathbb{R}^n$ are equivalent:*

(1) A discrete subgroup Γ of $\mathbb{R}^n$ is a lattice.

[5] According to the definition in Merriam-Webster Dictionary, a lattice is a regular geometrical arrangement of points or objects over an area or in space; specifically : the arrangement of atoms in a crystal.

(2) Γ *is of finite covolume, i.e.,* $vol(\Gamma\backslash\mathbb{R}^n) < +\infty$.
(3) Γ *is cocompact, i.e.,* $\Gamma\backslash\mathbb{R}^n$ *is compact.*
(4) Γ *is Zariski dense in* $\mathbb{R}^n$.

This is a special feature of lattices of $\mathbb{R}^n$ too. The equivalence between (2) and (3) is not true in general for all Lie groups.

Lattices in $\mathbb{R}^n$ have several rigidity properties, which are usually not emphasized. But they can help one to understand later rigidity results for lattices of semisimple Lie groups such as the Mostow strong rigidity and Margulis superrigidity.

Proposition 4.6. *Let* $\Gamma_1 \subset \mathbb{R}^{n_1}, \Gamma_2 \subset \mathbb{R}^{n_2}$ *be two lattices. Then any isomorphism* $\varphi : \Gamma_1 \to \Gamma_2$ *extends to an isomorphism* $\varphi : \mathbb{R}^{n_1} \to \mathbb{R}^{n_2}$.

The proof is simple since any isomorphism $\varphi : \Gamma_1 \to \Gamma_2$ must map a basis of a free abelian group Γ_1 to a basis of Γ_2, and the linear extension gives an isomorphism $\varphi : \mathbb{R}^{n_1} \to \mathbb{R}^{n_2}$.

But this is the simplest strong rigidity for lattices of Lie groups. The **basic point** is that *properties can be transferred between lattices and the ambient Lie groups.* Mostow strong rigidity is along the same spirit.

The following is a simple example of super-rigidity of lattices, or rather of showing an idea of the Margulis superrigidity.

Proposition 4.7. *Let* $\Lambda \subset \mathbb{R}^n$ *be a lattice, and* $\varphi : \Lambda \to \mathbb{R}^n$ *a group morphism, then it can be extended to a homomorphism* $\varphi : \mathbb{R}^n \to \mathbb{R}^m$.

There is also a simple characterization of lattices of $\mathbb{R}^n$.

Proposition 4.8. *An abstract group can be realized as a lattice of* $\mathbb{R}^n$ *if and only if it is a finitely generated torsion-free abelian group of rank* n.

The above results and discussions show that structures of lattices of $\mathbb{R}^n$ are simple, and classification of them as abstract groups is also simple since they are determined by the rank: *as abstract groups, all lattices of* $\mathbb{R}^n$ *are isomorphic.*

Question 4.9. *Is this the end of story about lattices in* $\mathbb{R}^n$*?*

The answer is a definite **No**! We can put additional structures on lattices and classify them in more refined ways.

We can put the standard Euclidean metric on $\mathbb{R}^n$ and endow lattices with the induced metric (or norm). Such enriched lattices are called *Euclidean lattices.*

We identify two Euclidean lattices if there is an isometry of $\mathbb{R}^n$ which moves one lattice to another one. (This is equivalent to that two lattices endowed with the norm of $\mathbb{R}^n$ are isometric as Euclidean lattices. The reason is that a basis of a lattice is mapped to a basis of the image lattice by an isomorphism. Since the isomorphism is an isometry, it is induced by an isometry of the whole ambient space $\mathbb{R}^n$.)

Then there are infinitely many isometry equivalence classes of Euclidean lattices. In fact, we can describe their isometric classes precisely.

Proposition 4.10. *The space of isometric classes of Euclidean lattices in $\mathbb{R}^n$ can be identified with a locally symmetric space $GL(n,\mathbb{Z})\backslash GL(n,\mathbb{R})/O(n)$.*

Proof. For any double coset $GL(n,\mathbb{Z})gO(n)$, define an Euclidean lattice $\mathbb{Z}^n g$. Since for every $\gamma \in GL(n,\mathbb{Z})$, $\mathbb{Z}^n\gamma = \mathbb{Z}^n$, it is easy to see that the isometry class of $\mathbb{Z}^n g$ is well-defined. Since every lattice $\Lambda \subset \mathbb{R}^n$ is of the form $\Lambda = \mathbb{Z}^n g$ for some $g \in GL(n,\mathbb{R})$, the proposition is proved.

In view of this proposition, the classification of Euclidean lattices up to isometry is too refined in some sense. For many applications such as crystals arranged along a lattice, the really important factor is the symmetry of the arrangement. Therefore, one natural way is to classify Euclidean lattices according to their symmetry group.

Given a Euclidean lattice $\Lambda \subset \mathbb{R}^n$, define its *isometry group* (or *symmetry group*)

$$Is(\Lambda) = \{g \in Is(\mathbb{R}^n) \mid g(\Lambda) = \Lambda\}.$$

If $Is(\Lambda)$ is bigger, then Λ is more symmetric. For example, it is visibly clear that the standard Euclidean square lattice $\mathbb{Z}^2 \subset \mathbb{R}^2$ is more symmetric than the rectangle lattice $\mathbb{Z}(1,0) + \mathbb{Z}(0,2)$.

Two Euclidean lattices $\Lambda_1, \Lambda_2 \subset \mathbb{R}^n$ are defined to be *equivalent* (or *of the same symmetry*) if their isometry groups $Is(\Lambda_1), Is(\Lambda_2)$ are isomorphic as abstract groups. This is the precise statement that two Euclidean lattices Λ_1, Λ_2 have the *same amount of symmetry.* It is clear that the square lattice $\mathbb{Z}^2$ and the rectangle lattice $\mathbb{Z}(1,0) + \mathbb{Z}(0,2)$ are not equivalent, since $\mathbb{Z}^2$ is more symmetric.

We summarize some basic facts on symmetry groups $Is(\Lambda)$ of lattices Λ. Recall that $Is(\mathbb{R}^n)$ is a Lie group given by the semi-direct product $\mathbb{R}^n \times O(n)$, where $O(n)$ acts on $\mathbb{R}^n$ by rotations.

Proposition 4.11. *For every Euclidean lattice $\Lambda \subset \mathbb{R}^n$, $Is(\Lambda)$ is a uniform lattice of $Is(\mathbb{R}^n)$.*

Proof. We note that $Is(\mathbb{R}^n)$ contains the normal cocompact subgroup $\mathbb{R}^n$ consisting of translations, and $Is(\Lambda)$ also contains the subgroup of translations isomorphic to Λ. Since Λ is a cocompact discrete subgroup of $\mathbb{R}^n$, $Is(\Lambda)$ is a cocompact subgroup of $Is(\mathbb{R}^n)$. We also note that since Λ is a discrete subgroup of $\mathbb{R}^n$, $Is(\Lambda)$ is a discrete subgroup of $Is(\mathbb{R}^n)$.

Proposition 4.12. *In every dimension, there are only finitely many equivalence classes of Euclidean lattices.*

This is related to finiteness of wallpaper patterns and symmetries of crystals in $\mathbb{R}^3$. It follows from some basic results of Bieberbach.

To state these results of Bieberbach, we need to consider one special class of groups.

Definition 4.13. *A discrete subgroup $\Gamma \subset Is(\mathbb{R}^n)$ is called a crystallographic group if it is a cocompact discrete subgroup of $Is(\mathbb{R}^n)$.*

As explained above, for any Euclidean lattice Λ, its isometry group $Is(\Lambda)$ is a crystallographic group. But not every crystallographic group is the isometry group of a lattice.

Besides clarifying structures of crystallographic groups, three basic results of Bieberbach from 1910s have also played an important role in proposing problems for lattices of Lie groups and understanding them. See [164] for detailed statements and proofs.

We note that these results of Bieberbach solved the first part of Problem 18 of Hilbert's famous list of 23 problems: *If there in n-dimensional euclidean space also only a finite number of essentially different kinds of groups of motions with a fundamental region?*

Theorem 4.14 (Bieberbach's first Theorem)**.** *Suppose that $\Gamma \subset Isom(\mathbb{R}^n)$ is a crystallographic group, then the intersection $\Gamma \cap \mathbb{R}^n$ is a lattice in $\mathbb{R}^n$ and has a finite index in Γ, and it is also the unique maximal normal abelian subgroup of Γ.*

If $\Gamma = Is(\Lambda)$, then $\Gamma \cap \mathbb{R}^n = \Lambda$.

Theorem 4.15 (Bieberbach's second Theorem)**.** *Two crystallographic groups of $\mathbb{R}^n$ are isomorphic as abstract groups if and only if they are conjugated in the affine group $Aff(\mathbb{R}^n) = GL(n, \mathbb{R}) \times \mathbb{R}^n$.*

The group $Aff(\mathbb{R}^n)$ consists of affine transformations of $\mathbb{R}^n$. This result reflects rigidity of crystallographic groups and is a precursor of the Mostow strong rigidity for lattices of semisimple Lie groups [132].

Theorem 4.16 (Bieberbach's third Theorem)**.** *In every dimension, there are only finitely many isomorphic classes of crystallographic groups.*

As mentioned before, this implies that there are only finitely many equivalence classes of Euclidean lattices in every dimension. An analogous finiteness result does not hold for lattices of general Lie groups. For example, for every semisimple Lie group G, there are infinitely many isomorphism classes of lattices Γ.

There is also a characterization of crystallographic groups as abstract groups [162].

Proposition 4.17 (Wolf, 1972). *An abstract group Γ is isomorphic to a crystallographic group in dimension n if and only if it contains a normal subgroup Γ' of finite index such that Γ' is isomorphic to $\mathbb{Z}^n$ and is a maximal abelian subgroup of Γ.*

As described before, crystallographic groups are natural from the isometry groups of lattices and enjoy good properties. But we can still ask the following question. As it is true in most cases, asking questions often leads to better understanding.

Question 4.18. *Why do we consider crystallographic groups? Are there other applications?*

We gave one answer earlier: to understand and classify symmetry groups of crystals. Of course, discrete subgroups of the isometry group $Is(\mathbb{R}^n)$ are interesting in themselves and describe discrete symmetries.

Another reason is to study flat manifolds. Recall that a Riemannian manifold is called *flat* if its sectional curvature is zero, i.e., it is locally isometric to $\mathbb{R}^n$. It is known that the universal covering space of a complete flat manifold is isomorphic to the Euclidean space $\mathbb{R}^n$.

Therefore, for any crystallographic group Γ of dimension n, $\Gamma\backslash\mathbb{R}^n$ is a compact flat manifold, and every compact flat manifold is of this form. The Bieberbach theorems provide the foundation to understand and classify compact flat manifolds. For example, the first Bieberbach theorem says that every compact flat manifold (or orbifold) has a finite cover which is a flat torus, the second theorem says that if two compact flat manifolds are homotopy equivalent, then they affinely equivalent, and the third theorem says that up to homotopy equivalence (and hence up to affine equivalence) there are only finitely many compact flat orbifolds.

Classification of compact flat manifolds is also one case of the so-called *space form problem*: to understand and classify manifolds of constant curvature. See the book [164] for detailed discussions.

Though crystallographic groups have been studied for a long time, the following problem is still open in higher dimensions.

Problem 4.19. *Classify and list all crystallographic groups in every dimension, i.e., classify and list all compact flat orbifolds in every dimension up to affine equivalence.*

In low dimensions, isomorphism classes of crystallographic groups have been listed. The number becomes very large in high dimensions and it seems too complicated already for $n > 8$.

Affine crystallographic groups

In the second Bieberbach theorem on rigidity of crystallographic groups, the group of affine transformations $Aff(\mathbb{R}^n) = \mathbb{R}^n \times GL(n, \mathbb{R}^n)$ of $\mathbb{R}^n$ appeared naturally. It is an important group by itself. A natural generalization of properties and classification of crystallographic groups concerns discrete subgroups of $Aff(\mathbb{R}^n)$. A discrete subgroup $\Gamma \subset Aff(\mathbb{R}^n)$ is called *an affine crystallographic group* if Γ acts properly on $\mathbb{R}^n$ and the quotient $\Gamma\backslash\mathbb{R}^n$ is compact.

Understanding affine crystallographic groups is still a very active area pursued by important mathematicians. A famous open problem is the Auslander conjecture.

Conjecture 4.20 (Auslander conjecture). *If $\Gamma \subset Aff(\mathbb{R}^n)$ is an affine crystallographic group, then Γ is virtually solvable, i.e., it contains a solvable subgroup of finite index.*

This corresponds to the first theorem of Bieberbach for crystallographic groups. We note that $\mathbb{R}^n$ is a normal subgroup of $Aff(\mathbb{R}^n)$ but is not a cocompact subgroup, since $GL(n, \mathbb{R})$ is not compact. Therefore, affine crystallographic groups are not lattices of $Aff(\mathbb{R}^n)$.

The quotient $\Gamma\backslash\mathbb{R}^n$ is a complete affine flat manifold. (Recall that an affine flat manifold is a manifold equipped with a flat, torsion-free connection.) Unlike Riemannian manifolds, a compact affine manifold may not be complete in the sense that every geodesic (with respect to the affine connection) can be extended infinitely in both directions.

The Auslander conjecture has been proved in dimensions up to 6. See [1] [31] and references there. But the classification of complete, compact affine flat manifolds seems to be a natural, really complicated.

Problem 4.21. *Understand and classify affine crystallographic groups and compact complete affine flat manifolds.*

Milnor raised a question on how to characterize affine crystallographic groups. He conjectured that any torsion-free polycyclic-by-finite group could be realized as an affine crystallographic group.

The answer turned out to be no. It might sound too strong. In the characterization of crystallographic groups, there is some strict condition on the finite index abelian subgroup. See [124] for references.

Case (2). *G is a nilpotent Lie group.*

In terms of commutativity of groups, the next class of Lie groups after abelian Lie groups consists of nilpotent groups.

Recall that a group G is called *nilpotent* if the central descending series $G^0 = G, G^1 = [G, G^0], \dots, G^i = [G, G^{i-1}], \dots,$ terminates at $\{1\}$ after finitely many steps. A Lie group is called *a nilpotent Lie group* if it is nilpotent as a group. For example, the group N_n consisting of $n \times n$ upper triangular matrices with 1s on the diagonal is nilpotent. This is a typical nilpotent Lie group in some sense. In fact, it is known that any simply connected nilpotent Lie group can be realized as a subgroup of N_n for some n.

A particularly important case is the Hisenberg group N_3 consisting of 3×3 upper triangular matrices with 1s on the diagonal.

Note that the higher dimensional Hisenberg group is also nilpotent and is a proper subgroup of the subgroup N_{n+2}. Specifically, it is the upper triangular matrices of size $(n+2) \times (n+2)$ corresponding to the block decomposition determined by the partition of $n+2$: $(1, n, 1)$. So it consists of matrices $\begin{pmatrix} 1 & b & c \\ 0 & I_n & b \\ 0 & 0 & 1 \end{pmatrix}$, where a, b are vectors with n entries, c is a scalar. The above example of the usual Heisenberg group corresponds to $n = 1$. They have important applications in quantum physics and is also useful in describing the boundary of the unit ball of $\mathbb{C}^2$ and hence in complex analysis of several variables.

For lattices of nilpotent Lie groups, there are important results due to Malcev [118]. We will concentrate on simply connected nilpotent Lie groups and their lattices. There are two reasons for this assumption:

(1) Simply connected nilpotent Lie groups can be realized as linear algebraic groups.
(2) The exponential map from the Lie algebra to a simply connected nilpotent Lie group is an isomorphism of algebraic manifolds. (They are also given by polynomials. This fact has many important consequences and is not true for other Lie groups. For a noncompact semisimple Lie group, the exponential map is not even surjective).

Proposition 4.22. *Suppose that Γ is a discrete subgroup of a simply connected nilpotent Lie group N. Then the following conditions are equivalent:*

(1) Γ is a lattice in N.
(2) Γ is a uniform lattice in N.
(3) Γ is dense in N with respect to the Zariski topology.

One source of lattices of nilpotent Lie groups comes from arithmetic subgroups of unipotent linear algebraic groups.

Recall that a linear algebraic group $\mathbf{G} \subset GL(n, \mathbb{C})$ is called *unipotent* if every element g of it is unipotent, i.e., $(g - Id)^k = 0$ for some integer k. Note that the notion of unipotence of a group depends on the matrix realization of the group, but the notion of nilpotence of it is intrinsic.

It is known that the real locus of a unipotent algebraic group is a nilpotent Lie group, and many important nilpotent Lie groups arise this way.

Proposition 4.23. *If N is the real locus of a unipotent linear algebraic group $\mathbf{N} \subset GL(n, \mathbb{C})$ defined over $\mathbb{Q}$, then its integral subgroup $\mathbf{N}(\mathbb{Z}) = \mathbf{N}(\mathbb{Q}) \cap GL(n, \mathbb{Z})$ is a uniform lattice in the Lie group $N = \mathbf{N}(\mathbb{R})$. Therefore, every arithmetic subgroup of $\mathbf{N}(\mathbb{Q})$ is a uniform lattice of N.*

The following result shows that this is the only way to construct lattices of simply connected nilpotent Lie groups.

Proposition 4.24. *Suppose that Γ is a lattice in a simply connected nilpotent Lie group N. Then there exists a linear algebraic group $\mathbf{N}$ defined over $\mathbb{Q}$ such that its real locus is equal to N and Γ is an arithmetic subgroup of $\mathbf{N}(\mathbb{Q})$.*

The combination of the above two propositions can be reformulated as follows.

Proposition 4.25. *A simply connected nilpotent Lie contains a lattice if and only if it (or equivalently its tangent algebra) admits the structure of a $\mathbb{Q}$-algebraic group (or a $\mathbb{Q}$-Lie algebra), and the lattice is an arithmetic subgroup with respect to the $\mathbb{Q}$-structure.*

Lattices of simply connected nilpotent Lie groups can be characterized as abstract groups as follows.

Proposition 4.26. *An abstract group Γ is isomorphic to a lattice of a simply-connected nilpotent Lie group if and only if Γ is nilpotent, finitely generated, and torsion-free.*

Lattices of nilpotent Lie groups also enjoy a strong rigidity property.

Proposition 4.27. *Suppose that Γ_1, Γ_2 are lattices in two simply connected nilpotent Lie groups N_1, N_2 respectively. Then every isomorphism $\varphi : \Gamma_1 \to \Gamma_2$ extends uniquely to an isomorphism $\varphi : N_1 \to N_2$ of the ambient Lie groups. Consequently, the homogeneous spaces $\Gamma_1 \backslash N_1$ and $\Gamma_2 \backslash N_2$ are isomorphic as homogeneous spaces and hence diffeomorphic.*

In literatures, a nilmanifold is a manifold which is a quotient of a nilpotent Lie group by a closed subgroup. This proposition can be reformulated as follows: *If two compact nilmanifolds $\Gamma_1 \backslash N_1, \Gamma_2 \backslash N_2$ are homotopy equivalent, then they are isomorphic as homogeneous manifolds and hence diffeomorphic.* This is called *the strong rigidity for compact nilmanifolds.*

An even stronger rigidity property holds.

Proposition 4.28. *Suppose that Γ is a lattice in a simply-connected nilpotent Lie group N, and N' is another simply connected nilpotent Lie group. Then any homomorphism $\varphi : \Gamma \to N'$ can be uniquely extended to a homomorphism $\varphi : N \to N'$.*

This is called the *super-rigidity* for lattices of nilpotent Lie groups. We will see below that the Margulis superrigidity is a generalization of this statement for lattices of semisimple Lie groups.

For references on results on lattices of nilpotent Lie groups discussed above, see [41, 118, 148].

Case (3). *G is solvable Lie group.*

After nilpotent Lie groups, the next less commutative class of Lie groups consists of solvable Lie groups.

Recall that a group is called *solvable* if the descending normal series reaches the trivial subgroup after finitely many steps, i.e., if we define $G^{(1)} = [G, G]$, and $G^{(i+1)} = [G^{(i)}, G^{(i)}]$, then $G^{(n)} = \{1\}$ for some n.

This is equivalent to the condition that there exists a subnormal series of subgroups $G_0 = \{1\} \subset G_1 \subset \cdots \subset G$ such that G_{j-1} is a normal subgroup of G_j and the quotient $G_{j-1}\backslash G_j$ is an abelian group.

A solvable Lie group is a Lie group which is also solvable as a group.

Proposition 4.29. *Let S be a solvable Lie group with finitely many connected components. Then the following conditions on a discrete subgroup $\Gamma \subset S$ are equivalent:*

(1) Γ is a lattice of S.
(2) Γ is a uniform lattice of S.

Unlike nilpotent linear algebraic groups or semisimple linear algebraic groups, arithmetic subgroups of solvable linear algebraic groups defined over $\mathbb{Q}$ are not necessarily lattices in the real locus of the algebraic groups. For example, let $\mathbf{S}$ be the linear algebraic subgroup of $GL(n, \mathbb{C})$ consisting of upper triangular matrices. Then $\mathbf{S}(\mathbb{Z})$ is not a lattice of $\mathbf{S}(\mathbb{R})$. This can be seen easily in the case $n = 2$.

Remark 4.30. For solvable linear algebraic groups $\mathbf{S}$, let $X(\mathbf{S})$ be the set of all characters of $\mathbf{S}$ defined $\mathbb{Q}$. Define S' by

$$S' = \{g \in \mathbf{S}(\mathbb{R}) \mid \chi(g) = 1, \chi \in X(\mathbf{S})\}.$$

Then $\mathbf{S}(\mathbb{Z})$ is contained in S' and it can be shown that $\mathbf{S}(\mathbb{Z})$ is a lattice of S'. This is a special case of the reduction theory for arithmetic subgroups due to Borel and Harish-Chandra.

Definition 4.31. *A group Γ is called polycyclic if there is a subnormal series of subgroups $\Gamma_0 = \{1\} \subset \Gamma_1 \subset \cdots \subset \Gamma_n$ such that Γ_{j-1} is a normal subgroup of Γ_i, and the quotient $\Gamma_{i-1}\backslash\Gamma_i$ is cyclic.*

If each quotient $\Gamma_{i-1}\backslash\Gamma_i$ is isomorphic to $\mathbb{Z}$, then Γ is called a strongly polycyclic group.

Relations between lattices of solvable Lie groups and polycyclic groups are described in the next proposition [148, p. 52, p. 68].

Proposition 4.32. *A lattice Γ of a simply connected solvable Lie group is strongly polycyclic. Conversely, every polycyclic group Γ admits a finite index normal subgroup which is isomorphic to a lattice of a simply connected solvable Lie group.*

If we do not require the solvable Lie group to be connected, then we have the following statement.

Proposition 4.33. *If Γ is a polycyclic group, then there exists a solvable Lie group G with finitely many connected components such that Γ is isomorphic to a lattice in G. Conversely, if G is a solvable Lie group with finitely many connected components, then every lattice of G is polycyclic.*

In general it is not true that we can find a connected solvable Lie group G in the above proposition.

An important step of proofs of the above theorem is an analogue of Bieberbach's first theorem.

Proposition 4.34. *If Γ is a lattice of a connected solvable Lie group S, and N is the nilradical of S (i.e., the largest connected nilpotent normal subgroup of S), then $\Gamma \cap N$ is a lattice in N, i.e., the lattice respects the nilpotent radical of S.*

Unlike the case of nilpotent Lie groups, there is no simple criterion on existence of lattices of solvable Lie groups.

It is also known that there exist lattices of simply connected Lie groups which are not arithmetic groups (see [162, p. 60]).

If Γ is a lattice of a connected solvable Lie group S, then the homogeneous space $\Gamma\backslash S$ is called a *compact solvmanifold.*

Solvmanifolds satisfy the following strong rigidity.

Proposition 4.35. *Let Γ_1, Γ_2 are lattices of two simply connected solvable Lie groups S_1, S_2 respectively. If Γ_1 is isomorphic to Γ_2 as abstract groups, then the solvmanifolds $\Gamma_1\backslash S_1$ and $\Gamma_2\backslash S_2$ are diffeomorphic as manifolds.*

The stronger statement that S_1, S_2 are isomorphic Lie groups is not true in general. See [162, p. 50]. If S is simply connected, then the fundamental group of the solvmanifold $\Gamma\backslash S$ is isomorphic to Γ. In this case, S is also contractible and the homotopy type of $\Gamma\backslash S$ is determined by Γ. Then the above proposition can be reformulated as follows.

Proposition 4.36. *Let S_1, S_2 are simply connected solvable Lie groups. If two compact solvmanifolds $\Gamma_1\backslash S_1$ and $\Gamma_2\backslash S_2$ are homotopy equivalent, then they are diffeomorphic.*

This is called the *Mostow strong rigidity* for solvmanifolds, which was proved by Mostow in 1954 and motivated the *Borel conjecture for aspherical manifolds*: if two aspherical manifolds are homotopy equivalent, then they are homeomorphic. This is still open and probably the most important conjecture in geometric topology.

The analogous superrigidity does not hold for lattices of solvable Lie groups. These results show that lattices of solvable Lie groups are less rigid than lattices of nilpotent Lie groups. See a recent paper [15] for more results and references.

We will see below that lattices of semisimple Lie groups, which are even less commutative Lie groups, enjoy important rigidity properties. If we put all Lie groups on a line, rigidity properties of their lattices have a dip at the solvable case.

Case (4). *G is semisimple.*

The class of semisimple Lie groups is probably the most important class of Lie groups in terms of structures and applications, and their lattices provide the most interesting and important discrete subgroups even among discrete subgroups of Lie groups. They have been applied to many different situations and their structures have been intensively studied. Their rich structures make them rigid in some sense and make detailed theories about them possible.

An important example is $G = SL(2, \mathbb{R})$, and discrete subgroups of $SL(2, \mathbb{R})$ are called *Fuchsian groups.* They have been studied intensively since the time of Gauss, and their action on the upper half plane gives rise to Riemann surfaces, which include important examples of Shimura varieties.

There are many special properties of Fuchsian groups among discrete subgroups of semisimple Lie groups. Probably the most striking feature is lack of rigidity properties. It is also amazing that this lack of rigidity motivated the rich and important theory of Teichmüller spaces and mapping class groups, a close analogy of symmetric spaces and arithmetic groups, which behave like "lattices of semisimple Lie groups of rank between 1 and 2". We note that lattices of semisimple Lie groups and their associated locally symmetric spaces have quite different properties when the real rank of the Lie groups is greater than or equal to 2. We will discuss them later in more detail.

Another more combinatorial property is that every finitely generated Fuchsian group is a hyperbolic group. This is not true for non-uniform lattices of any semisimple Lie group not locally isomorphic to $SL(2, \mathbb{R})$.

Discrete subgroups Γ of $SL(2, \mathbb{C})$ are called *Kleinian groups*, which include Fuchsian groups as special cases. They act isometrically and properly on the three-dimensional real hyperbolic space $\mathbf{H}^3$ and quotients $\Gamma \backslash \mathbf{H}^3$ are 3-dimensional hyperbolic manifold provide the most important class of 3-manifolds, which have played a crucial role in 3-dimensional topology and geometry, in particular the Thurston geometrization program.

$SL(2,\mathbb{R})$ and $SL(2,\mathbb{C})$ are semisimple Lie groups of rank 1. Results about Fuchsian, Kleinian groups and their associated hyperbolic manifolds have motivated problems and results for lattices of general semisimple Lie groups and locally symmetric spaces of real rank 1.

There are many properties of lattices of semisimple Lie groups and their associated locally symmetric spaces, and we will select a few of them here. For a more comprehensive list of properties and references, see [80].

Theorem 4.37 (Mostow strong rigidity [131]). *If $\Gamma_1 \subset G_1, \Gamma_2 \subset G_2$ be two irreducible lattices of semisimple Lie groups without center or compact factor, and not isomorphic to $PSL(2,\mathbb{R})$, then any isomorphism $\varphi : \Gamma_1 \to \Gamma_2$ extends to an isomorphism $\varphi : G_1 \to G_2$.*

Let $X_1 = G_1/K_1$ and $X_2 = G_2/K_2$ be the associated symmetric spaces. Then the above theorem can be reformulated as follows: if two locally symmetric spaces $\Gamma_1\backslash X_1, \Gamma_2\backslash X_2$ are homotopy equivalent and X_1, X_2 are not equal to the Poincaré upper half plane $\mathbf{H}^2$, then they are isometric after suitable scaling. For the exceptional case, the strong rigidity fails dramatically. There are continuous families of isomorphic Fuchsian groups whose hyperbolic surfaces are not isometric.

As mentioned before, every arithmetic subgroup of a semisimple Lie group is a lattice. One natural question is when lattices of semisimple Lie groups are arithmetic. This question was raised by Selberg and Piateskii-Shapiro.

Recall that a group Γ is called irreducible if Γ does not contain a subgroup Γ' of finite index such that Γ' splits as a product of two infinite subgroups.

Remark 4.38. The usual definition of an irreducible lattice Γ in a semisimple Lie group $G = G_1 \times G_2$, where G_1, G_2 are simple Lie groups, is that the projection of Γ in each factor G_1, G_2 is dense. It is known that this is equivalent to the above one using the normal subgroup theorem of Margulis for irreducible lattices and the notion of algebraic rank of groups. To show this equivalence more directly in the compact quotient case, we note that if up to finite index, $\Gamma = \Gamma_1 \times \Gamma_2$, then by the splitting theorem of closed manifolds of nonpositive curvature due to Gromoll-Wolf [69] and Lawson-Yau [105], the symmetric space G/K splits as a product $X = X_1 \times X_2$, which implies that G splits and we must have that $G = G_1 G_2$, $\Gamma_1 \subset G_1, \Gamma_2 \subset G_2$ are lattices, and hence Γ is not irreducible in the sense here. Conversely, if Γ is reducible, then by definition here, $\Gamma \cap G_1, \Gamma \cap G_2$ are lattices, and $\Gamma = \Gamma_1 \times \Gamma_2$ up to finite index.

An important theorem of Margulis [119] is the following one. His work on superrigidity was developed to prove this. Though the super-rigidity of lattices has its own interests, this application is undoubtedly the most interesting application.

Theorem 4.39 (Arithmeticity of lattices). *If Γ is an irreducible lattice of a semisimple Lie group G without compact factor and of rank at least 2, then Γ is an arithmetic subgroup of G with respect to a suitable $\mathbb{Q}$-structure on G.*

Case (5). *Lie groups which are none of the above cases.*

Though abelian, nilpotent, solvable and semisimple Lie groups are natural and important classes of Lie groups, their combinations give interesting and important Lie groups. For example, the isometry group $Is(\mathbb{R}^n) = \mathbb{R}^n \times O(n)$ does not belong to any groups of the above list. The Levi decomposition allows one to reduce a general Lie group to solvable and semisimple Lie groups, but a corresponding reduction for lattices is more complicated and subtle.

Much less work has been done for this class of Lie groups and their lattices.

Problem 4.40. *Study and understand structures and properties of lattices of general Lie groups and their associated locally homogeneous spaces by making use of results on lattices of solvable and nilpotent Lie groups and semisimple Lie groups.*

5 Actions of discrete subgroups of Lie groups

By definition, any discrete subgroup Γ of a Lie group G acts properly on the Lie group G by either the left multiplication or the right multiplication. The quotient $\Gamma\backslash G$ is a homogeneous space of G under the right multiplication.

For any closed subgroup $H \subset G$, the quotient G/H is a Hausdorff space, and the left G-multiplication gives a continuous action of G on G/H.

Proposition 5.1. *When H is a compact subgroup of G, then for any discrete subgroup $\Gamma \subset G$, Γ acts properly on G/H.*

One particularly important case is when G is a noncompact semisimple Lie group with finitely many connected components, and $H = K \subset G$ is a maximal compact subgroup. Then $X = G/K$ with an invariant Riemannian metric is a symmetric space of noncompact type, in particular it is simply connected and of nonpositive curvature.

Symmetric spaces of noncompact type are the most important examples of Hadamard manifolds, or more generally, CAT(0)-spaces. The action of Γ on X provides an important class of groups in geometric group theory.

Every discrete subgroup $\Gamma \subset G$ acts properly on X. Actions of these groups on symmetric spaces with suitable assumption on the quotient $\Gamma\backslash X$ imply some finiteness condition on the groups (for example, cohomological finiteness properties) and on the locally symmetric spaces.

One important way to understand the Γ-action (such as structures of orbits, quotient spaces etc) and to use it to prove properties of Γ is to find a good fundamental domain or a rough (or a coarse) fundamental domain.

Definition 5.2. *A domain Ω in X is called a globally finite coarse fundamental domain if the following conditions are satisfied:*

(1) $\cup_{\gamma\in\Gamma}\gamma\overline{\Omega} = X$.

(2) $\{\gamma \in \Gamma \mid \gamma\overline{\Omega} \cap \overline{\Omega} \neq \emptyset\}$ *is finite. In particular, the map* $\Omega \to \Gamma\backslash X$ *is finite to one.*

If we impose the condition that Γ*-translates of the domain* Ω *are disjoint, then we say that* Ω *is a globally finite fundamental domain.*

In general, it is not obvious why such a globally finite rough fundamental domain should exist for a discrete subgroup, or even for a lattice. One reason is that its existence implies various finiteness conditions of Γ.

A basic result in the reduction theory of arithmetic subgroups is a positive assertion on the existence of such fundemental domains.

Proposition 5.3. *Let* G *be a noncompact semisimple Lie group and* Γ *be an arithmetic subgroup. Then there exists a globally finite rough fundamental domain for the* Γ*-action on* X. *The rough fundamental domain has finite area and is a finite union of Siegel sets.*

The importance of the Siegel sets is that they can be described rather easily. For $G = SL(2,\mathbb{R})$ and $X = \mathbf{H}^2 = \{x + iy \mid x \in \mathbb{R}, y > 0\}$, it corresponds to a vertical strip $\{x+iy \mid x \in [a,b], y \geq c\}$, where $a, b \in \mathbb{R}, c > 0$, or its translates under elements of $SL(2,\mathbb{R})$. See [27] for a summary of reduction theory and references.

When H is a non-maximal compact subgroup, then G/H is a homogeneous Riemannian manifold but not a symmetric space. It is an important homogeneous space since period domains of Hodge structures appear as quotients of such spaces G/H for suitable choices of G and H.

If H is noncompact, then G/H is a homogeneous space which does not admit a G-invariant Riemannian metric, but it admits G-invariant pseudo-Riemannian metrics. If Γ is a lattice of G, then the action of Γ on G/H is not proper in general, and we can not take the usual quotients and study their properties.

In this case, one is interested in how severely Γ fails to act properly on G/H. For example, one can consider equivariant theory of G/H with respect to Γ, and also consider density of Γ-orbits and closures of Γ-orbits.

For a noncompact subgroup H, one natural problem is to find smaller discrete subgroups Γ' which are not lattices such that Γ' acts properly on G/H and the quotient $\Gamma'\backslash G/H$ is small in a suitable sense, for example, of finite volume or even compact. For example, one is interested in structures of the quotient $\Gamma'\backslash G/H$ as an orbifold, and the reduction theory for the Γ'-action. We will discuss more about such problems later.

6 Organization and brief summaries of open problems

In the previous sections, we gave some basic definitions and results on lattices of arithmetic subgroups and lattices of Lie groups. One important point we tried to emphasize is that properties of the ambient Lie groups and their lattices are closely related, and the existence of lattices imposes conditions on Lie groups.

Though some open problems were mentioned in the earlier sections, for example Problems 4.19–4.21 and Problem 4.40, we are now ready to discuss systematically open problems on discrete subgroups of Lie groups. We will provide some relevant definitions and results when it is convenient and important to do so, and will follow the following order:

1. Structures of subgroups of lattices: description and counting or distribution of various types of subgroups.
2. Computational aspects of arithmetic subgroups and discrete subgroups of Lie groups: both about the groups (construction of new groups and descriptions of existing ones) and their actions on the symmetric spaces.
3. Existence and construction of non-arithmetic lattices acting on the complex hyperbolic spaces, and understanding of internal structures of non-arithmetic groups.
4. Other natural discrete subgroups which are smaller than lattices and their associated locally symmetric spaces.
5. Characterization of lattices of semisimple Lie groups similar in spirit to those characterization for lattices of nilpotent Lie groups and solvable Lie groups, i.e., in terms of group theoretical properties.
6. Bounded generation of arithmetic subgroups and application to rigidity results.
7. Simplicial volumes of locally symmetric spaces: both vanishing or not and explicit evaluation of the simplicial volume.
8. Large scale geometry of lattices and geometric group theory: large scale geometric invariants and large scale geometric constructions such as the asymptotic dimension and asymptotic cone, and long time behaviors of random walks on groups.
9. Measure rigidity of lattices, and quantum unique ergodicity of locally symmetric spaces.
10. Rigidity of locally symmetric spaces and generalizations: characterizations of locally symmetric spaces and quotients of Euclidean buildings among CAT(0)-spaces.
11. Nonexistence of short geodesics of locally symmetric spaces of higher rank in contrast with the Margulis lemma for hyperbolic manifolds and rank one locally symmetric spaces.
12. Representation theories, cohomology variety of lattices of Lie groups, and associated rigidity properties of representations of lattices.
13. Counting of number of finite dimensional representations of lattices of Lie groups and other properties of representations of lattices.
14. Nonlinear actions of lattices of semisimple Lie groups of higher rank and the Zimmer program on rigidity and classification of lattice actions.

15. Families of discrete subgroups of semisimple Lie groups or families of locally symmetric spaces, not necessarily in towers, and asymptotic analytic and topological properties of these families of locally symmetric spaces.
16. Spectral analysis of locally symmetric spaces, and relation to the geometry of the locally symmetric spaces via more geometric versions of the Selberg trace formula.
17. Spectral rigidity of locally symmetric spaces among general Riemannian manifolds and deformation rigidity.
18. Moduli interpretation of locally symmetric spaces.
19. Cohomology of arithmetic groups: geometric construction of cycles.
20. Spines of symmetric spaces, i.e., equivariant deformation retractions of symmetric spaces, and finite classifying spaces of arithmetic groups.

7 Structures of subgroups of lattices of Lie groups

One deep insight of Lie is to study Lie groups through their Lie algebras. Because of this, Lie subgroups of Lie groups are relatively easy to be understood.

In some sense, the richness of Lie groups comes from a rich reply of subgroups and detailed descriptions of them. For example, if G is a semisimple Lie group, then parabolic subgroups P and maximal compact subgroups $K \subset G$ are important for understanding G and lattices of G, since G/P is a Furstenberg boundary, and $X = G/K$ is a symmetric space. For every reductive subgroup L of G that is stable under the Cartan involution associated with K, the orbit of L in X through the base point K in X is a totally geodesic subspace.

For a Lie algebra $\mathfrak{g}$, subalgebras are also fundamental. Besides the root spaces and roots, nilpotent and solvable subalgebras are also important for understanding Lie algebras and their representations.

If we pursue the analogy between Lie groups and their lattices, it is a natural problem to understand subgroups of lattices. Suppose that Γ is a lattice of a Lie group G. A naive and direct problem is to *classify all subgroups of* Γ. This is not doable in general.

One reason is that there might be too many finitely generated subgroups. For example, by the work of Mikhailova [123], there are continuously many non-isomorphic subgroups of the direct product $F_2 \times F_2$, where F_2 is the free group on two generators. It is expected that a lattice Γ of a semisimple Lie group G also contains many subgroups. In fact, the modular group $SL(2,\mathbb{Z})$ contains a copy of F_2 as a subgroup (by either explicitly writing down one or using the Tits alternative), and any semisimple linear algebraic group G that contains a copy of the product $SL(2,\mathbb{R}) \times SL(2,\mathbb{R})$ with the standard $\mathbb{Q}$-structure on $SL(2,\mathbb{R})$ will contain a copy of $F_2 \times F_2$. This implies that an arithmetic subgroup Γ of G contains at least continuously many nonisomorphic subgroups. By using the Tits alternative for $\mathbb{Q}$-rank 1 semisimple linear algebraic groups, we can see that if the $\mathbb{Q}$-rank of G is at least 3, then every arithmetic subgroup Γ of G contains a copy of $F_2 \times F_2$ and hence contains at least continuously many nonisomorphic subgroups.

It is natural to consider only finitely presented subgroups of Γ. For example, the class of finitely presented subgroups of $F_2 \times F_2$ can be described well in [17, Theorem 2]). On the other hand, it is conceivable that the class of finitely presented subgroups of a lattice Γ of G might still be too large and can not be classified and described easily.

It is perhaps helpful to point out that when we consider Lie subgroups (i.e., closed subgroups with finitely many connected components) of G, we only consider *certain special subgroups*.

Assume that G is a semisimple Lie group and $\Gamma \subset G$ is a lattice. A direct analogue of parabolic subgroups correspond to intersection $\Gamma \cap P$, where P is a parabolic subgroup of G. When Γ is an arithmetic subgroup of G with respect to a $\mathbb{Q}$-structure, and P is defined over $\mathbb{Q}$, the structure of $\Gamma \cap P$ is understood and basically clear and has played an important role in the study of the locally symmetric space $\Gamma\backslash X$ and automorphic forms on $\Gamma\backslash X$. For a general lattice Γ and a parabolic subgroup P, the situation is more complicated. For example, the question of when $\Gamma \cap P$ is infinite is not completely settled. We also note that if N_P is the nilpotent radical of P, then N_P is a nilpotent subgroup of G, but not all nilpotent subgroups of G are of this form. The intersection $\Gamma \cap N_P$ is also important for various applications.

For any compact subgroup $K \subset G$, $\Gamma \cap K$ is a finite subgroup. We will discuss more about finite subgroups of Γ below.

In view of the above discussions, the following problem seems natural.

Problem 7.1. *Let Γ be a lattice of a semisimple Lie group G. For suitable Lie subgroups G' of G, for example, reductive subgroups and nilpotent subgroups of G, understand the intersection of Γ with G', for example, decide when it is a lattice of G' or enjoy some other properties such as being infinite. A closely related problem is to describe and classify subgroups of Γ that are intersections of Γ with Lie subgroups G' of G.*

This problem is related to questions on how to produce subspaces of arithmetic locally symmetric spaces. See [40] for some applications of such related results.

One point of the above discussion is that *not all subgroups of Γ are equally interesting*, and we need to concentrate on those which are related to structures of the ambient Lie group G. Considering intersection with Lie subgroups of G is one way. Another way is to relate them to the geometry of the symmetric space X and the locally symmetric space $\Gamma\backslash X$.

For example, when G is a semisimple Lie group of rank 1, cyclic subgroups of Γ correspond to closed geodesics in the locally symmetric space $\Gamma\backslash X$, or loops around the cusps of $\Gamma\backslash X$. (It will depend on whether the elements in the cyclic group are semisimple or not).

When G is a semisimple Lie group of higher rank, we can consider abelian subgroups of higher rank of Γ. Some of them will correspond to flat torus in the locally symmetric space $\Gamma\backslash X$, and others involve horospheres near the infinity of $\Gamma\backslash X$.

In these cases, they correspond to closed geodesics and flat tori in the locally symmetric spaces. In terms of subgroups Γ' of Γ, these closed geodesics and flat tori of $\Gamma\backslash X$ can be obtained as follows: for each cyclic or abelian subgroup Γ', take the Zariski closure $\overline{\Gamma'}^Z$ of Γ' in G and consider quotient $\Gamma'\backslash\overline{\Gamma'}^Z$ of the Zariski closure by the subgroup Γ', or rather consider suitable orbits of $\overline{\Gamma'}^Z$ in X and their projections in $\Gamma\backslash X$.

We can apply the same procedure to other subgroups of Γ. The first natural question is

Question 7.2. *Besides the above abelian subgroups of* Γ*, what other non-abelian subgroups of* Γ *should be considered? For example, what kind of* $CAT(0)$*-groups can be embedded into* Γ *and how do they correspond to subspaces of the symmetric space* X*?*

The Tits alternative for Γ implies that every subgroup of Γ is either virtually solvable or contains a free subgroup F_2, and hence it gives us a lot of free subgroups of Γ. These subgroups might not be directly related to (or rather can be reflected by) structures or properties of the Lie group G or the locally symmetric space $\Gamma\backslash X$, and *the geometric and topological meaning* of each such subgroup might not be clear or understood yet. Maybe one version of the above problem is *to find some large subgroups of* Γ *which have some geometrical and topological interpretation.*

One natural class of groups that contain the free group F_2 consists of surface subgroups, i.e., the fundamental group of surfaces of negative Euler characteristic.

Problem 7.3. *Understand existence and structures of subgroups of* Γ *isomorphic to fundamental groups of surfaces.*

When $G = SL(2,\mathbb{C})$ and $X = G/K$ is the three dimensional real hyperbolic space, the existence of surface subgroups of Γ and the existence of embedded surfaces in the hyperbolic manifold $\Gamma\backslash X$ is important and complicated. This is related to classification of three dimensional manifolds. See recent breakthrough in [4, 5, 91]. See also [102].

The distribution of the lengths of closed geodesics and flat tori of locally symmetric space $\Gamma\backslash X$ has been intensively studied (see [81] for references and [56] for recent results).

A natural generalization concerns distribution of subgroups of Γ according to a suitable "volume". For example, for each subgroup $\Gamma' \subset \Gamma$, consider the quotient $\Gamma'\backslash\overline{\Gamma'}^Z$ and its associated subspace in $\Gamma\backslash X$.

Problem 7.4. *Fix a class of subgroups of Γ, and study the distribution of these subgroups.*

For a semisimple Lie group G, this might be difficult. If G is a nilpotent or solvable Lie group, this problem might be easier. It does not seem to have been systematically studied yet.

Cyclic subgroups and abelian subgroups of Γ are small subgroups of Γ. At the other end of the spectrum of subgroups of Γ are subgroups Γ' of finite index. Assume that G is a semisimple Lie group. For each such subgroup Γ', $\Gamma' \backslash X$ is a locally symmetric space of finite volume, and the volume of $\Gamma' \backslash X$ is proportional to the index $[\Gamma : \Gamma']$.

Problem 7.5. *Study the distribution of subgroups Γ' of finite index.*

This question has been studied by Lubotzky and his co-authors. See the book [112] for a summary and references.

A closely related problem is to study finite quotient groups.

Problem 7.6. *Given a lattice Γ of a Lie group G, study and understand finite quotient groups of Γ.*

When Γ is a lattice of a semisimple Lie group G, this is related to the congruence subgroup problem below. When Γ is virtually a solvable or nilpotent Lie groups, a lot of work has been done on problems related to the above. See the book [100] and references there.

A particularly important class of finite index subgroups of $\mathbf{G}(\mathbb{Z})$ consists of congruence subgroups. Recall that an arithmetic subgroup of $\mathbf{G}(\mathbb{Q})$ is called *a congruence subgroup* if it contains a *principal congruence subgroup of level N*, which is defined by

$$\Gamma(N) = \{\gamma \in \mathbf{G}(\mathbb{Z}) \mid \gamma = Id \mod N\}.$$

A famous problem, **the Congruence subgroup problem**, asks whether every arithmetic subgroup is a congruence subgroup. This is the congruence subgroup property in the strict sense. More generally, a relaxed condition is that *the congruence subgroup kernel is finite*, which is defined to measure the difference of two topologies on $G(\mathbb{Q})$, the congruence topology defined by the family of congruence subgroups as a basis of the identity element and the arithmetic subgroup topology which is defined by the family of arithmetic subgroups.

The answer to the general congruence subgroup problem is expected to be that it is positive, i.e., the congruence subgroup kernel is finite, in the higher rank case, and negative in the rank one case. This problem has been solved in many cases. For precise statements of the results, see [143] and [81] for references. The problem on distribution of congruence subgroups is easier and contributes to a lower order term among all arithmetic subgroups.

It is reasonable that congruence subgroups are easier to understand and hence the motivation for the congruence subgroup problem is also natural. For example, for the principal congruence subgroup $\Gamma(N)$ of $SL(n, \mathbb{Z})$, the quotient $SL(n, \mathbb{Z})/\Gamma(N)$ is isomorphic to $SL(n, \mathbb{Z}/N\mathbb{Z})$, and the associated locally symmetric spaces have lots of correspondences, for example, Hecke correspondences. Another reason is that locally symmetric spaces associated with congruence subgroups arise naturally as moduli spaces in number theory and algebraic geometry.

It is known that the answer to the congruence subgroup problem in the strict sense is positive for $SL(n, \mathbb{Z})$ when $n \geq 3$, and is negative for $n = 2$. Examples of non-congruence subgroups of $SL(2, \mathbb{Z})$ were known to Klein [96], Fricke [62], Pick [140] in 1880s, and also later in [135, 150]. See also the books by Fricke-Klein [63].

A generalization of the problem on surface subgroups of Γ is to consider maps from other discrete subgroups.

Problem 7.7. *Given another Lie group G' and a lattice or another discrete subgroup Γ' of G', study questions about homomorphisms from Γ' to Γ such as existence, distribution, structures on the set of all such homomorphisms (or monomorphisms, or epimorphisms etc).*

For example, it is true that if G' and G are semisimple Lie groups and are not isomorphic up to compact subgroups (i.e., there does not exist a compact normal subgroup $H \subset G$ such that there exists a Lie group homomorphism $\varphi : G' \to H\backslash G$ with a compact kernel and cokernel), then there is no isomorphism between lattices Γ' of G' and Γ of G. This follows from the Mostow strong rigidity. Specifically, if one of G and G' is not isomorphically to $SL(2, \mathbb{R})$ up to compact subgroups, then an isomorphism between Γ' and Γ extends to an isomorphism between G'/N' and G/N, where N', N are maximal compact normal subgroups. Otherwise, G', G are isomorphically to $SL(2, \mathbb{R})$ up to compact subgroups and hence isomorphic to each other up to compact subgroups.

In view of this, other discrete subgroups of G' are needed. One can also consider conditions on homomorphisms between Γ' and Γ which can be extended to homomorphisms between Lie subgroups of G' and G.

A closely related problem is to consider homomorphisms in the opposite direction in the above problem.

Problem 7.8. *Given another Lie group G', study homomorphisms from Γ to G'.*

This is a classical problem and many different aspects have been intensively studied. For example, when $G' = GL(n, \mathbb{C})$, the problem is concerned with finite dimensional representations of Γ. If we ask the question when a homomorphism $\Gamma \to G'$ can be extended to a Lie group homomorphism $G \to G'$, this amounts to the super-rigidity. If we consider the set of all such homomorphisms up to conjugacy, we obtain the character variety.

This opens up many problems. For example, geometry and topology of the whole character variety, and special homomorphisms and hence local geometry and topology of the character variety. A lot of work has been done when Γ is the fundamental group of a closed orientable surface of negative characteristic number, which is isomorphic to a uniform Fuchsian group. The recent active subject of higher Teichmüller spaces, maximal surface representations and Anosov representations of surface groups are all related to this. They study representations of surface groups with some special properties. See the papers [32, 33] for surveys and [73] for some most recent results. See also [15] for deformations of lattices of solvable groups.

In the earlier discussion, we implicitly assumed that subgroups of Γ under discussion were infinite. In studying locally symmetric space $\Gamma\backslash X$, finite subgroups of Γ are important and correspond to singular orbifold points of $\Gamma\backslash X$.

One basic property of arithmetic subgroups Γ of Lie groups G, or rather a consequence of the reduction theory, is that there are only finitely many conjugacy classes of finite subgroups.

Problem 7.9. *For specific examples of arithmetic groups such as $SL(n,\mathbb{Z})$, $Sp(n,\mathbb{Z})$, describe and determine explicitly the family of finite subgroups, and families of other subgroups, for example, virtually cyclic subgroups.*

Finite subgroups of $SL(n,\mathbb{Z})$ and their orders and related groups have been studied. See [125, 158] for some results and references. One reason for the above question is that finite subgroups of Γ correspond to orbifold singularities of the locally symmetric space $\Gamma\backslash X$, and resolutions of these singularities often contain valuable information too. For other families of subgroups, there are associated classifying spaces, which are important for computation of K-groups and surgery groups. A particularly important one is the family of virtually cyclic subgroups, which appears in the Farrell-Jones conjecture in geometric topology, which is closely related to the Borel conjecture on aspherical manifolds mentioned before. Understanding such families of subgroups are useful for understanding the corresponding classifying spaces. See [104] for some related results. Besides potential applications to classifying spaces, it is an interesting problem to understand families of virtually cyclic subgroups.

8 Computational aspects of arithmetic subgroups

In comparison with general groups, arithmetic groups and their actions on symmetric spaces admit explicit descriptions. A natural question is whether we can do some explicit computation with them. The general philosophy here is how some general theories and questions in combinatorial group theory can be applied to arithmetic subgroups.

Perhaps it is folklore that it is not easy to construct lattices and discrete subgroups of semisimple Lie groups G which are not isomorphic to $SL(2, \mathbb{R})$ up to compact subgroups. As recalled earlier, arithmetic subgroups of G are lattices, but arithmetic subgroups seem to provide the only systematic way to construct irreducible lattices of semisimple Lie groups of higher rank. This made Selberg and later Piateskii-Shapiro to conjecture that irreducible lattices of higher rank are arithmetic subgroups, which was proved to be true by Margulis [119].

All classical arithmetic subgroups such as $SL(n, \mathbb{Z})$ can be generated by some simple matrices. This leads to the following **dream**: *Is it possible to find conditions or algorithms which will allow us to decide whether a finite set of elements* $\gamma_1, \ldots, \gamma_m \in GL(n, \mathbb{Q})$ *generate an arithmetic subgroup of* $GL(n, \mathbb{R})$ *or an* S*-arithmetic subgroup of* $GL(n, \mathbb{Q})$.

When $n \geq 3$, this is probably beyond reach and not solvable. But any results in this direction will be interesting and valuable.

We can also consider elements of $\gamma_1, \ldots, \gamma_m \in GL(n, \mathbb{R})$ and ask for conditions that will determine whether they generate a discrete subgroup or a lattice.

It might be easier to decide when *a set of elements* $\gamma_1, \ldots, \gamma_m \in GL(n, \mathbb{Q})$ *generates a discrete subgroup of* $GL(n, \mathbb{R})$ *or a lattice of* $GL(n, \mathbb{R})/\mathbb{R}^\times$.

There are several variants of this. For example, let $\mathbf{G}$ be a linear algebraic group defined over $\mathbb{Q}$. We can consider when finitely many elements $\gamma_1, \ldots, \gamma_m \in \mathbf{G}(\mathbb{Q})$ generate an arithmetic subgroup of $\mathbf{G}(\mathbb{Q})$ or an S-arithmetic subgroup of $\mathbf{G}(\mathbb{Q})$. Naturally, one can also consider elements of the real locus $G = \mathbf{G}(\mathbb{R})$.

When $G = SL(2, \mathbb{R})$ or $SL(2, \mathbb{C})$, there have been a lot of work. For example, a classical result is the Poincaré polygon theorem for Fuchsian groups which gives a procedure on how to generate a Fuchsian group with an explicit fundamental domain by suitably chosen generators. For Kleinian groups, some work on discreteness of subgroups has been done, for example [89] (see also a recent preprint [107]). There are also results of Klein and Maskit on combination theorems, i.e., how to combine some discrete subgroups to produce bigger discrete subgroups [122]. For some more recent related work, see [22].

A natural generalization of Fuchsian and Kleinian groups are discrete groups acting isometrically on the real hyperbolic space $\mathbb{H}^n$ of dimension $n \geq 4$. Much less is known about them. For a systematic survey together with many open problems, see [93].

A generalization of the real hyperbolic space is the complex hyperbolic space. Even less results are known about them. See the book [68] and the survey paper [35].

For general semisimple Lie groups G, especially of the higher rank, there does not seem to be many results on the above problems. Probably one ideal model is the Poincaré polygon theorem. But such a direct analogue may not exist. If there is a good answer, this will allow us to produce more lattices of Lie groups in a systematic way. Probably it is helpful to keep in mind that given an arithmetic subgroup, for example, a congruence subgroup, it is not easy to find good generators and finite presentations. This makes a solution of the above problem more desirable.

The above discussion is related to the next problem.

Problem 8.1. *Given an arithmetic subgroup* Γ*, find generators with various desirable properties, for example, with the minimum number of generators, with the minimum number of relations, or other properties for special* Γ *(see Problem 8.2 below).*

This might not be a doable problem and is related to a general naive question: *what does it mean that a group is understood well (or to a certain extent)*, or equivalently, *what information about groups are important and useful.*

For a finite group, all the information of the group is contained in its finite multiplication table. But extracting valuable invariants about a group from its multiplication table is not easy. Similarly, a group is completely determined by any presentation. For some special classes groups, suitable presentations provide an effective method to describe them. (Examples probably include Coxeter groups, braid groups, Artin groups and knot groups.) For others, it is very difficult to understand groups from presentations. It might be sufficient to remind the reader difficulties in handling the three basic problems in combinatorial group theory posed by Dehn: the word problem, the conjugacy problem and the isomorphism problem. See the book [116] for more details about these points. It is also perhaps helpful to keep in mind that even though Coxeter groups have simple presentations, they provide a large class of important groups in geometric group theory, and their properties are not always obvious from the presentations [44].

Maybe one answer to the above naive question is that *we can understand a group if we can see it as a transformation group on a natural space.* This is vague. But the description of $SL(n, \mathbb{Z})$ as the group of all linear transformations of $\mathbb{R}^n$ that map the lattice $\mathbb{Z}^n$ to $\mathbb{Z}^n$ is important and helpful. From this, we can also interpret geometrically all the congruence subgroups of $SL(n, \mathbb{Z})$. More importantly, many basic properties of $SL(n, \mathbb{Z})$ and its action on the space of positive definite quadratic forms can be understood using the geometry of lattices of $\mathbb{R}^n$, i.e., the theory of geometry of numbers invented by Minkowski.

In spite of difficulties with using presentations of groups, it is still a natural problem to obtain explicit presentations of some concrete arithmetic groups. One part of the above question concerns how to define and construct "good and economic presentations". For a finite group, its multiplication table gives a finite presentation, which is certainly not economic. The requirement of minimal number of generators seems to be natural. In fact, for every finitely generated group Γ, *its rank* is defined to be the minimal number of generators. *One natural problem is to bound the rank of* Γ *in terms of other invariants of* Γ. An upper bound of the rank of a lattice of a semisimple Lie group is obtained in terms of the volume of the associated locally symmetric space in [66]. In view of the result in [155] (see also [111]) that under suitable conditions on the linear algebraic groups, non-uniform arithmetic subgroups always admit subgroups of finite index which are generated by at most three elements, the upper bound in [66] is not sharp and there is no chance to get a lower bound for the rank of a similar nature.

For a closed hyperbolic manifold $M = \Gamma\backslash\mathbf{H}^3$, the rank of Γ is bounded from above by the Heegaard genus of M, which is the minimal genus of all Heegaard surfaces of M. A conjecture from 1960s by Waldhausen that the rank of Γ is equal to Heegaard genus of M was disproved recently in [106].

It is perhaps helpful to consider the classical arithmetic subgroup $SL(n,\mathbb{Z})$. It is well-known and easy to see that $SL(2,\mathbb{Z})$ is generated by two elements (a translation and an inversion) by making use of the explicit fundamental domain on the upper half plane or the Gauss elimination. For general $n \geq 3$, there is no good explicit fundamental domain for $SL(n,\mathbb{Z})$ acting on the symmetric space $SL(n,\mathbb{R})/SO(n)$, but the Gauss elimination shows that $SL(n,\mathbb{Z})$ is generated by elementary matrices and hence is finitely generated. It turns out that $SL(n,\mathbb{Z})$ can be generated by two elements. Since it can not be generated by one element, the rank of $SL(n,\mathbb{Z})$ is equal to 2. For more details, see [158] [42, p. 85].

A naive question is which presentation of $SL(n,\mathbb{Z})$ of the two mentioned above is "better" and "more useful". For the second presentation with two generators, relations for the two generators are very complicated, in contrast with the simple relations for elementary matrices in the first more standard presentation. Probably there is no definite answer to this naive question.

It is perhaps helpful to point out that for uniform arithmetic subgroups of semisimple Lie groups, there are no known explicit examples whose presentations are known. This makes the above problem even more difficult or interesting.

The above problem can also be formulated as understanding what kind of generators and relations are considered "useful" or "practical" and constructing them for some natural and explicit groups. Perhaps it is helpful to keep in mind that the word problem is not solvable for finitely generated group, and one might not be able to get the "best" generators and relations. In this sense, this is a hopeless problem, but it is still interesting.

One reason for asking the above problem is that in combinatorial group theory, one often starts with a presentation of a group and then try to understand structures and properties of the group. It seems natural (and perhaps) naive that one should address the starting point: *finding a good presentation before anything else*. It seems fair to say that natural groups should have some natural presentations, and not all presentations are created equally.

Maybe one way is to consider the presentation complexes of groups. It is known that every presentation of Γ gives a 2-dimensional complex whose edges correspond to the elements of Γ (or its 1-skeleton is the Cayley graph) and the complex is simply connected. It is also known that every two presentation complexes of a group are quasi-isometric and share many global properties. Maybe one way to differentiate two presentations of a group is to consider more local and refined geometric properties of their presentation complexes. In the book [114, p. 129], a certain local property called *maximal principle* is used to show that the word problem is solvable.

One concrete result that shows a choice of generating sets matters concerns the value of the Kazhdan constant [67]. Motivated by this result, one natural problem is as follows:

Problem 8.2. *For a finitely generated subgroup Γ of a Lie group which has the Kazhdan property (T), for example, an irreducible lattice of a semisimple Lie group of rank at least 2, find generating sets whose associated Kazhdan constant of Γ is maximal among all generating sets, or understand how it depends on generating sets.*

It is probably true that for every such Γ, generating sets with maximal Kazhdan constant exists. We can ask whether for classical arithmetic subgroups such as $SL(n, \mathbb{Z})$, $n \geq 3$, the usually generators consisting of elementary matrices (or generating sets containing only two elements) will have this maximal property.

Given a general discrete group acting isometrically and properly on a Riemannian manifold, there is always a Dirichlet fundamental domain, which is not very useful except for some special cases.

Fundamental domains are useful for many purposes. For example, they can be used to study properties about the discrete subgroups, or its associated locally symmetric spaces. The model example is $\Gamma = SL(2, \mathbb{Z})$ acting on the upper half plane, and its structure and presentation can be derived from the known fundamental domain. For finitely generated Fuchsian groups, i.e., finitely generated discrete subgroups of $SL(2, \mathbb{R})$, their fundamental domains are well-understood. On the other hand, even for classical arithmetic groups such as $SL(n, \mathbb{Z})$, $n \geq 3$, a lot of work has been done on explicit fundamental domains, but finding good fundamental domains and describing their properties are not easy and not completely solved yet. See [80] for a summary of some results and references.

Problem 8.3. *Given an explicit arithmetic subgroup or a lattice Γ of a semisimple Lie group, find good explicit fundamental domains, or coarse fundamental domains with desirable properties, in the symmetric space $X = G/K$.*

Probably there are two general requirements: the domains have simple topology and boundary structures, in particular near infinity if the lattice is not cocompact. To get more precise conditions, consider the classical fundamental domain of $SL(2, \mathbb{Z})$ on the upper half plane. Near the infinity, it is bounded by two vertical rays $x = 0, x = 1$ and they are identified under a parabolic element of $SL(2, \mathbb{Z})$, and the translates of the vertical strip under a corresponding parabolic subgroup of $SL(2, \mathbb{Z})$ cover a horodisk of the infinity. This structure is convenient for analysis and automorphic forms, for example, the Fourier expansion of cusp forms. To study automorphic forms and the Arthur-Selberg trace, precise descriptions of the shape and geometry of fundamental domains near "cusps" are needed. This is one reason that Siegel sets are used in the reduction theory of arithmetic subgroups. But unfortunately, Siegel sets only give rough fundamental domains. It should also be pointed out that for different applications such as cohomology of arithmetic subgroups, requirements on fundamental domains might be different. There is no precise formulation of what gives a good fundamental domain. For some explicit computations, see the book [159] and references there.

9 Existence and construction of non-arithmetic lattices

It was known since Gauss (or even earlier) that $SL(2, \mathbb{Z})$ is a lattice of $SL(2, \mathbb{R})$. It was also known since Minkowski and Voronoi (or even earlier) that $SL(n, \mathbb{Z})$ is a lattice of $SL(n, \mathbb{R})$.

Due to the geometric construction of Fuchsian groups by Poincaré and the theory of Teichmüller space, there are many known lattices of $SL(2, \mathbb{R})$ which are not arithmetic.

A natural question is whether some lattices of Lie groups other than $SL(2, \mathbb{R})$ are not arithmetic subgroups.

For the positive direction, here is a list of Lie groups G for which every lattice is an arithmetic subgroup with respect to a suitable $\mathbb{Q}$-structure on the Lie group.

1. $G = \mathbb{R}^n$. Every lattice defines a $\mathbb{Z}$-structure and hence a $\mathbb{Q}$-structure on $\mathbb{R}^n$, and the lattice is arithmetic with respect to the $\mathbb{Q}$-structure.
2. $G = N$ is a simply connected nilpotent Lie group. Existence of a lattice in N implies that N is the real locus of a nilpotent linear algebraic group defined over $\mathbb{Q}$ and the lattice is an arithmetic subgroup with respect to this $\mathbb{Q}$-structure. Since $\mathbb{R}^n$ admits many different $\mathbb{Q}$-structures, it is natural that N also admits many different $\mathbb{Q}$-structures in general.
3. G is a semisimple Lie group of rank at least 2 and $\Gamma \subset G$ an irreducible lattice. Then Γ is an arithmetic subgroup with respect to a suitable $\mathbb{Q}$-structure on G. In general G admit many different $\mathbb{Q}$-structures. This is related to classification of linear algebraic groups whose real locus is a given Lie group.

There are several different ways to characterize arithmetic subgroups in comparison with general lattices. For example, for an arithmetic subgroup $\Gamma \subset \mathbf{G}(\mathbb{Q})$, its commensurator in G contains $\mathbf{G}(\mathbb{Q})$ and is dense in G. This corresponds to abundance of correspondences of the associated locally symmetric space $\Gamma\backslash G/K$.

After the initial work of Makarov [117] on construction of non-arithmetic lattices acting on the real hyperbolic spaces of dimension 3 using tessellation of the hyperbolic space, there has been a lot of work on constructing more not necessarily arithmetic lattices acting real hyperbolic spaces by Vinberg and others. One important method is to use reflections to generate groups. This method can only be applied to relatively low dimension (less than 30). See [145, 160, 161] (and also [115]) for a summary and references.

One more recent important method that constructs non-arithmetic lattices acting on the real hyperbolic spaces for all dimensions is due to Gromov and Piateskii-Shapiro [71]. They started out with two arithmetic subgroups and suitable fundamental domains, and then glued them up in a non-arithmetic way to get a non-arithmetic lattice.

Another recent construction of non-arithmetic lattices acting on the real hyperbolic spaces is given in [3]. It is a variation of the method of Gromov and Piateskii-Shapiro and makes use of arithmetic subgroups.

Problem 9.1. *Are there other ways to construct non-arithmetic hyperbolic manifolds besides the method of Gromov & Piateskii-Shapiro? In other words, can one always decompose a non-arithmetic hyperbolic manifold into pieces which are arithmetic, or whose structures are relatively simple (or simpler), or construct a non-arithmetic subgroup by combining some subgroups of arithmetic subgroups.*

Probably one example, or rather a non-example, is the pants decompositions of hyperbolic surfaces which decompose a general hyperbolic surfaces of finite area into finitely many simple pieces. Maybe one version of the above problem asks if there is a non-arithmetic hyperbolic manifold which is "truly" non-arithmetic, or non-arithmetic at different levels.

There are four types of rank 1 symmetric spaces of noncompact type. Only the real and complex hyperbolic spaces admit non-arithmetic lattices, since the Margulis super-rigidity and hence arithmeticity holds for lattices of the other two types.

There are some important work of Mostow [133], Mostow-Deligne [49] on construction of non-arithmetic lattices acting on the complex hyperbolic spaces of complex dimension 2 and 3. The results obtained so far are much less systematic than in the case of real hyperbolic spaces. For example, it is not known whether complex hyperbolic spaces of all dimension admit nonarithmetic lattices as in the case of real hyperbolic spaces, which were constructed by Gromov and Piateskii-Shapiro. The geometry of complex hyperbolic space is usually much more complicated than the real hyperbolic space.

Problem 9.2. *Prove or disprove that there exist nonarithmetic lattices acting on the complex hyperbolic spaces of complex dimension greater than 4.*

See the survey [139] for more results and references. See also [35, 68] for discrete groups acting on the complex hyperbolic space.

10 Other natural discrete subgroups of Lie groups

We have emphasized that arithmetic subgroups Γ of semisimple Lie groups G are lattices. One reason of their importance is that they enjoy many good group theoretical and cohomological properties, and the locally symmetric spaces $\Gamma\backslash X$ associated with them also enjoy many good geometric and topological properties thanks to the reduction theory for arithmetic subgroups and arithmeticity of lattices.

A natural question is what are other interesting discrete subgroups of Lie groups.

Problem 10.1. *Define and study finitely generated (or finitely presented) discrete subgroups Γ of Lie groups G which are not lattices but enjoy desirable group theoretic properties such as cohomological finiteness properties, or the associated locally symmetric spaces are topologically finite, admit nontrivial compactifications, and whose ends can be described and understood.*

The assumption of finite generation or finite presentation is important. Otherwise there will be too many groups. The case of $G = SL(2,\mathbb{R})$ is probably too simple. For a general semisimple Lie group G, this problem is probably too difficult. For example, when $G = SL(2,\mathbb{C})$, structure of finitely generated Kleinian groups and properties of their associated 3-dimensional hyperbolic manifolds are fairly compliacted and finally being understood.

For any discrete subgroup Γ of a Lie group G, and any compact subgroup $K \subset G$, the induced action of Γ on G/K is proper.

But if $H \subset G$ is a noncompact subgroup, then the action of Γ on G/H may not be proper anymore. One natural criterion on discrete subgroups Γ is that it acts properly on G/H. We can also impose that the quotient $\Gamma\backslash G/H$ is "small" or "finite" in a suitable sense. This is a well-studied problem. See the paper [98]. This provides a large class of noncofinite discrete subgroups Γ of G with some finiteness property.

For discrete subgroups Γ acting on $X = G/K$, we can also impose finiteness conditions on the quotient $\Gamma\backslash X$. For example, we can require that it is topologically finite, i.e., homeomorphic or diffeomorphic to the interior of a manifold with boundary. Or equivalently, $\Gamma\backslash X$ admits a compactification as a smooth manifold or some nice "singular" space.

One motivation is that for Fuchsian groups and Kleinian groups, there are many interesting discrete subgroups besides lattices, especially those geometrically finite ones.

For example, Fuchsian groups of the second kind are not lattices. For Fuchsian groups and Kleinian groups, there are both geometrical finiteness or topological finiteness imposed on groups. Finite generation of the groups is not sufficient for many applications. See the survey [93] for results and references.

The notion of geometrically finite groups acting on the real hyperbolic space of dimension $n \geq 3$ is complicated. See [29] for some discussions on several definitions. There are suitable generalizations of geometrically finite Kleinian groups to co-infinite volume discrete subgroups of semisimple Lie groups of rank 1. See [30].

The class of convex cocompact discrete subgroups of semisimple Lie groups of higher rank is quite restricted and is essentially reduced to the rank 1 case by [146] [97].

The following problem seems natural.

Problem 10.2. *Let X be a symmetric space of noncompact type. Define geometrically finite quotients $\Gamma\backslash X$ and describe end structures of such spaces.*

It is reasonable to expect that geometrically finite locally symmetric spaces $\Gamma\backslash X$ are topologically finite.

Recently, there is another approach to produce noncofinite discrete subgroups by considering representations (or homomorphisms) of a lattice of a "smaller" Lie group H into a "larger" semisimple Lie group. For example, for the fundamental group Π of a closed surface of negative Euler characteristic and a semisimple Lie group G, consider all homomorphisms $\Pi \to G$ and pick out some special ones such as the representations with some extremal property, for example the maximal representations in [32, 33], and the Anosov representations in [73].

Remark 10.3. These results on discrete subgroups of Lie groups seem to fit into the following general pattern: (1) Pick out the groups Γ which are known in some sense. (2) Map them into the given Lie groups G with some extremal properties. Then the problem is to understand properties of the quotient spaces $\Gamma\backslash G$, $\Gamma\backslash X$. This is different from the approach: for a discrete subgroup $\Gamma \subset G$, assume some information about the quotient $\Gamma\backslash G$ and derive properties of the group Γ.

The above problems are also related to the question: *What are other interesting locally symmetric spaces besides those of finite volumes?*

The meaning of "interesting" is vague. One reason is that both symmetric spaces and locally symmetric spaces of finite volume have been intensively studied and have many *interesting* applications. For example, locally symmetric spaces of finite volume can be used to construct finite classifying spaces of lattices of semisimple Lie groups. They also provide special Riemannian manifolds. Of course, the spectral theory of locally symmetric spaces is of essential importance in modern mathematics. Similarly, symmetric spaces are also important in geometry, topology and analysis.

Maybe a more precise question is what other quotients of symmetric spaces of noncompact type give rise to important examples in geometry, topology and analysis. For example, some interesting results on the L^2-Betti numbers, Novikov-Shubin invariants and L^2-torsions of arbitrary quotients of symmetric spaces are obtained in [137].

Infinite covers of locally symmetric spaces of finite volume certainly have infinite volume. But probably they do not enjoy nice finiteness properties. For example, they are usually not finitely generated. We also note that geometrically finite hyperbolic manifolds of infinite volume do not arise this way.

Usually we do not want the subgroups Γ to be too small in the Lie groups G. For example, cocompact lattices approximate the ambient Lie groups in terms of metric (or distance), and lattices approximate in terms of measure theory.

If G is a semisimple Lie group without compact factor and center, then any lattice Γ of G is Zariski dense when G is identified with the real locus of a linear algebraic group, which is also possible, for example through the adjoint representation. This is the Borel density theorem.

Therefore, a natural condition to impose on discrete subgroups of a Lie group is that they are Zariski dense, when the Lie group is a real linear algebraic group.

One natural problem is the next.

Problem 10.4. *If Γ is a finitely generated (or finitely presented) Zariski dense discrete subgroup Γ but is not a lattice, determine geometrical and topological restrictions it impose on the locally symmetric spaces $\Gamma\backslash X$, for example, the connectedness at infinity, number of ends, the volume growth of each end, quasi-isometry type of each end, and possible compactifications and their relations with harmonic functions.*

The class of finitely presented Zariski dense subgroups is still probably too large. Perhaps one more concrete version of this problem is to find *additional* algebraic conditions on the groups Γ so that the locally symmetric spaces $\Gamma\backslash X$ have some desirable geometric and topological properties. But it is not clear what are the natural conditions which are both broad and flexible enough. One natural class of groups consists of thin subgroups of lattices of semisimple Lie groups.

For some questions on structures of ends of locally symmetric spaces and their relations with harmonic functions and spectral theory, see the paper [86] and references there.

The geometry of locally symmetric spaces $\Gamma\backslash X$ of infinite volume can be very complicated. If Γ is a finitely generated Fuchsian group, the geometry and topology of the associated hyperbolic surface $\Gamma\backslash\mathbf{H}^2$ is well-understood, [6] but if

[6] We note that in contrast, if Γ is not required to be finitely generated, then $\Gamma\backslash\mathbf{H}^2$ can be very complicated since we can use a countably infinitely many pairs of pants to construct very complicated hyperbolic surfaces.

Γ is a finitely generated Kleinian group acting on $\mathbf{H}^3$, the geometry of the 3-hyperbolic manifold $\Gamma\backslash\mathbf{H}^3$ can be very complicated. Recent work by many people on the geometry of three dimensional hyperbolic manifolds of infinite volume could provide some guides to the above problem and also shows its difficulty. See [126] and references there.

Discrete subgroups Γ of G which are Zariski dense but not lattices are called *thin groups.* There are some open problems and many results about them from the perspective of expander graphs. See [152] for a fairly detailed survey. For some other results about Zariski dense subgroups, see [142] [16], but there are few results about the associated *locally symmetric spaces.*

One reason for posing the above problem is the belief in the general philosophy that *a group is interesting only when it acts on an interesting space, or an associated space is interesting.*

11 Characterization of lattices of semisimple Lie groups

As discussed in the earlier sections, there are characterizations of discrete subgroups as lattices of $\mathbb{R}^n$, of the isometry group $Is(\mathbb{R}^n)$, and of nilpotent and solvable Lie groups.

More specifically, discrete subgroups of $\mathbb{R}^n$ are characterized as free abelian groups of rank less than or equal to n, and they are lattices if and only if their rank is equal to n.

Structures of discrete subgroups of the isometry group $Is(\mathbb{R}^n)$ are more complicated. A particularly important class consists of those discrete subgroups of $Is(\mathbb{R}^n)$ that act cocompactly on $\mathbb{R}^n$. These are called *crystallographic groups*, and their structures are described by the three classical theorems of Bieberbach as mentioned before.

Structures and characterizations of lattices of nilpotent Lie groups are given satisfactorily by several theorems of Malcev [118], and discrete subgroups of nilpotent Lie groups can also be described. Lattices of solvable Lie groups can also be understood fairly well.

On the other hand, there is no characterization of lattices of semisimple Lie groups in terms of group theoretic properties or other related geometric properties. Therefore, an important open problem is the next.

Problem 11.1. *Give characterizations of lattices of semisimple Lie groups in the spirit of characterizations of lattices of nilpotent Lie groups and solvable Lie groups, maybe by combining group theoretical and geometric conditions.*

This is a vague conjecture and we will clarify the meaning of the group theoretical and geometric conditions below.

There has been a lot of work in directions related to this problem. Assume that a lattice Γ of a semisimple Lie group is torsion-free. Then it acts properly and fixed point freely on the symmetric space of noncompact type X, and Γ can be identified with the fundamental group of the manifold $\Gamma\backslash X$. $\Gamma\backslash X$ is a complete nonpositively curved Riemannian manifold of finite volume. A lot of work has been done on the questions:

1. Suppose M is a compact nonpositively curved Riemannian manifold or more generally a complete nonpositively curved Riemannian manifold of finite volume. How to characterize the fundamental group $\pi_1(M)$ combinatorially and intrinsically in terms of groups (in the sense of geometric group theory, for example)?
2. Among all complete nonpositively curved Riemannian manifolds of finite volume, how to characterize locally symmetric spaces?

A natural class of groups that contains the fundamental group of compact nonpositively curved Riemannian manifolds consists of CAT(0)-groups. We note that locally symmetric spaces of noncompact type with finite volume of rank at least 2 can be characterized among nonpositively curved irreducible[7] Riemannian manifolds in terms of higher rank, i.e., every geodesic is contained in an infinitesimal flat strip of dimension at least 2. This is the rank rigidity of nonpositively curved Riemannian manifolds due to Ballmann [10] and Burns-Spatzier [34].

Note that both symmetric spaces of noncompact type and Euclidean buildings provide important examples of CAT(0)-spaces, and groups acting properly on them and their products with compact quotients provide important examples of CAT(0)-groups, which are defined to be discrete groups that act properly on CAT(0)-spaces with compact quotients.

By combining these two results together, it is natural to propose the following conjecture.

Conjecture 11.2. *Suppose that Γ is an irreducible CAT(0)-discrete group, and acts properly and fixed point freely on a contractible Riemannian manifold $\tilde{M}^n$ with compact quotient, where n is equal to the virtual cohomology dimension of Γ. Assume that every element γ of Γ is contained in a free abelian group of rank at least 2. Then Γ is isomorphic to a uniform lattice of a semisimple Lie group G of rank at least 2.*

Recall that a group Γ is called irreducible if no finite index subgroup Γ' is the product of two infinite groups. This assumption is important. Otherwise, as explained below, the above conjecture will be false by taking products of fundamental group of compact negatively curved Riemannian manifolds which are not homeomorphic to locally symmetric spaces.

[7] A manifold is called irreducible if no finite cover splits isometrically as a product.

Proposition 11.3. *If M is a compact Riemannian manifold with strictly negative sectional curvature and is not homeomorphic to a locally symmetric space of negative curvature, then $\pi_1(M)$ is not isomorphic to a lattice of a semisimple Lie group.*

Proof. If $\pi_1(M)$ is isomorphic to a lattice Γ of a semisimple Lie group G, then M is homotopy equivalent to $\Gamma\backslash X$, where $X = G/K$ is a symmetric space of the noncompact type. Since M is aspherical, by the positive solution of the Borel conjecture for compact locally symmetric spaces in [58], M is homeomorphic to $\Gamma\backslash X$. Then the Mostow strong rigidity theorem implies that X is a rank 1 symmetric space, i.e., a symmetric space of negative curvature. This contradicts the assumption.

There have been a lot of work on compact Riemannian or Kahler manifolds which are negatively curved but not diffeomorphic (or biholomorphic) to locally symmetric spaces [50, 59, 72, 134]. The manifolds in [59] are homeomorphic to hyperbolic manifolds. It seems that the manifolds in [50, 72, 134] are probably not homeomorphic to locally symmetric spaces. If this is the case, they will provide examples that show that the irreducibility condition on Γ and the higher rank assumption in Conjecture 11.2 are necessary as asserted above. In any case, the recent paper [138, Corollary 5] contains closed strictly negatively curved Riemannian manifolds which are not homeomorphic to locally symmetric spaces and hence their fundamental group provides the desired examples.

To obtain a characterization of nonuniform lattices of semisimple Lie groups, we can replace in Conjecture 11.2 the compactness assumption of the quotient $\Gamma\backslash\tilde{M}^n$ by finite volume.

To explain Conjecture 11.2, we note the following result.

Proposition 11.4. *Under the assumption of Conjecture 11.2, assume further that the Riemannian manifold $\tilde{M}^n$ is nonpositively curved. Then Γ is isomorphic to a uniform lattice of a semisimple Lie group G of rank at least 2.*

Proof. This is a restatement of a result of [14, Theorem C]. In fact, the assumption of Conjecture 11.2 implies that the rank of Γ is at least 2. Then [14, Theorem C] implies that $\Gamma\backslash\tilde{M}^n$ is a locally symmetric space and hence Γ is isomorphic to a lattice in $Is(\tilde{M}^n)$.

In Conjecture 11.2, the condition that Γ is a CAT(0)-group is a group theoretical condition, since the class of CAT(0)-group is natural and has been intensively studied in geometric group theory. It is certainly a necessary condition but has a geometric flavor. It will be nice to replace it with a more algebraic condition. If we remove that assumption that Γ acts cocompactly on a contractible Riemann manifold, then lattices of p-adic semisimple Lie groups satisfy the other conditions in the conjecture but are not isomorphic to lattices of semisimple Lie groups.

Therefore, this condition is to exclude lattices of p-adic Lie groups.

It seems that Conjecture 11.2 imposes too many conditions on the group Γ. On the other hand, since lattices of semisimple Lie groups enjoy a lot of rigidity properties, it seems important or even necessary to impose more conditions as in the above conjecture in comparison with the characterizations of lattices of nilpotent and solvable Lie groups discussed above.

If we want to characterize lattices of products of semisimple real and p-adic Lie groups, we can weaken the condition on $\tilde{M}^n$ in Conjecture 11.2 as follows.

Conjecture 11.5. *Suppose that Γ is an irreducible $CAT(0)$-discrete group, and acts properly and fixed point freely on a contractible geodesic complete CAT(0)-space $\tilde{M}$ with compact quotient. Assume that every element γ of Γ is contained in a free abelian group of rank at least 2. Then Γ is isomorphic to a uniform lattice of a product of a semisimple Lie group and a p-adic Lie group.*

We note that this class of discrete groups of semisimple Lie groups and a p-adic Lie groups is very natural and arises from S-arithmetic subgroups of linear algebraic groups over number fields or over function fields.

A closely related problem concerns an extension of the rank rigidity of closed manifolds of nonpositive curvature to metric spaces of nonpositive curvature. Some results are obtained in the papers [12, 13]. For example, one result in [13] says: If X is a three-dimensional, geodesically complete, simply connected Euclidean polyhedron of nonpositive curvature which has rank at least 2 in the sense that every geodesic is contained in a half flat of dimension at least 2 and admits a properly discontinuous and cocompact action by a group Γ of isometries, then X is a Riemannian product or a thick Euclidean building of type $\tilde{A}_3$ or $\tilde{B}_3$. In particular, if Γ is irreducible, then Γ is isomorphic to a lattice of a Lie group over a locally compact group.

A conjecture characterizing compact quotients of Euclidean buildings in the higher dimension is formulated in [11]. The rank is defined in terms of half flats as above. Some recent results have been obtained in [36].

A slightly different formulation of the conjecture where the rank is defined in terms of finite width flat strips (i.e., every geodesic is contained in the middle of a flat strip) is formulated in [83], and this conjecture is true when the dimension of the space X is equal to the rank.

12 Bounded generation and rigidity results of arithmetic subgroups

Finite generation of a group is a basic property of discrete groups in the framework of geometric group theory. A closely related, but less known notion is the bounded generation of a group.

It is a *much stronger property* and has several important implications, including rigidity properties of the group.

Recall that a group Γ is called *boundedly generated* if there exists a finite subset of elements S and a positive integer m such that every element γ of Γ is of the form

$$\gamma = s_1^{k_1} \cdots s_m^{k_m}, \quad s_1, \ldots, s_m \in S.$$

Though the subscript m is bounded, the powers k_i are not when Γ is an infinite group.

It was first proved that $SL(n, \mathbb{Z})$ is boundedly generated when $n \geq 3$ in [37], motivated by the role of elementary matrices played in the congruence subgroup problem, in the work of Bass-Milnor-Serre. For example, one fact is that every element of $SL(n, \mathbb{Z})$ is generated by elementary matrices. The question in [37] asked whether there is a uniform bound on the number of elementary matrices which are needed in this statement. A positive answer was given in [37] (see [128] for an exposition).

Since then, the problem of bounded generation has been proved for a limited number of groups such as arithmetic subgroups of Chevalley groups of higher rank.

In the paper [16], the congruence subgroup property was used to prove superrigidity of lattices. The bounded generation of S-arithmetic groups implies the congruence subgroup problem for S-arithmetic subgroups under a suitable assumption (see [143, Theorem 5] for a statement and references.) The bounded generation of lattices is also closely related to Kazhdan's property [156]. See also the book [20] for more references.

The bounded generation property is not true for uniform lattices of rank 1 semisimple Lie groups. On the other hand, it is expected that the following problem might have a positive answer.

Problem 12.1. *Prove that an irreducible lattice Γ of a semisimple Lie group G of rank at least 2 is boundedly generated.*

As mentioned before, when G is the real locus of a Chevalley group and Γ is an arithmetic subgroup, then Γ is bounded generated. The other cases, in particular, when the lattice is uniform, are expected to be more complicated.

13 Simplicial volumes of locally symmetric spaces

The simplicial volume of oriented manifolds was introduced by Gromov to give a low bound of the minimal volume of the manifolds and to serve as a homotopy invariant for manifolds of odd dimension when the Euler characteristic is zero. For a complete hyperbolic manifold of finite volume, its simplicial volume is proportional to the hyperbolic volume, and this fact was used by Gromov to give a new proof of the Mostow strong rigidity of hyperbolic manifolds of dimension at least 3.

Briefly, if M^n is a compact oriented manifold, and $[M]$ is the fundamental class in $H_n(M, \mathbb{R})$, then the simplicial volume $||M||$ of M is defined to be the ℓ_1-norm of the class $[M]$, which is the infimum of ℓ_1-norm of all n-cocycles with coefficients in $\mathbb{R}$ that represent $[M]$. If M is noncompact, then we can use the locally finite homology group and the locally finite fundamental class. Since the ℓ_1-chains are dual to ℓ_∞-cochains, the simplicial volume is closely related to bounded cohomology.

In general it is not easy to compute the simplicial volume of manifolds. Among locally symmetric spaces $\Gamma\backslash X$ of noncompact type with finite volume, the following is known:

1. If $\Gamma\backslash X$ is a hyperbolic manifold, then the simplicial volume $||\Gamma\backslash X|| > 0$.
2. If $\Gamma\backslash X$ is compact, then the simplicial volume $||\Gamma\backslash X|| > 0$.
3. If $\Gamma\backslash X$ is noncompact and its $\mathbb{Q}$-rank is greater than or equal to 3, then the simplicial volume $||\Gamma\backslash X|| = 0$.

By the $\mathbb{Q}$-rank of $\Gamma\backslash X$, we meant the $\mathbb{Q}$-rank of the linear algebraic group $\mathbf{G}$ defined over $\mathbb{Q}$ which defines the arithmetic group Γ. For example, for arithmetic subgroups $\Gamma \subset SL(n, \mathbb{Z})$, the $\mathbb{Q}$-rank of $\Gamma\backslash SL(n, \mathbb{R})/SO(n)$ is equal to $n-1$. It is also known that when $\Gamma\backslash X$ is a certain $\mathbb{Q}$-rank 1 locally symmetric space of finite volume such as the Hilbert modular varieties, its simplicial volume $||\Gamma\backslash X|| > 0$.

See the article [110] for a nice introduction to simplicial volume and references on the above and some other known results.

On the other hand, no result is known about the simplicial volume of locally symmetric spaces of $\mathbb{Q}$-rank 2. The difficulty is that the infinity of $\Gamma\backslash X$ is connected in this case, but is not simply connected at infinity. Its topology is not as simple as in the $\mathbb{Q}$-rank 1 case either. (We note that when the $\mathbb{Q}$-rank of $\Gamma\backslash X$ is equal to 1, each end of $\Gamma\backslash X$ is homeomorphic to a cylinder.)

Problem 13.1. *Determine whether the simplicial volume of a $\mathbb{Q}$-rank 2 locally symmetric space $\Gamma\backslash X$ is positive or equal to 0.*

The answer is probably positive for the following consideration. For such a problem, the global topology near infinity is crucial and whether the arithmetic subgroup Γ or the locally symmetric space $\Gamma\backslash X$ is irreducible or not is probably not important. Suppose that $\Gamma\backslash X$ is the product of two $\mathbb{Q}$-rank 1 locally symmetric spaces of positive simplicial volume. Then the $\mathbb{Q}$-rank of $\Gamma\backslash X$ is equal to 2 and its simplicial volume is positive. A naturally related problem is to prove that the simplicial volume of a $\mathbb{Q}$-rank 1 locally symmetric space is positive.

It is known that the simplicial volume of a compact locally symmetric space $\Gamma\backslash X$ is proportional to the volume of $\Gamma\backslash X$, and the proportional constant only depends on the symmetric space X. When $X = \mathbf{H}^n$, the real hyperbolic space, the constant is equal to the volume of the ideal geodesic simplex in $\mathbf{H}^n$. A natural problem is to compute explicitly this constant in terms of the basic data that determine the Lie group $G = Is(X)$ such as its root system.

14 Large scale geometry of infinite discrete groups and lattices of Lie groups

A basic point of view of geometric group theory is to put metrics on groups and study groups as metric spaces.

For a finitely generated group Γ, each symmetric finite generating set S defines a left invariant metric d_S on Γ. It is known that two different generating sets give quasi-isometric metrics.

Due to this non-canonical nature of metrics, it is natural to consider large scale (or quasi-isometric) geometric invariants of groups endowed with word metrics, (Γ, d_S). This subject was initiated by Gromov (see [70]).

Among various notions, we mention the asymptotic dimension of (Γ, d_S) and the asymptotic cone $Cone_\infty(\Gamma, d_S)$.

Briefly, we recall that the definition of asymptotic dimension. First, given a non-compact metric space (M, d), its asymptotic dimension is defined as the smallest integer n, which could be ∞, such that for every $r > 0$, there exists a cover $\mathcal{C} = \{U_i\}_{i \in I}$ of M by uniformly bounded sets U_i with the r-multiplicity less than or equal to $n+1$, i.e., every ball in M of radius r intersects at most $n+1$ sets in $\mathcal{C}$. (Note that this is similar to the definition of dimension of a topological space via coverings of smaller and smaller sets). Clearly, this is a quasi-isometric invariant of (M, d). The asymptotic dimension of a finitely generated group Γ is defined to be the asymptotic dimension of (Γ, d_S).

One important application is that if the asymptotic dimension of Γ is finite, it implies several versions of the Novikov conjectures for Γ. See [82] for references.

For every $\varepsilon > 0$, scale the word metric d_S to get a new pointed metric space $(\Gamma, e, \varepsilon d_S)$, where e is the identity element of Γ. Then limits of the family of pointed metric spaces $\lim_{\varepsilon \to 0}(\Gamma, e, \varepsilon d_S)$, which always exist by taking limits over ultrafilters, are metric cones and called the *asymptotic cone* of Γ, denoted by $Cone_\infty(\Gamma, d_S)$. Asymptotic cones may not be unique and depend on the choice of the ultrafilters.

The asymptotic cone of groups was first used by Gromov to prove his celebrated characterization of virtual nilpotent groups via polynomial growth. The asymptotic cones of metric spaces can be defined similarly and have played an important role in understanding rigidity properties of symmetric spaces of noncompact types and Euclidean buildings. See [83] for references.

Naturally it is interesting and important to compute these large scale invariants for infinite discrete groups. It turns out that they are not completely known for lattices of Lie groups, or even for arithmetic subgroups of semisimple Lie groups.

Problem 14.1. *Compute the asymptotic dimension of a lattice Γ of a semisimple Lie group G.*

It can be shown easily that Γ is of finite asymptotic dimension, which is bounded by the dimension of the symmetric space $X = G/K$, where $K \subset G$ is a maximal compact subgroup. If the lattice Γ is uniform, then the asymptotic dimension of Γ is equal to the dimension of X. On the other hand, there is a low bound for it by the virtual cohomological dimension of Γ, which should be the right dimension. But proving the equality does not seem to be easy. See [82] for references.

Problem 14.2. *Determine the asymptotic cone of a lattice of a semisimple Lie group G.*

If the lattice Γ is uniform, then (Γ, d_S) is quasi-isometric to the symmetric space X associated with G, and $Cone_\infty(\Gamma, d_S)$ is equal to $Cone_\infty(X)$, which is a $\mathbb{R}$-building. According to [99, Theorem 1.1], when X contains an irreducible factor of rank at least 2, then the uniqueness of asymptotic cones $Cone_\infty(\Gamma, d_S)$ depends on the validity of the Continuum Hypothesis. On the other hand, if the rank X is equal to 1, then $Cone_\infty(\Gamma, d_S)$ is unique and equal to a homogeneous $\mathbb{R}$-tree of continuum degree at every point [54].

Let r be the $\mathbb{R}$-rank r of G, or equivalently the rank of the symmetric space X. Then every asymptotic cone $Cone_\infty(X)$ is a union of $\mathbb{R}^r$. One difference between the cases when $r \geq 2$ and $r = 1$ is that in the former case, $Cone_\infty(X)$ does not contain any *cut point*, i.e., a point whose complement is disconnected. On the other hand, when $r = 1$, every point of $Cone(X)$ is a cut point.

Therefore, for a *uniform lattice* Γ of a semisimple Lie group G, when the real rank $r \geq 2$, $Cone_\infty(\Gamma, d_S)$ does not contain any cut point, and when the real rank $r = 1$, every point of $Cone_\infty(\Gamma, d_S)$ is a cut point.

If Γ is a non-uniform lattice of G, structures of $Cone_\infty(\Gamma, d_S)$ are not clear. When $\Gamma \subset SL(2, \mathbb{R})$, Γ contains a free group as a subgroup of finite index, and $Cone_\infty(\Gamma, d_S)$ is a $\mathbb{R}$-tree of continuum degree at every point [54].

One natural problem is to study existence of cut points of $Cone_\infty(\Gamma, d_S)$ for nonuniform lattices Γ of G. When the real rank r of G is equal to 1, then Γ is a relatively hyperbolic group, and by [53], $Cone_\infty(\Gamma, d_S)$ is a tree-graded space and contains cut points. On the other hand, results of [52, Corollary 5.13, Theorem 6.1] show that when $r \geq 2$ and the $\mathbb{Q}$-rank of Γ is equal to 1, or when Γ is an arithmetic subgroup (or more generally S-arithmetic subgroup) of $SL(n, \mathbb{R})$ with respect to the standard $\mathbb{Q}$-structure and $n \geq 3$ (when the set of finite primes in S is nonempty, $n \geq 2$ is sufficient), then $Cone_\infty(\Gamma, d_S)$ has no cut point. Based on this, the following assertion is natural.

Problem 14.3. *If Γ is a lattice of a semisimple Lie group G of real rank at least 2, then $Cone_\infty(\Gamma, d_S)$ has no cut-point.*

Another natural problem is to understand local and global structures of $Cone_\infty(\Gamma, d_S)$. For example, when Γ is a uniform lattice, $Cone_\infty(\Gamma, d_S)$ is a $\mathbb{R}$-building and contains many Euclidean subspaces $\mathbb{R}^r$ which correspond to special subgroups of Γ, and intersections of these subspaces give a rich structure to $Cone_\infty(\Gamma, d_S)$. A non-uniform lattice Γ also contains many interesting subgroups. One can also try to understand influence of these subgroups on structures of $Cone_\infty(\Gamma, d_S)$.

More precise results on asymptotic cones are known for lattices of nilpotent Lie groups and solvable Lie groups. For example, a finitely generated group Γ is virtually a lattice of a nilpotent Lie group if and only if Γ has polynomial growth, and if and only if all asymptotic cones $Cone_\infty(\Gamma, d_S)$ are proper metric spaces. Furthermore, in this case, $Cone_\infty(\Gamma, d_S)$ is a graded nilpotent Lie group with a Carnot-Caratheodory metric. Results for solvable lattices are less complete. See [51, Sections 5, 6] for references. (It is known that if one asymptotic cone $Cone_\infty(\Gamma, d_S)$ is a proper metric space, then all other asymptotic cones $Cone_\infty(\Gamma, d_S)$ are also proper metric spaces and hence Γ is of polynomial growth. But this is a much more subtle and recent result due to Hrushovski [76] (or rather can be derived from it).)

Besides the large scale geometric properties of infinite groups, we can also consider large scale properties related to analysis.

For any finitely generated group Γ, we can define random walks and harmonic functions on the group Γ, and then the Martin boundary and Poisson boundary of the group. They depend on the choice of a probability measure whose support generates the group.

Problem 14.4. *Study the long term behaviors of random walks, identify the Martin boundary and the Poisson boundary of a lattice Γ of a semisimple Lie group G.*

The Martin boundary and Poisson boundary were originally first defined for domains in $\mathbb{R}^n$ and then for complete Riemannian manifolds. They are known for symmetric spaces of noncompact type.

If Γ is a uniform lattice of G, one can probably solve the above problems by using the known results for symmetric spaces.

But the situation is not so clear for non-uniform lattices. It seems that not much work has been done on them, especially for lattices of higher rank semisimple Lie groups. For free groups or hyperbolic groups, a lot of work has been done on the above problems and there are satisfactory answers. See the books [74, 163].

For Kleinian groups and discrete groups acting on rank 1 symmetric spaces of noncompact type, there is an important theory, called the Patterson-Sullivan theory, which relates measures on the limit set of the discrete group to other properties of the group and locally symmetric spaces. See the book [136] and the paper [147] for details and references.

Another related problem concerning the geometry and analysis at infinity is the following.

Problem 14.5. *Suppose X is a symmetric space of noncompact type and of rank at least 2 and Γ is a discrete group acting isometrically and properly on X. Pick a suitable boundary of X and establish an analogue of Patterson-Sullivan theory.*

Unlike the hyperbolic space or a rank 1 symmetric space, X has many different compactifications and hence admits several different boundaries. The first step is to choose the right boundary. It might not matter since the interesting Patterson-Sullivan measures might be supported only on certain parts of the boundaries. Some partial results have been obtained. See [6, 108, 109] and references therein.

There are other global aspects of arithmetic subgroups and lattices of Lie groups, such as the cohomological dimension and duality properties, that have motivated problems and results for other groups in geometric group theory such as the mapping class groups, the automorphism groups $Out(F_n)$, CAT(0)-groups etc.

Problem 14.6. *Formulate and solve the above problems and related ones for other important classes of groups in geometric group theory such as the mapping class group of surfaces and $Out(F_n)$.*

This is a vaguely stated problem on purpose. One concrete version asks about determination of the asymptotic cones of the mapping class groups of surfaces. Several results have been obtained in [18, 19]. It seems that not much results are known about the asymptotic cones of $Out(F_n)$.

In geometric group theory, the emphasis has been on the geometric and topological properties of groups. Notions related to analysis such as the Martin compactification, Poisson boundary and structures of harmonic functions are also worthwhile to be explored. The Poisson boundary of the mapping class groups is determined in [92]. The more refined Martin boundary and Martin compactification of mapping class groups seem to be unidentified.

It seems that the Poisson boundary of $Out(F_n)$ is not known, and the Martin compactification and Martin boundary have not been studied either.

The Poisson boundary and Martin boundary are known for Gromov hyperbolic groups [8]. Given what is known about the Poisson boundary and Martin boundary of symmetric spaces of noncompact type of higher rank, it is probably a subtle question to *determine the Poisson and Martin boundary of $CAT(0)$-groups.* We can also ask these questions for Coxeter groups, which have played an important role in geometric group theory and topology [44].

15 Measure rigidity of lattices, and quantum unique ergodicity

Homogeneous spaces provide a rich class of dynamical systems and results from ergodic theory can be applied effectively to solve problems in number theory.

Let $\Gamma \subset G$ be an arithmetic subgroup of a semisimple Lie group. Then the right G-multiplication on the homogeneous space $\Gamma\backslash G$ preserves the invariant measure. In particular, every subgroup H of G defines a dynamical system.

A general problem is to understand orbits and their closures of such actions. Since there are too many subgroups H of G, it is natural to assume that H is a closed subgroup.

Problem 15.1. *Identify and classify closure of the orbits of closed subgroups H in $\Gamma\backslash G$.*

This is a well-known problem and has been actively pursued after Margulis proved a conjecture of Raghunathan, which states that the closure of such an orbit Hx for some special subgroup H is equal to the orbit of another (larger) group. Such results for special subgroups H often have far reaching consequences. For example, Margulis proved the important Oppenheim conjecture in number theory in this way [121]. When H is a unipotent subgroup, this is the Ratner theory. See the book [129] for details and references.

For special subgroups H of G, another important, related known problem is to understand measures on $\Gamma\backslash G$ which are invariant under H.

Problem 15.2. *Understand conditions on subgroups $H \subset G$ so that H-invariant measures on $\Gamma\backslash G$ can be classified, i.e., they should correspond to the Haar measure on an orbit of an intermediate group I, $H \subset I \subset G$.*

One special case of this problem concerns the quantum unique ergodicity of the geodesic flow of the locally symmetric space $\Gamma\backslash G/K$.

Solutions to these problems have important applications, and they are being actively pursued by various people. For more precise and detailed description of the above problems and their applications, see [55, 120, 151].

16 Rigidity of locally symmetric spaces and generalizations

Locally symmetric spaces $\Gamma\backslash X$ of noncompact type with finite volume are quotients of symmetric spaces of noncompact type $X = G/K$ by lattices Γ in G. Such a symmetric space X is an important example of CAT(0)-spaces. Another important example of CAT(0)-space is an Euclidean building, in particular Bruhat-Tits buildings of p-adic semisimple algebraic groups.

As recalled before, a natural and important generalization of arithmetic subgroups consists of S-arithmetic subgroups. Let $\mathbf{G}$ be a linear semisimple algebraic group defined over $\mathbb{Q}$, and Γ_S be an S-arithmetic subgroup of $\mathbf{G}(\mathbb{Q})$. Then Γ_S acts properly on a product X_S of symmetric spaces and Bruhat-Tits buildings. The quotient of $\Gamma_S \backslash X_S$ is a natural and important space arising from a special CAT(0)-space.

The rank rigidity gives a nice characterization of locally symmetric spaces $\Gamma\backslash X$ of finite volume among complete nonpositively curved Riemannian manifolds of finite volume (see [157] for statements and references.) One important and natural problem is to characterize the above quotients $\Gamma_S\backslash X_S$ among quotients of CAT(0)-spaces.

Problem 16.1. *Characterize quotients $\Gamma_S\backslash X_S$ of products of symmetric spaces and Bruhat-Tits (or Euclidean) buildings by S-arithmetic subgroups among quotients of $CAT(0)$-spaces.*

One naive guess is that if M is an irreducible compact geodesically complete geodesic metric space such that its universal covering space $\tilde{M}$ is a CAT(0)-space and satisfies some higher rank assumption, then $\tilde{M}$ is a product of symmetric spaces and Euclidean buildings.

But there are subtle differences between contractible Riemannian manifolds and CAT(0)-geodesic metric spaces. For example, in a Riemannian manifold, two different geodesics can not coincide over any small interval. It seems that a direct generalization of the proof for the rank rigidity of locally symmetric space does not work.

There are many other types of rigidity properties associated with locally symmetric spaces (see the surveys [60, 157]).

The geodesic flow of manifolds is a classical and important dynamical system. For example, the geodesic flow of a compact hyperbolic surface $\Gamma\backslash\mathbf{H}^2$ had motivated many important concepts and results for general dynamical systems.

The geodesic flow of locally symmetric spaces $\Gamma\backslash X$ has been studied from various points of views, and the conjugacy rigidity of the geodesic flow has been proved (see [43] for a summary). A natural problem is to extend the geodesic flow to more general quotients of CAT(0)-spaces and prove the conjugacy rigidity.

Problem 16.2. *Let $\Gamma_S\backslash X_S$ be a quotient of a product of a symmetric space and Bruhat-Tits buildings. Define a natural class of geodesic metric spaces for which the geodesic flow is defined and show that for any space in this class, if its geodesic flow is conjugate to the geodesic flow of $\Gamma_S\backslash X_S$, then it is isomorphic to it. (Naturally it is assumed that the rank of X_S is at least 2 and Γ_S is irreducible.)*

It seems that if a geodesic metric space is locally isometric to a product of a Riemannian manifold and a polyhedral complex, then the geodesic flow is defined, and the above conjugacy rigidity of geodesic flow might hold.

When X is a Hermitian symmetric space, in particular, the unit disk in $\mathbb{C}^n$, or products of such disks, there are many other problems on rigidity of X and their quotients (Hermitian locally symmetric spaces). See the book [127] and the papers [77, 78] and references there.

17 Nonexistence of short geodesics and length spectrum rigidity

Geodesics as basic geometric and global invariants of Riemannian manifolds. They also correspond to orbits of the geodesic flow.

For compact Riemannian manifolds, lengths of closed geodesics provide important invariants of the manifolds. For example, the injectivity radius of a compact Riemannian manifold is equal to the half length of the shortest closed geodesics.

For hyperbolic surfaces, the collar theorem describes the geometry of the surface near a simple closed geodesic well and is a basic tool in understanding the geometry of the surface.

The collar theorem for hyperbolic surfaces is a special case of the Margulis Lemma for negatively curved manifolds. As it is known, the Margulis lemma is a fundamental tool in describing the geometry and topology of nonpositively curved Riemannian manifolds through the thick and thin decomposition of the manifolds (especially the detailed description of the thin part).

It is also known that the moduli space of hyperbolic Riemann surfaces of genus at least 2 is noncompact and the reason is the existence of hyperbolic surfaces with arbitrarily short closed geodesics. On the other hand, by the Mostow strong rigidity, the moduli space of compact (or finite volume) locally symmetric spaces is trivial.

Margulis proved in [119, (4.21), p. 322] that *if G is a semisimple Lie group without compact factor, then there exists a neighborhood U of the identity element of G such that for every irreducible non-cocompact lattice $\Gamma_0 \subset G$, the intersection $\Gamma_0 \cap U$ consists of unipotent elements.* Then he made the following conjecture [119, (4.21)(B), p. 322].

Conjecture 17.1. *Let G is a semisimple Lie group of real rank at least 2 (i.e., its associated symmetric space $X = G/K$ of noncompact type has rank at least 2) and does not contain any compact factor. Then there exists a neighborhood U of the identity element of G such that for every irreducible cocompact lattice $\Gamma_0 \subset G$, the intersection $\Gamma_0 \cap U$ consists of elements of finite order.*

When Γ is a torsion-free lattice, this conjecture implies the following: *If X is a symmetric space of noncompact type of rank at least 2, and there exists a positive constant $c = c(G)$ such that for every torsion-free irreducible cocompact* *lattice acting on X, the locally symmetric space $\Gamma\backslash X$ does not admit a closed geodesic of length less than c.*

This conjecture of Margulis is related to and stronger than the Kazhdan-Margulis Theorem on uniform low bounds on volumes of locally symmetric spaces.

As mentioned earlier, it is easy to construct hyperbolic surfaces with arbitrarily short geodesics. The cusp closing theorem of Thurston allows one to construct via Dehn surgery compact hyperbolic 3-manifolds with arbitrarily short geodesics from complete hyperbolic 3-manifolds of finite volume. Compact hyperbolic 4-manifolds with short geodesics are constructed in [3].

It is perhaps helpful to point out that for noncompact locally symmetric spaces of finite volume, its injectivity radius is equal to 0, and it always has a thin part near infinity, but this does not imply that there are short closed geodesics. The above conjecture essentially says that all thin parts of irreducible locally symmetric spaces of rank at least 2 are contained in neighborhoods of the infinity.

Besides the shortest length of closed geodesics, lengths of other geodesics are also important and interesting. They form the length spectrum. Many questions about them are still open.

It is known that the length spectrum of a locally symmetric space does not determine it up to isometry.

On the other hand, the following problem might have a positive solution.

Problem 17.2. *Suppose that M_t, $t \in [0,1]$, is a family of compact Riemannian manifolds with the same length spectrum, and M_0 is a locally symmetric space $\Gamma\backslash X$. Prove that M_t are isometric to M_0 for all t.*

One reason for believing in this assertion is that the length spectrum of locally symmetric spaces are special. For example, for a hyperbolic surface $\Gamma\backslash\mathbf{H}^2$, its length spectrum defines the Selberg zeta function, which satisfies a functional equation. This is certainly not conceivable for a compact surface of variable negative curvature.

This leads one to believe in the following assertion.

Problem 17.3. *If a compact Riemannian manifold M has the same length spectrum as a locally symmetric space $\Gamma\backslash X$, then M is a locally symmetric space.*

As mentioned before, there are nonisometric locally symmetric spaces which have the same length spectrum [144]. Therefore, the above assertion is the optimal. Combining with the deformation rigidity of locally symmetric spaces which are not hyperbolic surfaces and the only finitely many hyperbolic surfaces can have the same length spectrum, one can see that the assertion in Problem 17.2 is reasonable.

18 Representation theory of lattices of Lie groups and cohomology varieties

One reason why groups are useful is that they appear in representation theory, i.e., they act as linear transformations on vector spaces.

Representation theory of finite groups was first studied by the founders of representation theories of groups. Later, representations of compact Lie groups were studied, for example, one central result is the Peter-Weyl theorem. Representations of both finite groups and compact Lie groups have many important applications.

Representations of noncompact Lie groups have been intensively studied but are still not completely understood.

One natural problem is to *study representations of infinite discrete groups, in particular, of lattices of semisimple Lie groups.*

This is in general a very difficult problem even when we restrict to finite dimensional representations or unitary representations.

If Γ is a lattice of a semisimple Lie group G, we naturally hope to reduce or relate representations of Γ to representations of G. If this is true, then we can use the better understood representations of Lie groups, and the problem has become easier. In some sense, this is the basic spirit of the super-rigidity of lattices of semisimple Lie groups.

Remark 18.1. Super-rigidity of lattices started with results for lattices of nilpotent Lie groups and solvable groups. The notion of algebraic hulls for such lattices allows one to pass from lattices to Lie groups. See [118] for algebraic hulls of nilpotent groups and [131] for lattices of solvable Lie groups.

Another instance which shows close relation between representations of a lattice Γ and its ambient Lie group G is concerned with the Kazhdan property (T).

Basically, a group Γ satisfies property (T) if the trivial representation is isolated in the space of all unitary representations of Γ which is equipped with the Fell topology. This means that if Γ acts on a Hilbert space unitarily with an almost fixed point, then it has a fixed point. It is known that a lattice Γ of G has property (T) if and only if the Lie group G has property (T) (see the book [20] for detailed definitions of property (T) and many applications).

One notion concerning finite dimensional representations of a finitely generated group Γ is the SS-rigidity. By definition, Γ is called *SS-rigid if for every integer n, there are only finitely many equivalent classes of representations of Γ of dimension n.* It is known that some arithmetic subgroups Γ are SS-rigid.

Proposition 18.2. *If a group Γ is boundedly generated and satisfies the Kazhdan property (T), then it is SS-rigid.*

Proof. By [149, Theorem 2 in Section 3], if a group Γ is boundedly generated and satisfies the condition that for every subgroup $\Gamma_1 \subset \Gamma$ of finite index, the abelianization $\Gamma_1/[\Gamma_1, \Gamma_1]$ is finite, then Γ is SS-rigid. Since this latter condition is satisfied if Γ satisfies the Kazhdan property (T), the proposition is proved.

A natural problem is the next one.

Problem 18.3. *Prove that lattices of semisimple Lie groups are SS-rigid.*

When Γ is an irreducible lattice of a semisimple Lie group of rank at least 2, then it satisfies the Kazhdan property (T). As mentioned before in Problem 12.1, Γ is also expected to be boundedly generated, and hence Γ is expected to be SS-rigid in this case. The situation when Γ is a lattice of a semisimple Lie group of rank 1 is less clear, and the answer could be no.

For finite groups, there are close connections between representation theories and cohomology groups. One basic result for finite groups is the Venkov-Evens theorem which says that *the cohomology ring of a finite group is finitely generated,* and this result is the foundation of the cohomology variety of finite groups and support variety of modules of finite groups, which are all defined in terms of the ring of cohomology of finite groups. Cohomology varieties and support varieties are basic constructions of representation theories of finite groups. See the book [23] for details.

For a large class of infinite discrete subgroups Γ including lattices of semisimple Lie groups, the cohomology rings $H^i(\Gamma, \mathbb{Z})$ are finitely generated (see [87] for a precise statement, related finiteness results and references).

Problem 18.4. *Develop representation theories of lattices of semisimple Lie groups by following the representation theories of finite groups, in particular, the notions of cohomology varieties and support varieties for representation theories of finite groups.*

One reason for posing this problem is that infinite groups such as arithmetic subgroups share various common things with finite groups. For example, starting with a linear algebraic group $\mathbf{G} \subset GL(n, \mathbb{C})$ defined over $\mathbb{Z}$, we can get the Lie group $G = G(\mathbb{R})$, an arithmetic subgroup $\mathbf{G}(\mathbb{Z})$ and finite groups $\mathbf{G}(\mathbb{Z}/p\mathbb{Z})$. As it is known that most finite simple groups arise from $\mathbf{G}(\mathbb{Z}/p\mathbb{Z})$ and their variants. As we emphasized before, the embedding of $\mathbf{G}(\mathbb{Z})$ into G has played an important role in understanding $\mathbf{G}(\mathbb{Z})$. There is a natural map $\mathbf{G}(\mathbb{Z}) \to \mathbf{G}(\mathbb{Z}/p\mathbb{Z})$, whose kernel gives the congruence subgroups of $\mathbf{G}(\mathbb{Z})$.

Some results in this direction for representations of infinite discrete groups have been obtained and discussed in [24].

There are another classes of problems related to representations of Γ in connection with character varieties. As mentioned before, for any discrete group Γ and a Lie group G, we can consider all homomorphisms $\Gamma \to G$. A lot of work has been done. For example, the uniformization theorem of Riemann surfaces allows one to study Teichmüller spaces of surfaces by the character variety for $G = SL(2, \mathbb{R})$. There is also a lot of recent work on higher Teichmüller theories. Many problems are still open. See the discussion related to Problem 7.8 above and the papers [93] [33] for more details and references.

19 Counting representations of lattices of Lie groups

For a compact semisimple Lie group, its finite dimensional representations are classified by the theory of highest weights. Their dimension can be read off from the highest weights. If one is interested in counting the number of representations in terms of their dimensions, its asymptotic behavior can be obtained.

Suppose that Problem 18.3 on SS-rigidity for lattices of Lie groups has a positive solution. (As explained there, it is natural to expect that an irreducible lattice of a semisimple Lie group of higher rank is SS-rigid.) Then there are only finitely many non-isomorphic representations in each dimension, and a natural problem is to count representations of a lattice according to dimension. One can count them by defining a Dirichlet series using the number of representations in each dimension.

Problem 19.1. *Count finite dimensional representations of lattices of semisimple Lie groups.*

It is natural to form a Dirichlet series using the number of representations in each dimension and to understand its properties, such as whether it converges when the imaginary part of the complex variable is large enough, and whether it admits a meromorphic continuation. If the answers to the above questions are positive, we can also ask whether it satisfies a functional equation.

Lubotzky has some recents on this problem for higher rank arithmetic subgroups such as $SL(n, \mathbb{Z})$.

One can also ask similar questions for other important infinite groups.

Problem 19.2. *Prove the SS-rigidity and count finite dimensional representations of mapping class groups of surfaces and $Out(F_n)$.*

This might be more difficult or false. The mapping class groups and $Out(F_n)$ are not boundedly generated [57, Theorem 1.3, Theorem 3.5].

20 Nonlinear actions of lattices and the Zimmer program

A finite dimensional representation of a group Γ, $\rho : \Gamma \to GL(n, \mathbb{R})$, is an action of Γ on the vector space $\mathbb{R}^n$ by linear transformations.

Then a natural extension of representations of Γ consists of actions of Γ on manifolds. Corresponding to classification of representations of Γ, *a natural problem is to understand and classifications of* Γ *on manifolds, in particular on compact manifolds.*

If Γ is a lattice of a semisimple Lie group $\mathbf{G}$, then it acts naturally on homogeneous spaces of G. If Γ is an arithmetic subgroup, then there is another type of actions. For example, $SL(n, \mathbb{Z})$ acts on the flat torus $\mathbb{Z}^n \backslash \mathbb{R}^n$ via the standard action of $SL(n, \mathbb{Z})$ on $\mathbb{R}^n$. Then representations of $\mathbf{G}$ defined over $\mathbb{Q}$ induce action of arithmetic subgroups of $\mathbf{G}(\mathbb{Q})$ on torus.

The Zimmer program aims to understand and classify actions of arithmetic subgroups of higher rank semisimple Lie groups from many different perspectives.

Basically, there are two types of problems: (1) classification of ergodic actions on compact manifolds preserving some geometric structures such as affine connections and volume forms etc in terms of known algebraic actions, (2) rigidity of algebraic actions of lattices on suitable geometric structures.

Depending the geometric structures that are preserved by there actions, there could be restrictions on the dimension of the space relative to the rank of the Lie group.

In the simplest formulation, the general problem can be formulated *roughly* as follows: *For a lattice* Γ *of a semisimple Lie group* G*, determine or classify spaces where* Γ *acts nontrivially. For known actions of* Γ *on manifolds, understand their special properties.*

As stated, this general problem is too vague and too broad. To make it more concrete and precise, we state several sample conjectures in some special cases in the Zimmer program, following [61, pp. 76–77].

The first conjecture concerns the question of what kind of spaces a lattice can act *nontrivially* (to be interpreted suitably).

Conjecture 20.1. *Assume that* $n > 2$ *and* $\Gamma \subset SL(n, \mathbb{R})$ *is a lattice. For every smooth manifold* M *with* $\dim M < n - 1$*, every homomorphism* $\rho : \Gamma \to \mathrm{Diff}(M)$ *has finite image.*

The next two conjectures concern a classification of actions of lattices on manifolds which preserve some rigid geometric structures. Imposing the condition of preserving additional geometric structures is important. Otherwise, classification of actions is impossible.

Conjecture 20.2. *Let $\Gamma \subset SL(n, \mathbb{R})$ be a lattice. Then there is a classification of actions of Γ on compact manifolds that preserve both a volume form and an affine connection in terms of some algebraically defined actions (for example, the action of $SL(n, \mathbb{Z})$ on $\mathbb{Z}^n \backslash \mathbb{R}^n$ is an algebraically defined).*

Conjecture 20.3. *Let $\Gamma \subset SL(n, \mathbb{R})$ be a lattice. Then there is a classification of actions of Γ on compact manifolds that preserve both a volume form and where one element of Γ acts as an Anosov diffeomorphism in terms of some algebraically defined actions.*

The Zimmer program has been actively studied by many people. There are many conjectures and problems, and precise formulations for general lattices are complicated too. For example, in discussing rigidity of actions, one natural problem requires clarification of what geometric structures are rigid. For detailed descriptions of the many conjectures and open problems, and some known results, see [61, 167].

21 Family of discrete groups, not necessarily in towers, or family of locally symmetric spaces

Locally symmetric spaces $\Gamma \backslash X$ are either rigid or have a rich deformation theory. The former corresponds to the Mostow strong rigidity, and the latter corresponds to the Teichmüller space or moduli spaces of hyperbolic (or Riemann surfaces).

For hyperbolic surfaces, understanding behaviors of the geometry and analysis of hyperbolic surfaces when the surfaces go to the boundary of the moduli spaces is an important and interesting problem.

For rigid locally symmetric spaces and also for non-rigid hyperbolic surfaces, another way to producing varying families of locally symmetric spaces is to consider towers of them.

More specifically, for a fixed symmetric space X of noncompact type, consider a shrinking family of lattices $\Gamma_1 \supset \Gamma_2 \cdots \supset \Gamma_n \supset \cdots$. Then their corresponding locally symmetric spaces $\Gamma_1 \backslash X, \dots, \Gamma_n \backslash X$ form a tower of covering spaces.

Problem 21.1. *Assume that the tower is infinite, i.e., there is an infinite sequence of subgroups $\Gamma_1 \supset \cdots \supset \Gamma_n \supset \cdots$, understand limits of their geometric and analytic invariants as $n \to +\infty$.*

We can also consider a family of locally symmetric spaces which are quotients of X but do not necessarily form a tower.

There are many questions we can ask, for example, about the growth of the Betti number with respect to the volume, the growth of the size of torsion elements in homology groups with respect to the volume, the convergence of the spectral measure, convergence of the heat kernel, geometric conditions to ensure that the first positive eigenvalue of the Laplace operator has a uniform lower bound. We can also ask questions on comparing properties of two different towers. These questions and results will shed light on both locally symmetric spaces $\Gamma\backslash X$ and the symmetric space X. See [2, 25, 113, 154] for example, for some results and references.

Another type of problems concerns comparison of locally symmetric spaces by going to finite covers. The philosophy is that this extra flexibility of going to many possible finite covers will allow one to gain more control on geometry and topology.

Problem 21.2. *Given two locally symmetric spaces $\Gamma_1\backslash X, \Gamma_2\backslash X$, decide whether there are finite covers which are close to each other in terms of geometrical, topological, or analytical properties.*

This is a vague question and needs clarification. More precisely, one can ask if they admit finite covers which are close in the sense of Gromov-Hausdorff distance or quasi-isometry. One can also ask whether the Betti numbers of their covers are close to each other in a suitable sense, for example, the ratio of the difference over the volumes. In terms of analysis, one can ask about behaviors of the small eigenvalues and the trace of the heat kernel. Since the spectrum of $\Gamma'\backslash X$ will become dense in the continuous spectrum of the symmetric space X when Γ' becomes smaller in a sequence of subgroups that approaches the trivial group (for example, a decreasing sequence of lattices that converges to the trivial group), one can also ask how the distribution of eigenvalues of finite covers of $\Gamma_1\backslash X, \Gamma_2\backslash X$ compare.

One reason for believing in this assertion is that they have the same limit when considering coverings, i.e., the same universal covering space, and that they should have some close finite covers before reaching the limit. See [90] for some *positive* results when $\Gamma\backslash X$ is a hyperbolic surface.

Due to the Mostow strong rigidity for other locally symmetric spaces of finite volume, the answer is expected to be negative when Γ_1, Γ_2 are not commensurable up to conjugation. One possible argument is as follows: suppose that $\Gamma_1\backslash X, \Gamma_2\backslash X$ are two compact locally symmetric spaces, which are not hyperbolic surfaces. Suppose that for every positive constant ε, there exist finite covers $\Gamma_1'\backslash X, \Gamma_2'\backslash X$ such that the Hausdorff distance between them is less than ε. Then it should be true that $\Gamma_1'\backslash X$ and $\Gamma_2'\backslash X$ are homotopy equivalent and the Mostow strong rigidity implies that $\Gamma_1'\backslash X, \Gamma_2'\backslash X$ are isometric.

One motivation for posing the above problem is that locally symmetric spaces which are quotients of the same symmetric spaces need to be studied together. This is of course in the spirit of Hecke correspondences in number theory and automorphic forms.

22 Analysis of locally symmetric spaces, and relation to the geometry of the locally symmetric spaces

Among all Riemannian manifolds, locally symmetric spaces are special both in terms of the geometry of the space and the spectral theory of the Laplace operator. For example, for a noncompact finite volume complete Riemannian manifold with sectional curvature bounded, there is in general little information available about its spectrum, for example, whether there is nonempty continuous spectrum. On the other hand, for a locally symmetric space $\Gamma\backslash X$ of finite volume, its continuous spectrum is described satisfactorily by the theory of Eisenstein series.

One natural and important problem is to understand relations between the rich geometry and spectral theory of locally symmetric spaces. One important example is the Selberg trace formula, which can be considered as a non-abelian generalization of the Poisson summation formula. In fact, the Poisson summation formula can be considered as the trace formula for the simplest 1-dimensional compact Riemannian manifold, $\ell\mathbb{Z}\backslash\mathbb{R}$, where ℓ is a positive number and is equal to the length of the shortest geodesic of the manifold $\ell\mathbb{Z}\backslash\mathbb{R}$. Specifically, the Poisson summation formula is an identity between quantities depending on the lengths $n\ell$ of closed geodesics of the circle $\mathbb{R}/\ell\mathbb{Z}$ and quantities depending on the eigenvalues $\frac{4}{\ell^2}\pi^2 n^2$ of the Laplace operator of the circle $\mathbb{R}/\ell\mathbb{Z}$.

For hyperbolic surfaces $\Gamma\backslash\mathbf{H}^2$ of finite area, the Selberg trace formula relates the lengths of closed geodesics of the hyperbolic surface to the eigenvalues of the Laplace operator of the hyperbolic surface $\Gamma\backslash\mathbf{H}^2$.

Such identities have far reaching consequences for both the geometry and spectral theory of the hyperbolic surfaces, for example, the distribution of the length spectrum and of the eigenvalues.

A natural problem is to establish similar identities for locally symmetric spaces of finite volume, in particular, those of higher rank.

There are indeed a version of Selberg trace formula by Arthur for $\Gamma\backslash X$ when Γ is a congruence subgroup, which was designed mainly for understanding the Langlands program, in particular the principle of functoriality (see [9] for a summary). The geometric side is in terms of orbital integrals, and their geometric meaning in terms of the locally symmetric space is not so clear.

After the initial work of Arthur on the trace formula, one major effort was to stabilize the trace formula so that one can compare the geometric side of trace formula of two groups in order to transfer the spectral side, hence proving the principle of functoriality. The recent breakthrough of Ngo provides the foundation for the stabilization of the trace formula (see [39] for details and references). This is of course very important. But information about the geometry of $\Gamma\backslash X$ is completely lost, and it is difficult to use the stable trace formula to understand the geometry of $\Gamma\backslash X$, for example, the distribution of flat tori and other submanifolds according to their volume.

Problem 22.1. *Obtain more geometric versions of the Arthur-Selberg trace formula, maybe more in the style of the original Selberg trace formula, and derive geometric consequences, such as distribution of special submanifolds, about locally symmetric spaces from them.*

There have been some work on the geometry of locally symmetric spaces in the spirit of Selberg trace formula and Selberg zeta function, but nothing comparable with the Selberg trace formula for hyperbolic surfaces $\Gamma\backslash\mathbf{H}^2$ has been obtained. See the papers [46, 47, 64, 65, 75, 130] and references therein.

The spectral theory of locally symmetric spaces, or the spectral theory of automorphic forms, is a huge and important subject. A lot is known about them, but there are also many other outstanding problems. For example, explicitly computing (or obtaining good bounds on) eigenvalues of $\Gamma\backslash X$ is difficult and important. We can mention the Selberg $\frac{1}{4}$-conjecture for congruence subgroups of $SL(2,\mathbb{Z})$ and the Ramanujan conjecture for the general case. Constructing locally symmetric spaces with desirable spectral data is also difficult and important. See [82] for more references and questions.

There is another generalization of the Poisson summation formula in the direction of Poisson relation, which relates the singularities of the Fourier transform of the spectral measure to the structure of closed geodesics. One such generalization of the Poisson relation for $\mathbb{Q}$-rank 1 locally symmetric spaces was obtained in [88]. One important geometric notion needed is the so-called sojourn time of scattering geodesics, which are certain geodesics going to infinity in both directions (more precisely, both ends are distance minimizing eventually), and the sojourn time can be considered as a normalized length of such geodesics. One natural problem is to generalize it to the higher $\mathbb{Q}$-rank locally symmetric spaces.

Problem 22.2. *For a locally symmetric space of $\mathbb{Q}$-rank at least 2, define scattering flats, generalized vector-valued sojourn times, and establish a Poisson relation.*

Products of two $\mathbb{Q}$-rank 1 locally symmetric spaces might suggest a correct definition of sojourn times and a right formulation of the Poisson relation.

23 Spectral rigidity of locally symmetric spaces

The spectrum of the Laplace operator of locally symmetric spaces $\Gamma\backslash X$ can be studied via representation theories of G and automorphic forms, and hence it has rich structures.

Given the close relation between the length spectrum and the spectrum of compact locally symmetric spaces, and rigidity problems for locally symmetric spaces from the perspective of lengths of closed geodesics, a natural type of question concerns how special the eigenvalues of locally symmetric spaces are.

There are known nonisometric locally symmetric spaces which are isospectral, i.e., have the same spectrum [143]. On the other hand, the next assertion seems possible.

Problem 23.1. *If a compact Riemannian manifold M is isospectral to a locally symmetric space $\Gamma\backslash X$, i.e., they have the same spectrum of the Laplace operator, then M is also a locally symmetric space, which is a quotient of the same symmetric space X.*

Some results on this problem has been obtained in [166]. If the above problem has a positive answer, then the next problem seems reasonable.

Problem 23.2. *Suppose that a family of compact Riemannian manifolds $M_t, t \in [0,1]$, has the same spectrum, and M_0 is a locally symmetric space $\Gamma\backslash X$, then for all t, M_i is a locally symmetric space and is isomorphic to $M_0 = \Gamma\backslash X$.*

For a general compact Riemannian manifold M, the Laplace operator has a discrete spectrum, and a zeta function can be defined in terms of these eigenvalues. Unlike the zeta functions that arise from number theory and arithmetic algebraic geometry, these zeta functions usually do not enjoy many good properties, for example, they do not satisfy functional equations.

For hyperbolic surfaces $\Gamma\backslash\mathbf{H}^2$, the Selberg zeta function can also be expressed in terms of the eigenvalues of the hyperbolic surfaces, and this shows that the zeta function defined by the eigenvalues of hyperbolic surfaces enjoys good properties.

Problem 23.3. *Study properties of the zeta functions defined in terms of eigenvalues of compact locally symmetric spaces.*

24 Moduli interpretation of locally symmetric spaces

An important application of locally symmetric spaces is that some important moduli spaces in algebraic geometry and number theory are given by them, usually by arithmetic Hermitian locally symmetric spaces, also called arithmetic locally symmetric varieties.

For example, the moduli space of elliptic curves (or Riemann surfaces of genus 1) is equal to $SL(2,\mathbb{Z})\backslash\mathbb{H}^2$, and this gives it the name "the modular curve". The moduli space of Abelian varieties of dimension g is equal to $Sp(2g,\mathbb{Z})\backslash\mathfrak{h}_g$, where $\mathfrak{h}_g$ is the Siegel upper half space of genus g and is an important Hermitian symmetric space of noncompact type.

Another important locally symmetric space which can be realized as a moduli space is the space $SL(n,\mathbb{Z})\backslash SL(n,\mathbb{R})/SO(n)$, which is the moduli space of equivalence classes of positive definite quadratic forms in n variables of determinant 1, or equivalently, the moduli space of compact flat manifolds of dimension n of volume 1. These moduli interpretations of $SL(n,\mathbb{Z})\backslash SL(n,\mathbb{R})/SO(n)$ are the basic setup of the geometry of numbers. The philosophy is clear: once every point of a locally symmetric space $\Gamma\backslash X$ can be interpreted as an equivalence class of some geometric or algebraic objects, we can use these objects to understand structures of the locally symmetric space. For example, we can give invariants to each point in the locally symmetric space. Conversely, structures of locally symmetric spaces can be used to study these objects.

A very important problem is the next one. We note that when Γ is an arithmetic subgroup of a semisimple Lie group G, the associated locally symmetric space $\Gamma\backslash X$ is called an arithmetic locally symmetric space.

Problem 24.1. *Given an arithmetic locally symmetric space $\Gamma\backslash X$, interpret it as the moduli space of some geometric or algebraic objects.*

In the above problem, it is natural to expect that Γ should be a congruence subgroup, since in most known examples, this is the case, and the level of the congruence subgroup corresponds to some level or other structures on the objects under consideration.

Besides the above moduli interpretation of the locally symmetric space $SL(n,\mathbb{Z})\backslash SL(n,\mathbb{R})/SO(n)$ in terms of compact flat manifolds and the Siegel modular variety $Sp(2g,\mathbb{Z})\backslash\mathfrak{h}_g$ in terms of principally polarized abelian varieties of dimension g, there are many others. For example, the book [79] discusses moduli spaces of several interesting configurations in algebraic geometry and their identification with arithmetic locally symmetric varieties.

According to [48, p. 248, the third paragraph], the above problem is related to a dream of Deligne. In his own words, "To interpret more complex Hodge structures, one would like to replace abelian varieties by suitable 'motives', but this is still a dream." (This French translation was provided by Eduard Looijenga.) Basically, moduli spaces of abelian varieties or abelian varieties satisfying certain conditions (for example, with some restrictions on their endomorphism rings) will not exhaust all locally symmetric spaces, and we need to extend the objects. Deligne suggested that moduli spaces of suitable motives might give moduli interpretation of all arithmetic locally symmetric varieties.

When $\Gamma\backslash X$ is noncompact, it admits many natural compactifications such as the Satake compactifications, the Borel-Serre compactification, which is a real analytic manifold with corners. When $\Gamma\backslash X$ is also Hermitian, it admits more compactifications: the Baily-Borel compactification, which is a normal projective variety, and the toroidal compactification, which is an algebraic variety with mild singularities (many of them are smooth and hence give explicit resolutions of the singularities of the Baily-Borel compactification).

Another related problem is to give a moduli interpretation of such compactifications.

Problem 24.2. *Suppose* $\Gamma\backslash X$ *is a noncompact arithmetic locally symmetric space which is a moduli space of some classes of objects. For a natural compactification of* $\Gamma\backslash X$*, identify its boundary points with some degenerated objects, i.e., give a modular interpretation of the boundary points.*

See the paper [7] for some results and references.

25 Cohomology of arithmetic groups and geometric cycles

Cohomology groups of locally symmetric spaces $\Gamma\backslash X$ have been intensively studied by many people. They are closely related to existence of and other properties of automorphic forms and automorphic representations. See the book [28] and the survey article [153] for overviews and references.

The connection with automorphic forms gives rich structures to cohomology groups of $\Gamma\backslash X$. For example, the theory of Eisenstein series gives rise to Eisenstein cohomology classes.

One natural and basic problem is

Problem 25.1. *Understand and construct cohomology classes of* $\Gamma\backslash X$ *geometrically.*

One natural geometric construction is to take cycles generated by sub-locally symmetric spaces. It is not necessarily true that they represent nontrivial cycles, or cohomology classes. But these cycles do not seem to provide enough classes to generate cohomology groups.

For algebraic varieties, the Hodge conjecture gives a clean characterization of cohomology classes coming from subvarieties. One natural problem is to characterize the cohomology classes of $\Gamma\backslash X$ coming from sub-locally symmetric spaces or the subspaces spanned by them.

Vanishing and nonvanishing of cohomology and homology of locally symmetric spaces have important implications. As mentioned before, they often imply existence or nonexistence of certain automorphic forms, which is one of the major and subtle problems in automorphic forms. But there are other applications. For example, the infinitesimal rigidity of a locally symmetric space $\Gamma\backslash X$ is related to vanishing of the first cohomology of $\Gamma\backslash X$ with suitable coefficients. (The reason is that the tangent space of an associated character variety is computed by such a cohomology group. This is also reasonable in view of the Kodaira-Spencer deformation theory for complex manifolds).

If $\Gamma\backslash X$ is a hyperbolic 3-manifold $\Gamma\backslash\mathbf{H}^3$, an important conjecture of Thurston, which is recently proved, asserts that a finite cover has the nonzero first Betti number in order to get suitable surfaces inside the manifold to make it a Haken manifold. One can also ask about nonvanishing of the first Betti number of finite covers of rank 1 locally symmetric spaces. We note that for locally symmetric spaces of higher rank, the Kazhdan property (T) implies the vanishing of the first Betti number.

26 Spines of symmetric spaces and classifying spaces of arithmetic groups

If Γ is a torsion-free lattice of a semisimple Lie group G, then Γ acts properly and fixed point freely on the symmetric space $X = G/K$. Since X is contractible, X is the universal space for proper and fixed point free actions of Γ, i.e., the universal covering space $E\Gamma$ of the classifying space $B\Gamma$, and $\Gamma\backslash X$ is a classifying space of Γ:

$$\pi_1(\Gamma\backslash X) = \Gamma, \quad \pi_i(\Gamma\backslash X) = \{1\}, \text{ for } i \geq 2.$$

We note that models of $B\Gamma$ are unique up to homotopy equivalence. Explicit models of $B\Gamma$ are important for many applications. For example, cohomology of Γ can be computed by

$$H^i(\Gamma) = H^i(B\Gamma).$$

For many applications, it is important to find compact models of $B\Gamma$, or those models which are given by finite CW-complexes, called finite models. It is also a natural problem to construct models of $B\Gamma$ of the smallest possible dimension, which is equal to the cohomological dimension of Γ.

If $\Gamma\backslash X$ is noncompact, then $\Gamma\backslash X$ is not a finite model of $B\Gamma$. There are two ways to construct finite models. One way is to construct a compactification $\overline{\Gamma\backslash X}$ of $\Gamma\backslash X$ such that the inclusion $\Gamma\backslash X \to \overline{\Gamma\backslash X}$ is a homotopy equivalence. The Borel-Serre compactification satisfies this property. But in this case, the cohomological dimension of Γ is equal to $\dim \Gamma\backslash X - r_{\mathbb{Q}}$, where $r_{\mathbb{Q}}$ is the $\mathbb{Q}$-rank of the linear algebraic group $\mathbf{G}$ which defines the $\mathbb{Q}$-structure, and is hence strictly smaller than $\dim \Gamma\backslash X$.

Another approach is to find a submanifold $(\Gamma\backslash X)_T$ such that the inclusion $(\Gamma\backslash X)_T \to \Gamma\backslash X$ is a homotopy equivalence. When $\Gamma\backslash X$ is a hyperbolic surface, then $(\Gamma\backslash X)_T$ can be obtained by cutting off cusp neighborhoods. For a general locally symmetric space, the truncated subspace $(\Gamma\backslash X)_T$ can also be constructed. In this case, $(\Gamma\backslash X)_T$ is a compact submanifold of codimension 0, whose dimension is strictly greater than the cohomological dimension of Γ. See [80] for references on these results.

One natural problem is the following.

Problem 26.1. *Find a compact subspace S of $\Gamma\backslash X$ which is a deformation retract of $\Gamma\backslash X$ and is of dimension equal to* $\dim \Gamma\backslash X - r_{\mathbb{Q}}$.

Such a subspace S is called *a spine of $\Gamma\backslash X$ of optimal dimension.* In terms of the symmetric space X, the inverse image of S in X is a Γ-equivariant deformation retraction of X and is also called a *spine* (or an equivariant spine) of X.

When $X = \mathbf{H}^2$, such spines are known and they are given by embedded trees invariant under Γ. They also exist for special symmetric spaces such as linear symmetric spaces (a typical example of linear symmetric space is $GL(n,\mathbb{R})/O(n)$, the space of positive definite symmetric matrices, or $SL(n,\mathbb{R})/SO(n)$), and spines can be constructed by the well-rounded deformation retraction or linear simplicial decomposition, and for symmetric spaces X when the $\mathbb{Q}$-rank of $\Gamma\backslash X$ is equal to 1. See [165] for more information and references.

If Γ contains nontrivial torsion elements, then $\Gamma\backslash X$ is not a model of classifying space of Γ. For example, when $\Gamma = SL(2,\mathbb{Z})$ and $X = \mathbb{H}^2$, $SL(2,\mathbb{Z})\backslash\mathbf{H}^2$ is simply connected. In fact, according to a general result, every model of $B\Gamma$ is infinite dimensional whenever Γ contains nontrivial torsion elements.

In this case, X is a universal space for proper actions of Γ, i.e, Γ acts properly on X and for every finite subgroup $F \subset \Gamma$, the set of fixed points X^F is nonempty and contractible. In other words, X is a model of $\underline{E}\Gamma$. Good models of $\underline{E}\Gamma$ are important for understanding many problems in geometric topology about Γ such as the Novikov conjectures and the Baum-Connes conjecture. One useful criterion for a good model of $\underline{E}\Gamma$ is that $\underline{E}\Gamma$ is a Γ-CW complex such that $\Gamma\backslash\underline{E}\Gamma$ is a finite CW-complex. Such models are called *cofinite models* of $\underline{E}\Gamma$. The Borel-Serre partial compactification of X is a cofinite model $\underline{E}\Gamma$ for arithmetic subgroups Γ (see [84] for more information and references).

As in the special torsion-free cases mentioned above, we can also find Γ-equivariant deformation retractions W of X such that $\Gamma\backslash W$ is compact, which are called *equivariant spines.* The construction by truncation X_T or $(\Gamma\backslash X)_T$ gives spines of codimension 0. The smallest possible dimension of equivariant spines W of X is the virtual cohomology dimension of Γ, which is the cohomological dimension of a finite index torsion-free subgroup of Γ.

A natural and largely open problem for arithmetic groups and symmetric spaces is the following.

Problem 26.2. *For every lattice Γ of a semisimple Lie group G which may contain torsion elements, find a Γ-equivariant spine W of X whose dimension is equal to the virtual cohomological dimension of Γ.*

As in the previous case for torsion-free arithmetic subgroups Γ, such spines of optimal dimension exist for some special symmetric spaces. But it is not known whether they exist for general symmetric spaces.

For a torsion free discrete group Γ, there always exists model of $B\Gamma$ of dimension equal to the cohomological dimension of Γ unless it is equal to 2. But a similar result is not known for models of $\underline{E}\Gamma$ when the groups Γ contain torsion elements. This shows some subtle difference between these two cases.

References

[1] H. Abels, G.A. Margulis, G.A. Soifer, *On the Zariski closure of the linear part of a properly discontinuous group of affine transformations*, J. Differential Geom. 60 (2002), 315–344.

[2] M. Abert, N. Bergeron, I. Biringer, T. Gelander, N. Nikolov, J. Raimbault, I. Samet, *On the growth of Betti numbers of locally symmetric spaces*, arXiv:1104.5559.

[3] I. Agol, *Systoles of hyperbolic 4-manifolds*, arXiv:math/0612290.

[4] I. Agol, *The virtual Haken conjecture*, arXiv:1204.2810.

[5] I. Agol, *Criteria for virtual fibering*, J. Topol. 1 (2008), no. 2, 269–284.

[6] P. Albuquerque, *Patterson-Sullivan theory in higher rank symmetric spaces*, Geom. Funct. Anal. 9 (1999), no. 1, 1–28.

[7] V. Alexeev, *Complete moduli in the presence of semiabelian group action*, Ann. of Math. (2) 155 (2002), 611–708.

[8] A. Ancona, *Théorie du potentiel sur les graphes et les variétés*, 1–112, Lecture Notes in Math., 1427, Springer, Berlin, 1990.

[9] J. Arthur, *An introduction to the trace formula*, in *Harmonic analysis, the trace formula, and Shimura varieties*, 1–263, Clay Math. Proc., 4, Amer. Math. Soc., Providence, RI, 2005.

[10] W. Ballmann, *Nonpositively curved manifolds of higher rank*, Ann. of Math. (2) 122 (1985), no. 3, 597–609.

[11] W. Ballmann, S. Buyalo, *Periodic rank one geodesics in Hadamard spaces*, in *Geometric and probabilistic structures in dynamics*, 19–27, Contemp. Math., 469, Amer. Math. Soc., Providence, RI, 2008.

[12] W. Ballmann, M. Brin, *Orbihedra of nonpositive curvature*, Inst. Hautes Études Sci. Publ. Math. No. 82 (1995), 169–209.

[13] W. Ballmann, M. Brin, *Rank rigidity of Euclidean polyhedra*, Amer. J. Math. 122 (2000), no. 5, 873–885.

[14] W. Ballmann, P. Eberlein, *Fundamental groups of manifolds of nonpositive curvature*, J. Differential Geom. 25 (1987), no. 1, 1–22.

[15] O. Baues, B. Klopsch, *Deformations and rigidity of lattices in solvable Lie groups*, arXiv:1111.5589.

[16] G. Baumslag, A. Myasnikov, V. Shpilrain, *Open problems in combinatorial group theory*. Second edition, *Combinatorial and geometric group theory*

(New York, 2000/Hoboken, NJ, 2001), 1–38, Contemp. Math., 296, Amer. Math. Soc., Providence, RI, 2002.

[17] G. Baumslag, J. Roseblade, *Subgroups of direct products of free groups*, J. London Math. Soc. (2) 30 (1984), no. 1, 44–52.

[18] J. Behrstock, C. Drutu, M. Sapir, *Median structures on asymptotic cones and homomorphisms into mapping class groups*, Proc. Lond. Math. Soc. (3) 102 (2011), no. 3, 503–554.

[19] J. Behrstock, Y. Minsky, *Dimension and rank for mapping class groups*, Ann. of Math. (2) 167 (2008), no. 3, 1055–1077.

[20] B. Bekka, P. de la Harpe, A. Valette, *Kazhdan's property (T)*, New Mathematical Monographs, 11, Cambridge University Press, Cambridge, 2008. xiv+472 pp.

[21] Y. Benoist, *Five lectures on lattices in semisimple Lie groups*, in *Géométries á courbure négative ou nulle, groupes discrets et rigidités*, 117–176, Sémin. Congr., 18, Soc. Math. France, Paris, 2009.

[22] Y. Benoist, H. Oh, *Discreteness criterion for subgroups of products of SL(2)*, Transform. Groups 15 (2010), no. 3, 503–515.

[23] D. Benson, *Representations and cohomology. II. Cohomology of groups and modules*, Second edition. Cambridge Studies in Advanced Mathematics, 31, Cambridge University Press, Cambridge, 1998, xii+279 pp.

[24] D. Benson, *Complexity and varieties for infinite groups. I, II*, J. Algebra 193 (1997), no. 1, 260–287, 288–317.

[25] N. Bergeron, A. Venkatesh, *The asymptotic growth of torsion homology for arithmetic groups*, arXiv:1004.1083.

[26] A. Borel, *Compact Clifford-Klein forms of symmetric spaces*, Topology 2 (1963), 111–122.

[27] A. Borel, L. Ji, *Compactifications of symmetric and locally symmetric spaces*, Mathematics: Theory & Applications, Birkhäuser Boston, Inc., Boston, MA, 2006, xvi+479 pp.

[28] A. Borel, N. Wallach, *Continuous cohomology, discrete subgroups, and representations of reductive groups*, Annals of Mathematics Studies, 94, Princeton University Press, Princeton; University of Tokyo Press, Tokyo, 1980, xvii+388 pp.

[29] B. Bowditch, *Geometrical finiteness for hyperbolic groups*, J. Funct. Anal. 113 (1993), no. 2, 245–317.

[30] U. Bunke, M. Olbrich, *Scattering theory for geometrically finite groups*, in *Geometry, analysis and topology of discrete groups*, 40–136, Adv. Lect. Math. (ALM), 6, Int. Press, Somerville, MA, 2008.

[31] D. Burde, K. Dekimpe, S. Deschamps, *The Auslander conjecture for NIL-affine crystallographic groups*, arXiv:math/0409476.

[32] M. Burger, A. Iozzi, F. Labourie, A. Wienhard, *Maximal representations*

of surface groups: symplectic Anosov structures, Pure Appl. Math. Q. 1 (2005), no. 3, 543–590.

[33] M. Burger, A. Iozzi, A. Wienhard, *Higher Teichmüller spaces: from* $SL(2,\mathbb{R})$ *to other Lie groups*, arXiv:1004.2894.

[34] K. Burns, R. Spatzier, *Manifolds of nonpositive curvature and their buildings*, Inst. Hautes Études Sci. Publ. Math. No. 65 (1987), 35–59.

[35] A. Cano, J. Seade, *An overview on complex Kleinian groups*, arXiv:1110.2674.

[36] P.E. Caprace, M. Sageev, *Rank rigidity for* $CAT(0)$ *cube complexes*, Geom. Funct. Anal. 21 (2011), no. 4, 851–891.

[37] D. Carter, G. Keller, *Bounded elementary generation of* $SL_n(\mathcal{O})$, Amer. J. Math. 105 (1983), no. 3, 673–687.

[38] B. Chandler, W. Magnus, *The history of combinatorial group theory. A case study in the history of ideas*, Studies in the History of Mathematics and Physical Sciences, 9, Springer-Verlag, New York, 1982. viii+234 pp.

[39] L. Clozel, M. Harris, J. P. Labesse, B. C. Ngo, *On the stabilization of the trace formula. Stabilization of the Trace Formula, Shimura Varieties, and Arithmetic Applications, 1*, International Press, Somerville, MA, 2011, xiv+527 pp.

[40] L. Clozel, T.N. Venkataramana, *Restriction of the holomorphic cohomology of a Shimura variety to a smaller Shimura variety*, Duke Math. J. 95 (1998), no. 1, 51–106.

[41] L. Corwin, F. Greenleaf, *Representations of nilpotent Lie groups and their applications. Part I. Basic theory and examples*, Cambridge Studies in Advanced Mathematics, 18, Cambridge University Press, Cambridge, 1990, viii+269 pp.

[42] H.S.M. Coxeter, W.O.J. Moser, *Generators and relations for discrete groups*, Fourth edition. Ergebnisse der Mathematik und ihrer Grenzgebiete, 14, Springer-Verlag, Berlin-New York, 1980. ix+169 pp.

[43] C. Croke, *Rigidity theorems in Riemannian geometry*, in *Geometric methods in inverse problems and PDE control*, pp. 47–72, IMA Vol. Math. Appl., 137, Springer, New York, 2004.

[44] M. Davis, *The geometry and topology of Coxeter groups*, London Mathematical Society Monographs Series, 32, Princeton University Press, Princeton, NJ, 2008. xvi+584 pp.

[45] Y. de Cornulier, *Dimension of asymptotic cones of Lie groups*, J. Topol. 1 (2008), no. 2, 342–361.

[46] A. Deitmar, *Geometric zeta functions of locally symmetric spaces*, Amer. J. Math. 122 (2000), no. 5, 887–926.

[47] A. Deitmar, W. Hoffmann, *Asymptotics of class numbers*, Invent. Math. 160 (2005), no. 3, 647–675.

[48] P. Deligne, *Variétés de Shimura: interprétation modulaire, et techniques de construction de modéles canoniques*, in *Automorphic forms, representations and L-functions*, Proc. Sympos. Pure Math. XXXIII, Oregon State Univ., Corvallis, Ore., 1977, Part 2, pp. 247–289, Amer. Math. Soc., Providence, R.I., 1979.

[49] P. Deligne, G. Mostow, *Monodromy of hypergeometric functions and non-lattice integral monodromy*, Inst. Hautes Études Sci. Publ. Math. No. 63 (1986), 5–89.

[50] M. Deraux, *A negatively curved Kähler threefold not covered by the ball*, Invent. Math. 160 (2005), no. 3, 501-525.

[51] C. Drutu, *Quasi-isometry invariants and asymptotic cones*, Internat. J. Algebra Comput. 12 (2002), 99–135.

[52] C. Drutu, S. Mozes, M. Sapir, *Divergence in lattices in semisimple Lie groups and graphs of groups*, Trans. Amer. Math. Soc. 362 (2010), no. 5, 2451–2505.

[53] C. Drutu, M. Sapir, *Tree-graded spaces and asymptotic cones of groups, With an appendix by Denis Osin and Sapir*, Topology 44 (2005), no. 5, 959–1058.

[54] A. Dyubina, I. Polterovich, *Structures at infinity of hyperbolic spaces*, Russian Math. Surveys 53 (1998), no. 5, 1093–1094

[55] M. Einsiedler, *Applications of measure rigidity of diagonal actions*, Proceedings of the International Congress of Mathematicians. Volume III, 1740–1759, Hindustan Book Agency, New Delhi, 2010

[56] M. Einsiedler, E. Lindenstrauss, P. Michel, A. Venkatesh, *Distribution of periodic torus orbits and Duke's theorem for cubic fields*, Ann. of Math. (2) 173 (2011), 815–885.

[57] B. Farb, A. Lubotzky, Y. Minsky, *Rank-1 phenomena for mapping class groups*, Duke Math. J. 106 (2001), no. 3, 581–597.

[58] F.T. Farrell, L.E. Jones, *Rigidity for aspherical manifolds with* $\pi_1 \subset GL_m(\mathbb{R})$, Asian J. Math. 2 (1998), no. 2, 215–262.

[59] F.T. Farrell, L.E. Jones, *Negatively curved manifolds with exotic smooth structures*, J. Amer. Math. Soc. 2 (1989), 899–908.

[60] D. Fisher, *Local rigidity of group actions: past, present, future*, in *Dynamics, ergodic theory, and geometry*, 45–97, Math. Sci. Res. Inst. Publ., 54, Cambridge Univ. Press, Cambridge, 2007.

[61] D. Fisher, *Groups acting on manifolds: around the Zimmer program*, in *Geometry, rigidity, and group actions*, 72–157, Chicago Lectures in Math., Univ. Chicago Press, Chicago, IL, 2011

[62] R. Fricke, *Ueber die Substitutionsgruppen, welche zu den aus dem Legendre'schen Integralmodul* $k^2(\omega)$ *gezogenen Wurzeln gehören*, Math. Ann. 28 (1887), 99–118.

[63] R. Fricke, F. Klein, *Vorlesungen über die Theorie der automorphen Funktionen. Band 1: Die gruppentheoretischen Grundlagen. Band II: Die funktionentheoretischen Ausführungen und die Andwendungen*, Bibliotheca Mathematica Teubneriana, Bände 3, 4, Johnson Reprint Corp., New York; B. G. Teubner Verlagsgesellschaft, Stuttg art 1965, Band I: xiv+634 pp.; Band II: xiv+668 pp.

[64] R. Gangolli, G. Warner, *Zeta functions of Selberg's type for some noncompact quotients of symmetric spaces of rank one*, Nagoya Math. J. 78 (1980), 1–44.

[65] R. Gangolli, G. Warner, *Zeta functions of Selberg's type for compact space forms of symmetric spaces of rank one*, Illinois J. Math. 21 (1977), no. 1, 1–41.

[66] T. Gelander, *Volume vs. rank of lattices*, J. Reine Angew. Math. 661 (2011), 237–248.

[67] T. Gelander, A. Zuk, *Dependence of Kazhdan constants on generating subsets*, Israel J. Math. 129 (2002), 93–98.

[68] W. Goldman, *Complex hyperbolic geometry,* Oxford Mathematical Monographs, Oxford Science Publications, The Clarendon Press, Oxford University Press, New York, 1999. xx+316 pp.

[69] D. Gromoll, J. A. Wolf, *Some relations between the metric structure and the algebraic structure of the fundamental group in manifolds of nonpositive curvature*, Bull. Amer. Math. Soc. 77 (1971), 545–552.

[70] M. Gromov, *Asymptotic invariants of infinite groups*, in *Geometric group theory, Vol. 2*, pp. 1–295, London Math. Soc, Lecture Note Ser., vol. 182, Cambridge Univ. Press, 1993.

[71] M. Gromov, I. Piateskii-Shapiro, *Nonarithmetic groups in Lobachevsky spaces*, Inst. Hautes Études Sci. Publ. Math. No. 66 (1988), 93–103.

[72] M. Gromov, W. Thurston, *Pinching constants for hyperbolic manifolds*, Invent. Math. 89 (1987), no. 1, 1–12.

[73] O. Guichard, A. Wienhard, *Topological invariants of Anosov representations*, Xiv:0907.0273.

[74] Y. Guivarc'h, L. Ji, J.C. Taylor, *Compactifications of symmetric spaces*, Progress in Mathematics, 156, Birkhäuser Boston, Inc., Boston, MA, 1998. xiv+284 pp.

[75] W. Hoffmann, *Geometric estimates for the trace formula*, Ann. Global Anal. Geom. 34 (2008), no. 3, 233–261.

[76] E. Hrushovski, *Stable group theory and approximate subgroups*, arXiv:0909.2190.

[77] X. Huang, *On some problems in several complex variables and CR geometry*, in *First International Congress of Chinese Mathematicians* (*Beijing, 1998*), 383–396, AMS/IP Stud. Adv. Math., 20, Amer. Math. Soc., Providence, RI,

2001.

[78] X. Huang, *On a linearity problem for proper holomorphic maps between balls in complex spaces of different dimensions*, J. Differential Geom. 51 (1999), no. 1, 13–33.

[79] B. Hunt, *The geometry of some special arithmetic quotients*, Lecture Notes in Mathematics, 1637, Springer-Verlag, Berlin, 1996, xiv+332 pp.

[80] L. Ji, *A tale of two stories: arithmetic groups and mapping class groups*, in *Handbook of Teichmüller Theory*, volume 3, 157–295, Eur. Math. Soc., Zürich, 2012.

[81] L. Ji, *Arithmetic groups and their generalizations. What, why, and how.* AMS/IP Studies in Advanced Mathematics, 43, American Mathematical Society, Providence, RI; International Press, Cambridge, MA, 2008, xviii+259 pp.

[82] L. Ji, *Large scale geometry, compactifications and the integral Novikov conjectures for arithmetic groups*, in *Third International Congress of Chinese Mathematicians*, Part 1, 2, 317–344, AMS/IP Stud. Adv. Math., 42, pt. 1, 2, Amer. Math. Soc., Providence, RI, 2008.

[83] L. Ji, *Buildings and their applications in geometry and topology*, in Differential Qeometry: under the inpluence of S.-S. Chern, ed. by Y. Shen, Z. Shen, S.-T. Yau, ALM 22, pp. 89–210, HZP/IP, 2012.

[84] L. Ji, *Integral Novikov conjectures and arithmetic groups containing torsion elements*, Comm. Anal. Geom. 15 (2007), no. 3, 509–533.

[85] L. Ji, *A summary of the work of Gregory Margulis*, Pure Appl. Math. Q. 4 (2008), no. 1, Special Issue: In honor of Grigory Margulis. Part 2, 1–69.

[86] L. Ji, P. Li, J.P. Wang, *Ends of locally symmetric spaces with maximal bottom spectrum*, J. Reine Angew. Math. 632 (2009), 1–35.

[87] L. Ji, Z. Lin, *Finite generation properties of cohomology rings for infinite groups*, Journal of pure and applied algebra 216 (2012), 1118–1133.

[88] L. Ji, M. Zworski, *Scattering matrices and scattering geodesics of locally symmetric spaces*, Ann. Sci. École Norm. Sup. (4) 34 (2001), no. 3, 441–469.

[89] T. Jorgensen, *On discrete groups of Möbius transformations*, Amer. J. Math. 98 (1976), no. 3, 739–749.

[90] J. Kahn, V. Markovic, *The good pants homology and a proof of the Ehrenpreis conjecture*, Ann. of Math 175 (2012), 1127–1190.

[91] J. Kahn, V. Markovic, *Immersing almost geodesic surfaces in a closed hyperbolic three manifold*, arXiv:0910.5501.

[92] V. Kaimanovich, H. Masur, *The Poisson boundary of the mapping class group*, Invent. Math. 125 (1996), no. 2, 221–264.

[93] M. Kapovich, *Kleinian groups in higher dimensions*, in *Geometry and dynamics of groups and spaces*, 487–564, Progr. Math., 265, Birkhäuser,

Basel, 2008.

[94] M. Kapovich, *Hyperbolic manifolds and discrete groups*, Reprint of the 2001 edition, Modern Birkhäuser Classics, Birkhäuser Boston, Inc., Boston, MA, 2009, xxviii+467 pp.

[95] S. Katok, *Fuchsian groups*, Chicago Lectures in Mathematics, University of Chicago Press, Chicago, IL, 1992, x+175 pp.

[96] F. Klein, *Zur Theorie der elliptischen Modulfunctionen*, Math. Ann, 17 (1880), 62–70.

[97] B. Kleiner, B. Leeb, *Rigidity of invariant convex sets in symmetric spaces*, Invent. Math. 163 (2006), no. 3, 657–676.

[98] T. Kobayashi, T. Yoshino, *Compact Clifford-Klein forms of symmetric spaces–revisited*, Pure Appl. Math. Q. 1 (2005), no. 3, part 2, 591–663.

[99] L. Kramer, S. Shelah, K. Tent, S. Thomas, *Asymptotic cones of finitely presented groups*, Adv. Math. 193 (2005), no. 1, 142–173.

[100] L. Kurdachenko, J. Otal, I. Subbotin, *Groups with prescribed quotient groups and associated module theory*, Series in Algebra, 8, World Scientific Publishing Co., Inc., River Edge, NJ, 2002. xvi+227 pp.

[101] M. Lackenby, *Finite covering spaces of 3-manifolds*, in *Proceedings of the International Congress of Mathematicians*, Volume II, 1042–1070, Hindustan Book Agency, New Delhi, 2010

[102] M. Lackenby, *Surface subgroups of Kleinian groups with torsion*, Invent. Math. 179 (2010), no. 1, 175–190.

[103] M. Lackenby, *Expanders, rank and graphs of groups*, Israel J. Math. 146 (2005), 357–370.

[104] J. Lafont, I. Ortiz, *Relative hyperbolicity, classifying spaces, and lower algebraic K-theory*, Topology 46 (2007), no. 6, 527–553.

[105] J. Lawson, S. T. Yau, *Compact manifolds of nonpositive curvature*, J. Differential Geometry 7 (1972), 211–228.

[106] T. Li, *Rank and genus of 3-manifolds*, arXiv:1106.6302.

[107] C. Li, X. Zhang, *Test map and discreteness criteria for Subgroups in* $PU(1,n;\mathbb{C})$, arXiv:1111.5125.

[108] G. Link, *Hausdorff dimension of limit sets of discrete subgroups of higher rank Lie groups*, Geom. Funct. Anal. 14 (2004), no. 2, 400–432.

[109] G. Link, *Ergodicity of generalised Patterson-Sullivan measures in higher rank symmetric spaces*, Math. Z. 254 (2006), no. 3, 611–625.

[110] C. Löh, *Simplicial volume*, http://www.map.him.uni-bonn.de/Simplicial-volume.

[111] D.D. Long, A. Reid, *Small subgroups of* $SL(3,\mathbb{Z})$, Exp. Math. 20 (2011), no. 4, 412–425.

[112] A. Lubotzky, D. Segal, *Subgroup growth*, Progress in Mathematics, 212, Birkhäuser Verlag, Basel, 2003, xxii+453 pp.

[113] W. Lück, *Approximating L^2-invariants by their finite-dimensional analogues,* Geom. Funct. Anal. 4 (1994), 455–481.

[114] R. Lyndon, P. Schupp, *Combinatorial group theory*, Reprint of the 1977 edition, Classics in Mathematics. Springer-Verlag, Berlin, 2001, xiv+339 pp.

[115] C. Maclachlan, A. Reid, *The arithmetic of hyperbolic 3-manifolds*, Graduate Texts in Mathematics, 219, Springer-Verlag, New York, 2003, xiv+463 pp.

[116] W. Magnus, A. Karrass, D. Solitar, *Combinatorial group theory: Presentations of groups in terms of generators and relations*, Interscience Publishers, New York-London-Sydney, 1966, xii+444 pp.

[117] V.S. Makarov, *Geometric methods of construction of discrete groups of motions of a Lobachevskii space, Itogi Naukii Tekhniki, Seriya Problemy Geometrii*, Vol. 15, 3–59. Moscow, 1983, English translation: J. Sov. Math. 29 (1985), 1519–1550.

[118] A.I. Malcev, *On a class of homogeneous spaces*, Amer. Math. Soc. Translation 1951, (1951), no. 39, 33 pp.

[119] G. Margulis, *Discrete subgroups of semisimple Lie groups*, Ergebnisse der Mathematik und ihrer Grenzgebiete (3), 17, Springer-Verlag, Berlin, 1991, x+388 pp.

[120] G. Margulis, *Problems and conjectures in rigidity theory*, in *Mathematics: frontiers and perspectives*, 161–174, Amer. Math. Soc., Providence, RI, 2000.

[121] G. Margulis, *Oppenheim conjecture*, in *Fields Medallists' lectures*, 272–327, World Sci. Ser. 20th Century Math., 5, World Sci. Publ., River Edge, NJ, 1997.

[122] B. Maskit, *Kleinian groups*, Grundlehren der Mathematischen Wissenschaften, 287, Springer-Verlag, Berlin, 1988. xiv+326 pp.

[123] K.A. Mikhailova, *The occurrence problem for direct products of groups*, Dokl. Akad. Nauk SSSR 119 (1958) 1103–1105.

[124] J. Milnor, *On fundamental groups of complete affinely flat manifolds*, Advances in Math. 25 (1977), no. 2, 178–187.

[125] J. Milnor, *Introduction to algebraic K-theory,* Annals of Mathematics Studies 72, Princeton, N. J., Princeton University Press and University of Tokyo Press, XIII, 1971. xiii+184 pp.

[126] Y. Minsky, *End invariants and the classification of hyperbolic 3-manifolds*, in *Current developments in mathematics*, 2002, 181–217, Int. Press, Somerville, MA, 2003.

[127] N. Mok, *Metric rigidity theorems on Hermitian locally symmetric manifolds*, Series in Pure Mathematics, 6, World Scientific Publishing Co., Inc., Teaneck, NJ, 1989. xiv+278 pp.

[128] D. Morris, *Bounded generation of $SL(n, A)$ (after D. Carter, G. Keller,*

and E. Paige). New York J. Math. 13 (2007), 383–421.

[129] D. Morris, *Ratner's theorems on unipotent flows*, Chicago Lectures in Mathematics, University of Chicago Press, Chicago, IL, 2005, xii+203 pp.

[130] H. Moscovici, R. Stanton, *R-torsion and zeta functions for locally symmetric manifolds*, Invent. Math. 105 (1991), no. 1, 185–216.

[131] G. Mostow, *Representative functions on discrete groups and solvable arithmetic subgroups*, Amer. J. Math. 92 (1970), 1–32.

[132] G. Mostow, *Strong rigidity of locally symmetric spaces*, Annals of Mathematics Studies, No. 78, Princeton University Press, Princeton, N.J.; University of Tokyo Press, Tokyo, 1973. v+195 pp.

[133] G. Mostow, *On a remarkable class of polyhedra in complex hyperbolic space*, Pacific J. Math. 86 (1980), no. 1, 171–276.

[134] G. Mostow, Y.T. Siu, *A compact Kähler surface of negative curvature not covered by the ball*, Ann. of Math. (2) 112 (1980), no. 2, 321–360.

[135] M. Newman, *Normal subgroups of the modular group which are not congruence subgroups*, Proc. Amer. Math. Soc. 16 (1965), 831–832.

[136] P. Nicholls, *The ergodic theory of discrete groups*, London Mathematical Society Lecture Note Series, 143, Cambridge University Press, Cambridge, 1989, xii+221 pp.

[137] M. Olbrich, *L^2-invariants of locally symmetric spaces*, Doc. Math. 7 (2002), 219–237.

[138] P. Ontaneda, *Pinched Smooth Hyperbolization*, arxiv.org/pdf/1110.6374.pdf.

[139] J. Parker, *Complex hyperbolic lattices*, in *Discrete groups and geometric structures*, pp. 1–42, Contemp. Math., 501, Amer. Math. Soc., Providence, RI, 2009.

[140] G. Pick, *Ueber gewisse ganzzahlige lineare Substitutionen, welche sich nitch durch algebraische Congruenzen erkll'aren lassen*, Math. Ann 28 (1887), 119–124.

[141] H. Poincaré, *Papers on Fuchsian functions*, Translated from the French and with an introduction by John Stillwell, Springer-Verlag, New York, 1985, ii+483 pp.

[142] G. Prasad, A. Rapinchuk, *Zariski-dense subgroups and transcendental number theory*, Math. Res. Lett. 12 (2005), 239–249.

[143] G. Prasad, A. Rapinchuk, *Developments on the congruence subgroup problem after the work of Bass, Milnor and Serre*, in *Collected Papers of John Milnor: V. Algebra*, AMS, 2010, pp. 307–326.

[144] G. Prasad, A. Rapinchuk, *Number-theoretic techniques in the theory of Lie groups and differential geometry*, in *Fourth International Congress of Chinese Mathematicians*, 231–250, AMS/IP Stud. Adv. Math., 48, Amer. Math. Soc., Providence, RI, 2010.

[145] M.N. Prokhorov, *Absence of discrete groups of reflections with a noncompact fundamental polyhedron of finite volume in a Lobachevskii space of high dimension*, Math. USSR-Izv. 28 (1987), 401–411.

[146] J.F. Quint, *Groupes convexes cocompacts en rang supérieur*, Geom. Dedicata 113 (2005), 1–19.

[147] J.F. Quint, *An overview of Patterson-Sullivan theory*, Lecture Notes posted at http://www.math.univ-paris13.fr/~quint/publications/courszurich.pdf.

[148] M.S. Raghunathan, *Discrete subgroups of Lie groups*, Ergebnisse der Mathematik und ihrer Grenzgebiete, Band 68, Springer-Verlag, New York-Heidelberg, 1972, ix+227 pp.

[149] A.S. Rapinchuk, *On SS-rigid groups and A. Weil's criterion for local rigidity. I*, Manuscripta Math. 97 (1998), no. 4, 529–543.

[150] I. Reiner, *Normal subgroups of the unimodular group*, Illinois J. Math. 2 (1958), 142–144.

[151] P. Sarnak, *Recent progress on the quantum unique ergodicity conjecture*, Bull. Amer. Math. Soc. (N.S.) 48 (2011), no. 2, 212–228.

[152] P. Sarnak, *Notes on thin groups*, MSRI Lecture notes, February 2012.

[153] J. Schwermer, *Geometric cycles, arithmetic groups and their cohomology*, Bull. Amer. Math. Soc. (N.S.) 47 (2010), no. 2, 187–279.

[154] P. Sarnak, X. Xue, *Bounds for multiplicities of automorphic representations*, Duke Math. J. 64 (1991), 207–227.

[155] R. Sharma, T. Venkataramana, *Generations for arithmetic groups*, Geom. Dedicata 114 (2005), 103–146.

[156] Y. Shalom, *Bounded generation and Kazhdan's property (T)*, Inst. Hautes Etudes Sci. Publ. Math. No. 90 (1999), 145–168 (2001).

[157] R. Spatzier, *An invitation to rigidity theory*, in *Modern dynamical systems and applications*, 211–231, Cambridge Univ. Press, Cambridge, 2004.

[158] C. Soulé, *An introduction to arithmetic groups*, in *Frontiers in number theory, physics, and geometry. II*, 247–276, Springer, Berlin, 2007.

[159] A. Terras, *Harmonic analysis on symmetric spaces and applications. II*, Springer-Verlag, Berlin, 1988, xii+385 pp.

[160] E. Vinberg, *Absence of crystallographic groups of reflections in Lobachevskii spaces of large dimension*, Trans. Moscow Math. Soc 1985 (1985), 75–112.

[161] E. Vinberg, O. Shvartsman, *Discrete groups of motions of spaces of constant curvature,* in *Geometry, II*, pp. 139–248, Encyclopaedia Math. Sci., 29, Springer, Berlin, 1993.

[162] E. Vinberg, V. Gorbatsevich, O. Shvartsman, *Discrete subgroups of Lie groups,* in *Lie groups and Lie algebras. II*, Encyclopaedia of Mathematical Sciences, 21, Springer-Verlag, Berlin, 2000, 1–124.

[163] W. Woess, *Random walks on infinite graphs and groups.* Cambridge Tracts in Mathematics, 138, Cambridge University Press, Cambridge, 2000,

xii+334 pp.

[164] J. Wolf, *Spaces of constant curvature*, Sixth edition. AMS Chelsea Publishing, Providence, RI, 2011. xviii+424 pp.

[165] D. Yasaki, *On the existence of spines for $\mathbb{Q}$-rank 1 groups*, Selecta Math. (N.S.) 12 (2006), 541–564.

[166] S.K. Yeung, *Isospectral problem of locally symmetric spaces*, Int. Math. Res. Not. IMRN (2011), no. 12, 2810–2824

[167] R. Zimmer, D. Morris, *Ergodic theory, groups, and geometry*, CBMS Regional Conference Series in Mathematics, 109, American Mathematical Society, Providence, RI, 2008, x+87 pp.

Open Problems and Surveys of Contemporary Mathematics
SMM 6, pp. 325–342

Motives — Grothendieck's Dream

James S. Milne*

Abstract

Grothendieck introduced the notion of a "motif" in a letter to Serre in 1964. Later he wrote that, among the objects he had been privileged to discover, they were the most charged with mystery and formed perhaps the most powerful instrument of discovery.[1] In this article, I shall explain what motives are, and why Grothendieck valued them so highly.

2000 Mathematics Subject Classification: 11G09, 14G10

Keywords and Phrases: Motives, Weil conjectures, Standard conjectures

Contents

---

*The origin of this article is a lecture I gave at the University of Michigan on February 3, 2009. A Chinese translation of it (by Xu Kejian) was published in Mathematical Advances in Translation, Vol.28, No.3, 193–206, 2009. I thank Xu Kejian for his comments and for suggesting the present title. Email: `jmilne@umich.edu`

[1] Parmi toutes les chose mathématiques que j'avais eu le privilège de découvrir et d'amener au jour, cette réalité des motifs m'apparaît encore comme la plus fascinante, la plus chargée de mystère — au coeur même de l'identité profonde entre la "géométrie" et l' "arithmétique". Et le "yoga des motifs" ... est peut-être le plus puissant instrument de décourverte que j'aie dégagé dans cette première période de ma vie de mathématicien.

Grothendieck, Récoltes et Semailles, Introduction.

1 Cohomology in topology

Attached to a compact manifold X of dimension $2n$, there are cohomology groups

$$H^0(X,\mathbb{Q}),\ldots,H^{2n}(X,\mathbb{Q}),$$

which are finite-dimensional $\mathbb{Q}$-vector spaces satisfying Poincaré duality (H^i is dual to H^{2n-i}), a Lefschetz fixed point theorem, and so on. There are many different ways of defining them — singular cochains, Čech cohomology, derived functors — but the different methods all give exactly the same groups, provided they satisfy the Eilenberg-Steenrod axioms. When X is a complex analytic manifold, there are also the de Rham cohomology groups $H^i_{\text{dR}}(X)$. These are vector spaces over $\mathbb{C}$, but they are not really new because $H^i_{\text{dR}}(X) \simeq H^i(X,\mathbb{Q}) \otimes_{\mathbb{Q}} \mathbb{C}$.[2]

[2] The symbol $\simeq$ denotes a *canonical* isomorphism.

2 Cohomology in algebraic geometry

Now consider a nonsingular projective algebraic variety X of dimension n over an algebraically closed field k. Thus X is defined by polynomials over k, and the conditions mean that, when $k = \mathbb{C}$, the points $X(\mathbb{C})$ of the variety form a compact manifold of dimension $2n$.

Weil's work on the numbers of points on algebraic varieties with coordinates in finite fields led him in 1949 to make his famous "Weil" conjectures "concerning the number of solutions of equations over finite fields and their relation to the topological properties of the varieties defined by the corresponding equation over the field of complex numbers". In particular, he found that the numbers of points seemed to be controlled by the Betti numbers of a similar algebraic variety over $\mathbb{C}$. For example, for a curve X of genus g over the field $\mathbb{F}_p = \mathbb{Z}/p\mathbb{Z}$ of p elements, the number $|X(\mathbb{F}_p)|$ of points on the curve satisfies the inequality

$$||X(\mathbb{F}_p)| - p - 1| \leqslant 2gp^{\frac{1}{2}}, \quad g = \text{genus of } X. \tag{2.1}$$

Weil was able to predict the Betti numbers of certain hypersurfaces over $\mathbb{C}$ by counting the numbers of points on a hypersurface of the same dimension and degree over $\mathbb{F}_p$, and his predictions were confirmed by Dolbeault. It was clear that most of Weil's conjectures would follow from the existence of a cohomology theory for algebraic varieties with good properties ($\mathbb{Q}$ coefficients, correct Betti numbers, Poincaré duality theorem, Lefschetz fixed point theorem, and so on). In fact, as we shall see, no such cohomology theory exists with $\mathbb{Q}$ coefficients, but in the following years attempts were made to find a good cohomology theory with coefficients in some field of characteristic zero other than $\mathbb{Q}$. Eventually, in the 1960s Grothendieck defined étale cohomology and crystalline cohomology, and showed that the algebraically defined de Rham cohomology has good properties in characteristic zero. The problem then became that we had too many good cohomology theories!

Besides the usual valuation on $\mathbb{Q}$, there is another valuation for each prime number ℓ defined by

$$|\ell^r \tfrac{m}{n}| = 1/\ell^r, \quad m, n \in \mathbb{Z} \text{ and not divisible by } \ell.$$

Each valuation makes $\mathbb{Q}$ into a metric space, and on completing it we obtain fields $\mathbb{Q}_2$, $\mathbb{Q}_3$, $\mathbb{Q}_5$, ..., $\mathbb{R}$. For each prime number ℓ distinct from the characteristic of k, étale cohomology gives cohomology groups

$$H^0_{\text{et}}(X, \mathbb{Q}_\ell), \ldots, H^{2n}_{\text{et}}(X, \mathbb{Q}_\ell)$$

which are finite-dimensional vector spaces over $\mathbb{Q}_\ell$ and satisfy Poincaré duality, a Lefschetz fixed point formula, and so on. Also, there are de Rham groups $H^i_{\text{dR}}(X)$, which are finite-dimensional vector spaces over k, and in characteristic $p \neq 0$, there are crystalline cohomology groups, which are finite-dimensional vector spaces over a field of characteristic zero (field of fractions of the ring of Witt vectors with coefficients in k).

These cohomology theories can't be the same, because they give vector spaces over very different fields. But they are not unrelated, because, for example, the trace of the map $\alpha^i\colon H^i(X) \to H^i(X)$ defined by a regular map $\alpha\colon X \to X$ of algebraic varieties is a *rational* number independent of the cohomology theory[3]. Thus, in many ways, they behave as though they all arose from an algebraically defined $\mathbb{Q}$ cohomology, but we know they don't.

3 Why is there is no algebraically defined $\mathbb{Q}$ cohomology?

I give two explanations, the first of which applies in nonzero characteristic, and the second in all characteristics.

3.1 First explanation

An elliptic curve E is a curve of genus 1 with a chosen point (the zero for the group structure). Over $\mathbb{C}$, $E(\mathbb{C})$ is isomorphic to the quotient of $\mathbb{C}$ by a lattice Λ (thus, topologically it is a torus). The endomorphisms of E are the complex numbers α such that $\alpha\Lambda \subset \Lambda$, from which it follows easily that either[4] $\mathrm{End}(E)_{\mathbb{Q}} = \mathbb{Q}$ or $\mathrm{End}(E)_{\mathbb{Q}}$ is a field of degree 2 over $\mathbb{Q}$. The cohomology group $H^1(X(\mathbb{C}), \mathbb{Q})$ has dimension 2 as a $\mathbb{Q}$-vector space, and so in the second case it has dimension 1 as an $\mathrm{End}(E)_{\mathbb{Q}}$-vector space.

In characteristic $p \neq 0$, there is a third possibility, namely, $\mathrm{End}(E)_{\mathbb{Q}}$ can be a division algebra (noncommutative field) of degree 4 over $\mathbb{Q}$. The smallest $\mathbb{Q}$-vector space such a division algebra can act on has dimension 4. Thus there is no $\mathbb{Q}$-vector space $H^1(E, \mathbb{Q})$ such that $H^1(E, \mathbb{Q}_\ell) \simeq H^1(E, \mathbb{Q}) \otimes_{\mathbb{Q}} \mathbb{Q}_\ell$ as an $\mathrm{End}(E)_{\mathbb{Q}}$-module.

3.2 Second explanation

Let X be a nonsingular projective variety over an algebraically closed field k of characteristic zero (and not too big). When we choose an embedding $k \to \mathbb{C}$, we get a complex manifold $X(\mathbb{C})$ and it is known that

$$H^i_{\mathrm{et}}(X, \mathbb{Q}_\ell) \simeq H^i(X(\mathbb{C}), \mathbb{Q}) \otimes \mathbb{Q}_\ell$$
$$H^i_{\mathrm{dR}}(X) \otimes_k \mathbb{C} \simeq H^i(X(\mathbb{C}), \mathbb{Q}) \otimes_{\mathbb{Q}} \mathbb{C}.$$

In other words, each embedding $k \hookrightarrow \mathbb{C}$ *does* define a $\mathbb{Q}$-structure on the different cohomology groups. However, different embeddings give different $\mathbb{Q}$-structures, demonstrating that they don't come from an algebraically defined $\mathbb{Q}$ cohomology.

To see this, note that because, because X is defined by finitely many polynomials having only finitely many coefficients, it has a model X_0 over a subfield k_0 of k such that k is an infinite Galois extension of k_0 — let Γ be the Galois group of k

[3]At present, the proof of this in nonzero characteristic requires Deligne's results on the Weil conjectures.

[4]I write $M_{\mathbb{Q}}$ for $M \otimes \mathbb{Q}$.

over k_0. The choice of the model determines an action of Γ on $H^i_{\text{et}}(X, \mathbb{Q}_\ell)$. If the different embeddings of k into $\mathbb{C}$ over k_0 gave the same subspace $H^i(X(\mathbb{C}), \mathbb{Q})$ of $H^i_{\text{et}}(X, \mathbb{Q}_\ell)$, then the action of Γ on $H^i_{\text{et}}(X, \mathbb{Q}_\ell)$, would stabilize $H^i(X, \mathbb{Q})$. But infinite Galois groups are uncountable and $H^i(X, \mathbb{Q})$ is countable, and so this would imply that Γ acts through a finite quotient on $H^i_{\text{et}}(X, \mathbb{Q}_\ell)$. This is known to be false in general.[5]

Thus there is *not* a $\mathbb{Q}$ cohomology theory underlying the different cohomology theories defined by Grothendieck. So how are we going to express the fact that, in many ways, they behave as if there were? Grothendieck's answer is the theory of motives. Before discussing it, I need to explain algebraic cycles.

4 Algebraic cycles

4.1 Definitions

Let X be a nonsingular projective variety of dimension n over a field k. A prime cycle on X is a closed algebraic subvariety Z of X that can not be written as a union of two proper closed algebraic subvarieties. Its codimension is $n - \dim Z$. If Z_1 and Z_2 are prime cycles, then

$$\operatorname{codim}(Z_1 \cap Z_2) \leqslant \operatorname{codim}(Z_1) + \operatorname{codim}(Z_2),$$

and when equality holds we say that Z_1 and Z_2 intersect properly.

The group $C^r(X)$ of algebraic cycles of codimension r on X is the free abelian group generated by the prime cycles of codimension r. Two algebraic cycles γ_1 and γ_2 are said to intersect properly if every prime cycle in γ_1 intersects properly with every prime cycle in γ_2, in which case their intersection product $\gamma_1 \cdot \gamma_2$ is well-defined — it is a cycle of codimension $\operatorname{codim}\gamma_1 + \operatorname{codim}\gamma_2$. For example:

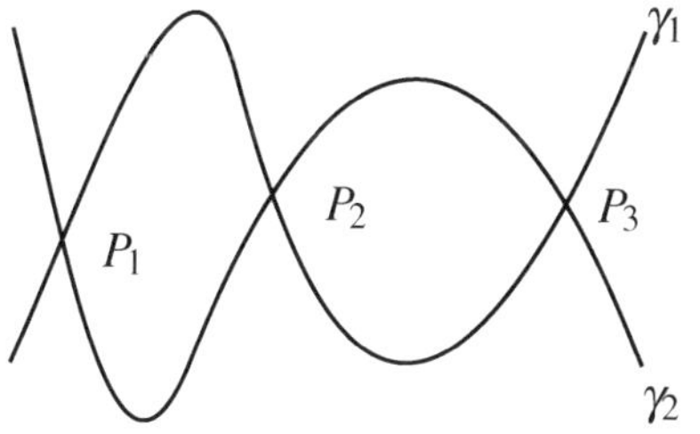

$\gamma_1 \cdot \gamma_2 = P_1 + P_2 + P_3$

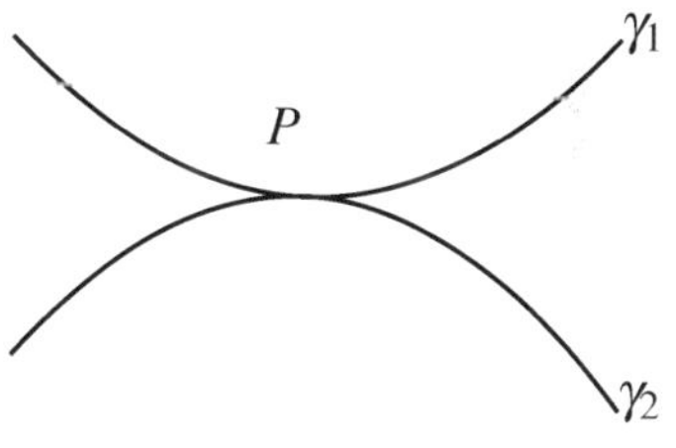

$\gamma_1 \cdot \gamma_2 = 2P$

In this way, we get a partially defined map

$$C^r(X) \times C^s(X) - - \to C^{r+s}(X).$$

[5] In fact, roughly speaking the Tate conjecture predicts that, when k_0 is finitely generated over $\mathbb{Q}$, the image of the Galois group is as large as possible subject only to the constraints imposed by the existence of algebraic cycles.

In order to get a map defined on the whole of a set, we need to be able to move cycles. Two cycles γ and γ' on X are said to be rationally equivalent if there exists an algebraic cycle on $X \times \mathbb{P}^1$ having $\gamma - \gamma'$ as its fibre over one point of $\mathbb{P}^1$ and 0 as its fibre over a second point. Any two algebraic cycles γ_1 and γ_2 are rationally equivalent to algebraic cycles γ_1' and γ_2' that intersect properly, and then the rational equivalence class of $\gamma_1' \cdot \gamma_2'$ is independent of the choice of γ_1' and γ_2'. Therefore on passing to the quotients by rational equivalence, we obtain a well-defined bi-additive map

$$C^r_{\text{rat}}(X) \times C^s_{\text{rat}}(X) \longrightarrow C^{r+s}_{\text{rat}}(X). \tag{4.1}$$

Let $C^*_{\text{rat}}(X) = \bigoplus_{r=0}^{\dim X} C^r_{\text{rat}}(X)$. This is a $\mathbb{Q}$-algebra, called the Chow ring of X.

For example, the prime cycles of codimension 1 on the projective plane $\mathbb{P}^2$ are just the curves defined by irreducible homogeneous polynomials, and two such cycles are rationally equivalent if and only if they are defined by polynomials of the same degree. Therefore $C^1_{\text{rat}}(\mathbb{P}^2) \simeq \mathbb{Z}$ with basis the class of any line in $\mathbb{P}^2$.

A prime cycle of codimension 1 in $\mathbb{P}^1 \times \mathbb{P}^1$ is a curve defined by an irreducible polynomial $P(X_0, X_1; Y_0, Y_1)$ separately homogeneous in each pair of symbols (X_0, X_1) and (Y_0, Y_1). The rational equivalence class of the cycle is determined by the pair of degrees. The group $C^1_{\text{rat}}(\mathbb{P}^1 \times \mathbb{P}^1) \simeq \mathbb{Z} \times \mathbb{Z}$ with basis the classes of $\{0\} \times \mathbb{P}^1$ and $\mathbb{P}^1 \times \{0\}$. The diagonal $\Delta_{\mathbb{P}^1}$ is rationally equivalent to $\{0\} \times \mathbb{P}^1 + \mathbb{P}^1 \times \{0\}$.

Rational equivalence is the finest equivalence relation on algebraic cycles giving a well defined map (4.1) on equivalence classes and satisfying certain natural conditions. The coarsest such equivalence relation is numerical equivalence: two algebraic cycles γ and γ' are numerically equivalent if $\gamma \cdot \delta = \gamma' \cdot \delta$ for all algebraic cycles δ of complementary dimension for which the intersection numbers are defined. The numerical equivalence classes of algebraic cycles form a ring $C^*_{\text{num}} = \bigoplus_{r=0}^{\dim X} C^r_{\text{num}}(X)$ which is a quotient of the Chow ring.

From now on, $\sim =$ rat or num.

4.2 Cycle maps

For all the cohomology theories we are interested in, there is a cycle class map

$$\text{cl}\colon C^*_{\text{rat}}(X)_{\mathbb{Q}} \to H^*(X) \stackrel{\text{def}}{=} \bigoplus_{r=0}^{2\dim X} H^r(X)$$

that doubles degrees and sends intersection products to cup products.

4.3 Correspondences

We are only interested in cohomology theories that are contravariant functors, i.e., such that a regular map $f\colon Y \to X$ of algebraic varieties defines homomorphisms $H^i(f)\colon H^i(X) \to H^i(Y)$. However, this is a weak condition, because there are typically not many regular maps from one algebraic variety to a second. Instead, we should allow "many-valued maps", or, more precisely, correspondences.

The group of correspondences of degree r from X to Y is defined to be

$$\mathrm{Corr}^r(X,Y) = C^{\dim X + r}(X \times Y).$$

For example, the graph Γ_f of a regular map $f\colon Y \to X$ lies in $C^{\dim X}(Y \times X)$, and its transpose Γ_f^t lies in $C^{\dim X}(X \times Y) = \mathrm{Corr}^0(X,Y)$. In other words, a regular map from Y to X defines a correspondence of degree zero from X to Y.[6]

A correspondence γ of degree 0 from X to Y defines a homomorphism $H^*(X) \to H^*(Y)$, namely,

$$x \mapsto q_*(p^*x \cup \mathrm{cl}(\gamma)).$$

Here p and q are the projection maps

$$X \xleftarrow{p} X \times Y \xrightarrow{q} Y.$$

The map on cohomology defined by the correspondence Γ_f^t is the same as that defined by f.

We use the notations:

$$\mathrm{Corr}^r_\sim(X,Y) = \mathrm{Corr}^r(X,Y)/\!\sim, \quad \mathrm{Corr}^r_\sim(X,Y)_\mathbb{Q} = \mathrm{Corr}^r_\sim(X,Y) \otimes_\mathbb{Z} \mathbb{Q}.$$

5 Definition of motives

Grothendieck's idea was that there should be a universal cohomology theory taking values in a $\mathbb{Q}$-category of motives $\mathcal{M}(k)$.

Thus, $\mathcal{M}(k)$ should be a category similar to the category $\mathsf{Vec}_\mathbb{Q}$ of finite-dimensional $\mathbb{Q}$-vector spaces (but not too similar!). Specifically:

- Homs should be $\mathbb{Q}$-vector spaces (preferably finite-dimensional);
- $\mathcal{M}(k)$ should be an abelian category;
- even better, $\mathcal{M}(k)$ should be a tannakian category over $\mathbb{Q}$ (see below).

There should be a universal cohomology theory

$$X \rightsquigarrow hX\colon (\text{nonsingular projective varieties}) \to \mathcal{M}(k).$$

Specifically:

- each algebraic variety X should define a motive hX, and each correspondence of degree zero from X to Y should define a homomorphism $hX \to hY$ (in particular, a regular map $Y \to X$ should define a homomorphism $hX \to hY$).
- every good[7] cohomology theory should factor uniquely through $X \rightsquigarrow hX$.

[6] The switching of directions is unfortunate, but we have to do it somewhere, and I'm following Grothendieck and most subsequent authors.

[7] The technical term is Weil cohomology theory.

5.1 First attempt

We can simply define $\mathcal{M}_\sim(k)$ to be the category with one object hX for each nonsingular projective variety X over k, and with the morphisms defined by

$$\operatorname{Hom}(hX, hY) = \operatorname{Corr}^0_\sim(X, Y)_\mathbb{Q}.$$

Correspondences compose, and so this is a category. However, it is clearly deficient. For example, an endomorphism e of a $\mathbb{Q}$-vector space V such that $e^2 = e$ decomposes the vector space into its 0 and 1 eigenspaces

$$V = \operatorname{Ker}(e) \oplus eV,$$

and if (W, f) is a second such pair, then

$$\operatorname{Hom}_{\mathbb{Q}\text{-linear}}(eV, fW) \simeq f \circ \operatorname{Hom}_{\mathbb{Q}\text{-linear}}(V, W) \circ e \qquad (\text{inside } \operatorname{Hom}_{\mathbb{Q}\text{-linear}}(V, W)).$$

A similar statement holds in every abelian category, and so, if we want $\mathcal{M}_\sim(k)$ to be abelian, we should at least add the images of idempotents in $\operatorname{End}(hX) \stackrel{\text{def}}{=} \operatorname{Corr}^0_\sim(X, X)_\mathbb{Q}$.

5.2 Second attempt

We now define $\mathcal{M}_\sim(k)$ to be the category with one object $h(X, e)$ for each pair with X as before and e an idempotent in the ring $\operatorname{Corr}^0_\sim(X, X)_\mathbb{Q}$. Morphisms are defined by

$$\operatorname{Hom}(h(X, e), h(Y, f)) = f \circ \operatorname{Corr}^0_\sim(X, Y)_\mathbb{Q} \circ e$$

(subset of $\operatorname{Corr}^0_\sim(X, Y)_\mathbb{Q}$). That's it! This is the category of effective motives for rational or numerical equivalence depending on the choice of $\sim$, which we should denote $\mathcal{M}^{\text{eff}}_\sim(k)$. It contains the preceding category as the full subcategory of objects $h(X, \Delta_X)$.

For example, the earlier discussion shows that $\operatorname{End}(h\mathbb{P}^1, \Delta_{\mathbb{P}^1}) = \mathbb{Z} \oplus \mathbb{Z}$ with $e_0 \stackrel{\text{def}}{=} (1, 0)$ represented by $\{0\} \times \mathbb{P}^1$ and $e_2 \stackrel{\text{def}}{=} (0, 1)$ represented by $\mathbb{P}^1 \times \{0\}$. Corresponding to the decomposition $\Delta_{\mathbb{P}^1} \sim e_0 + e_2$, we obtain a decomposition

$$h(\mathbb{P}^1, \Delta_{\mathbb{P}^1}) = h^0\mathbb{P}^1 \oplus h^2\mathbb{P}^1 \tag{5.1}$$

with $h^i\mathbb{P}^1 = h(\mathbb{P}^1, e_i)$ (this is true both in $\mathcal{M}^{\text{eff}}_{\text{rat}}(k)$ and in $\mathcal{M}^{\text{eff}}_{\text{num}}(k)$). We write $\mathbb{1} = h^0\mathbb{P}^1$ and $\mathbb{L} = h^2\mathbb{P}^1$.

For some purposes, the category of effective motives is the most useful, but generally we would prefer a category in which objects have duals.

5.3 Third attempt

This can be achieved quite easily simply by inverting $\mathbb{L}$. The objects of $\mathcal{M}_\sim(k)$ are now triples $h(X, e, m)$ with X and e as before, and with $m \in \mathbb{Z}$. Morphisms are defined by

$$\operatorname{Hom}(h(X, e, m), h(Y, f, n)) = f \circ \operatorname{Corr}^{n-m}_\sim(X, Y)_\mathbb{Q} \circ e.$$

This is the category of motives over k. It contains the preceding category as the full subcategory of objects $h(X, e, 0)$.

Sometimes $\mathcal{M}_{\mathrm{rat}}(k)$ is called the category of Chow motives and $\mathcal{M}_{\mathrm{num}}(k)$ the category of Grothendieck (or numerical) motives.

6 What is known about the categories of motives

6.1 Properties of the category $\mathcal{M}_{\sim}(k)$

- The Hom sets are $\mathbb{Q}$-vector space, which are finite-dimensional if $\sim$=num (but not usually otherwise).
- Direct sums of motives exist, so $\mathcal{M}_{\sim}(k)$ is an additive category. For example,
$$h(X, e, m) \oplus h(Y, f, m) = h(X \sqcup Y, e \oplus f, m).$$
- An idempotent f in the endomorphism ring of a motive M decomposes the motive into a direct sum of the kernel and image of f, so $\mathcal{M}_{\sim}(k)$ is a pseudo-abelian category. For example, if $M = h(X, e, m)$, then
$$M = h(X, e - efe, m) \oplus h(X, efe, m).$$
- The category $\mathcal{M}_{\mathrm{num}}(k)$ is abelian and semisimple, but $\mathcal{M}_{\mathrm{rat}}(k)$ is not even abelian, except perhaps when k is algebraic over a finite field.
- There is a good tensor product structure on $\mathcal{M}_{\sim}(k)$, which is defined by
$$h(X, e, m) \otimes h(Y, f, n) = h(X \times Y, e \times f, m + n).$$
Denote $h(X, \Delta_X, 0)$ by hX. Then $hX \otimes hY = h(X \times Y)$, i.e., the Künneth formula holds for the functor $X \rightsquigarrow hX$.
- The above statements hold also for the category of effective motives, but in both $\mathcal{M}_{\mathrm{num}}(k)$ and $\mathcal{M}_{\mathrm{rat}}(k)$ objects have duals. This means that for each motive M there is a dual motive $M^{\vee}$ and an "evaluation map" $\mathrm{ev}\colon M^{\vee} \otimes M \to \mathbb{1}$ having a certain universal property. For example,
$$h(X, e, m)^{\vee} = h(X, e^t, \dim X - m)$$
if X is connected.

I should stress that, although $\mathcal{M}_{\mathrm{rat}}(k)$ is not abelian, it is still a very important category. In particular, it contains more information than $\mathcal{M}_{\mathrm{num}}(k)$.

6.2 Is $X \rightsquigarrow hX$ a universal cohomology theory?

Certainly, the functor $X \rightsquigarrow hX$ sending a X to its Chow motive is universal. This is almost a tautology: good cohomology theories are those that factor through $\mathcal{M}_{\mathrm{rat}}(k)$.

With $\mathcal{M}_{\mathrm{num}}(k)$ there is a problem: a correspondence numerically equivalent to zero will define the zero map on motives, but we don't know in general that it defines the zero map on cohomology. In order for a good cohomology theory to factor through $\mathcal{M}_{\mathrm{num}}(k)$, it must satisfy the following conjecture:

Conjecture D. *If an algebraic cycle is numerically equivalent to zero, then its cohomology class is zero.*

In other words, if $\mathrm{cl}(\gamma) \neq 0$ then γ is not numerically equivalent to zero. Taking account of Poincaré duality, we can restate this as follows: if there exists a cohomology class γ' such that $\mathrm{cl}(\gamma) \cup \gamma' \neq 0$, then there exists an algebraic cycle γ'' such that $\gamma \cdot \gamma'' \neq 0$. Thus, the conjecture is an existence statement for algebraic cycles. Unfortunately, we have no method for proving the existence of algebraic cycles. More specifically, when we expect that a cohomology class is algebraic, i.e., the class of an algebraic cycle, we have no way of going about proving that it is. This is a major problem, perhaps the major problem, in arithmetic geometry and in algebraic geometry.

In characteristic zero, Conjecture D is known for abelian varieties and it is implied by the Hodge conjecture.

6.3 Is hX graded?

When we assume Conjecture D, every good cohomology theory H does factor through $X \rightsquigarrow hX$. This means that there is a functor from ω from $\mathcal{M}_{\mathrm{num}}(k)$ to the category of vector spaces over the ground field for H such that

$$\omega(hX) = H^*(X) \stackrel{\mathrm{def}}{=} \bigoplus_{i=0}^{2\dim X} H^i(X).$$

Clearly there should be a decomposition of hX that underlies the decomposition of $H^*(X)$ for every good cohomology theory. For $\mathbb{P}^1$ we saw in (5.1) that that this is true. The following was conjectured by Grothendieck.

Conjecture C. *In the ring* $\mathrm{End}(hX) = C^{\dim X}_{\mathrm{num}}(X \times X)$, *the diagonal* Δ_X *has a canonical decomposition into a sum of orthogonal idempotents*

$$\Delta_X = \pi_0 + \cdots + \pi_{2\dim X}. \tag{6.1}$$

Such an expression defines a decomposition

$$hX = h^0 X \oplus h^1 X \oplus \cdots \oplus h^{2\dim X} X, \tag{6.2}$$

with $h^i X = h(X, \pi_i, 0)$, *and this decomposition should have the property that it is mapped to the decomposition*

$$H^*(X) = H^0(X) \oplus H^1(X) \oplus \cdots \oplus H^{2\dim X}(X)$$

by every good cohomology theory for which Conjecture D holds.

Again the conjecture is an existence statement for algebraic cycles, and hence is hard. It is known for all nonsingular projective varieties over finite fields (here certain polynomials in the Frobenius map can be used to decompose the motive) and for abelian varieties in characteristic zero (by definition abelian varieties have a group structure, which is commutative, and the maps $m \colon X \to X$, $m \in \mathbb{Z}$, can be used to decompose hX).

When Conjecture C is assumed, it becomes possible to speak of the weights of a motive. For example, the motive h^iX has weight i, and $h(X, \pi_i, m)$ has weight $i - 2m$. A motive is said to be pure if it has a single weight. Every motive is a direct sum of pure motives.

Until Conjectures C and D are proved, Grothendieck's dream remains unfulfilled.

Aside: Murre conjectured that a decomposition (6.1) exists with certain properties even in $C_{\text{rat}}^{\dim X}(X \times X)$. It has been shown that his conjecture is equivalent to the existence of an interesting filtration on the Chow groups, which had been conjectured by Beilinson and Bloch.

6.4 What is a tannakian category?

By an affine group, I mean a matrix group (possibly infinite dimensional)[8]. For such a group G over $\mathbb{Q}$, the representations of G on finite-dimensional $\mathbb{Q}$-vector spaces form an abelian category $\text{Rep}_{\mathbb{Q}}(G)$ with tensor products and duals, and the forgetful functor is an exact faithful functor from $\text{Rep}_{\mathbb{Q}}(G)$ to $\mathsf{Vec}_{\mathbb{Q}}$ preserving tensor products.

A neutral tannakian category T over $\mathbb{Q}$ is an abelian category with tensor products and duals for which there exist exact faithful functors to $\mathsf{Vec}_{\mathbb{Q}}$ preserving tensor products; the tensor automorphisms of such a functor ω form an affine group G, and the choice of such functor ω determines an equivalence of categories $\mathsf{T} \to \text{Rep}_{\mathbb{Q}}(G)$. Thus, a neutral tannakian category is an abstract version of the category of representations of an affine group that has no distinguished "forgetful" functor (just as a vector space is an abstract version of k^n that has no distinguished basis).

A tannakian category T over $\mathbb{Q}$ (not necessarily neutral) is an abelian category with tensor products and duals for which there exists an exact faithful tensor functor to the category of vector spaces over some field of characteristic zero (not necessarily $\mathbb{Q}$).[9] The choice of such a functor defines an equivalence of T with the category of representations of an affine groupoid.

6.5 Is $\mathcal{M}_{\text{num}}(k)$ tannakian?

No, it isn't. In an abelian category T with tensor products and duals it is possible to define the trace of an endomorphism of an object. This is preserved by every exact faithful tensor functor $\omega\colon \mathsf{T} \to \mathsf{Vec}_{\mathbb{Q}}$, and so, for the identity map u of an object M,

$$\text{Tr}(u|M) = \text{Tr}(\omega(u)|\omega(M)) = \dim_{\mathbb{Q}} \omega(M)$$

[8]More precisely, an affine group is an affine group scheme over a field (not necessarily of finite type). Such a group is an inverse limit of affine algebraic group schemes, each of which can be realized as a subgroup of some GL_n.

[9]We also require that $\text{End}(\mathbb{1}) = \mathbb{Q}$ where $\mathbb{1}$ is an object such that $\mathbb{1} \otimes X \simeq X \simeq X \otimes \mathbb{1}$, all X in T.

which, being the dimension of a vector space, is a nonnegative integer. For the identity map u of a variety X, $\mathrm{Tr}(u|hX)$ turns out to be the Euler-Poincaré characteristic of X (alternating sum of the Betti numbers). For example, if X is a curve of genus g, then

$$\mathrm{Tr}(u|hX) = \dim H^0 - \dim H^1 + \dim H^2 = 2 - 2g,$$

which may be negative. This proves that there does not exist an exact faithful tensor functor $\omega\colon \mathcal{M}_{\mathrm{num}}(k) \to \mathsf{Vec}_F$ for any field F.

To fix this we have to change the inner workings of the tensor product structure. Assume Conjecture C, so that every motive has a decomposition (6.2). When we change the sign of the "canonical" isomorphism

$$h^iX \otimes h^jX \simeq h^jX \otimes h^iX$$

for ij odd, then $\mathrm{Tr}(u|h(X))$ becomes the sum of the Betti numbers of X rather than the alternating sum. Then $\mathcal{M}_{\mathrm{num}}(k)$ becomes a tannakian category (neutral if k has characteristic zero, but not otherwise). Thus, when k is algebraic over a finite field, $\mathcal{M}_{\mathrm{num}}(k)$ is known to be a nonneutral tannakian category (but, lacking Conjecture D, we don't know that the standard cohomologies factor through it).

7 The Weil conjectures revisited

7.1 Zeta functions

Let X be a projective nonsingular variety over $\mathbb{F}_p$, and fix an algebraic closure $\mathbb{F}$ of $\mathbb{F}_p$. For each m, there is exactly one subfield $\mathbb{F}_{p^m}$ of $\mathbb{F}$ of with p^m elements. Let $X(\mathbb{F}_{p^m})$ denote the set of points on X with coordinates in $\mathbb{F}_{p^m}$. This set is finite, and the zeta function $Z(X,t)$ of X is defined by

$$\log Z(X,t) = \sum\nolimits_{m\geqslant 1} |X(\mathbb{F}_{p^m})| \frac{t^m}{m}.$$

For example, let $X = \mathbb{P}^0 =$ point. Then $|X(\mathbb{F}_{p^m})| = 1$ for all m, and so

$$\log Z(X,t) = \sum\nolimits_{m\geqslant 1} \frac{t^m}{m} = \log \frac{1}{1-t};$$

thus

$$Z(X,t) = \frac{1}{1-t}.$$

As our next example, let $X = \mathbb{P}^1$. Then $|X(\mathbb{F}_{p^m})| = 1 + p^m$, and so

$$\log Z(X,t) = \sum (1+p^m)\frac{t^m}{m} = \log \frac{1}{(1-t)(1-pt)};$$

thus

$$Z(X,t) = \frac{1}{(1-t)(1-pt)}.$$

7.2 Weil's fundamental work

In the 1940s, Weil proved that for a curve X of genus g over $\mathbb{F}_p$,

$$Z(X,t) = \frac{P_1(t)}{(1-t)(1-pt)}, \quad P_1(t) \in \mathbb{Z}[t], \tag{7.1a}$$

$$P_1(t) = (1-a_1t)\cdots(1-a_{2g}t) \quad \text{where } |a_i| = p^{\frac{1}{2}}. \tag{7.1b}$$

In particular, this says that

$$|X(\mathbb{F}_p)| = 1 + p - \textstyle\sum_{i=1}^{2g} a_i$$

and so

$$||X(\mathbb{F}_p)| - p - 1| = |\textstyle\sum_{i=1}^{2g} a_i| \leqslant 2gp^{\frac{1}{2}}.$$

One of Weil's proofs of these statements makes use of the Jacobian variety of the curve. For a curve X of genus g over $\mathbb{C}$, $X(\mathbb{C})$ is a Riemann surface of genus g, and so the holomorphic differentials on $X(\mathbb{C})$ form a complex vector space $\Omega^1(X)$ of dimension g and the homology group $H_1(X(\mathbb{C}),\mathbb{Z})$ is a free $\mathbb{Z}$-module of rank $2g$. An element γ of $H_1(X(\mathbb{C}),\mathbb{Z})$ defines an element $\omega \mapsto \int_\gamma \omega$ of the dual vector space $\Omega^1(X)^\vee$ of $\Omega^1(X)$. It has been known since the time of Abel and Jacobi that this map realizes $H_1(X(\mathbb{C}),\mathbb{Z})$ as a lattice Λ in $\Omega^1(X)^\vee$, and so the quotient $J(X) = \Omega^1(X)^\vee/\Lambda$ is complex torus — the choice of a basis for $\Omega^1(X)$ defines an isomorphism $J(X) \approx \mathbb{C}^g/\Lambda$. The endomorphisms of $J(X)$ are the linear endomorphisms of $\Omega^1(X)^\vee$ mapping Λ into itself, from which it follows that $\mathrm{End}(J(X))$ is a finitely generated $\mathbb{Z}$-module. Therefore $\mathrm{End}(J(X))_{\mathbb{Q}}$ is a $\mathbb{Q}$-algebra of finite rank. A polarization of $J(X)$ defines an involution $\alpha \mapsto \alpha^\dagger$ of $\mathrm{End}(J(X))_{\mathbb{Q}}$ which is positive definite in the sense that the trace $\mathrm{Tr}(\alpha\alpha^\dagger) > 0$ for all nonzero α.

The complex torus $J(X)$ is an algebraic variety. When Weil was working on these questions in the 1940s, it was not known how to define the Jacobian variety of a curve over a field of nonzero characteristic. In fact, the foundations of algebraic geometry at the time were inadequate for this task, and so, in order to give substance to his proof of (7.1a,7.1b) he had first to rewrite the foundations of algebraic geometry, and then develop the theory of Jacobian varieties over arbitrary fields.

For any variety X over $\mathbb{F}_p$ there is a regular map $\pi\colon X \to X$ (called the Frobenius map) that acts on points as $(a_0 : \cdots : a_n) \mapsto (a_0^p : \cdots : a_n^p)$ and which has the property that the fixed points of π^m acting on $X(\mathbb{F})$ are exactly the elements of $X(\mathbb{F}_{p^m})$. Weil proved a fixed-point formula which allowed him to show that, for a curve X over $\mathbb{F}_p$, $Z(X,t) = P_1(t)/(1-t)(1-pt)$ with $P_1(t)$ equal to the characteristic polynomial π acting on $J(X)$, which he knew to have $\mathbb{Z}$-coefficients. A polarization of $J(X)$ defines an involution on $\mathrm{End}(J(X))_{\mathbb{Q}}$, which Weil proved to be positive definite. From this he deduced the inequality $|a_i| < p^{\frac{1}{2}}$.

7.3 Statement of the Weil conjectures

Weil's results on curves and other varieties suggested the following conjectures: for a nonsingular projective variety X of dimension n over $\mathbb{F}_p$,

$$Z(X,t) = \frac{P_1(t) \quad \cdots \quad P_{2n-1}(t)}{(1-t)P_2(t)\cdots P_{2n-2}(t)(1-p^n t)}, \quad P_i(t) \in \mathbb{Z}[t], \tag{7.2a}$$

$$P_i(t) = (1-a_{i1}t)\cdots(1-a_{ib_i}t) \quad \text{where } |a_{ij}| = p^{i/2}; \tag{7.2b}$$

moreover, if X arises by reduction modulo p of a variety $\tilde{X}$ over $\mathbb{Q}$, then the b_i (degrees of the P_i) should be the Betti numbers of the complex manifold $\tilde{X}(\mathbb{C})$.

7.4 The standard conjectures and the Weil conjectures

When Grothendieck defined his étale cohomology groups, he and his collaborators (Artin and Verdier) proved a fixed-point theorem which allowed them to show that $Z(X,t)$ can be expressed in the form (7.2a) with P_i replaced by the characteristic polynomial $P_{i,\ell}$ of the Frobenius map π acting in $H^i_{\text{et}}(X,\mathbb{Q}_\ell)$. However, rather than having coefficients in $\mathbb{Z}$, the polynomials P_i have coefficients in $\mathbb{Q}_\ell$, and it could not be excluded that they might depend on ℓ.

In 1968 Grothendieck announced two conjectures, known respectively as the Lefschetz standard conjecture and the Hodge standard conjecture, whose proof would allow one to extend Weil's proof of the Weil conjectures from curves to varieties of arbitrary dimension by replacing the Jacobian of the curve with the motive of the variety.

Our Conjecture C is a weak form of the Lefschetz standard conjecture. As we have seen, together with the folklore Conjecture D, it implies that there is a good theory of motives, and this implies that (7.2a) holds with $P_i(t)$ the characteristic polynomial of π acting on the motive h^iX. Now $P_i(t)$ has coefficients in $\mathbb{Q}$, and an elementary argument shows that it has coefficients in $\mathbb{Z}$.

The Hodge standard conjecture is a positivity statement which implies that the endomorphism algebra of every motive admits a *positive definite* involution. Assuming this, Weil's argument then proves (7.2b).

In characteristic zero, the Hodge standard conjecture can be proved by analytic methods, but in nonzero characteristic it is known for very few varieties. However, it is implied by the Hodge and Tate conjectures.

In 1973, Deligne succeeded in completing the proof of the Weil conjectures by means of a very clever argument not involving the standard conjectures. However, Grothendieck's statement[10]

> Alongside the problem of resolution of singularities [in nonzero characteristic], the proof of the standard conjectures seems to me to be the most urgent task in algebraic geometry

remains valid.

[10]Grothendieck, A., Standard conjectures on algebraic cycles. 1969 Algebraic Geometry (Internat. Colloq., Tata Inst. Fund. Res., Bombay, 1968) pp. 193–199 Oxford Univ. Press, London, p.198.

8 Zeta functions of motives

8.1 Zeta functions of varieties over $\mathbb{Q}$

Let X be a projective nonsingular variety over $\mathbb{Q}$. When we scale the polynomials defining X so that they have integer coefficients, and then look at the equations modulo a prime number p, we obtain a projective variety X_p over $\mathbb{F}_p$. We call p "good" if X_p is again nonsingular. All but finitely many primes are good, and we define the zeta function of X to be[11]

$$\zeta(X,s)=\prod\nolimits_{p \text{ good}} Z(X_p,p^{-s}).$$

For example, when $X=\mathbb{P}^0=$ point,

$$\zeta(X,s)=\prod\nolimits_p \frac{1}{1-p^{-s}},$$

which is the Riemann zeta function $\zeta(s)$, and when $X=\mathbb{P}^1$,

$$\zeta(X,s)=\prod\nolimits_p \frac{1}{(1-p^{-s})(1-p^{1-s})}=\zeta(s)\zeta(s-1).$$

Consider an elliptic curve E over $\mathbb{Q}$. For a good p,

$$Z(E_p,t)=\frac{(1-a_pt)(1-\bar{a}_pt)}{(1-t)(1-pt)},\quad a_p+\bar{a}_p\in\mathbb{Z},\quad a_p\bar{a}_p=p,\quad |a_p|=p^{1/2}$$

(see 7.1a,7.1b). Therefore

$$\zeta(E,s)=\frac{\zeta(s)\zeta(s-1)}{L(E,s)}$$

where

$$L(E,s)=\prod\nolimits_p \frac{1}{(1-a_pp^{-s})(1-\bar{a}_pp^{-s})}.$$

8.2 Zeta functions of motives

We first consider motives over $\mathbb{F}_p$. We can't define the zeta function of a motive M over $\mathbb{F}_p$ in terms of the points of M with coordinates in the fields $\mathbb{F}_{p^m}$ because they are not defined. However, we do know that the category $\mathcal{M}(\mathbb{F}_p)$ is tannakian, and in any tannakian category endomorphisms of objects have characteristic polynomials. We define the zeta function $Z(M,t)$ of a pure motive M over $\mathbb{F}_p$ of weight i to be the characteristic polynomial of the Frobenius map of M if i is odd, or its reciprocal if i is even. The characteristic polynomial has coefficients in $\mathbb{Q}$, and even in $\mathbb{Z}$ if M is effective. For motives M_1 and M_2 of the same weight

$$Z(M_1\oplus M_2,t)=Z(M_1,t)\cdot Z(M_2,t), \tag{8.1}$$

[11] We should also include factors for the "bad" primes and one for the real numbers. In what follows, I am ignoring a finite number of factors.

and we use this formula to extend the definition to all motives.

How does this relate to the zeta functions of varieties? Let X be a smooth projective variety of dimension n over $\mathbb{F}_p$. As we noted earlier, Grothendieck and his collaborators showed that $Z(X,t) = P_1(t)\cdots P_{2n-1}(t)/P_0(t)\cdots P_{2n}(t)$ where $P_i(t)$ is the characteristic polynomial of the Frobenius map of X acting on the étale cohomology group $H^i_{\mathrm{et}}(X_{\mathbb{F}}, \mathbb{Q}_\ell)$ (any prime $\ell \neq p$; *a priori* $P_i(t)$ may depend on ℓ). Now assume that Conjecture D holds for ℓ-adic étale cohomology. Then there exists a functor $\omega\colon \mathcal{M}(\mathbb{F}_p) \to \mathsf{Vec}_{\mathbb{Q}_\ell}$ such that $\omega(h^iX) = H^i_{\mathrm{et}}(X_{\mathbb{F}}, \mathbb{Q}_\ell)$. The functor preserves characteristic polynomials, and this shows[12] that $Z(h^iX,t) = P_i(X,t)^{(-1)^{i+1}}$. Thus,

$$Z(X,t) = Z(h^0X,t)\cdots Z(h^{2n}X,t).$$

From (6.2) and (8.1), we see that the right hand side equals $Z(hX,t)$, and so $Z(X,t) = Z(hX,t)$.

We now consider motives over $\mathbb{Q}$. Such a motive M is described by a projective smooth variety X over $\mathbb{Q}$, an algebraic cycle γ on $X \times X$, and an integer m. For all but finitely many prime numbers p, X and γ will reduce well to give a motive M_p over $\mathbb{F}_p$, and we can define

$$\zeta(M,s) = \prod\nolimits_{p \text{ good}} Z(M_p, p^{-s}).$$

For example,

$$\zeta(h^0(\mathbb{P}^1)) = \zeta(s), \quad \zeta(h^2(\mathbb{P}^1)) = \zeta(s-1).$$

For an elliptic curve E,

$$hE = h^0E \oplus h^1E \oplus h^2E$$

and so

$$\begin{aligned}\zeta(hE,s) &= \zeta(h^0E,s)\cdot\zeta(h^1E,s)\cdot\zeta(h^2E,s)\\ &= \zeta(s)\cdot L(E,s)^{-1}\cdot\zeta(s-1).\end{aligned}$$

Notice that, without assuming any unproven conjectures, we have defined a category of motives over $\mathbb{Q}$, and we have attached a zeta function to each object of the category. This is a function of the complex variable s, which conjecturally has many wonderful properties. The functions that arise in this way are called motivic L-functions. On the other hand, there is an entirely different method of constructing functions $L(s)$ from modular forms, automorphic forms, or, most generally, from automorphic representations — these are called automorphic L-functions. Their definition does not involve algebraic geometry. The following is a fundamental guiding principle in the Langlands program.

Big Modularity Conjecture. *Every motivic L-function is an alternating product of automorphic L-functions.*

Let E be an elliptic curve over $\mathbb{Q}$. The (little) modularity conjecture says that $\zeta(h^1E,s)$ is the Mellin transform of a modular form. The proof of this by Wiles et al. was the main step in the proof of Fermat's Last Theorem.

[12] In particular, $P_i(X,t)$ is a polynomial with $\mathbb{Z}$ coefficients, independent of ℓ. This shows that Conjectures C and D imply (7.2a), independently of the work of Deligne.

9 The conjecture of Birch and Swinnerton-Dyer, and some mysterious squares

Let E be an elliptic curve over $\mathbb{Q}$. Beginning about 1960, Birch and Swinnerton-Dyer used one of the early computers (EDSAC 2) to study $L(E,s)$ near $s = 1$. These computations led to their famous conjecture: let $L(E,1)^*$ denote first nonzero coefficient in the expansion of $L(E,s)$ as a power series in $s-1$; the conjecture states that

$$L(E,1)^* = \{\text{terms understood}\}\,\{\text{mysterious term}\}.$$

The mysterious term is conjectured to be the order of the Tate-Shafarevich group of E, which (if finite) is known to be a square.

About the same time, they studied

$$L_3(E,s) = \prod\nolimits_p \frac{1}{(1-a_p^3 p^{-s})(1-\bar{a}_p^3 p^{-s})}$$

near $s = 2$, and they found (computationally) that

$$L_3(E,1)^* = \{\text{terms understood}\}\{\text{mysterious square}\}.$$

The mysterious square can be quite large, for example 2401. What is it?

As we noted above, $L(E,s)^{-1} = \zeta(h^1 E, s)$. We can regard the conjecture of Birch and Swinnerton-Dyer as a statement about the motive $h^1 E$. The conjecture has been extended to all motives over $\mathbb{Q}$. One can show that

$$h^1(E) \otimes h^1(E) \otimes h^1(E) = 3h^1(E, \Delta_E, -1) \oplus M$$

for a certain motive M, and that

$$\zeta(M,s) = L_3(E,s)^{-1}.$$

Thus, the mysterious square is conjecturally the "Tate-Shafarevich group" of the motive M.

10 Final note

Strictly, $\mathcal{M}(k)$ should be called the category of *pure* motives. It is attached to the category nonsingular projective varieties over k. Grothendieck also envisaged a category *mixed* motives attached to the category of *all* varieties over k. It should no longer be semisimple, but each mixed motive should have a filtration whose quotients are pure motives. There is at present no definition of a category of mixed motives, but several mathematicians have constructed triangulated categories that are candidates to be its derived category; it remains to define a t-structure on one of these categories whose heart is the category of mixed motives itself.

Literature

Grothendieck himself published nothing on motives, but he mentioned them frequently in his letters to Serre [1].The first expositions of the theory, based on lectures of Grothendieck, were those of Demazure [2] and Kleiman [3]. The proceedings of the 1991 motives conference [4] survey what was known about motives at the time. In particular, the article of Kleiman discusses Grothendieck's standard conjectures, that of Scholl describes in detail the construction of the category of motives, and the first article of the author (especially 2.48) explains the application of the standard conjectures to the Weil conjectures. At present, the only book on motives are those of André [5] and Murre, Nagel, and Peters [6].

References

[1] Pierre Colmez and Jean-Pierre Serre (Editors), Correspondance Grothendieck-Serre. Edited by. Documents Mathématiques (Paris), 2. Société Mathématique de France, Paris, 2001. (English translation published by the AMS, 2004).

[2] Demazure, M., Motifs des variétés algébriques, Séminaire Bourbaki 1969/70, Exposé 365, 20pp.

[3] Kleiman, S. L., Motives. Algebraic geometry, Oslo 1970 (Proc. Fifth Nordic Summer-School in Math., Oslo, 1970), pp. 53–82. Wolters-Noordhoff, Groningen, 1972.

[4] Uwe Jannsen, Steven Kleiman, and Jean-Pierre Serre (Editors), Motives. Proceedings of the AMS-IMS-SIAM Joint Summer Research Conference held at the University of Washington, Seattle, Washington, July 20–August 2, 1991. Proceedings of Symposia in Pure Mathematics, 55. American Mathematical Society, Providence, RI, 1994.

[5] André, Y., Une introduction aux motifs (motifs purs, motifs mixtes, périodes), Panoramas et Synthèses, 17, Société Mathématique de France, Paris, 2004.

[6] Murre, J. P., Nagel, J., and Peters, Chris A. M., Lectures on the theory of Pure motives, University Lecture Series, 61. American Mathematical Society, Providence, RI, 2013.

Open Problems and Surveys of Contemporary Mathematics
SMM 6, pp. 343–378

Representations are Everywhere*

Nanhua Xi†

Contents

*Translated by Xu Fan, from the Chinese version of the draft prepared for the talk "Talks in Mathematics Center" in Mathematical Sciences Center, Tsinghua university.

†Academy of Mathematics and System Science, Chinese Academy of Sciences, China.
Email: nanhua@math.ac.cn

Introduction

Today, I will talk about representations, which appear almost everywhere in mathematics. You may have heard of representation theory before or had some feeling about what this subject is about, but perhaps not everyone of you understand it well.

Representation theory is a very active research field, it attracts a lot of top mathematicians in the world. But in China, there are very few people working on it.

Sometimes we may feel that representation theory is very abstract and not easy to understand, or even not understandable. In fact representations are so common in mathematics and appear everywhere, such as polynomial rings, square integrable functions in analysis, cohomology theory and K-theory in topology and so on. All of them carry very rich structures in representation theory.

Let us see what I. M. Gelfand said: All of mathematics is some kind of representation theory. I. M. Gelfand is a great mathematician. There were few mathematicians who could be as important as him for the extensity and intensity of research in the last fifty years. You would agree with me about this if you read his three collected works. The second volume of his collected works is very popular. I tried to buy one from Amazon and found that it was sold out. This also indicates that in some sense representation theory is getting much more attention and believed to be very important among the math community. When talking about Gelfand's contribution to representation, his student A. Kirillov said the following words in the preface of this volume: "Representation theory has been 'very lucky': such perspective and powerful mathematician as E. Cartan, H. Weyl, Harish-Chandra, A. Weil, R. Langlands, and P. Deligne all worked in the field. But even against this background, I. M. Gelfand's contributions know no equal in range, depth of approach, and beauty of results". Intentionally or not, the student did not mention Lusztig, who is also an expert in representation theory.

It is worth thinking about why the great mathematician Gelfand would say the previous words above. How should we understand his words?

First, let us explain the terminology. Literally, it is "expression" in Chinese, "appearance" in Japanese and "representation" in English. Its stem in English is "present", which has the meanings "appearance/behavior/exhibit/expression/ display/performance". The prefix "re" has the meanings "again/once more".

We often hear words like this: "You should show what you mean", "His performance is very good". If you pay excessive attention to the wording here, you will see that the Japanese translation is more accurate, as its meaning is closer to what the word "presentation" in English means.

So literally, "representation" is reappearance of some properties or structures of some subjects on another. With this understanding, we could understand Gelfand's words more easily now. For example, a function over the real number field can be thought as a representation of the field. From this perspective, not merely mathematics, any science, even all of the knowledge about the world can be seen as some kinds of representations theory. This is similar to a viewpoint in Buddhism: everything consists of its appearances. Of course this is going too far. This viewpoint of representation may be interesting in philosophy, but useless for mathematics. Mathematics needs a more precise definition of "representation".

Representation in mathematics is the reappearance of an abstract algebraic structure in terms of some more explicit objects which consist of linear transformations or matrices. The algebraic structures we usually use are groups, rings and fields etc. What are the algebraic structures that representation theory cares about? Groups, some special rings which are called algebras, and some special noncommutative rings which are called Lie (super)algebras are the focus here. These are the main algebraic structures that representation theory deals with.

It has been known that an algebraic structure is determined by operations. In general, different objects are related to using mappings between them. Some maps are easy to identify, but others are not. We focus on those maps which preserve the operations and thus reflect some relations between the algebraic structures. We call these maps homomorphisms. Then a representation is a special kind of homomorphism, whose image consists of some linear transformations. With this understanding, we can begin our adventure in representation theory now.

1 A rough partition of representation theory

Roughly speaking, representation theory consists of three parts: the representation theory of groups, the representation of (associative) algebras, and the representation of Lie algebras.

All invertible linear transformations on a linear space V form a group under the composition between maps. A representation of a group G is just a homomorphism from this group G to the group $GL(V, F)$:

$$\text{Group} \xrightarrow{\text{group homomorphism}} \{\text{invertible linear transformation on a linear space}\}$$

The space of all linear transformations $\text{End}(V, F)$ on a linear space V is an algebra since it has addition, composition as multiplication, and could be multiplied by elements in the field F. A representation of an algebra is just a homomorphism from this algebra to the algebra $\text{End}(V, F)$:

$$\text{Algebra} \xrightarrow{\text{algebra homomorphism}} \{\text{linear transformation on a linear space}\}$$

The algebra structure of all linear transformations $GL(V, F)$ on a linear space V induces a Lie algebra structure over $\text{End}(V, F)$, it is called the general linear Lie algebra. A representation of a Lie algebra is just a homomorphism from this Lie algebra to the general linear Lie algebra $\text{End}(V, F)$:

$$\text{Lie algebra} \xrightarrow{\text{homomorphism of Lie algebras}} \{\text{linear transformation on a linear space}\}$$

There are many connections among these three parts of representation theories. However there are also many differences among the methods and idea used in the specific research on them.

A linear space V has finite or infinite dimension. If the dimension is finite, the corresponding representation is called a finite dimensional representation. For a fixed basis of V, a linear transformation could be represented by a matrix. Hence, there is another form for a finite dimensional representation:

Representation of groups:

$$\text{Group} \xrightarrow{\text{homomorphism between groups}} \{n \times n \text{ invertible matrix over some field } F\}$$

Representation of algebras:

$$\text{Algebra} \xrightarrow{\text{homomorphism between algebras}} \{n \times n \text{ matrix on some field } F\}$$

Representation of Lie algebras:

$$\text{Lie algebra} \xrightarrow{\text{homomorphism between Lie algebras}} \{n \times n \text{ matrix on some field } F\}$$

(Here the Lie bracket is defined by $[X, Y] = XY - YX$, for any two $n \times n$ matrices X, Y over F.)

2 Examples of representations: One dimensional case

The simplest representations are the one dimensional ones.

In these cases, a representation of a group G is given by a homomorphism from this group to the group F^* which consists of all nonzero elements in the field F. Such a representation is also called a character of this group.

The determinant is denoted by det. It is something we are very familiar with. In fact it is also a one-dimensional representation of the group of all invertible matrices. This can be seen as follows. The determinant gives a homomorphism from the group of all invertible matrices to F^*: the determinant of an invertible matrix is not zero, the determinant of the product of two matrices is the product of the determinants of them. This representation can be expressed as follows:

$$GL_n(F) \to F^*, \quad A \mapsto \det A,$$

where $F^* = F - \{0\}$ and $GL_n(F) = \{n \times n \text{ invertible matrix over } F\}$. We have learnt this during studying the first course of linear algebra.

Many other examples of one dimensional representations come from the number theory. The Legendre symbol $\left(\frac{x}{p}\right)$ in quadratic reciprocity law is a character of a cyclic group with order $(p-1)$. The quadratic reciprocity law is a pearl in number theory. It can be used to compute the Legendre symbol and gives the solvability conditions for quadratic equations modulo prime numbers. It also reveals the relation among equations modulo different prime numbers.

The first rigorous proof of the quadratic reciprocity law was given by Gauss and published in 1801 in his famous book *Disquisitiones Arithmeticae* (*Research in Number Theory* or *Research of Arithmetics* in its Chinese translation). In the book the quadratic reciprocity law was called "The Fundamental Theorem".

As we all know, the Gaussian sum is very important in solving polynomial equations modulo some prime number and polynomial equations over finite fields. It is also used in the proof of the quadratic reciprocity law, cubic reciprocity law and quartic reciprocity law. One dimensional representations (character) are essential here. The Gaussian sum is given by

$$G(\chi, \sigma) = \sum \chi(t)\sigma(t), \quad t \in \mathbb{F}_p^*,$$

where χ and σ are the characters of the additive group $\mathbb{Z}/p\mathbb{Z} = \mathbb{F}_p$ and the multiplicative group $\mathbb{F}_p^* = \mathbb{F}_p \setminus \{0\}$, respectively.

Let us take a look at an example from analysis. A periodic function with period 2π can be viewed as a function on the circle $S = \{e^{ix} | x \in \mathbb{R}\}$. All the square-integrable functions on S form a Hilbert space H, with the inner product defined by:

$$\langle f, g\rangle = \frac{1}{2\pi}\int_0^{2\pi} f(x)\overline{g(x)}dx.$$

The complex characters of S are the homomorphisms $e^{ix} \to e^{inx}, n \in \mathbb{Z}$. It is obvious that

$$\frac{1}{2\pi}\int_0^{2\pi} e^{inx}e^{imx} = \delta_{nm}.$$

Hence all the characters of S form an orthonormal basis of H. Essentially, this is Fourier analysis: any square-integrable functions on S can be expanded as a linear combination of the orthonormal basis. Notice that the exponential function e^{inx} is an eigenfunction of the differential operator $\frac{d}{dx}$. Perhaps you feel that there is a close relation between representation theory and differential equations. This is indeed the case.

3 The language of module

Before going further, let us introduce another language—the language of module. Let us exam the three components of a representation. It contains an algebraic structure A, a homomorphism from A to another similar algebra structure consisting of all linear transformations on a linear space V. Certainly the space V plays an important role.

We call V an A-module, we also say V is a representation of A. This actually oversimplifies things. Essentially what we need is the homomorphism. However, it is convenient to say that V is a representation of A. Certainly, when we say this, actually we also mean there is a linear action of A on V, in other words, the elements of A act on V linearly. However it is often convenient to say V is a representation of A.

4 Examples of representations: Higher dimensional case

Now we can look at some examples of higher dimensions.

Example 4.1. *All invertible linear transformations on a linear space V is denoted by $GL(V)$. This is a group, called the general linear group. The identity map $GL(V) \to GL(V)$ is of course a representation of $GL(V)$. It is a very simple and natural representation.*

It is also simple to say this using the language of module. Since $GL(V)$ acts on V naturally, V is a module of the group $GL(V)$.

In this language, it is convenient to see that $V \otimes V$ is also a representation of $GL(V)$, with the group action defined by:

$$g(u \otimes v) = gu \otimes gv,$$

where $g \in GL(V), u, v \in V$. In general, if U, V are G-modules, then $U \otimes V$ is also a G-module. Thus the nth tensor product $V^{\otimes n}$ is also a representation of $GL(V)$.

Example 4.2. *There is an addition on the space of all linear transformations on a vector space V. A linear transformation can also be multiplied by elements in the ground field. Further, we define the Lie bracket as: $[f, g] = f \circ g - g \circ f$, where f, g are linear transformations on the linear space V and $\circ$ is the composition. Then the set of all linear transformations of V becomes a Lie algebra, it is called the general linear Lie algebra and denoted by $gl(V)$. The identity map $gl(V) \to gl(V)$ is a representation of $gl(V)$. This is the most natural one.*

It is also easy to rephrase this in the language of module. V is a $gl(V)$-module since $gl(V)$ act on V linearly. In this language, it is clear that $V \otimes V$ is also a representation of $gl(V)$, with the linear action defined by:

$$a(u \otimes v) = au \otimes v + u \otimes av,$$

where $a \in gl(V), u, v \in V$. In general, if U, W are $gl(V)$-modules, then $U \otimes W$ is also a $gl(V)$-module. Hence $V^{\otimes n}$ is also a representation of $gl(V)$. Comparing an action of a group with that of a Lie algebra on the tensor product, we can see that the action of the Lie algebra on a certain space is like differentiation on the group, both of them satisfy the Leibnitz rule.

From the above two examples, we can see that the language of module is very convenient and useful.

Example 4.3. *The polynomial ring $F[x_1, \ldots, x_n]$ over a field F is a module of the general linear group $GL_n(F)$.*

There are several ways to define an action of $GL_n(F)$ on $F[x_1, \ldots, x_n]$.

The n-tuple $(x_1, \ldots, x_n)$ can be viewed as a vector. We denote it by x. For any element g in $GL_n(F)$, we can define a transformation $x \to xg^{-1}$. This sends every x_i to a degree 1 polynomial in $F[x_1, \ldots, x_n]$, denoted by $\sigma_g(x_i)$. Then we define a linear action of g on $F[x_1, \ldots, x_n]$ as follows:

$$\sigma(g) : \sum \xi_{a_1 \cdots a_n} x_1^{a_1} \cdots x_n^{a_n} \to \sum \xi_{a_1 \cdots a_n} \sigma_g(x_1)^{a_1} \cdots \sigma_g(x_n)^{a_n},$$

where $\xi_{a_1 \cdots a_n} \in F$ and $a_i \in \mathbb{Z}_{\geq 0}$. In this way we get a $GL_n(F)$-module structure on $F[x_1, \ldots, x_n]$. This representation is denoted by σ.

For any element g in $GL_n(F)$, we denote g^t to be the transport of g. By the transformation $x \to xg^t$, we can also get a $GL_n(F)$-module structure on $F[x_1, \ldots, x_n]$. It is denoted by τ.

Similarly, using the transformation $x^t \to gx^t$, *we can get a* $GL_n(F)$*-module structure on* $F[x_1, \ldots, x_n]$. *It is denoted by* ρ.

Regarding the x_i *as the coordinate functions of the linear space* F^n, *then* $F[x_1, \ldots, x_n]$ *is the set of all the polynomial functions on* F^n. *For* $p \in F[x_1, \ldots, x_n]$ *and* $g \in GL_n(F)$, *we can define:*

$$\lambda_g(p)(u) = p(g^{-1}u),$$

$$\mu_g(p)(u) = p(g^t u),$$

and

$$\phi_g(p)(v) = p(vg).$$

where $u, v \in F^n$, u *is thought of as a column vector and* v *a row vector. In this way, we get three* $GL_n(F)$*-module structures on* $F[x_1, \ldots, x_n]$. *They are denoted by* λ, μ *and* ϕ, *respectively.*

It is easy to see that essentially σ *and* λ *are the same as representations, while* τ, ρ, μ *and* ϕ *are the same. Also, there is a close relation between the representations* σ *and* τ. *There is actually a duality between them.*

Example 4.4. *The polynomial ring* $F[x_{11}, x_{12}, \ldots, x_{n,n-1}, x_{n,n}]$ *has a natural* $GL_n(F)$*-module structure.*

Indeed, the indeterminates $x_{11}, x_{12}, \ldots, x_{n,n-1}, x_{n,n}$ *can be arranged into an* $n \times n$ *matrix*

$$x = (x_{ij})_{1 \le ij \le n}.$$

For $g \in GL_n(F)$, *consider the transformation*

$$x \to gx.$$

This transformation sends every x_{ij} *to a degree one polynomial of* $x_{11}, x_{12}, \ldots, x_{n,n-1}, x_{n,n}$. *As in the example mentioned above, we can get a* $GL_n(F)$*-module structure on* $F[x_{11}, x_{12}, \ldots, x_{n,n-1}, x_{n,n}]$. *We denote it by* α.

One can also consider the transformation

$$x \to xg^{-1}.$$

Similarly, we get a $GL_n(F)$*-module structure on* $F[x_{11}, x_{12}, \ldots, x_{n,n-1}, x_{n,n}]$. *It is denoted by* β.

We can consider the transformation

$$x \to xg^t.$$

Then we get a $GL_n(F)$*-module structure on* $F[x_{11}, x_{12}, \ldots, x_{n,n-1}, x_{n,n}]$. *It is denoted by* γ.

Let $M_n(F)$ *denote all the* $n \times n$ *matrices over* F. *It is a linear space and contains* $GL_n(F)$ ($GL_n(F)$ *is open in* $M_n(F)$ *under the Zariski topology*). *Viewed as*

$x_{11}, x_{12}, \ldots, x_{n,n-1}, x_{n,n}$ as coordinate functions of the linear space $M_n(F)$, then $F[x_{11}, x_{12}, \ldots, x_{n,n-1}, x_{n,n}]$ is the ring of all the polynomial functions on $M_n(F)$. As before, we can define a linear action of $GL_n(F)$ on $F[x_{11}, x_{12}, \ldots, x_{n,n-1}, x_{n,n}]$ as follows:

$$\theta_g(p)(u) = p(g^{-1}u),$$

$$\kappa_g(p)(u) = p(g^t u),$$

and

$$\eta_g(p)(v) = p(vg),$$

where $p \in F[x_{11}, x_{12}, \ldots, x_{n,n-1}, x_{n,n}]$, $g, u \in GL_n(F)$.

This way, we give three $GL_n(F)$-module structures on $F[x_{11}, x_{12}, \ldots, x_{n,n-1}, x_{n,n}]$. They are denoted by θ, κ and η respectively.

It is easy to see that essentially β and θ are the same representations, while α, γ, κ and η are also the same. Also, α and β are dual to each other.

As we have seen, there are very rich representation structures on the polynomial rings. These structures are important in representation theory, algebraic geometry, and combinatorics. Many important problems can be solved if the representations of the polynomial ring are understood clearly. Unfortunately people are not able to do that now.

Example 4.5. *The tangent space of a Lie group (algebraic group) at the identity element gives a representation of the Lie group (algebraic group), induced by the conjugate action on itself. It is called the adjoint representation of the Lie group (algebraic group).*

We know that the tangent bundle of a differential manifold is important in understanding differential manifolds. Similarly, as can be inferred, the adjoint representation may also be important for Lie groups. Indeed, if one is asked to pick out the most important one among representations of a Lie group, the first choice would be the adjoint representation.

The adjoint representation contains very rich information. It is important in the study of the structure and representation of Lie groups. For example, through the adjoint representation, we can define the root system, whose significance is never over-emphasized in understanding the structures and representations of Lie groups.

The next example comes from topology and algebraic geometry.

Example 4.6. *A filtration of subspaces in $\mathbb{C}^n$ consists of the following data*

$$v_1 \subset V_2 \subset \cdots \subset V_n = \mathbb{C}^n, \quad \dim V_i = i.$$

All the filtrations form a set denoted by X. This is a smooth complex manifold, called a flag manifold or flag variety, which is important in topology, algebraic *geometry and representation theory. If $\mathcal{L}$ is a holomorphic line bundle, then the cohomology group $H^i(X,\mathcal{L})$ is a $GL_n(\mathbb{C})$-module. It is a very interesting object in mathematics. Of course if we consider a holomorphic vector bundle instead of a holomorphic line bundle, the same thing happens.*

To state more general results, we need to use the language of algebraic groups.

Assume G is a connected reductive algebraic group over some algebraic closed field k, B is its Borel subgroup. Then the homogeneous space G/B is a smooth variety called a flag variety. (For example, when $G = GL_n(\mathbb{C})$, one can take B to be the set of all the invertible upper triangular matrices, then G/B is isomorphic to the flag manifold X defined above). If E is a rational B-module, then $G\times^B E$ is vector bundle over G/B its cohomology groups are G-modules. The study of the G-module structure of these cohomology groups is important in representation theory.

Now we turn to some infinite dimensional modules.

Example 4.7. *Denote $SL_2(\mathbb{C})$ to be the set of all 2×2 complex matrices with determinant 1. It is a group under matrix multiplication. The set $L^2(\mathbb{C})$ of all square-integrable functions over complex number field $\mathbb{C}$ is an $SL_2(\mathbb{C})$-module.*

In fact, $L^2(\mathbb{C})$ has infinitely many $SL_2(\mathbb{C})$-module structures. Let $GL(L^2(\mathbb{C}))$ be the group of all the invertible linear transformations on $L^2(\mathbb{C})$. For any integer k and real number v, we can define a representation $\mathcal{P}^{k,iv}: SL_2(\mathbb{C})\to GL(L^2(\mathbb{C}))$ as follows:

$$\mathcal{P}^{k,iv}\begin{pmatrix} a & b \\ c & d\end{pmatrix} f(z) = |-bz+d|^{-2-iv}\left(\frac{-bz+d}{|-bz+d|}\right)^{-k} f\left(\frac{az-c}{-bz+d}\right),$$

where $z\in\mathbb{C}$, $f\in L^2(\mathbb{C})$.

These representations are called principal series. They are unitary representations. That is, for every $g\in SL_2(\mathbb{C})$ the linear transformation $\mathcal{P}^{k,iv}(g)$ is a unitary operator, i.e. it is invertible and preserves the inner product on $L^2(\mathbb{C})$.

More generally, for any number $w = u+iv$, consider the measure $(1+|z|)^{\operatorname{Re} w}dxdy$ and the space $L^2_w(\mathbb{C})$ of all square-integrable functions with respect to this measure. For any integer k, we can define a representation $\mathcal{P}^{k,w}: SL_2(\mathbb{C})\to GL(L^2_w(\mathbb{C}))$ as follows:

$$\mathcal{P}^{k,w}\begin{pmatrix} a & b \\ c & d\end{pmatrix} f(z) = |-bz+d|^{-2-w}\left(\frac{-bz+d}{|-bz+d|}\right)^{-k} f\left(\frac{az-c}{-bz+d}\right),$$

where $z\in\mathbb{C}$, $f\in L^2_w(\mathbb{C})$. It is unitary if and only if w is an imaginary number.

If $k=0$, w is a real number and $0<w<2$, then $\mathcal{P}^{k,w}$ is a unitary representation with respect to the following inner product:

$$\langle f,g\rangle = \int_{\mathbb{C}}\int_{\mathbb{C}}\frac{f(z)\overline{g\zeta}dzd\zeta}{|2-\zeta|^{2-w}}.$$

These unitary representations are called complementary series.

Example 4.8. *Let $SL_2(\mathbb{R})$ be the set of all the 2×2 real matrices with determinant 1. It is a group under the matrix multiplication. All square-integrable function $L^2(\mathbb{C})$ over the field of real numbers $\mathbb{R}$ is an $SL_2(\mathbb{R})$-module.*

Similarly, $L^2(\mathbb{R})$ has infinite many $SL_2(\mathbb{R})$-module structures. For any real number v, we can define a representation $\mathcal{P}^{\pm,iv} : SL_2(\mathbb{R}) \to GL(L^2(\mathbb{R}))$ as follows:

$$\mathcal{P}^{\pm,iv}\begin{pmatrix} a & b \\ c & d \end{pmatrix} f(x) = \begin{cases} |-bx+d|^{-1-iv} f(\frac{ax-c}{-bx+d}), & \text{for } + \\ \operatorname{sgn}(-bx+d)|-bx+d|^{-1-iv} f(\frac{ax-c}{-bx+d}), & \text{for } - \end{cases}$$

where $x \in \mathbb{R}$, $f \in L^2(\mathbb{R})$.

These representations are called principal series. They are unitary representations.

For any real number $0 < u < 1$, consider the Hilbert space

$$H_u = \left\{ f : \mathbb{R} \to \mathbb{C} \middle| \|f\|^2 = \int_{-\infty}^{\infty}\int_{-\infty}^{\infty} \frac{f(x)\overline{f(y)}dxdy}{|x-y|^{1-u}} \right\},$$

and the following action of $SL_2(\mathbb{R})$:

$$\mathcal{C}^u \begin{pmatrix} a & b \\ c & d \end{pmatrix} f(x) = |-bx+d|^{-1-u} f\left(\frac{ax-c}{-bx+d}\right).$$

These unitary representations are called complementary series.

For any integer $n \geq 2$, consider the Hilbert space

$$L_n = \left\{ f \text{ is analytic on } \operatorname{Im} z > 0 \middle| \|f\|^2 = \int\int_{\operatorname{Im} z>0} |f(z)|^2 y^{n-2} dxdy < 0 \right\},$$

with the following action of $SL_2(\mathbb{R})$:

$$\mathcal{D}_n^+ f(z) = |-bz+d|^{-n} f\left(\frac{az-c}{-bz+d}\right).$$

This is a unitary representation. Under complex conjugation, it produces another unitary representation $\mathcal{D}_n^-$. These representations are called discrete series. The representations $\mathcal{D}_1^+$ and $\mathcal{D}_1^-$ defined below are their limits. Their definitions are similar to those of the discrete series, but with the following norm

$$\|f\|^2 = sup_{y>0} \int_{-\infty}^{\infty} |f(x+iy)|^2 dx.$$

Example 4.9. *Assume G is a Lie group, Γ is a discrete subgroup of G such that the volume of G/Γ is finite. For example, G is the real number field and Γ is the set $\mathbb{Z}$ of all integers. G acts on the coset space G/Γ by right multiplication, thus acting linearly on the space $L^2(G/\Gamma)$ of square-integrable functions on G/Γ. Then $L^2(G/\Gamma)$ is a representation of G. It is significant in number theory.*

At this point it should be mentioned that there is a famous Selberg trace formula for the character. It has many applications in number theory, representation theory and even in physics.

We just mentioned physics, let us talk about it a little more. In fact group theory is very important for physics. In the single particle model, the wave function of a single electron generates a representation of the group $SO(3)$. If you consider the spin-wave function of a spinor, you will get a representation of $SU(2)$.

Using the ten-dimensional representation of $SU(3)$, Gell-Mann predicted the existence of the Ω-particle. This was proved by experiments later, in 1964.

So far we mainly talked about the representations of groups and Lie algebras. Now let us take a look at some representations of path algebras, it would be simpler to describe the path algebras by quivers. Remember that the modality should serve for context.

We consider a quiver Γ with n vertices:

$$\circ \longrightarrow \circ \longrightarrow \circ \longrightarrow \cdots \longrightarrow \circ \longrightarrow \circ$$

A representation of Γ is just a set of vector spaces and linear maps between them. One assigns a linear space to a vertex and a linear map to an arrow:

$$\{V_i, f_j | 1 \leq i \leq n, 1 \leq j \leq n-1\}$$

where V_i is the linear space over F corresponding to the vertex i and $f_j : V_j \to V_{j+1}$ is a linear transformation corresponding to $j \to j+1$.

This definition seems complicated and formal, it is different from the definitions of representations we mentioned before. However, a representation of the quiver Γ is actually equivalent to that of the path algebra $F[\Gamma]$. The representations of path algebras is an important class of representations in representation theory.

It is convenient to express the representations of path algebras in terms of representations of quivers. Moreover, this will give unexpected connections to some other things in representation theory. This will be talked later.

5 Basic thought of representation theory

We have seen many examples and digested the terminologies. We can see that the basic thought of representations consists of two things: symmetry and linearization.

The algebraic structure reflects the symmetry. This is especially easy to understand for groups. We know that a square is symmetric , while a circle is more symmetric. From the perspective of the corresponding groups, the group preserving the circle is much larger than that preserving the square. To see this, we put the circle and the square into $\mathbb{R}^2$ such that their common centers coincide with the origin. All the reflections, revolutions, and the identity map are invertible linear transformations. They are closed under composition and thus form a group. All the reflections and revolutions preserve the circle. However there are very few transformations which preserve the square, only two reflections and four revolutions.

A representation of an algebraic structure gives the corresponding linearization. It also reflects some symmetry of the associated linear space. This is a mutual benefit, on one hand, it gives a concrete representation of the algebraic structure, on the other hand it associates some symmetry to the linear space.

6 Fundamental problems in representation theory

Now that representations exist commonly in so many branches of mathematics in so interesting ways, it is necessary for us to study them systematically. Then there are some basic problems:

- ⋆ What kind of representations are the most basic ones?
- ⋆ How does one build general representations from the basic ones?
- ⋆ How does one construct the most basic representations?
- ⋆ What are the properties of the most basic representations, such as classification, dimension, character?
- ⋆ What are the properties of some naturally constructed representations?

7 The most basic representations

Due to the limited time in this lecture, we shall only discuss the first problem.

To proceed, we need to introduce a new concept. We consider an algebraic structure A (group, Lie algebra, and so on) and a representation of A. Then A acts on the corresponding linear space V linearly. If a subspace of V is invariant under the action of A, we call this subspace a subrepresentation of V.

Let us take a look at some examples first. The symmetric group S_n acts naturally on the linear space $V = \mathbb{C}^n$ by

$$\sigma(a_1, \ldots, a_n) = (a_{\sigma(1)}, \ldots, a_{\sigma(n)}),$$

where $\sigma \in S$ and $(a_1, \ldots, a_n) \in \mathbb{C}^n$. Obviously, $\{(a, \ldots, a) | a \in \mathbb{C}\}$ is a subrepresentation of V.

We have known that $F[x_{11}, x_{12}, \ldots, x_{n,n-1}, x_{n,n}]$ has a natural $GL_n(F)$-module structure. The determinant spans a one-dimensional subspace of this representation. It is easy to see that this one-dimensional subspace is a subrepresentation of $F[x_{11}, x_{12}, \ldots, x_{n,n-1}, x_{n,n}]$. Also the space of all homogeneous polynomials with degree i gives a subrepresentation of $F[x_{11}, x_{12}, \ldots, x_{n,n-1}, x_{n,n}]$.

Every representation has two trivial subrepresentations: the null space and itself.

What are the most basic representations? They are the irreducible representations, which are also called simple representations. If a representation has no other subrepresentations besides the null representation and itself, it is then called an irreducible representation. Sometimes people also call it a simple representation, an irreducible module or a simple module, etc.

The simplest irreducible representations are one dimensional representations, which cannot have subrepresentations besides the null representation and itself.

Let us see more examples. Recall that the set $GL(V)$ of all the invertible linear transformations on a linear space V form a group. Under invertible linear transformations, a nonzero vector can be transformed to any other nonzero vector. So as a representation of $GL(V)$, V can not have any nontrivial subrepresentations, thus it is irreducible. Similarly, V is an irreducible representation of the Lie algebra $gl(V)$.

The linear space $\mathbb{C}^n$ is an irreducible representation of $SL_n(\mathbb{C})$.

The polynomial ring $F[x_1, \ldots, x_n]$ mentioned before is a representation of the general linear group $GL_n(F)$. Assume $F = \mathbb{C}$, collecting all degree i homogeneous polynomials with n variables, one then gets a representation of the general linear group $GL_n(F)$. It is irreducible. This tells us that there are many irreducible representations.

Let $sl_n(\mathbb{C})$ denote all the $n \times n$ matrices with trace zero. Note that in matrix theory, conjugation is frequently used. The trace of an invertible matrix will not change under conjugation. Under conjugation, $GL_n(\mathbb{C})$ acts on $sl_n(\mathbb{C})$ and then $sl_n(\mathbb{C})$ is an irreducible representation of $GL_n(\mathbb{C})$. Also $GL_n(\mathbb{C})$ acts on $gl_n(\mathbb{C})$ in the same way. But $gl_n(\mathbb{C})$ is not irreducible, since all the scalar matrices form a one dimensional representation. Now $gl_n(\mathbb{C})$ is the direct sum of two irreducible representations. In addition, $V \otimes V$ is not irreducible as long as V is not of dimension one.

$L^2(\mathbb{C})$ is an irreducible representation of $SL_2(\mathbb{C})$. In fact, there are infinitely many irreducible representation structures of $SL_2(\mathbb{C})$ on $L^2(\mathbb{C})$.

Similar to the case of algebraic structures, when we try to study the relation between different representations, we use the terms homomorphism and isomorphism. A homomorphism between two representations is a linear map between then which preserves the action. In other words, for two representations(modules) U, V of some algebraic structure, a linear map $\phi : U \to V$ is called a homomorphism if it satisfies the condition $\phi(au) = a\phi(u), u \in U, a \in A$. A homomorphism is called an isomorphism if it is invertible and its inverse is also a homomorphism. Isomorphic representations are viewed as the same.

For irreducible representations, we have the following famous lemma.

Lemma 7.1. (*Schur Lemma*) *A nonzero homomorphism between two irreducible representations is an isomorphism.*

This lemma seems very simple, as well as its proof. But it is extremely useful. For example, suppose we consider two irreducible representations with a linear map between them. In the case of compact Lie groups, we get a homomorphism between them by integration. If the two representations are not isomorphic, then the homomorphism must be zero.

We can get a lot of useful information if the homomorphism is zero, or if the homomorphism is an isomorphism. From Schur Lemma, it is easy to get the orthogonal relation among the characters of irreducible representations of compact Lie groups.

In the case of the unit circle, this orthogonal relation is the familiar equality $\int_S e^{inx}dx = 0$, where $n \in \mathbb{Z}$. Here we used the fact that the irreducible representations of S are $e^{ix} \to e^{inx}, n \in \mathbb{Z}$. In other cases, the orthogonal relation may be not easy to identify at first glance.

8 Classification of irreducible modules

To classify irreducible modules is an important problem in mathematics. Sometimes classification could be simple, but it can also be very complicated however. It often depends on the ground field of representations.

Theorem 8.1. *Let G be a finite group and the spaces of representations be complex linear spaces. Then*
(1) *The number of irreducible representations is equal to the number of the conjugaey classes in G;*
(2) *Every representation is a direct sum of some irreducible representations;*
(3) *The sum of squares of the dimensions of all irreducible representations (up to isomorphism) is equal to the order of the group.*

This theorem is important, but does not give much information about the classification. For some important groups, the classification of irreducible representations is clear.

Example 8.1. *Symmetric group.*

- ⋆ *The number of complex irreducible representations of S_n is equal to the number $P(n)$ of partitions of n.*
- ⋆ *The dimension of an irreducible representation is computable.*

★ *The classification of irreducible representations is known if the characteristic p of the ground field F satisfies $p \leq n$.*

★ *But we know little for other properties, such as the dimensions and characters of irreducible representations. This is an important problem. Unexpectedly, there are deep connections between this natural and seeming elementary question, and affine Lie algebras, affine Schubert varieties. There is also a close relation with the representations of the general linear group.*

Let us continue to consider the classification of irreducible representations.

We are already very familiar with the unit circle. This is a topological group. In this lecture we shall only consider the unitary representations for topological groups. That is, we require that the representation space to be a Hilbert space and the elements of the group act as bounded unitary operators on the space. For the unit circle, we have the following theorem.

Theorem 8.2. *Let S be the unit circle.*

$$\begin{aligned} \{\textit{irreducible unitary representations of } S\} &\overset{1-1}{\longleftrightarrow} \mathbb{Z} \\ e^{inx} &\leftrightarrow n \\ \{\textit{irreducible unitary representations of } \mathbb{Z}\} &\overset{1-1}{\longleftrightarrow} S \end{aligned}$$

In general, all the irreducible representations of a commutative topological group over complex number field are one dimensional. If G is a connected commutative topological group, then the isomorphism classes of its irreducible representations form a discrete commutative topological group A. There is a one to one correspondence between the isomorphism classes of the irreducible unitary representations of A and those of G. This is the Pontryagin duality theorem. Therefore, the properties of irreducible representations for commutative topological groups are clear.

Pontryagin was a Soviet mathematician. He lost his eyesight when he was 15 years old. Despite of that, he did many important work in mathematics with highly perseverance. He wrote a famous book *Continuous Groups* which has been translated into Chinese. The Pontryagin classes introduced by him is important in differential topology.

For the representations of compact Lie groups, the work of Weyl is far-reaching.

Theorem 8.3. (*Peter-Weyl*) *For a compact Lie group G,*
(1) *The subspace spanned by the entries of the matrix coefficients of all finite dimensional unitary representations is dense in $L^2(G)$.*
(2) *Every irreducible complex representation is a unitary representation and finite dimensional.*
(3) *Every unitary representation is completely reducible.*

(4) *The multiplicity of every irreducible representation in* $L^2(G)$ *is equal to the dimension of this representation.*

Theorem 8.4. (*Weyl*) *The formula for the characters of irreducible representations of compact connected semisimple Lie groups.*

The Weyl character formula is very famous. To prove this formula, he combined techniques and knowledge from different branches in mathematics.

We have mentioned the general linear group several times. Its representation theory has very rich contents. Now we consider the classification of its irreducible representations. We shall restrict ourselves to finite dimensional representations.

Consider the general linear groups over the complex number field. There is a one-to-one correspondence between the isomorphism classes of its finite dimensional irreducible representations and those n-tuples of integers satisfying the condition that a component is not less than the component following it.

Theorem 8.5.

$$\{\text{ finite dimensional irreducible representations of } GL_n(\mathbb{C})\}$$
$$\downarrow 1-1$$
$$\{(a_1, a_2, \ldots, a_n) \in \mathbb{Z}^n | a_1 \geq a_2 \geq \cdots \geq a_n\}$$

The character (including dimension) of an irreducible representation of $GL_n(\mathbb{C})$ is given by the Weyl character formula. The maximal compact subgroup of $GL_n(\mathbb{C})$ is the unitary group $U(n)$. A finite dimensional representation of $GL_n(\mathbb{C})$ is irreducible if and only if it is irreducible as a representation of $U(n)$. This is the famous unitary trick.

The general linear Lie algebra $gl_n(\mathbb{C})$ is the tangent space of $GL_n(\mathbb{C})$ at the unit element. Hence a finite dimensional representation of $GL_n(\mathbb{C})$ induces a finite dimensional representation of $gl_n(\mathbb{C})$ by differentiation.

Theorem 8.6.

$$\{\text{finite dimensional irreducible representations of } gl_n(\mathbb{C})\}$$
$$\downarrow 1-1$$
$$\{(a_1, a_2, \ldots, a_n) \in \mathbb{Z}^n | a_1 \geq a_2 \geq \cdots \geq a_n\}$$

The character(including dimension) of an irreducible representation of $gl_n(\mathbb{C})$ is also given by the Weyl character formula.

All the linear transformations on a linear space V is denoted by $\text{End}(V)$. This is an algebra and has a Lie algebra structure. When the dimension of V is finite, $\text{End}(V)$ can be identified with the matrix algebra $M_n(F)$. The corresponding Lie algebra is denoted by $gl_n(F)$, where F is the ground field. We have seen that the representation theory of $gl_n(F)$ has rich structures. However, $\text{End}(V)$ has only one irreducible representation V. Indeed, $\mathrm{End}(V)$ is a simple algebra. Similarly, $M_n(F)$ has only one irreducible representation, the n dimensional linear space F^n.

Note that a set could be an algebra and a Lie algebra at the same time. The content of its representation theory when regarded as an algebra is much less than the content of its representation theory when viewed as a Lie algebra. As you can see, all $n \times n$ complex matrices viewed as an algebra, denoted by $M_n(\mathbb{C})$, has only one irreducible representation $\mathbb{C}^n$. However when it is regarded as a Lie algebra denoted by $gl_n(\mathbb{C})$, it has infinitely many isomorphism classes of irreducible representations, whose content is richer. The latter has deep connections with other branches of mathematics, and also has wide applications in physics. Hence one would believe that the representation theory of Lie groups and Lie algebras is richer, more sophisticated and more interesting.

Let us pay more attention to the irreducible representations of the general linear group. Now the ground field is $\bar{\mathbb{F}}_p$, where p is a prime number. It is the algebraic closure of $\mathbb{F}_p$.

One considers polynomials in one variable with coefficients in $\mathbb{F}_p$. Collecting the roots of these polynomials, one then gets $\bar{\mathbb{F}}_p$. The classification of finite dimensional irreducible representations of $GL_n(\bar{\mathbb{F}}_p)$ is the same as the classification of finite dimensional irreducible representations of $GL_n(\mathbb{C})$, both corresponding to some n tuples of integers:

Theorem 8.7.

$$\{\textit{finite dimensional irreducible representations of } GL_n(\bar{\mathbb{F}}_q)\}$$
$$\downarrow 1-1$$
$$\{(a_1, a_2, \ldots, a_n) \in \mathbb{Z}^n | a_1 \geq a_2 \geq \cdots \geq a_n\}$$

But the characters of irreducible representations become more complicated. So far, we even do not know the formula of the dimension in general. Lusztig's conjecture claims that there is connection between the finite dimensional irreducible representations and the singularities of some affine Schubert variety when p is not too small. Now this conjecture has been proved when p is very large.

Over the field of complex numbers, the finite dimensional representation theory of the general linear group $GL_n(\mathbb{C})$ is the same as that of $gl_n(\mathbb{C})$. But they are totally different when the field is $\bar{\mathbb{F}}_p$. Every irreducible representation of $gl_n(\bar{\mathbb{F}}_q)$ is finite dimensional. However, the classification is complicated, and has a close relation with Springer fibers. The dimensions are not known in general cases. Lusztig has also given some conjectures about them.

Theorem 8.8. *Every irreducible representation of $gl_n(\bar{\mathbb{F}}_q)$ is finite dimensional.*

Part of irreducible representations coincide with the representations of $GL_n(\bar{\mathbb{F}}_p)$, which is difficult to study. The other part of irreducible representations does not have direct connection with those of $GL_n(\bar{\mathbb{F}}_p)$.

The classification of irreducible representations of a quiver is simple. The classification of indecomposable representations of a quiver is more interesting. It is one of main topics in representation theory of algebras.

Consider the quiver Γ

$$\circ \longrightarrow \circ \longrightarrow \cdots \longrightarrow \circ \longrightarrow \circ$$

Γ has n irreducible representations over F up to isomorphisms, corresponding to the n vertices of Γ. Denote by E_i the irreducible representation corresponding to the i-th vertex. The linear space associated to the vertex $j \neq i$ is 0 and is F for the vertex i. All the linear transformations corresponding to the arrows are 0. We can then use the following simple notation

$$E_i = (0, \dots, 0, \underset{i\text{-th}}{F}, 0, \dots, 0)$$

There are $n(n+1)/2$ in total indecomposable representations:

$$\{\text{indecomposable representations of } \Gamma\} \xrightarrow{1-1} \{\text{positive roots of } GL_{n+1}(\mathbb{C})\}.$$

The relation between these indecomposable representations and positive roots is interesting. When F is a finite field, we can construct corresponding Hall algebra.

In the end of the eighties in the last century, Ringel found a connection between the Hall algebras and quantum groups. This connection led to the discovery of canonical bases of quantum groups by Lusztig. The canonical bases have nice properties and have deep connections with algebraic geometry and representation theory. Since a quantum group is a deformation of the universal enveloping algebra of a Lie algebra and becomes the universal enveloping algebra when the parameter is taken to be 1, the canonical basis of a quantum group gives the canonical basis of the universal enveloping algebra of a Lie algebra. It is surprising that the canonical basis cannot be found in a Lie algebra without the help of quantum groups.

Let us consider the classification of infinite dimensional irreducible representations of Lie groups. We consider some simple examples such as $SL_2(\mathbb{C})$ and $SL_2(\mathbb{R})$. Since a Lie group has a topological structure, we mainly concern about unitary representations. By the following theorem, we can see that the classification of infinite dimensional irreducible representations of Lie groups is not elementary.

Theorem 8.9. *The classification of infinite dimensional irreducible unitary representations of $SL_2(\mathbb{C})$ is as follows:*
(1) *The trivial representation;*
(2) *Unitary principal series: $\mathcal{P}^{k,iy}$ $(k \in \mathbb{Z},\ y \in \mathbb{R})$;*
(3) *Complementary series: $\mathcal{C}^x$, where $0 < x < 2$.*
The isomorphisms are $\mathcal{P}^{k,iy} \cong \mathcal{P}^{-k,-iy}$.

From the above theorem, we know that $SL_2(\mathbb{C})$ has a large number of infinite dimensional irreducible unitary representations. Roughly speaking, it mainly consists of unitary principal series and complementary series. A representation in unitary principal series has two arguments: the integer k and the real number y. A representation in complementary series has only one argument: the real number x which is between 0 and 2.

The group $SL_2(\mathbb{C})$ has many irreducible finite dimensional representations, which are not unitary except the trivial one.

Theorem 8.10. *The classification of infinite dimensional representations of $SL_2(\mathbb{R})$ is given as follows:*
(1) The trivial representation;
(2) Discrete series: $\mathcal{D}_n^{\pm}, n \geq 2$ and the limits $\mathcal{D}_1^{\pm}$;
(3) Unitary principal series: $\mathcal{P}^{+,iy}$ and $\mathcal{P}^{-,iz}$ $(y \in \mathbb{R},\ 0 \neq z \in \mathbb{R})$;
(4) Complementary series: $\mathcal{C}^x$, where $0 < x < 1$.
The isomorphisms are $\mathcal{P}^{+,iy} \cong \mathcal{P}^{-,-iy}$, $\mathcal{P}^{-,iy} \cong \mathcal{P}^{+,-iy}$.

From the theorem, we see that the classification of irreducible unitary representations of $SL_2(\mathbb{R})$ is similar to that of $SL_2(\mathbb{C})$, but is a little more complicated and has a new set of representations: the discrete series.

The problem of classifying of irreducible unitary representations of a general Lie group has not been solved yet, whether the Lie group is real or complex. The classification is clear only for some groups.

We now stop talking about classification of irreducible representations here.

9 Research methods

Now let us talk about the methods used in the study of representation theory. We have seen that representations appear naturally in several branches of mathematics such as algebra, analysis and geometry. Hence the methods used in classification of irreducible representations vary from ideas and techniques in algebra, analysis, differential geometry, algebraic geometry and topology, etc.

Let us see how some methods are used in representation theory.

9.1 Algebraic methods

Algebraic method is one of the main methods used in the study of representation theory. After all, we are considering homomorphisms between algebraic structures. In representation theory of finite groups, Lie (super) algebras, Lie groups, algebraic groups, quantum groups, and algebras etc, the algebraic methods are extremely important.

Let us give more details about the algebraic methods.

Representation theory of finite groups. Let G be a finite group, F be a field. We consider the group algebra $F[G]$, which is a vector space over F with the elements of G being a basis. The multiplication in G induces the multiplication in $F[G]$:

$$\sum_{g\in G} a_g g \cdot \sum_{h\in G} a_h h = \sum_{g,h\in G} a_g a_h gh.$$

The group algebra is then naturally a G-module. A main problem representation theory of finite groups is to consider the properties of $F[G]$, such as semisimplicity, how to decompose it into the direct sum of small modules and the properties of the indecomposable factors, etc.

Another important problem is to consider induced representations. Let H be a subgroup of G, V be an H-module. Then the tensor product

$$Ind_H^G V = F[G] \otimes_{F[H]} V$$

is a representation of G. We want to know whch irreducible modules could be produced by this module.

Character is an important notion in representation theory. For representation theory of finite groups, the definition of character is simple when the characteristic of the ground field F is zero. If $\phi : G \to GL_n(F)$ is such a representation of G, then the character of ϕ is the trace function induced by ϕ:

$$ch\phi : G \to F, \;\; g \mapsto \text{the trace of the matrix}\phi(g), \;\forall g \in G$$

Since similar matrices have the same trace, isomorphic representations have the same character. More interestingly, if the characters of two irreducible representations are the same, then the two representations are isomorphic.

Representation theory of Lie (super)algebras and Lie groups. The tangent space of a Lie group at the unit element is a Lie algebra, so a representation of a Lie group gives naturally a representation of its Lie algebra. Conversely, a representation of Lie algebra often has a structure of a representation of Lie group.

The bracket of a Lie algebra is not associative. This is not convenient. For this reason, people constructed an associative algebra associated to the Lie algebra, called the universal enveloping algebra. A representation of a Lie algebra is equivalent to a representation of its universal enveloping algebra. Using the universal enveloping algebra it is convenient to define the induced representations of a Lie subalgebra.

Assume $\mathfrak{g}$ is a Lie algebra, $\mathfrak{h}$ is a Lie subalgebra. Their universal enveloping algebras are denoted by $U(\mathfrak{g})$ and $U(\mathfrak{h})$, respectively. Let V be a representation of $\mathfrak{h}$. Then it is a $U(\mathfrak{h})$-module. So the tensor product

$$Ind_{\mathfrak{h}}^{\mathfrak{g}} V = U(\mathfrak{g}) \otimes_{U(\mathfrak{h})} V$$

is naturally a $U(\mathfrak{g})$-module, i.e., it is a representation of $\mathfrak{g}$.

Universal enveloping algebra and induced representation are important concepts in representation theory of Lie (super)algebras and Lie groups. In representation theory of semisimple Lie algebras, the notion of character is also important, whose definition is similar to that of the finite groups.

Representation theory of algebraic groups and quantum groups. Algebraic methods are also important in the representation theory of algebraic groups and quantum groups. Suppose G is an algebraic group over an algebraic closed field k. $k[G]$ is the algebra consisting of the all rational functions on G, it is also called the coordinate algebra. The coordinate algebra is actually a Hopf algebra, whose coproduct is induced by the multiplication in G. For example, when G is the general linear group $GL_n(k)$, $k[G]$ is the ring $k[x_{11}, x_{12}, \ldots, x_{nn}, \det^{-1}]$, whose coproduct is defined as:

$$\Delta : k[G] \to k[G] \otimes k[G], \quad x_{ij} \to \sum_{1 \leq k \leq n} x_{ik} \otimes x_{kj},$$

where $x_{11}, x_{12}, \ldots, x_{nn}$ are the coordinate functions of G and det is the determinant of the matrix $(x_{ij})_{1 \leq i,j \leq n}$.

Any rational representation of an algebraic group G is naturally a comodule of the coordinate algebra $k[G]$ and any comodule of the coordinate algebra $k[G]$ has a structure of a rational representation of the algebraic group G.

If you are not familiar with the notion comodule, you can use the hyperalgebra U_G of $k[G]$ to think about it. This super algebra is an associative algebra. A structure of rational G-module on a finite dimensional vector space V is equivalent to a U_G-module structure on V.

A quantum group is a deformation of a universal enveloping algebra. Hence in many ways representation theory of quantum groups is parallel with representation theory of Lie algebras and algebraic groups (partially). It borrows many ideas and skills from them.

Certainly, algebraic methods are the main methods used in representation theory of algebra.

9.2 Analytical methods

The analytical methods are mainly used in representation theory of topological groups and Lie groups. Our main interest there is the representation of locally compact groups. For example, the unit circle S is a locally compact group and the additive group $\mathbb{R}$ is also locally compact.

Suppose G is a locally compact topological group, X is a locally compact Hausdoff space. G acts on X continuously and there is a G-invariant positive measure on X. A central problem in the representation theory of topological groups is to decompose $L^2(X)$ into a direct sum of irreducible representations. Since $L^2(X)$ is a Hilbert space, functional analysis is a main tool which is widely used.

Analytic methods are also important in representation theory of algebraic groups over p-adic fields. Assume G is such an algebraic group (such as $GL_n(\mathbb{Q}_p)$), F is a field (often $\mathbb{C}$). Consider the space

$$\mathcal{C}^\infty(G, F) = \{\text{locally constant function } G \to F\}.$$

It is a representation of G naturally. The research of this representation is an important subject.

9.3 Differential geometric methods

In representation theory, the differential geometric methods are mainly used in the researches of representation theory of Lie groups, since Lie groups are manifolds. Let G be a Lie group, H its closed subgroup. Then the geometry on homogeneous space G/H has deep relation with representation theory of the Lie group G.

9.4 Algebraic geometric methods

Algebraic geometric methods have significant influences on several aspects of representation theory and have brought tremendous progresses to representation theory of algebraic groups, finite groups and Lie algebras.

Since an algebraic group is a variety itself, algebraic geometry plays an important role in representation theory of algebraic groups naturally. We will mainly consider the linear algebraic groups. Several familiar groups such as $GL_n(F)$, $SL_n(F)$, $Sp_{2n}(F)$ and $SO_n(F)$ are linear algebraic groups, where F is algebraically closed field such as $\mathbb{C}$, the algebraic closure $\bar{\mathbb{F}}_p$ of finite field $\mathbb{F}_q$ and so on. One would be able to understand the algebraic groups if one could understand these groups. However, for these groups, there are many problems which have not been solved.

Let X be a variety with an action of G on it, $\mathcal{L}$ a G-equivariant vector bundle over X. Then the cohomology group $H^i(X, \mathcal{L})$ is a rational G-module. The cohomology groups are important objects in representation theory of algebraic groups.

One problem is to determine when these cohomology groups are not zero. When G is a reductive algebraic group over $\mathbb{C}$, X is a flag variety and $\mathcal{L}$ is a G-equivariant line bundle over X, the famous Borel-Weil-Bott Theorem says that there exists at most one i such that $H^i(X, \mathcal{L})$ is not zero and in that case it is an irreducible G-module.

If the character of the field F is greater than 0, things become very complicated. In these cases the Borel-Weil-Bott Theorem does not hold. But for the

0th order cohomology group of the line bundle over the flag variety, a theorem of Kempf tells us when it is zero. If the 0th order cohomology group of the line bundle is not zero, its structure as a G-module is clear. By Serre duality, the highest cohomology group is also understood. However, little is known about other cohomology groups.

Algebraic geometry also plays a vital role in the representation theory of finite groups of Lie type. A finite groups of Lie type is a linear group over a finite field, such as $GL_n(\mathbb{F}_q)$, $SL_n(\mathbb{F}_q)$, $Sp_{2n}(\mathbb{F}_q)$, $SO_n(\mathbb{F}_q)$ and so on, where $q = p^a$ is power of some prime number p. The importance of finite groups of Lie type also lies in the following fact: finite simple groups, except alternating group, finite commutative groups of prime order and 26 discrete simple groups, are finite groups of Lie type.

A finite group of Lie type is a special subgroup of an algebraic group, it is the fixed point locus of the Frobenius map. Then algebraic group theory comes into play. In 1976, Deligne and Lusztig constructed some representations of finite groups of Lie type by the l-adic cohomology in algebraic geometry. Soon, Lusztig built the character theory of finite groups of Lie type. This might be the deepest theory in representation theory of finite groups.

Algebraic geometric methods also have wonderful applications in the representation theory of Lie algebras. A complex semisimple Lie algebra is the tangent space of some complex semisimple Lie group at the unit element. Finite dimensional representations of a complex semisimple Lie algebra are all highest weight modules. They could be obtained from the finite dimensional representations of complex semisimple Lie groups by differentiation. Hence the characters are given by the Weyl character formula, which was obtained in the 1920s. However for an infinite dimensional irreducible representation, people could not determine the character for quite a long time, even the great mathematicians like I. M. Gelfand could not give a conjecture about it.

Then a breakthrough was made by Kazhdan and Lusztig. In 1979, they gave the following conjecture;

$$chL_w = \sum_{y \leq w} (-1)^{l(w)-l(y)} P_{y,w}(1) chM_y.$$

In this conjecture, the key is the term $P_{y,w}(1)$, other terms could be understood easily.

Let us explain the notations: y, w are elements in the Weyl group, $P_{y,w}(1)$ is the value of Kazhdan-Lusztig polynomial $P_{y,w}$ at 1, M_y is some Verma module, L_w is some irreducible highest weight module which is in general infinite dimensional, chL_w and chM_y are the characters of M_y and L_w, respectively.

The importance of the Kazhdan-Lusztig polynomial $P_{y,w}$ is that it computes the intersection cohomology of a Schubert variety. So it reflects the singularity of a Schubert variety.

Many people were surprised by it after this conjecture is made since they never thought that there would be connections between the representation theory of Lie algebras and the singularity of a Schubert variety, and would be related by such a complicated expression through intersection cohomology.

Soon, Kazhdan-Lusztig conjecture was proved by Bernstein-Beilinson and Kashiwara-Brylinski. Their proofs used the D-module theory and the perverse sheaf theory on flag varieties. There was also a novelty in the method they used in their proof. They did not prove the conjecture directly, instead they transformed the conjecture from the category of representation of Lie algebras to the category of D-modules and then to the category of perverse sheaves by Riemann-Hilbert correspondence. In the category of perverse sheaves, the Kazhdan-Lusztig polynomial has a proper interpretation. Hence the conjecture became easy to understand and prove.

9.5 Topological methods

Topological methods are mainly used in representation theory of Lie groups and algebraic groups. The tools used are fiber bundles, characteristic classes, cohomology, K-theory and so on.

As an example, let us take a look at the representations of an affine Hecke algebra. Let G be the general linear group $GL_n(\mathbb{C})$ over $\mathbb{C}$. Let X be the flag variety, that is, X consists of all the subspace filtrations in $\mathbb{C}^n$:

$$V_1 \subset V_2 \subset \cdots \subset V_n = \mathbb{C}^n, \dim V_i = i.$$

Let $\mathcal{N}$ be the set of all the nilpotent linear transformations on $\mathbb{C}^n$. Define

$$Z = \{(\xi, x_1, x_2) | \xi(x_i) - x_i\} \subset \mathcal{N} \times X \times X.$$

This is an algebraic variety, called Steinberg triples. The group $G \times \mathbb{C}^*$ acts on Z naturally. Hence the Grothendieck group of the category of $G \times \mathbb{C}^*$-equivariant coherent sheaves on Z is the K-equivariant group $K^{G\times\mathbb{C}^*}(Z)$. In this K-group, a product can be defined by convolution. Then it becomes an associative ring. An unexpected thing is that the associative ring structure on this K-group is isomorphic to an affine Hecke algebra of type A defined purely algebraically. Similar results also hold for other reductive algebraic groups.

A result obtained by Borel in year 1976 shows that there are connections between representations of affine Hecke algebras and the representations of p-adic groups. Indeed, part of the interesting representations of p-adic groups can be reduced to those of the affine Hecke algebra. Then for this part of representations of p-adic groups, a conjecture given by Langlands and Deligne can be restated using the language of affine Hecke algebras. Since Langlands and Deligne conjecture was not precise, Lusztig gave a modified statement. In 1987, Kazhdan and Lusztig used exactly the connection between affine Hecke algebras and K-groups to prove the refined Langlands and Deligne conjecture.

A related variety is also interesting. It is the set of all the nilpotent elements in $gl_n(\mathbb{C})$. Every reductive Lie algebra over the complex number field gives rise to a similar variety which has singular points. The famous Springer resolution of these singularities is in fact the cotangent bundle of the flag variety. The fiber of this resolution, now called the Springer fiber, is very useful in representation theory.

Kazhdan and Lusztig introduced affine Springer fibers. It played a key role in the proof of the fundamental lemma. In 2010, Bao-Châu Ngô received Fields Medal for his proof of the fundamental lemma.

In representation theory, you may find that some aspects, which seem to be unrelated, in fact have deep connections with each other. You will often feel that there are so many magical things in representation theory and all kinds of methods work beautifully, no matter algebraic, or analytic, or algebro-geometric, differential geometric, or topological.

10 History

We have acquainted ourselves with the concepts, examples, methods and basic properties of representation theory, to some extent.

Now let us trace the history of representation theory so that we should be able to see it better and more deeply.

From the historical point of view, representation theory has several origins. At the beginning, it started from the study of characters in number theory. Then various representation theories showed up: representations of finite commutative groups, representations of finite groups, finite dimensional representations of Lie groups and Lie algebras, representations of algebras, infinite dimensional representations of Lie groups and Lie algebras. In this process the study of invariants also enhanced the study of representation theory. We will mainly talk about the history of the representation theory of groups, since it is the most vital one in its wide and deep extensions and influences in applications.

Characters in Number Theory In number theory, characters appeared long time ago, such as the Gauss sum, which appeared in Gauss' book *Disquisitiones Arithmetica* in 1802. We have seen that in Gauss sum two characters of commutative groups appear. *Disquisitiones Arithmeticae* was finished in 1798. Gauss was 21 and then 24 when the book was published. This book had great influences in mathematics. You will admire his talent if you do research in number theory.

Dirichlet was a diligent student of Gauss. He studied this book carefully. In

1837, using the characters of groups, he defined a series called the L-series:

$$L(\chi, s) = \sum_{n=1}^{\infty} \frac{\chi(n)}{n^s},$$

where χ comes from a character ϕ of a commutative group $(\mathbb{Z}/a\mathbb{Z})^*$ and a is an integer greater than 1. The definition of χ is as follows: $\chi(n) = 0$ if the common factor of n and a is greater than 1, and $\chi(n) = \phi(\bar{n})$ where $\bar{n} = n + a\mathbb{Z}$ is the residue class of n modulo a.

As we all know, there are infinitely many primes. Using the L-series, Dirichlet proved a stronger result: there are infinite primes in some arithmetic series. That is, if two integer a and m are coprime, then there are infinitely many primes in the arithmetic series $a + m, a + 2m, \ldots, a + km, \ldots$.

It is fair to say that the L-series has deep influences. Let us say more about this.

The Dirichlet series converges in the area $\mathrm{Re}\, s > 1$ and can be analytically extended to the whole complex plane. If χ is nontrivial, this analytic extension is a holomorphic function. Later in the research of representations of Galois groups of finite extension fields of a field, Artin also defined a kind of L-series and extended them to L-functions now called the Artin L-functions, by analytical continuation.

Using these L-functions, Artin proved the famous Artin reciprocity law in abelian global class field theory. This is the case of finite groups. Then Langlands wanted to generalize Artin's work to the case of non-abelian global class field theory. Jacquet and Langlands also defined L-functions for the irreducible representations of reductive algebraic groups over the p-adic field and the automorphic representations of reductive algebraic groups over a global field. Langlands gave a series of conjectures which built a deep connection between number theory and representation theory of algebraic groups. This is the hot topic Langlands program. It can be seen that there are connections between the ideas of Langlands and Dirichlet's work. This shows that the connections between representation theory and number theory have a long history and will continue to play their important roles.

10.1 Representations of commutative groups

The characters of finite commutative groups have appeared in the work of Gauss and Dirichlet essentially. However the concept was developed only after decades. It was introduced by Dedekind and Webber in 1878 and 1881 respectively. Webber was the first one who gave the definition of characters of abstract finite commutative groups. Dedekind was the last student of Gauss. He also studied under the direction of Dirichlet for some time. He collected Dirichlet's lecture notes on number theory. Since the characters of some commutative groups were used in the work of Dirichlet in various ways, Dedekind mentioned in the appendix of the published lecture notes (1984) that he hoped some people could study the characters of general commutative groups. In his book *Lehrbuch der Algebra* (1986), Webber gave a general description of the characters of commutative groups.

10.2 Representations of finite groups

Representation theory of finite groups began with the work of Dedekind, which was not published. Dedekind considered the decomposition of the group determinants of finite commutative groups, and mentioned in a letter to Frobenius in 1896. The definition of group determinant of a finite commutative group is simple and natural. Let G be a finite group with elements $g_1, g_2, \ldots, g_n$. For any element g_i, we associate an indeterminate x_{g_i}. Then consider an $n \times n$ matrix, whose ij entry is $x_{g_i g_j^{-1}}$. Hence the matrix could be written as $(x_{g_i g_j^{-1}})_{a \leq i,j \leq n}$. The determinant of the group G is the determinant $D = |x_{g_i g_j^{-1}}|$ of the matrix. It is obvious that D is a polynomial of the indeterminate $x_{g_1}, x_{g_2}, \ldots, x_{g_n}$ with integer coefficients and different orderings of the elements in G give rise to group determinants which differ at most by sign.

Dedekind also wanted to decompose a polynomial into the product of irreducible polynomials. One can decompose the polynomial with two indeterminates into the product of two linear polynomials. But for general commutative groups it can also be decomposed into a product of linear polynomials. Then something interesting happens. In the linear polynomials, the coefficients are just a character of this commutative group. Hence, from these linear polynomials, we can get the characters of the group G.

The determinant can be decomposed into the product of n linear polynomials. Hence we can get all the n characters of the group. That is, when G is a commutative group,

$$D = \prod_{\chi} (\chi(g_1)x_{g_1} + \chi(g_2)x_{g_2} + \cdots + \chi(g_n)x_{g_n})$$

where χ runs over all the characters of G.

As a simple example, we take $n = 2$. Assume g_1 is the identity element. Then

$$x_{g_i g_j^{-1}} = \begin{cases} x_{g_1} & \text{if } 1 \leq i = j \leq 2, \\ x_{g_2} & \text{if } 1 \leq i \neq j \leq 2. \end{cases}$$

Hence the group determinant is $x_{g_1}^2 - x_{g_2}^2$ which can be decomposed as $(x_{g_1} + x_{g_2})(x_{g_1} - x_{g_2})$. The case $n = 3$ can also computed easily. The general case is not difficult too, which can be reduced to the case of cyclic groups.

Dedekind also wanted to know what would happen for noncommutative groups. He computed some examples and found that the degrees of some factors were greater than 1. As an example, one can compute the group determinant of the permutation group of three words. This is the smallest noncommutative group, which contains six elements. Its determinant has an irreducible factor of degree two and multiplicity 2. It also has two irreducible factors of degree one.

He wrote a letter to Frobenius about this. Inspired by this letter, Frobenius picked this up and kept on studying the problem with his intelligence and enthusiasm. In the year the letter was written, Frobenius published three papers on the characters of finite groups. These papers established the theory of character of finite groups, giving the decomposition of the determinant. Several results in these paper now become standard in this field. One of them is the orthogonal relation between the characters. For two characters χ and ψ corresponding to two irreducible representations, we have

$$\frac{1}{|G|}\sum_{g\in G}\chi(g)\psi(g^{-1}) = \begin{cases} 1 & \text{if } \chi = \psi \\ 0 & \text{if } \chi \neq \psi. \end{cases}$$

He found that there was a one to one correspondence between the irreducible factors of the group determinant of G and the irreducible characters of G, with the degrees and the multiplicities of the irreducible factors equal to those of the corresponding irreducible characters respectively. Then the order of G is equal to the sum of the squares of the degrees of all irreducible characters:

$$|G| = \sum_{\chi}(deg\chi)^2.$$

This is a beautiful and simple result.

A character of a commutative group is a homomorphism from the group to the multiplicative group of complex numbers, but a character of a noncommutative group is only a function from the group to the complex number field. This was noticed by Frobenius. In 1897, he gave the definition of a representation of a group: a homomorphism from this group to the general linear group. This was where the concept and terminology "representation" came from. Frobenius also found that a character was just the trace function of the representation. In other words, if $\rho : G \to GL_d(\mathbb{C})$ is a representation, then its trace function $\chi : G \to \mathbb{C}, g \to tr(\rho(g))$ is a character. Using this notion, we can see that the definition of characters of commutative groups and noncommutative groups get unified.

The concept of induced representation was also given by Frobenius. He studied the relation between representations of small groups and large groups and found the famous Frobenius reciprocity law. Frobenius was a mathematician full of diligence and fecundity. He wrote 20 papers and brought many contributions to representation theory from 1896 to 1901, he not only built representation theory of finite groups but also computed the characters of many groups such as symmetric groups, alternative groups, the projective special linear group $PSL_2(\mathbb{F}_p), p =$ *odd prime* and so on.

Then W. Burnside came into the history and played an important role. He was born in 1851 and died in 1927. He actually did not contribute much new in representation theory. But in those years he often gave different proofs quickly after Frobenius gave a new result. This made Frobenius feel unhappy since he was afraid that the person who was following him so closely would give better results anytime. However Burnside himself cared more about the structure theory of finite groups and the applications of representation theory in the structure theory of groups.

In 1904, Burnside gave an amazing result using the characters of finite groups. Assume that p, q are primes. Then the finite groups of order $p^a q^b$ is solvable. Only until the 1970s, people could find a proof without the help of characters. Even so up to now the proof using character is still the simplest one. Burnside wrote a famous book *Theory of Groups of Finite Order.* He was especially interested in the classification of simple groups and was always trying to find some simple groups and to determine which groups were solvable. In 1911, he published the second edition of this book, where he gave a conjecture that the finite groups of odd order were solvable. This conjecture had great influences and attracted many people. In 1957, Suzuki made progress in a paper in which he used many interesting theories on characters of finite groups. In 1963, this conjecture was completely solved by Feit and Thompson, whose paper was about 250 pages and occupied a single issue of *Pacific Journal of Math.* In their proof, character theory plays an important role. Until now, there is no proof without using the character theory. Needless to say, Burnside made great contribution to the development of finite group theory in the history.

Schur changed the appearance of representation theory of finite groups. The Schur Lemma we mentioned earlier was obtained by him. Using this lemma, he gave very simple proofs of several results of Frobenius and brought a new perspective to look at things. He also introduced the projective representation (1907):

$$G \to GL_n(\mathbb{C})/\{\text{scalar matrices}\}$$

and studied its properties, such as the Schur index. It was said that Schur's lectures were very popular among students. Sometimes there were more than 400 students in his class. Due to the large amount of people, students sitting in the back had to use opera glasses to see the blackboard clearly.

The modern theory of finite group representation theory began with Noether. She used very convenient language by introducing the group algebra and the language of module. In this way, the description of the results of Frobenius is then very clear and natural.

The next great surge was leaded by Brauer. Brauer was a student of Schur and got his PhD degree in 1926. At that time Frobenius, Burnside, Schur only considered representation theory over complex number field. But Brauer thought that representation over arbitrary fields should also be considered. When the order of a group is not divisible by the characteristic of the field, the finite group representation theory over this field has no difference with the theory over the complex field. When the order of the group is divisible by the characteristic of the field, they have a big difference. We can see this from the example of symmetric group discussed earlier. In the latter case, the representation theory over such fields is called the modular representation theory of the finite groups.

Brauer built the modular representation theory of finite groups. He proved that the number of the irreducible modular representations of a group was the same as the number of certain conjugacy classes of that group, and found the relation between the modular representations and the representations of finite groups over complex number field. He also obtained a beautiful relation between the decomposition matrix and Cartan matrix. Brauer introduced p-block theory and the concept of defect group. Until now, they are still the basis concepts in modular representation theory of finite groups.

Brauer had good relationship with the development of representation theory in China. He had two Chinese students: Xuefu Duan and Xihua Cao. Professor Xuefu Duan and Professor Xihua Cao made great contributions to the development of representation theory of finite groups in China.

You may find that in the first thirty or forty years in the history of representation theory, the main figures were all German except Burnside, who was an English. So the first few decades in the research of representation of finite groups carries a strong German mark.

Finite Dimensional Representations of Lie Groups and Lie Algebras
At the same time, the representation theory of Lie groups and Lie algebras were being developed too. The first things which were considered were the representations of special linear group $SL_2(\mathbb{C})$ and the special linear Lie algebra $sl_2(\mathbb{C})$. We know $SL_2(\mathbb{C})$ is the set of all 2×2 complex matrices with determinant 1 and $sl_2(\mathbb{C})$ is its tangent space at the unit matrix, consisting of all the 2×2 complex matrices with trace 0.

In 1893, S. Lie determined their irreducible representations. He also introduced the polynomial representations, and proved that the homogeneous parts of those polynomials with two indeterminate over complex number field gave all the finite dimensional irreducible representations of the Lie group and its Lie algebra. Therefore, there was a correspondence between the set of finite dimensional irreducible representations and the set of natural numbers.

Then Cartan came into play. Cartan's work had a close relation with Lie groups. He proved the complete reducibility of the finite dimensional representations, i.e., every finite dimensional representation is a direct sum of some irreducible representations. Fano, an algebraic geometer, also proved the complete reducibility from the algebro-geometric point of view.

Various methods came in. In 1897, Hurwitz introduced the unitarian trick, which had very strong influences in representation theory. Let us still take the special linear group $SL_2(\mathbb{C})$ for example. Hurwitz found a unitary subgroup $G_u =$

$SU(2) = S^3$. The starting point of Hurwitz was the invariant theory. Note that G_u is a compact subgroup. Given a representation $\sigma : G \to GL_n(\mathbb{C})$ of $G = SL_2(\mathbb{C})$. Then by σ the group G acts on $\mathbb{C}^n$. For any polynomial function $P : \mathbb{C}^n \to \mathbb{C}$, Hurwitz wanted to get a G-invariant polynomial function. The standard method is averaging. But $SL_2(\mathbb{C})$ is not compact, thus we could only do that on the compact subgroups of G. Consider the following integration on $G_u = SU(2) = S^3$:

$$\hat{P}(x) = \frac{1}{8\pi} \int_{C_u} P(g^{-1}.x)dv \in \mathbb{C}[\mathbb{C}^n]^G,$$

where we use the standard measure on S^3. It is now good to see that $\hat{P}(x)$ is G-invariant polynomial, that is $\hat{P}(g.x) = \hat{P}(x)$ for all $g \in G$. Using this trick, Hurwitz proved that the algebra consisting of these G-invariant polynomials was finitely generated. As we know, in the invariant theory, whether the algebra consisting of these invariant polynomials is finitely generated or not is an important problem. Hurwitz then told us it was finitely generated in many cases. His results were important for both commutative algebra and algebraic geometry.

This unitarian trick could also be used to study the representations of groups. Let us still consider the group $G = SL_2(\mathbb{C})$. Given a representation $\sigma : G \to GL_n(\mathbb{C})$, we have an action of G on the linear space $\mathbb{C}^n$. For any positive definite nondegenerate Hermite form $H(,)$, by the average

$$\hat{H}(x, y) = \int_{G_u} H(g^{-1}.x, g^{-1}.y)dv, \quad (x, y) \in \mathbb{C}^n,$$

Hurwitz got a G_u-invariant nondegenerate Hermite form $\hat{H}(,)$. For a G_u-invariant subspace of $\mathbb{C}^n$, the complement space under $\hat{H}(,)$ is also G_u-invariant. But the fact that a subspace is G_u-invariant is equivalent to that it is G-invariant. Therefore, all the finite dimensional representations of G are completely reducible. In fact, Hurwitz also studied the higher dimension cases. Using the unitarian trick, Hurwitz proved the complete reducibility for $SL_n(\mathbb{C})$ and $SO_n(\mathbb{C})$. However, these important work did not get much attention. Hurwitz's work also gave a proof of E. Study conjecture for $SL_n(\mathbb{C})$ and $SO_n(\mathbb{C})$ which first appeared in a letter written by him to S. Lie. In that letter, he claimed that for general semisimple groups, finite dimensional representations were completely reducible. Hurwitz's paper was silent for 25 years.

In 1913, Cartan constructed all finite dimensional irreducible representations of complex semisimple Lie algebras. In 1922, Schur found the paper written by Hurwitz in 1987. Following the method used in that paper, he extended the character theory of complex representations for finite groups to SU_n and SO_n and got the irreducible characters of unitary groups and orthogonal groups.

Hurmann Weyl combined the previous results. From 1924 to 1926, combining the work of Cartan, Hurwitz, Schur and using his own creativity, he proved the complete reducibility of complex semisimple Lie groups, complex semisimple Lie algebras and compact connected semisimple Lie groups, and gave the characters of representations. The proofs of these results were the sign of the beginning of the global theory of Lie groups, which also had deep influences on E. Cartan. From then on, Cartan often considered problems from the viewpoint of the global properties of Lie groups, such as his work on symmetric spaces. As you know, symmetric spaces are important in differential geometry.

In 1927, Weyl and his student Peter considered the analysis on compact Lie groups. Their work was the foundation of harmonic analysis on compact topological groups and the beginning of harmonic analysis on groups. The Peter-Weyl theorem we mentioned before is a classic result in representation theory.

Around 1927, Weyl was interested in the applications of representation theory of groups to quantum mechanics. In 1928, he published a famous book *Gruppentheorie und Quantenmechanik*, which had great influences on both mathematics and physics. It is fair to say that it was Weyl and the physicist Wigner who introduced representation theory to quantum mechanics.

The most surprising result inspired by this book came from the physicist Casimir. You have seen the Casimir operator when studying Lie algebras. Casimir was a physicist and his starting point was also the applications of representation theory of groups in quantum mechanics. He found the Casimir operator for $sl_2(\mathbb{C})$, using it he proved the complete reducibility of finite dimensional representations of $sl_2(\mathbb{C})$. Now we know that this operator also appears in the center of the universal enveloping algebra. Physicists like using the Casimir operator, since its eigenvalues have physical meanings. In 1931, Casimir introduced the Casimir operator for any complex semisimple Lie algebra. Using this operator, van der Waerden gave the first proof of the complete reducibility of finite dimensional representations of complex semisimple Lie algebras. The proof was published in 1935.

In the mid 1930s, Pontryagin and van Kampen studied locally compact commutative topological groups. The famous Pontryagin duality theorem says the space of irreducible unitary representations of locally compact connected commutative topological group G is a discrete topological group $\hat{G}$, and the space of irreducible unitary representations of $\hat{G}$ is G.

In 1936, Brauer gave another algebraic proof of the complete reducibility of finite dimensional representations of complex semisimple Lie algebras. This work did not get much attention for 25 years as well. In the 1950s, The Bourbaki school wanted to rewrite the whole system of mathematics. When trying to write a book on Lie groups and Lie algebras, they were not satisfied with the proof of complete reducibility of finite dimensional representation of complex semisimple Lie algebras given by van der Waerden. So they tried a new algebraic proof. That book was published in 1961. After that, they found their proof was actually the same as the one given by Brauer. Good work is not sometimes noticed.

10.3 Infinite dimensional representations of Lie groups and Lie algebras

When we said that Weyl considered the application of representation theory of groups in quantum mechanics, we also mentioned Wingner, who was a physicist and also a mathematician. He got the Nobel Prize in 1963. For the need of physics, this Hungarian studied the infinite dimensional representations of groups. There are some important groups in relativity, such as Lorentz groups, homogeneous and nonhomogeneous Poincaré groups. In 1939, Wingner studied the irreducible (projective) unitary representations of Poincaré groups.

Soon, Dirac noticed the significance of this work. He thought that group theory was a wide field and was important. He also studied the irreducible representations of Lorentz groups. Then he asked an Indian called Harish-Chandra to consider this problem. Harish-Chandra finished it and got his PhD degree. Later, Harish-Chandra found that his proof was not rigorous. He was worried about that and told Dirac that the proof in his paper was not good. But Dirac said was as follows: I don't care whether the proof is good or not. I only care what the nature does. Harish-Chandra felt perplexed about this. During the years studying under the direction of Dirac, Harish-Chandra thought he was lacking the sixth sense as a physicist and could only be a mathematician. Then he switched to representation theory and reshaped representation theory of Lie groups. I will not talk about his great work here.

The German V. Bargmann was also a physicist and mathematician. In 1947, he studied the irreducible unitary representations of Lorentz groups. The motivation came from physics. Bargmann was also famous among physicists, he received the Planck prize and the Wigner prize.

Then the Soviet school began. In 1947, I. M. Gelfand and M. A. Naimark gave the classification of the irreducible unitary representations of $SL_2(\mathbb{C})$, which was quite complicated. Gelfand and his collaborators then did a lot of work on this. After that, the development of representation is flourished and many great figures appeared.

We have seen that before 1950 several great mathematicians, such as E. Cartan, H. Weyl, did important work in representation theory. After that, representation theory was developed more rapidly. In this field, many great figures came in: I. M. Gelfand, M. A. Maimark, Harish-Chandra, Selberg, A. Weil, Grothendieck, Borel, Langlands, Deligne, Kazhdan, Drinfeld, Lafforgue, Bernstein, Beilinson, Kac, Kashiwara, Lusztig... Certainly, the development and influence of representation theory is more intense than before, the methods and perspectives of research are also various.

11 Conclusion

In this talk, you have seen that representations are everywhere in mathematics and are colorful. There are many first-class mathematicians working in this field and it is always attracting the best mathematicians. Unfortunately, you can not find any Chinese in the list of the top ones. I sincerely hope that someone of you would get interested in this field and become one member of this great community.

12 Excursus

This article is a revised version of the draft prepared for the talk "Talks in Mathematics Center" on May 05, 2011.

The content mentioned in that talk are well known. For the history part, we followed the following references:

[1] C. Curtis, Representation Theory of Finite groups: from Frobenius to Brauer. The Mathematical Intelligencer, Vol. 14, No. 4, 1992, 48-57.

[2] A. Borel, Essays in the History of Lie Groups and Algebraic Groups. American Mathematical Society and London Mathematical Society, 2001.

[3] G. Mackey, Unitary Group Representations in Physics, Probability and Number Theory. Benjamin, 1978.

[4] G. Mackey, The Scope and History of Commutative and Noncommutative Harmonic Analysis. American Mathematical Society and London Mathematical Society, 1992.

For the other parts, there are too many references to list here.

For beginners, the following five books give good introductions to representation theory of finite groups, representation theory of compact Lie groups, representation theory of semisimple Lie groups, representation theory of semisimple Lie algebras, representation theory of algebras.

[5] J. P. Serre, Linear Representations of Finite Groups. Graduate Texts in Mathematics 42, Springer-Verlag, 1977.

[6] T. Bröcker, T. tom Dieck, Representations of Compact Lie Groups. Graduate Texts in Mathematics 98, Springer-Verlag, 1985.

[7] A. W. Knapp, Representation Theory of Semisimple Groups, an Overview Based on Examples. Princeton University Press, 1986.

[8] J. E. Humphreys, Introduction to Lie Algebras and Representation Theory. Graduate Texts in Mathematics 9, Springer-Verlag, 1972.

[9] P. Gabriel, A. V. Roiter, Representation of Finite-dimensional Algebras. Springer-Verlag, 1997.

The research on automorphic forms is a very active field. The collection *Representation Theory and Automorphic Forms* (American Mathematical Society, 1993) edited by P. Sally and N. R. Wallach includes eleven interesting papers published in *Bulletin of the AMS* between 1955 and 1984. It is worth reading.

If possible, it is pleasant and beneficial to study carefully or read freely the collected papers and books of mathematicians such as I. M. Gelfand, Harish-Chandra, Selberg, Borel, Langlands, Lusztig. You will learn a lot about many research fields.

Open Problems and Surveys of Contemporary Mathematics
SMM 6, pp. 379–381

A Note on The Distribution of Critical Points of Eigenfunctions*

Shing-Tung Yau†

This author has always been curious about the distributions of critical points of eigenfunctions when the eigenvalue λ grows to infinity. Perhaps its growth is at least as large as $\sqrt{\lambda}$. In order to gain some information, we study the two dimensional case where we prove that for certain eigenvalue λ related to the curvature of the surface, there is always a critical point of the eigenfunction with some control on the value of this eigenfunction at this point.

Let φ be an eigenfunction of the operator $\Delta - V$ on the compact surface M. If $\partial M \neq \phi$, we assume ∂M is concave and we take either Dirichlet or Neumann conditions. Let $-\lambda$ be the eigenvalue. We define

$$u = |\nabla\varphi|^2 + c\varphi^2 \tag{1}$$

where c is a constant to be chosen.

The function u achieves its infimum at some point in the interior of M. (This follows from the sharp maximum principle and the boundary conditions.) At this point,

$$\begin{aligned} 0 &\le \Delta u \\ &= 2\sum \varphi_{ij}^2 + 2\sum \varphi_i(\Delta\varphi)_i + 2K|\nabla\varphi|^2 \\ &\quad +2c\varphi(\Delta\varphi) + 2c|\nabla\varphi|^2 \\ &= 2\sum \varphi_{ij}^2 + 2\sum \varphi_i(V\varphi - \lambda\varphi)_i + 2K|\nabla\varphi|^2 \\ &\quad +2c\varphi(V\varphi - \lambda\varphi) + 2c|\nabla\varphi|^2. \end{aligned} \tag{2}$$

*Supported by NSF Grant No. DMS-9206938

†Department of Mathematics, Harvard University, Cambridge, MA 02138, USA. Email: yau@math.harvard.edu

If $\nabla\varphi \neq 0$ at this point, we can assume $\varphi_1 \neq 0$ and $\varphi_2 = 0$. Then $u_1 = u_2 = 0$ implies that

$$\varphi_{11} + c\varphi = 0 \tag{3}$$

$$\varphi_{12} = 0. \tag{4}$$

Hence

$$\begin{aligned}\varphi_{22} &= \Delta\varphi - \varphi_{11} \\ &= (V - \lambda_1 + c)\varphi.\end{aligned} \tag{5}$$

Putting these equations into (2), we obtain

$$\begin{aligned}0 &\leq c^2\varphi^2 + (V - \lambda + c)^2\varphi^2 \\ &\quad +(V-\lambda)|\nabla\varphi|^2 + \varphi\sum \varphi_i V_i + K|\nabla\varphi|^2 \\ &\quad +cV\varphi^2 - c\lambda\varphi^2 + c|\nabla\varphi|^2 \\ &= \left[2c^2 - 3c\lambda + \lambda^2 + V^2 + (3c - 2\lambda)V\right]\varphi^2 \\ &\quad +(K + V + c - \lambda)|\nabla\varphi|^2 + \varphi\sum \varphi_i V_i \\ &\leq \left[(2c-\lambda)(c-\lambda) + V^2 + (3c-2\lambda)V + \frac{|\nabla V|}{2}\right]\varphi^2 \\ &\quad + \left[K + V + c - \lambda + \frac{|\nabla V|}{2}\right]|\nabla\varphi|^2.\end{aligned} \tag{6}$$

If we can find a constant c so that

$$(2c-\lambda)(c-\lambda) + V^2 + (3c-2\lambda)V + \frac{|\nabla V|}{2} \leq 0 \tag{7}$$

and

$$K + V + c - \lambda + \frac{|\nabla V|}{2} \leq 0. \tag{8}$$

Then the sharp minimum principle will given rise to a contradiction. Hence the assumption that $\nabla\varphi \neq 0$ is violated. In other words, if (7) and (8) hold, the minimum of the function $|\nabla\varphi|^2 + c\varphi^2$ occurs at a critical point of φ.

Theorem 1. *Let φ be an eigenvalue function of $\Delta - V$ with eigenvalue $-\lambda$ on a two-dimensional compact surface M. Assume that for some constant $c > 0$, (7) and (8) hold. Then there is a critical point x_0 of φ so that*

$$|\nabla\varphi|^2(x) + c\varphi^2(x) \geq c\varphi^2(x_0) \tag{9}$$

for all $x \in M$.

By integrating (9), we obtain the following

Theorem 2. *If φ is normalized so that $\int_M \varphi^2 = 1$, then φ has a critical point x_0 where*

$$\varphi^2(x_0) \leq \frac{1}{\text{Area}(M)} + \frac{\lambda}{c\text{Area}(M)}.$$

By following the shortest path from x_0 to the nodal line of φ, we obtain the following

Theorem 3. *Let φ be an eigenfunction of the operator $\Delta - V$ on the compact surface with eigenvalue $-\lambda$. Let c be a constant so that (7) and (8) hold. Then there is a critical point x_0 of φ so that the distance from x_0 to the nodal line of φ is not greater than $\frac{\pi}{2\sqrt{c}}$.*

NOTE. If $V = 0$ and $K \leq 0$, we can choose $c = \lambda$. If $V = 0$ and $K \leq 1$, we can choose $c = \frac{\lambda}{2}$ and (7) and (8) are equivalent to $K \leq \frac{\lambda}{2}$. It can be seen that the estimate is optimal for the sphere.

Reference

[1] R. Schoen and S. T. Yau, *Lecture on Differential Geometry*, International Press, 1995.

Open Problems and Surveys of Contemporary Mathematics
SMM 6, pp. 383–396

Sobolev Inequality for Measure Space*

Shing-Tung Yau†

This note is an outgrowth of a student seminar held in Taiwan in 1991. It should be considered as a classroom note on a fundamental work of Michael and Simon [1].

Let (M, Ω) be a measure space with the following structure:

1. For each $x \in M$, there exists a function ρ_x and a measurable transformation F_x defined on the set $\{\rho_x \le \rho_0\}$ such that $F_x^*\Omega \ge \Omega$ and $\rho_x(F_x) \le 2\rho_x$.
2. Let $g_x(\rho) = \sup_{\rho_x \le \rho} \left[(\rho_x(F_x) - \rho_x) + \frac{F_x^*(\Omega)}{F_x^*(\Omega)-\Omega}\right]$.

We assume that there is a nonincreasing positive function $G_x(\rho)$ so that $-G'_x(\rho) \le \frac{G_x(\rho)}{g_x(\rho)}$ for $\rho \le \rho_0$ and that for a given non-decreasing function k, there are two non-decreasing functions H and $\tilde{k}$ such that for all $a > 0$

$$k'(a) \sup_x \int_0^{4G_x^{-1}(\frac{1}{5t})} \frac{G_x(\frac{\tau}{4})}{G_x(\tau)} d\tau \le H(\tilde{k}(a)t).$$

If $g_x(\rho) \ge \frac{\rho}{m}$, we can choose $G_x(\rho) = \rho^{-m}$, $k(t) = \tilde{k}(t) = t^{\frac{m}{m-1}}$ and $H(t)$ a constant multiple of $t^{\frac{1}{m}}$.

Theorem 1. *Assume*

$$\lim_{\epsilon \to 0} G_x(\epsilon) \left(\int_{\substack{\rho_x < \epsilon \\ or\ \rho_x(F_x) < \epsilon}} (F_x^*\Omega - \Omega) \right) \ge 1$$

and the measure of the support of a non-negative function h is not greater than $\frac{1}{5} \sup_x \frac{1}{G_x(\frac{\rho_0}{4})}$, then for any nondecreasing function k, the integral of $k(h)$ with

*Research partially supported by NSF Grant No. DMS 92-06938

†Department of Mathematics, Harvard University, Cambridge, MA 02138, USA. Email: yau@math.harvard.edu

respect to the measure $\inf_x(F_x^*\Omega - \Omega)$ *over the set* $\{h \geq \delta\}$, *with* $\delta > 0$, *is not greater than*

$$H\left(\int_M \tilde{k}(h+\delta)\right)\left\{\int \sup_x \frac{h - h(F_x)}{(\rho_x - \rho_x(F_x))_+}(F_x^*\Omega - \Omega) + \delta^{-2}\int_M \sup_x(\delta + h - h(F_x))\frac{h - h(F_x)}{(\rho_x - \rho_x(F_x))_+}(F_x^*\Omega - \Omega)\right\}.$$

If we have a one-parameter family of measurable transformations $F_x(t)$ satisfying the same hypothesis as before so that $\lim_{t\to 1} F_x(t)(y) = y$ for all y, then we can apply the above Theorem for each $F_x(t)$ and let $t \to 1$.

Theorem 2. *Assume* $\lim_{\epsilon\to 0} G_x(\epsilon)\int_{\rho_x<\epsilon}\lim_{t\to 1}\frac{(F_x^*(t)\Omega-\Omega)}{t-1} \geq 1$ *and the measure of the support of* h *is not greater than* $\frac{1}{5}\sup_x \frac{1}{G_x(\frac{\rho_0}{4})}$. *Then*

$$\int_M k(h) \leq H\left(\int_M \tilde{k}(h)\right)\int_M \limsup_{t\to 1}\sup_x\left(\frac{|h - h(F_x(t))|}{(\rho_x - \rho_x(F_x(t)))_+}\right)\lim_{t\to 1}\frac{(F_x^*(t)\Omega - \Omega)}{t-1}.$$

A major example is given by assuming $g_x(\rho) \geq \frac{\rho}{m}$. In this case, we take $G_x(\rho) = c^{-1}\rho^{-m}$ and we assume $\int_{\rho_x\leq\rho}\lim_{t\to 1}\left(\frac{F_x^*(t)\Omega-\Omega}{t-1}\right) \leq c\rho^n$. Since

$$\int_0^{4G_x^{-1}(\frac{1}{5t})}\frac{G_x(\frac{\tau}{4})}{G_x(\tau)} = 4^{m+1}(\frac{5}{c})^{\frac{1}{m}}t^{\frac{1}{m}},$$

we take $k(t) = t^{\frac{m}{m-1}}$, $H(t) = 4^{m+1}(\frac{5}{c})^{\frac{1}{m}}t^{\frac{1}{m}}$ and $\tilde{k}(t) = t^{\frac{m}{m-1}}$.

Theorem 3. *Let* (M,Ω) *be a measure space equipped with the following structure.*

1. *For each* $x \in M$, *there exists a function* ρ_x *and a continuous family of transformations* $F_x(t)$ *defined on* $\{\rho_x \leq \rho_0\}$ *where* $F_x^*(t)\Omega > \Omega$ *and* $\lim_{t\to 1}F_x(t)(y) = y$ *for* $\rho_x(y) \leq \rho_0$.
2. *Assume that*

$$\sup_{\rho_x\leq\rho}\lim_{t\to 1}\left[(\rho_x(F_x(t)) - \rho_x)_+\frac{F_x^*(t)(\Omega)}{F_x^*(t)(\Omega) - \Omega}\right] \leq \frac{\rho}{m}\Omega.$$

3. $$\lim_{\epsilon\to 0}\epsilon^{-m}\int_{\rho_x<\epsilon}\lim_{t\to 1}\left(\frac{F_x^*(t)(\Omega) - \Omega}{t-1}\right) \geq c.$$

Then for any nonnegative function h *whose support has* $\sup_x \lim_{t\to 1}\frac{F_x^*(t)\Omega-\Omega}{t-1}$ *measure less than* $(\frac{4}{5})^m c\rho_0^m$,

$$\left(\int h^{\frac{m}{m-1}}\right)^{\frac{m-1}{m}} \leq \left(\frac{m-1}{m}\right)^m 4^{m+1}\left(\frac{5}{c}\right)^{\frac{1}{m}} \times \int_M \lim_{t\to 1}\frac{|h - h(F_x(t)|}{(\rho_x - \rho_x(F_x(t)))}\lim_{t\to 1}\left(\frac{F_x^*(t)(\Omega) - \Omega}{t-1}\right).$$

A corollary for Theorem 3 is the Sobolev inequality for $\mathbf{R}^m$. We set $F_x(t)(y) = y + (t-1)(y-x)$ and $\Omega =$ the standard euclidean volume form. In this case $F_x^*(t)\Omega - \Omega = (t^m - 1)\Omega$ so that

$$\lim \lim_{t\to 1} (\rho_x(F_x(t)y) - \rho_x(y))_+ \frac{F_x^*(t)(\Omega)}{F_x^*(t)(\Omega) - \Omega} = \frac{\rho}{m}\Omega.$$

In the proof of Theorem 1, we prove the following proposition which has independent interest.

Proposition. *Let h be a nonnegative function. Let A be the set of points x so that*

$$h(x) \geq \left[\lim_{\epsilon\to 0} G_x(\epsilon) \int_{\substack{\rho_x < \epsilon \\ \rho_x(F_x) < \epsilon}} (F^*\Omega - \Omega)\right]^{-1}.$$

Then the measure of A measured with respect to $\inf_x(F_x^*\Omega - \Omega)$ can be estimated by

$$2\left(\int_0^{\rho_0'} \frac{G_x(\frac{t}{4})}{G_x(t)} dt\right) \int_M \sup_x \frac{h - h(F_x)}{(\rho_x - \rho_x(F_x))_+} (F_x^*\Omega - \Omega)$$

if

$$\int_M h \leq \inf_x \left(\frac{1}{G_x(\frac{\rho_0'}{4})}\right).$$

In the last section, we also discuss Poincaré inequality with no compact support. Hence we can derive Harnack inequality.

1 Proof of the main theorems

Let λ be any increasing such that $\lambda(t) = 0$ for $t \leq 0$. Then for any measurable function h

$$\begin{aligned} 0 &= \int \lambda(\rho - \rho_x(F_x))h(F_x)F_x^*\Omega \qquad (1.1)\\ &\quad - \int \lambda(\rho - \rho_x)h\Omega \\ &= \int [\lambda(\rho - \rho_x(F_x))h(F_x) - \lambda(\rho - \rho_x)h]F_x^*\Omega \\ &\quad + \int \lambda(\rho - \rho_x)h(F_x^*\Omega - \Omega) \\ &= \int (\lambda(\rho - \rho_x(F_x)) - \lambda(\rho - \rho_x))hF_x^*\Omega \\ &\quad + \int \lambda(\rho - \rho_x(F_x))(h(F_x) - h)F_x^*\Omega \\ &\quad + \int \lambda(\rho - \rho_x)h(F_x^*\Omega - \Omega). \end{aligned}$$

Since

$$\lambda(\rho-\rho_x(F_x))-\lambda(\rho-\rho_x)=\int_0^1\frac{d}{dt}\lambda(\rho-t\rho_x(F_x)-(1-t)\rho_x) \tag{1.2}$$
$$=-\left[\int_0^1\lambda'(\rho-t\rho_x(F_x)-(1-t)\rho_x)\right](\rho_x(F_x)-\rho_x)$$

we find

$$\int\lambda(\rho-\rho_x)h(F_x^*\Omega-\Omega)\le\int\left(\int_0^1\lambda'(\rho-t\rho_x(F_x)-(1-t)\rho_x)\right) \tag{1.3}$$
$$\times(\rho_x(F_x)-\rho_x)hF_x^*\Omega$$
$$-\int\lambda(\rho-\rho_x(F_x))(h-h(F_x))F_x^*(\Omega).$$

Since

$$\sup_{\rho_x\le\rho}\left[(\rho_x(F_x)-\rho_x)+\frac{F_x^*(\Omega)}{F_x^*\Omega-\Omega}\right]\le g_x(\rho) \tag{1.4}$$

we obtain

$$\int\lambda(\rho-\rho_x)h(F_x^*\Omega-\Omega)\le\int\left(\int_0^1\lambda'(\rho-t\rho_x(F_x)-(1-t)\rho_x)\right) \tag{1.5}$$
$$\times g_x(\rho)hF_x^*(\Omega-\Omega)$$
$$+\int\lambda(\rho-\rho_x(F_x))(h-h(F_x))F_x^*\Omega.$$

Note that $\lambda'\ge 0$, we need only to restrict the integral of the first term on the right to be over the set where $\rho_x(F_x)\ge\rho_x$.

Multiplying the inequality by $\frac{G_x(\rho)}{g_x(\rho)}$, where $G_x(\rho)$ is a non-negative non-increasing function and integrating

$$\int\left(\int_{\sigma_1}^{\sigma_2}\frac{G_x(\rho)}{g_x(\rho)}\lambda(\rho-\rho_x)d\rho\right)h(F_x^*\Omega-\Omega)$$
$$=\int\int_0^1\left(\int_{\sigma_1}^{\sigma_2}\lambda'(\rho-t\rho_x(F_x)-(1-t)\rho_x)G_x(\rho)d\rho\right)h(F_x^*\Omega-\Omega) \tag{1.6}$$
$$+\int\int_{\sigma_1}^{\sigma_2}\frac{G_x(\rho)}{g_x(\rho)}\lambda(\rho-\rho_x(F_x))(h-h(F_x))F_x^*\Omega.$$

Integrating by parts we obtain

$$\begin{aligned}
&\int\int_{\sigma_1}^{\sigma_2}\frac{G_x(\rho)}{g_x(\rho)}\lambda(\rho-\rho_x)h(F_x^*\Omega-\Omega)\\
&\leq -\int\int_0^1\int_{\sigma_1}^{\sigma_2}G_x'(\rho)\lambda(\rho-t\rho_x(F_x)-(1-t)\rho_x)h(F_x^*\Omega-\Omega) \qquad (1.7)\\
&\quad+\int\int_0^1\lambda(\sigma_2-t\rho_x(F_x)-(1-t)\rho_x)G_x(\sigma_2)h(F_x^*\Omega-\Omega)\\
&\quad-\int_{\rho_x(F_x)\geq\rho_x}\int_0^1\lambda(\sigma_1-t\rho_x(F_x)-(1-t)\rho_x)G_x(\sigma_1)h(F_x^*\Omega-\Omega)\\
&\quad+\int\int_{\sigma_1}^{\sigma_2}\frac{G_x(\rho)}{g_x(\rho)}\lambda(\rho-\rho_x(F_x))\frac{|(h-h(F_x))|}{(\rho_x(F_x)-\rho_x)_+}g_x(\rho)(F_x^*\Omega-\Omega).
\end{aligned}$$

Since for $\rho_x(F_x)\geq\rho_x$, $\rho-t\rho_x(F_x)-(1-t)\rho_x\leq\rho-\rho_x$; we conclude that $-G_x'(\rho)\lambda(\rho-t\rho_x(F)-(1-t)\rho_x)\leq G(\rho)\lambda(\rho-\rho_x)g_x^{-1}(\rho)$ if λ is chosen (by approximation) to be the step function with $\lambda(t)=0$ for $t\leq 0$ and $\lambda(t)=1$ for $t>0$. Hence the first term will cancel the left hand side, (Recall that the integral there can be restricted to the set where $\rho_x(F_x)\geq\rho_x$.)

Hence

$$\begin{aligned}
&G_x\ (\sigma_1)\int\int_0^1\lambda(\sigma_1-t\rho_x(F_x)-(1-t)\rho_x)h(F_x^*\Omega-\Omega)\\
&\leq G_x(\sigma_2)\int\int_0^1\lambda(\sigma_2-t\rho_x(F_x)-(1-t)\rho_x)h(F_x^*\Omega-\Omega) \qquad (1.8)\\
&\quad+\int\int_{\sigma_1}^{\sigma_2}G_x(\rho)\lambda(\rho-\rho_x(F_x))\frac{(h-h(F_x))}{(\rho_x(F_x)-\rho_x)_+}(F_x^*\Omega-\Omega).
\end{aligned}$$

Define

$$\begin{aligned}
\varphi_x(\rho)&=\int_{\rho_x(F_x)\geq\rho_x}\int_0^1\lambda(\rho-t\rho_x(F)-(1-t)\rho_x)h(F_x^*\Omega-\Omega)\\
\varphi_x(\rho)&=\int\lambda(\rho-\rho_x(F_x))\frac{h-h(F_x)}{(\rho_x(F_x)-\rho_x)_+}(F_x^*\Omega-\Omega) \qquad (1.9)
\end{aligned}$$

where $\lambda(t)=0$ for $t\leq 0$ and $\lambda(t)=1$ for $t>0$.

Then

$$G_x(\sigma_1)\varphi_x(\sigma_1)\leq G_x(\sigma_2)\varphi_x(\sigma_2)+\int_{\sigma_1}^{\sigma_2}G_x(\rho)\psi_x(\rho) \qquad (1.10)$$

for all $0<\sigma_1<\sigma_2<\rho$.

Lemma 1.1. *Suppose for some $0<\rho_0'<\rho_0$,*

$$\sup_{0<\sigma<\rho_0}G_x(\rho)\varphi_x(\sigma)>G_x(\frac{\rho_0'}{4})\varphi_x(4\rho_0')\left[2+3\rho_0'\left(\int_0^{\rho_0'}\frac{G_x(\frac{t}{4})}{G_x(t)}\right)^{-1}\right]. \qquad (1.11)$$

Then there exists $\rho(x)$, $0 < \rho(x) < \rho_0'$ *so that*

$$\varphi_x(4\rho(x)) \le \psi_x(\rho(x)) \left(2\int_0^{\rho_0'} \frac{G_x(\frac{t}{4})}{G_x(t)} \right). \tag{1.12}$$

Proof. If the lemma was not true,

$$\psi_x(\sigma) \le \varphi_x(4\sigma) \left(2\int_0^{\rho_0'} \frac{G_x(\frac{t}{4})}{G_x(t)} dt \right)^{-1} \tag{1.13}$$

for $o < \sigma < \rho_0$.

Hence

$$\begin{aligned}
&\int_0^{\rho_0'} G_x(\sigma)\psi_x(\sigma) \qquad (1.14)\\
&\le \left(\int_0^{\rho_0'} G_x(\sigma)\varphi_x(4\sigma) \right) \left(2\int_0^{\rho_0'} \frac{G_x(\frac{t}{4})}{G_x(t)} \right)^{-1}\\
&\le \left(\int_0^{4\rho_0'} G_x(\frac{\sigma}{4})\varphi_x(\sigma) \right) \left(2\int_0^{\rho_0'} \frac{G_x(\frac{t}{4})}{G_x(t)} \right)^{-1}\\
&\le \left[\int_0^{\rho_0'} \frac{G_x(\frac{\sigma}{4})}{G_x(\sigma)} G_x(\sigma)\varphi_x(\sigma) + \int_{\rho_0'}^{4\rho_0'} G_x(\frac{\sigma}{4})\varphi_x(\sigma) \right] \left(2\int_0^{\rho_0'} \frac{G_x(\frac{t}{4})}{G_x(t)} \right)^{-1}\\
&\le \frac{1}{2} \sup_{0 \le \sigma \le \rho_0} G_x(\sigma)\varphi_x(\sigma)\\
&\quad + \frac{3\rho_0'}{2} \left(\int_0^{\rho_0'} \frac{G_x(\frac{t}{4})}{G_x(t)} \right)^{-1} G_x(\frac{\rho_0'}{4})\varphi_x(4\rho_0').
\end{aligned}$$

Combining this inequality with (1.10), we obtain

$$\begin{aligned}
&\frac{1}{2} \sup_{0 < \sigma < \rho_0} G_x(\sigma)\varphi_x(\sigma)\\
&\quad \le G_x(\rho_0')\varphi_x(\rho_0') + \frac{3\rho_0'}{2} G_x(\frac{\rho_0'}{4}) \left(\int_0^{\rho_0'} \frac{G_x(\frac{t}{4})}{G_x(t)} \right)^{-1} \varphi_x(4\rho_0')
\end{aligned}$$

then

$$\sup_{0<\sigma<\rho_0'} G_x(\rho)\varphi_x(\sigma) \le G_x(\frac{\rho_0'}{4})\varphi_x(4\rho_0') \left[2 + 3\rho_0' \left(\int_0^{\rho_0'} \frac{G_x(\frac{t}{4})}{G_x(t)} \right)^{-1} \right] \tag{1.15}$$

which contradicts the hypothesis of the lemma. $\square$

Since G is decreasing, $G_x(\frac{\sigma}{4}) \ge G_x(\sigma)$ and $\rho_0' \left(\int_0^{\rho_0'} \frac{G_x(\frac{\sigma}{4})}{G_x(\sigma)} dt \right)^{-1} \le 1$. Hence we can replace the hypothesis by

$$\sup_{0<\sigma<\rho_0'} G_x(\sigma)\varphi_x(\sigma) > 5G_x(\frac{\rho_0'}{4})\varphi_x(4\rho_0'). \tag{1.16}$$

Let A be the set of points x so that $h(x) \geq 1$. We shall assume that ρ_0' is chosen such

$$1 \geq 5G_x(\frac{\rho_0'}{4})\varphi_x(4\rho_0'). \tag{1.17}$$

Then for all $x \in A$, there exists $0 < \rho(x) < \rho_0'$ such that

$$\varphi_x(4\rho(x)) \leq \psi_x(\rho(x)) \left(2\int_0^{\rho_0'} \frac{G_x(\frac{t}{4})}{G_x(t)}\right). \tag{1.18}$$

Let $A(i) = \{x \in A | \varphi_x(4\rho) \leq \psi_x(\rho)\left(2\int_0^{\rho_0'} \frac{G_x(\frac{t}{4})}{G_x(t)}\right)$ for some $0 \leq (\frac{4}{3})^{-i-1}\rho_0 < \rho < (\frac{4}{3})^{-i}\rho_0\}$.

Then define B_i inductively so that

$$B_k = A^{(i)} - \bigcup_{i=0}^{k-1}\bigcup_{x\in F_i}\{\rho_x < (\frac{4}{3})^{1-i}\rho_0\} \tag{1.19}$$

where F_i is a finite subset of B_i so that $B_i \subset \bigcup_{x\in F_i}\{\rho_x < (\frac{4}{3})^{1-i}\rho_0'\}$ and the sets $\{\rho_x < (\frac{4}{3})^{-i}\rho_0'\}$ with $x \in F_i$ are pairwise disjoint.

Therefore $A \subset \bigcup_{i=1}^{\infty}\bigcup_{x\in F_i}\{\rho_x < (\frac{4}{3})^{1-i}\rho_0'\}$

For each $x \in F_i$, there exists $\rho(x)$ so that

$$\varphi_x(4\rho(x)) \leq \psi_x(\rho(x))\left(2\int_0^{\rho_0}\left(\frac{G_x(\frac{t}{4})}{G_x(t)}\right)dt\right).$$

Hence

$$\varphi_x(4(\frac{4}{3})^{-i}\rho_0') \leq \psi_x((\frac{4}{3})^{-i}\rho_0')\left(2\int_0^{\rho_0}\frac{G_x(\frac{t}{4})}{G_x(t)}\right). \tag{1.20}$$

Summing up, we obtain

$$\sum_i\sum_{x\in F_i}\varphi_x(4(\frac{4}{3})^{-i}\rho_0') \leq \left(2\int_0^{\rho_0'}\frac{G_x(\frac{t}{4})}{G_x(t)}\right)\sum_i\sum_{x\in F_i}\psi_x((\frac{4}{3})^{-i}\rho_0'). \tag{1.21}$$

Since the sets $\{\rho_x < (\frac{4}{3})^{-i}\rho_0'\}$ are disjoint, the sets $\{\rho_x(F_x) < (\frac{4}{3})^{-i}\rho_0'\}$ are also disjoint. Hence

$$\sum_{x\in F_i}\psi_x((\frac{4}{3})^{-i}\rho_0') \leq \int \sup_x\left(\frac{|h - h(F_x)|}{(\rho_x(F_x) - \rho_x)_+}\right)(F_x^*\Omega - \Omega). \tag{1.22}$$

The set A is covered by the sets $\{\rho_x(y) \leq (\frac{4}{3})^{1-i}\rho_0'\}$. Since $\rho_x(F_x) \leq 2\rho_x$, $\lambda(4(\frac{4}{3})^{-i}\rho_0' - t\rho_x(F_x) - (1-t)\rho_x) \geq 1$ on the set $\{\rho_x \leq (\frac{4}{3})^{1-i}\rho_0'\}$. Hence we conclude that the $F_x^*\Omega - \Omega$ measure of A is dominated by

$$2\left(\int_0^{\rho_0}\sup_x\frac{G_x(\frac{t}{4})}{G_x(t)}\right)\left(\sum_i\sum_{x\in F_i}\psi_x((\frac{4}{3})^{-i}\rho_0')\right). \tag{1.23}$$

We shall apply this assertion to the function $\tilde{\lambda}(h-\alpha)$ where $\tilde{\lambda}(t)=0$ for $t<0$ and $\tilde{\lambda}(t)=1$ for $t\geq\delta$. Then we conclude that the $\inf(F_x^*\Omega-\Omega)$ measure of the set $\{x|h(x)\geq\alpha\}$ is dominated by

$$2\left(\sup\int_0^{\rho_0'}\frac{G_x(\frac{t}{4})}{G_x(t)}\right)\sum_i\sum_{x\in F_i}\psi_x((\frac{4}{3})^{-i}\rho_0') \tag{1.24}$$

where ψ_x is defined using the function $\tilde{\lambda}(h-\alpha)$ and $\rho_0'=4G_x^{-1}\left[\frac{1}{5\int_M\tilde{\lambda}(h-\alpha)}\right]\leq\rho_0$.

Now

$$\begin{aligned}
&\tilde{\lambda}(h-\alpha)-\tilde{\lambda}(h(F_x)-\alpha)\\
&\quad=\left[\int_0^1\tilde{\lambda}'(th+(1-t)h(F_x)-\alpha)\right](h-h(F_x))\\
&\quad=(\tilde{\lambda}'(h-\alpha))(h-h(F_x))\\
&\qquad-\left\{\int_0^1\int_0^1(1-t)\tilde{\lambda}''[h-(1-t)\tau(h-h(F_x))-\alpha]\right\}(h-h(F_x)).
\end{aligned} \tag{1.25}$$

Therefore

$$\begin{aligned}
&\sum_i\sum_{x\in F_i}\psi_x\left((\frac{4}{3})^{-i}\rho_0'\right)\\
&\qquad\leq\int_M\tilde{\lambda}'(h-\alpha)\sup_x\frac{|(h-h(F_x))|}{(\rho_x-\rho_x(F_x))_+}(F_x^*\Omega-\Omega)\\
&\qquad\quad+\int_M\sup_x\int_0^1\int_0^1(1-t)|\lambda''[h-(1-t)\tau(h-h(F_x))-\alpha]|\\
&\qquad\quad\times\frac{(h-h(F_x))^2}{(\rho_x-\rho_x(F_x))_+}(F_x^*\Omega-\Omega).
\end{aligned} \tag{1.26}$$

Let k be any increasing function of t. Then the integral of $k(\tilde{\varphi})$ with respect to the measure $\inf_x(F_x^*\Omega-\Omega)$ is given by the integral of $k'(\alpha)$ multiplied with the measure of $\{\tilde{\varphi}\geq\alpha\}$. Hence it is dominated by the integral of

$$\begin{aligned}
&2\left(\sup_x\int_0^{\rho_0'}\frac{G_x(\frac{t}{4})}{G_x(t)}\right)k'(\alpha)\left\{\int_M\tilde{\lambda}'(h-\alpha)\sup_x\frac{h-h(F_x)}{(\rho_x-\rho_x(F_x))_+}(F_x^*\Omega-\Omega)\right.\\
&+\int_M\sup_x\left|\int_0^1\int_0^1(1-t)\tilde{\lambda}''[h-(1-t)\tau(h-h(F_x))-\alpha]\right|\\
&\left.\times\frac{(h-h(F_x))^2}{(\rho_x-\rho_x(F_x))_+}(F_x^*\Omega-\Omega)\right\}.
\end{aligned} \tag{1.27}$$

We assume

$$k'(a)\sup_x\int_0^{4G_x^{-1}(\frac{1}{t})}\frac{G_x(\frac{\tau}{4})}{G_x(\tau)}d\tau\leq\tilde{G}(\tilde{k}(a)t). \tag{1.28}$$

When $\tilde{G}$ and $\tilde{k}$ are non-decreasing functions then

$$2\sup_x \left(\int_0^{\rho_0'} \frac{G_x(\frac{t}{4})}{G_x}\right) k'(\alpha) \leq \tilde{G}(\tilde{k}(\alpha)\mu\{h \geq \alpha\}) \qquad (1.29)$$
$$\leq \tilde{G}(\int_M \tilde{k}(h)).$$

Note that the right hand side is independent of α.

The integral of $k(h)$ with respect to the $\inf_x(F_x^*\Omega - \Omega)$ measure is therefore dominated by

$$\begin{aligned}
&\tilde{G}(\int_M \tilde{k}(h))\Big\{\int_M \sup_x \frac{h - h(F_x)}{(\rho_x - \rho_x(F_x))_+}(F_x^*\Omega - \Omega) \\
&+ \int_{\alpha=0}^{\infty}\int_M \sup_x \Big|\int_0^1\int_0^1 (1-t)\lambda''[h - (1-t)\tau(h - h(F_x)) - \alpha]\Big| \\
&\frac{(h - h(F_x))^2}{(\rho_x - \rho_x(F_x))_+}(F_x^*\Omega - \Omega)\Big\} \\
\leq\ &\tilde{G}(\int_M \tilde{k}(h))\Big\{\int_M \sup_x \frac{h - h(F_x)}{(\rho_x - \rho_x(F_x))_+}(F_x^*\Omega - \Omega) \qquad (1.30)\\
&+ \sup(|\lambda''|)\int_M (|h - h(F_x)| + \delta)\frac{(h - h(F_x))^2}{(\rho_x - \rho_x(F_x))_+}(F_x^*\Omega - \Omega)\Big\}.
\end{aligned}$$

Theorems 1, 2, and 3 follow quite easily from (1.30).

2 Generalization

In Section 1, it is sometimes difficult to verify the condition

$$\lim_{\epsilon\to 0} G_x(\epsilon)\int_{\rho_x<\epsilon} \lim_{t\to 1}\frac{F_x^*(t)\Omega - \Omega}{t-1} \geq 1. \qquad (2.31)$$

It would be easier if we do not have to take the limit for $\epsilon \to 0$. To achieve this, we notice that Lemma 1.1 can be modified to

Lemma 2.1. *Suppose for some* $0 < \rho_0' < \rho_0$*;*

$$\sup_{\epsilon\leq\sigma\leq\rho_0'} G_x(\sigma)\varphi_x\sigma > G_x(\frac{\rho_0'}{4})\varphi_x(4\rho_0')\left[2 + 3\rho_0'\left(\int_0^{\rho_0'}\frac{G_x(\frac{t}{4})}{G_x(t)}\right)^{-1}\right]. \qquad (2.32)$$

Then there exists $\rho(x)$, $\epsilon < \rho(x) < \rho_0's$ *so that*

$$\varphi_x(4\rho(x)) \leq \psi_x(\rho(x))\left(2\int_0^{\rho_0'}\frac{G_x(\frac{t}{4})}{G_x(t)}\right).$$

Based on this lemma, we define

$$\bar{h}_\epsilon(x) = \int\int_0^1 \lambda(\epsilon - t\rho_x(F_x) - (1-t)\rho_x)h(F_x^*\Omega - \Omega). \tag{2.33}$$

Then for these points x so that $\bar{h}_\epsilon(x) \geq 1$, we have

$$\sup_{\epsilon \leq \sigma \leq \rho_0'} G_x(\sigma)\varphi_x(\sigma) \geq G_x(\epsilon).$$

Hence if ρ_0' is chosen so that

$$G_x(\epsilon) > 5G_x(\frac{\rho_0'}{4}) \int_M^0 \int_0^1 \lambda(4\rho_0' - t\rho_x(F_x) - (1-t)\rho_x)h(F_x^*\Omega - \Omega) \tag{2.34}$$

there is $\epsilon < \rho(x) < \rho_0'$, so that

$$\varphi_x(4\rho(x)) \leq \psi_x(\rho(x)) \left(2\int_0^{\rho_0'} \frac{G_x(\frac{t}{4})}{G_x(t)}\right). \tag{2.35}$$

We are going to replace the function h by $\bar{h}$. Hence we shall reinterpret (2.4) and (2.5).

In fact,

$$\int_M \bar{h}_\epsilon \geq \left[\int h(F_x^*\Omega - \Omega)\right] \left[\inf_y \int_M \int_0^1 \lambda(\epsilon - t\rho_y(F_y) - (1-t)\rho_y)(F_y^*\Omega - \Omega)\right]. \tag{2.36}$$

Hence if

$$\begin{aligned} G_x(\epsilon) \inf_y &\left[\int_M \int_0^1 \lambda(\epsilon - t\rho_y(F_y) - (1-t)\rho_y)(F_y^*\Omega - \Omega)\right] \\ &\geq 5G_x(\frac{\rho_0'}{4}) \int_M \bar{h}_\epsilon. \end{aligned} \tag{2.37}$$

then the conclusion of the lemma holds. Here ρ_0' will be the smallest number so that (2.7) holds. Note that

$$\begin{aligned} \int_{\rho_x < 2\rho(x)} \bar{h}_\epsilon \inf_x (F_x^*\Omega - \Omega) \leq \sup_y &\left[\int_M \int_0^1 \lambda(\epsilon - t\rho_y(F_y) - (1-t)\rho_y)(F_y^*\Omega - \Omega)\right] \\ &\int h\lambda(4\rho(x) - t\rho_x(F_x) - (1-t)\rho(x))(F_x^*\Omega - \Omega). \end{aligned} \tag{2.38}$$

We conclude therefore that the $\inf_x(F_x^*\Omega - \Omega)$ measure of the set $\{\bar{h}_\epsilon \geq 1\}$ can be estimated by

$$\sup_y \left(\int_M \int_0^1 \lambda(\epsilon - t\rho_y(F_y) - (1-t)\rho_y)(F_y^*\Omega - \Omega) \right) \tag{2.39}$$
$$\left(2 \sup_x \int_0^{\rho_0'} \frac{G_x(\frac{t}{4})}{G_x(t)} \right) \int_M \sup_x \frac{|h - h(F_x)|}{(\rho_x - \rho_x(F_x))_+} (F_x^*\Omega - \Omega)$$

where ρ_0' is defined by (2.7).

With this modification, most of the theorems in Section 1 can be generalized. We shall just state the generalization of Theorem 3. When we have a family of maps $F_x(t)$, we define

$$h_\epsilon(x) = \int_{\rho_x \leq \epsilon} \left(\lim_{t \to 1} \frac{F_x^*(t)\Omega - \Omega}{t-1} \right) h. \tag{2.40}$$

Theorem 4. *In Theorem 3, we can drop the assumption that* $\lim_{\epsilon \to 0} G_x(\epsilon)$ $\lim_{t\to 1} \frac{F_x^*(t)\Omega - \Omega}{t-1} \geq 1$ *if we modify the Sobolev inequality to be*

$$\int_M k(h_\epsilon) \leq H \left(\int_M \tilde{k}(k'(h_\epsilon)) \right) \int_M \limsup_{t\to 1} \sup_x \frac{h - h(F_x(t))}{(\rho_x - \rho_x(F_x(t)))_+} \lim_{t \to 1} \frac{F_x^*(t)\Omega - \Omega}{t-1}.$$

We have also to assume the measure of the support of h_ϵ is not greater than

$$\frac{1}{5}\left[\inf_x \frac{1}{G_x(\frac{\rho_x}{4})} \right] \sup_y \left\{ G_y(\epsilon) \int_{\rho_y \leq \epsilon} \lim_{t\to 1} \frac{F_x^*(t)\Omega - \Omega}{t-1} \right\}.$$

For example, when $g_x(\rho) \geq \frac{\rho}{m}$, we can take $G_x(\rho) = c^{-1}\rho^{-m}$ and all the assumptions will hold as long ρ_0 is large.

3 Some applications

The assumptions that we made on ρ_x and F_t are required to satisfy within the sets $\{\rho_x \leq \rho_0\}$ only. As a result, the Sobolev inequality depends on the assumption that the support of h has measure less than a constant C_{ρ_0}. In order to apply this theorem, we assume that there is a set of non-negative function $\{\varphi_\alpha\}$ so that $\sum_\alpha \varphi_\alpha \geq 1$ and the measure of the support of φ_α is not greater than C_{ρ_0}. Then we can apply the theorem to $\varphi_\alpha h$.

For simplicity, we shall restrict ourselves to the case when $g(t) = \frac{t}{m}$. In this case,

$$C_{\rho_0} = \left(\frac{4}{5}\right)^m C\rho_0^m$$

where

$$\lim_\epsilon \epsilon^{-m} \int_{\rho_x < \epsilon} \lim_{t\to 1} \left(\frac{F_x^*(t)\Omega - \Omega}{t-1} \right) \geq C.$$

We can apply the Holder's inequality

$$\left(\int_M h^{\frac{m}{m-1}}\right)^{\frac{m-1}{m}} \leq \left(\int_M (\sum \varphi_\alpha h)^{\frac{m}{m-1}}\right)^{\frac{m-1}{m}} \tag{3.41}$$

and the Sobolev inequality to $\varphi_\alpha h$.

Such an inequality can be used to prove some versions of Sobolev embedding theorems.

Let us assume that there is a bilinear form $\langle\ ,\ \rangle$ defined on the subspace of L^2 functions so that

$$\langle h, h\rangle \geq \int_M \sup_x \lim_{t\to 1} \frac{|h - h(F_x(t))|^2}{|\rho_x - \rho_x(F_x(t))|^2} \lim_{t\to 1}\left(\frac{F_x^*(t)\Omega - \Omega}{t-1}\right). \tag{3.42}$$

The subspace of L^2 consists of h so that $\langle h, h\rangle < \infty$ is therefore an analogue of the L_1^2 space. When the total measure is finite, the embedding of it into L^2 is precompact. This can be proved in the following way. We replace h by h^2 in the above inequality and obtain the statement that $\left(\int h^{\frac{2m}{m-1}}\right)^{\frac{m-1}{m}}$ can be estimated by

$$\sum_\alpha \int_M \lim_{t\to 1} \frac{|\varphi_\alpha(F_x(t)) - \varphi_\alpha(x)|}{|\rho_x(F_x(t)) - \rho_x|}, \quad \int h^2 \quad \text{and} \quad \langle h, h\rangle.$$

Hence if the above three quantities are uniformly bounded, we can bound the L^2 norm of h on any set S in terms of $(\mu(S))^{\frac{1}{m}}$. If $\{h_i\}$ is a sequence with $\int h_i^2 \leq 1$ and $\langle h_i, h_i\rangle \leq 1$, we can find a uniform convergent subsequence outside a set S with small measure. In this set, we can apply the above statement to extract a strong convergent subsequence of $\{h_i\}$ in L^2 norm.

Once we have the Sobolev embedding theorem, we have the discreteness of the spectrum of the self-adjoint operator A defined by $\langle\ ,\ \rangle$. In particular, we can form the integral kernel of the operator A. The standard Nash argument would then imply that the integral kernel of A can be estimated by $t^{-\frac{m}{2}}$ when t is not large.

Theorem 5. *Let M be a finite measure space which supports structure given by ρ_x and F_x. Assume that there is a set of function $\{\varphi_\alpha\}$ which satisfies the property that $\sum_\alpha \varphi_\alpha \geq 1$, suppose φ_α has $\sup_x \lim_{t\to 1} \frac{F_x^*(t)\Omega - \Omega}{t-1}$ measure less than $\left(\frac{4}{5}\right)^m C\rho_0^m$ and*

$$\sum_\alpha \int_M \sup_x \lim_{t\to 1} \frac{|\varphi_\alpha(F_t) - \varphi_\alpha|}{|\rho_x(F_t) - \rho_x|} < \infty.$$

Then for any self adjoint operator A defined by a pairing $\langle\ ,\ \rangle$ stronger than right hand side of (3.2), it has discrete spectrum whose integral kernel can be estimated.

Note that if $\rho_0 = \infty$, the set $\{\varphi_\alpha\}$ is not necessary and no assumption needs to be made, if the total measure is not finite, the spectrum need not be discrete.

On the other hand, the Nash argument can still be applied to estimate the heat kernel.

Once we know the Sobolev inequality, one can apply the De Giorgi-Moser argument to prove Harnack inequality as long as we have an inequality of the type

$$\inf_{\alpha} \int_{\rho_x \leq \gamma} |h - \alpha| \leq \frac{c}{\gamma} \int_{\rho_x \leq \gamma} \sup_x \lim_{t \to 1} \frac{|h - h(F_x(t))|}{|\rho_x - \rho_x(F_x(t))|}.$$

In order to be able to find such a Poincaré inequality, we generalize a method of Korevare and Schoen. We shall assume that there is a family of measurable transformations $F_x(\frac{k}{2m})$ defined on the set $\rho_x \leq \gamma$ where $k = 0, \dots, 2m$. We require that

(1) $F_x(0)(z) = x \quad \text{for} \quad \rho_x(z) \leq \gamma.$

(2) $F_x(1)(y) = y.$

(3) $F_x(\frac{k}{2m})(y) = F_y(\frac{2m-k}{2m})(x).$ (3.43)

Under condition (3), for any function h

$$h\left[F_x(\frac{k}{2m})(y)\right] - h\left[F_x(\frac{k-1}{2m})(y)\right]$$
$$= h(a)\left[F_y(\frac{2m-k}{2m})(x)\right] - h\left[F_y(\frac{2m-k+1}{2m})(x)\right]. \qquad (3.44)$$

Hence

$$|h(x) - h(y)| \leq \sum_{k=1}^{2m} \left| h\left[F_x(\frac{k}{2m})(y)\right] - h\left[F_x(\frac{k-1}{2m})(y)\right] \right| \qquad (3.45)$$
$$= \sum_{k=m+1}^{2m} \left| h\left[F_y(\frac{k}{2m})(x)\right] - h\left[F_y(\frac{k-1}{2m})(x)\right] \right|$$
$$+ \sum_{k=m+1}^{2m} \left| h\left[F_x(\frac{k}{2m})(y)\right] - h\left[F_x(\frac{k-1}{2m})(y)\right] \right|.$$

$$\int_{\rho_0(y) \leq \gamma} \int_{\rho_0(x) \leq \gamma} |h(x) - h(y)| \Omega_x \Omega_y$$
$$\leq \sum_{k=m+1}^{2m} \int_{\rho_0(y) \leq \gamma} \left\{ \int_{\rho_0(x) \leq \gamma} |h[F_y(\frac{k}{2m})(x)] - h[F_y(\frac{k-1}{2m})(x)]| \Omega_x \right\} \Omega_y$$
$$+ \sum_{k=m+1}^{2m} \int_{\rho_0(x) \leq \gamma} \left\{ \int_{\rho_0(y) \leq \gamma} |h[F_x(\frac{k}{2m})(y)] - h[F_x(\frac{k-1}{2m})(y)]| \Omega_y \right\} \Omega_x.$$

We shall assume that for some $C > 0$,

$$\sum_{k=m+1}^{2m} \left|\rho_x[F_x(\frac{k}{2m})F_x^{-1}(\frac{k-1}{2m})(y)] - \rho_x(y)\right| (F_x^{-1})^*(\frac{k-1}{2m})\Omega_y$$
$$\leq C\rho_x(y)\Omega_y. \qquad (3.46)$$

Then (3.6) shows that

$$\int_{\rho_0(y)\le\gamma}\int_{\rho_0(x)\le\gamma}|h(x)-h(y)|\le 2C\int_{\rho_0(x)\le\gamma}\gamma\mathcal{D}_x(\gamma) \tag{3.47}$$

where

$$\mathcal{D}_x(\gamma)=\int_{\rho_0(y)\le\gamma}\sup_{k>m}\frac{|h(F_x(\frac{k}{2m})F_x^{-1}(\frac{k-1}{2m})(y))-h(y)|}{|\rho_x(F_x(\frac{k}{2m})F_x^{-1}(\frac{k-1}{2m})(y))-\rho_x(y)|}\Omega_y. \tag{3.48}$$

We conclude the following Poincaré inequality

$$\inf_a\int_{\rho_0(x)\le\gamma}|h(x)-a|\le 2C\gamma\sup_{\rho_0(x)\le\gamma}\mathcal{D}_x(\gamma). \tag{3.49}$$

Theorem 6. *Suppose then for each point x with $\rho_0(x)\le\gamma$, we can find a family of measurable transformations $F_x(\frac{k}{2m})$ defined on the set $\rho_x\le\gamma$ with the properties defined by* (3.3) *and* (3.7)*. Then* (3.10) *holds.*

In particular, if the pairing $\langle\ ,\ \rangle$ has the property that $\langle h,h\rangle$ dominates

$$\int_{\rho_0(x)\le\gamma}\sup_{k>m}\frac{|h(F_x(\frac{k}{2m})F_x^{-1}(\frac{k-1}{2m})(y))-h(y)|^2}{|\rho_x(F_x(\frac{k}{2m})F_x^{-1}(\frac{k-1}{2m})(y))-\rho_x(y)|^2}$$

then the operator A will have the Harnack property as long as the hypothesis of Theorem 5 and (3.9) hold. In this case, there is no nonconstant bounded solution for the equation $Ah=0$.

Reference

[1] J. H. Michael and L. M. Sobolev, *Sobolev and mean-value inequalities on generalized submanifolds of* $\mathbf{R}^n$, Comm. Pure and Appl. Math., 26, 1973, pp. 361-379.

Open Problems and Surveys of
Contemporary Mathematics
SMM 6, pp. 397–477

Open Problems in Differential Geometry

Shing-Tung Yau

Hui Ma, Chung-Jun Tsai, Mu-Tao Wang, En-Tao Zhao

Contents

1 Projective geometry

1.1 History

Note: here we refer to [102].

Felix Klein stated in his famous Erlangen Program that geometry is the study of invariants under the action of a group of transformations. With this point of view, projective geometry is characterized by invariants under transformations of the projective group $\mathrm{PGL}(n)$. It started from great geometers such as Pappus and also perspective geometry developed by Italian painters.

Klein's Erlangen program was rather influenced by Sophus Lie. Lie is famous for his theory of Lie groups but he never considered himself as a group theorist; he always considered himself as a geometer. He did major contribution for affine geometry and projective geometry.

Klein was a student of Plücker. In Plücker's paper, geometry is the study of families of points or of spheres etc. To explain this we consider two points in $\mathbb{R}^k$ with homogeneous coordinates

$$x = [x_1, \cdots, x_{k+1}], \quad y = [y_1, \cdots, y_{k+1}]$$

and their Plücker coordinates

$$p_{ij} = x_i y_j - y_i x_j.$$

Up to a factor the p_{ij}'s describe the line through x and y. In the case of $k = 3$, in terms of homogeneous coordinates of two given points, lines through these two points can be described by six Plücker coordinates satisfying the following equation

$$P = p_{12}p_{34} + p_{13}p_{42} + p_{14}p_{23} = 0.$$

Here $[p_{ij}] \subseteq \mathbb{RP}^5$ is defined by a quadric. So the space of lines in $\mathbb{R}^3$ becomes a 4-dimensional space, sitting in $\mathbb{RP}^5$, defined by a quadric. There is an important duality: two lines intersect if the corresponding points on the quadric lie on the same line in the quadric. Curves, surfaces and three-dimensional bodies in point geometry are dual to their respective analogs in line geometry, called line surfaces, line congruences and line complexes. For example, $(p_{12}, p_{13}, p_{14}, p_{34}, p_{42}, p_{23}) =$

$(t, 0, 1, t, 0, -t^2)$ describes one of the families of lines for a hyperboloid. It is a line surface. The set of tangents to a surface is a line complex.

By using Plücker's line geometry, S. Lie defined sphere geometry, which is an important field. Consider a sphere in $\mathbb{R}^3$ defined by

$$x^2 + y^2 + z^2 - 2ax - 2by - 2cz + d = 0,$$

where a, b, c, and d are real parameters. In addition, we introduce the fifth quantity, the radius r of the sphere given by

$$r^2 = a^2 + b^2 + c^2 - d.$$

Here a, b, c, d and r are all real numbers, one immediately can complexify them. Introduce homogeneous coordinates by

$$a = \frac{\xi}{\nu},\ b = \frac{\eta}{\nu},\ c = \frac{\zeta}{\nu},\ r = \frac{\lambda}{\nu},\ d = \frac{\mu}{\nu}.$$

Then

$$\Phi := \xi^2 + \eta^2 + \zeta^2 - \lambda^2 - \mu\nu = 0.$$

Note that once using homogenous coordinates, $\lambda = 0$ corresponds to a point and $\nu = 0$ corresponds to the plane at ∞. In this terminology we have points and planes. S. Lie introduced the following correspondence

$$\begin{aligned} p_{12} &= \xi + i\eta, \quad p_{34} = \xi - i\eta, \\ p_{13} &= \zeta + \lambda, \quad p_{42} = \zeta - \lambda, \\ p_{14} &= \mu, \quad p_{23} = -\nu, \end{aligned}$$

which defines linear transformations on Plücker coordinates. p_{ij} describes coordinates of lines in $\mathbb{R}^3$ and $(\xi, \eta, \zeta, \lambda, \mu, \nu)$ associates a sphere. Then $\Phi = 0$ implies $P = 0$. So we get a map from the space of spheres to the space of lines. This is Lie's famous line-sphere transformation.

Now we complexify spheres to complex spheres and lines to complex lines. $\mathrm{SL}(3, \mathbb{C})$ acts via projective transformations on the space of lines in $\mathbb{C}^3$ and $\mathrm{SO}(6, \mathbb{C})$ acts on the projective bundle associated to the tangent bundle of $T\mathbb{C}^3$. These two groups are local isomorphic under Lie line-sphere transformation. This is a remarkable fact, particularly in general relativity.

Thus one can associate the set of tangent lines to a given surface $S \subset \mathbb{R}^3$ to the set of spheres tangent to another surface S', that is, lines tangent to the surface S at a point M correspond to the set of spheres tangent to S' at M'. Then an asymptotic line tangent to S at a point M will be mapped to a line of curvature of S' at M'. If one takes Kummer surface whose degree is 4, then under this correspondence, asymptotic curves of Kummer surface correspond to algebra curves of degree 16. This is one of important statements that inspired Picard to prove it independently. I mention this because we haven't paid enough attention on this geometry.

1.2 Projective flat structure

1.2.1 Projective flat structures on surfaces

Recall that a Riemann surface admits a *real* or *complex projective flat structure* if there exists an atlas such that the coordinate transformation (or gluing map) on overlapping coordinate charts are in $\mathrm{PGL}(3, \mathbb{R})$ or $\mathrm{PGL}(2, \mathbb{C})$, respectively. The former one has $\mathbb{RP}^2$ as its model space while the latter one has $\mathbb{CP}^1$ as its model space. Gunning's notes [97] is an excellent introduction to the complex projective structure. The structure corresponds to a $\pi_1(\Sigma_g)$ representation in $\mathrm{PGL}(2, \mathbb{C})$.

The projective flat structure of a Riemann surface Σ_g of genus $g > 1$ is related to the affine structure in the following way. On a convex domain $\Omega \subset \mathbb{RP}^n$, one solves the equation

$$\det(\frac{\partial^2 u}{\partial x^i \partial x^j}) = (-\frac{1}{u})^{n+2}$$

subject to the boundary condition $u = 0$ on $\partial\Omega$ for a convex, negative function u. Then $\mathrm{d}s^2 = -\frac{1}{u}\sum \frac{\partial^2 u}{\partial x^i \partial x^j}\mathrm{d}x^i \mathrm{d}x^j$ defines a metric on Ω that is invariant under projective transformations which leaves Ω invariant. In the case of a ball, this is simply the Hilbert metric on which all geodesics are straight segments. The existence of such solutions is proved in $n = 2$ case by Loewner–Nirenberg [133], and in $n \geq 3$ case by Cheng–Yau [52]. Thus for a subgroup $\Gamma \subset \mathrm{PGL}(3, \mathbb{R})$, the quotient space Ω/Γ carries a real projective flat structure. The deformation space of real projective structures on a compact orientable Riemann surface Σ_g of $g > 1$ is a countable disjoint union of open cells of dimension $16(g - 1)$. This moduli space identifies with a holomorphic vector bundle over Teichmüller space whose fiber is the space of holomorphic cubic differentials on Σ_g, which is parametrized by $(\Sigma_g, H^0(K^3))$, i.e. a conformal structure on Σ_g and a holomorphic cubic differential. These are called *convex* projective flat structure as we require Ω to be convex. When $\partial\Sigma_g \neq 0$, we require that

(1) The boundary $\partial\Sigma_g$ consists of closed geodesics with geodesically convex collar neighborhoods.
(2) The holonomy representation along a closed geodesic is an element in $\mathrm{SL}(3, \mathbb{R})$ $\subset \mathrm{PGL}(3, \mathbb{R})$ of the form $T = \begin{bmatrix} a & 0 & 0 \\ 0 & b & 0 \\ 0 & 0 & c \end{bmatrix}$ with $a > b > c > 0$. In this case, the model is the group generated by T acting on the first octant $\{(x, y, z) | x > 0, y > 0, z > 0\}$. T has 3 fixed points. The projection onto the 2 dimensional space defines a triangle Δ in $\mathbb{RP}^2$. $\Delta/\langle T\rangle$ is an annuls.

Choi in his thesis [57, 58, 59] proved the following:

Theorem 1.1. *Let $g > 1$ and suppose Σ_g has a real flat projective structure, then there exists a collection of disjoint simple closed geodesics $C_1, \cdots, C_k$ on Σ_g such that $\Sigma_g / \cup C_i$ consists of*

- *an annulus covered by an affine half-space;*
- *the interior of a compact convex real flat projective surface with negative Euler characteristic.*

Furthermore, each C_i bounds either one or two annuli. Such a decomposition is unique.

Question 1.2. *How about 3-dimensional case?*

It is far more complicated.

Choi's approach is topological/combinatorial, and we ask:

Question 1.3. *Is there any analytic proof? Does there exist a geometric flow to accomplish such a decomposition?*

For Choi's theorem, we need to study convex projective structure with geodesic boundary. Note that geodesics in projective structure are straight lines. We need to solve the above Monge–Ampère equation on a convex domain with some straight line boundaries. The boundary values of u are imposed so these boundaries become geodesics with respect to the metric $\mathrm{d}s^2$. One should try to write down these conditions explicitly.

About the geometric flow approach to Choi's theorem, recall Hamilton's Ricci flow in 3-dimension: Along the Ricci flow, a 3-manifold decompose as hyperbolic components, the Seifert fibered components, and quotients of 3-sphere. We are now dealing with projective flat structure, there should not be any spherical component. Analogously, the hyperbolic component is now the convex real projective flat structure, and the Seifert fibered component is the annulus. Under the flow, the annulus should collapse, and we capture the convex projective component. Here are two possible scenarios for its boundary: The first one is that the flow terminate at some t_0, and the boundaries become geodesics. The second possibility is that the boundary goes to infinity, and becomes a thin end. In Hamilton's work, the boundary of the hyperbolic components are some tori, and goes to infinity along the Ricci flow.

We also need a metric to start the evolution. A possible candidate is a metric with vector potential, $\mathrm{d}s^2 = \sum(\frac{\partial v^i}{\partial x^i} + \frac{\partial v^j}{\partial x^j})\mathrm{d}x^i\mathrm{d}x^j$. The idea is based on the Hessian type metric and the Kähler potential. Such a metric with a vector potential may always exist on a real flat projective surface. Also, how to characterize metrics of this form?

Another possible approach is the continuity method: Solve u for

$$\det(\varepsilon g_{ij} + u_{ij}) = (-\frac{1}{u})^4.$$

Let us say for $\varepsilon >> 1$, the equation has a solution. One can try to show that the equation has a solution as $\varepsilon \to 0$ by continuity argument.

Let us make a final remark on the boundary condition. For the triangle $\triangle$ as described above, Calabi [34] wrote down an explicit solution u with $u = 0$ on $\partial\triangle$. There might be solutions with other boundary condition, and it is helpful to understand more about this case.

Choi's idea is also used to study the complex projective flat structure. Suppose Σ_g has a complex projective flat PGL$(2,\mathbb{C})$ structure. Then there is a unique (up to conjugation) reduction of the monodromy representation $\pi_1(\Sigma_g) \to$ SL$(2,\mathbb{C})$. We recall that a homomorphism $\rho : \pi(\Sigma) \to \mathrm{SL}(2,\mathbb{C})$ is *non-elementary* if $\rho(\pi_1)$ does not fix any point or geodesic in $\mathbb{H}^3 \cup \bar{\mathbb{C}}$. Note that SL$(2,\mathbb{C})$ acts on $\mathbb{H}^3$ by isometry and on the boundary $\bar{\mathbb{C}}$ by Möbius transformation. It is well-known that if $g \geq 2$ and ρ is the monodromy representation of a projective flat structure on Σ_g, then ρ is non-elementary, see [97]. In 1995, Kapovich [115] proved the converse statement: If $\rho : \pi_1(\Sigma) \to \mathrm{SL}(2,\mathbb{C})$ is non-elementary, then there exists a PGL$(2,\mathbb{C})$ structure on Σ_g with ρ as its monodromy representation. Kapovich used a continuity argument that starts with a simpler complex structure. Perhaps one can use the equivariant harmonic map approach as Donaldson [64] and Corlette [60].

1.2.2 Higher dimensional projective flat structure

There are very few known interesting complex projective flat structures in higher dimensions. Here we look for a real projective structure with group PGL$(n+1,\mathbb{R})$. The general question is whether there exists a projective flat structure on a manifold with negative sectional curvature. We start with the convex or strictly convex case which is in terms of the developing map from the universal covering $\tilde{M}$ to $\mathbb{RP}^n$. Also we note that the domain with convex projective structure is not smooth in general. In fact, if Ω/Γ is compact and $\partial\Omega \neq S^{n-1}$, then $\partial\Omega$ is not smooth (C^1 but not $C^{1+\epsilon}$ for $\epsilon > 0$).

Thus, let M be a compact orbifold with a convex projective flat structure. Benoist [13] showed that the structure is strictly convex if and only if the fundamental group $\Gamma = \pi_1(M)$ is Gromov hyperbolic (i.e. the word metric on the group satisfies the triangle inequality for a hyperbolic space). There are several examples of strictly convex projective flat structure by gluing convex cones with infinite number of faces, for example, by Vinberg–Kac [193] and Vinberg [194].

In 2007, Kapovich [116] constructed convex projective structures on Gromov–Thurston [94] manifolds. These are manifolds whose sectional curvature K is pinched by $-1-\epsilon \leq K \leq -1$, and not homotopic to a hyperbolic manifold. He found a totally geodesic hypersurface and showed that the complement always admits strictly convex projective flat structure. In general, the Hilbert metric on a general domain Ω is not quasi-isometric to the one on a ball.

1.2.3 Complex projective flat structure over complex manifolds

A theorem of Cheng–Yau [51] shows that a Kähler manifold with complex projective flat structure is a locally Hermitian symmetric space. Thus we look at the non-Kähler case.

2 Affine geometry

2.1 Background

In the old days, S. Lie looked at differential equations and the group transformations which leave the equations invariant. He studied linearization of differential equations obtained Lie algebras of vector fields. This is the very beginning of the development of Lie theory. Here is more complicated than linear theory since the equation is nonlinear. Affine geometry has been important ever since up to now the study on Monge-Ampère equation. A couple years after Lie discussed Lie sphere theorem, Klein announced his Erlangen Program in 1872.

In the spirit of Klein's Erlangen program, affine geometry is the study of invariants under the affine group. To each geometry, we wish to associate an elliptic equation or a general partial differential equation. The important equation associated with affine geometry is the following Monge-Ampère equation

$$\det(u_{ij}) = F. \tag{2.1}$$

The affine group leaves this equation invariant.

Cheng and I studied the equation of the following general type ([49])

$$\det(u_{ij}) = F(x, \nabla u, u).$$

One can solve the Dirichlet problem. The Legendre transform can often be useful to simplify the above equation.

There are other reasons to consider the Monge-Ampère equation. Gauss curvature of a graph given by

$$K = \frac{\frac{\partial^2 u}{\partial x^2}\frac{\partial^2 u}{\partial y^2} - (\frac{\partial^2 u}{\partial x \partial y})^2}{(1 + |\nabla u|^2)^2}.$$

is naturally related to Monge-Ampère equation. Monge looked at this formula and found affine group acting on it. The numerator is

$$M(u) := \frac{\partial^2 u}{\partial x^2}\frac{\partial^2 u}{\partial y^2} - (\frac{\partial^2 u}{\partial x \partial y})^2 = F.$$

Take a new function $\tilde{u}(x) = u \circ (Ax)$, where A is a linear transformation. Then

$$M(\tilde{u}) = (\det A)^2 M(u).$$

Hence Monge-Ampère equation is invariant under affine transformations up to a constant factor. And $M(\tilde{u}) = M(u)$ if $\det A = \pm 1$. Thus the differential equation is invariant under $\mathrm{SL}(2, \mathbb{R})$. Geometric invariants under affine group thus enter the study of Monge-Ampère equation.

2.2 Hypersurfaces in affine geometry

Recall that in the study of a hypersurface M in $\mathbb{R}^n$ and its invariants with respect to $G \subseteq \mathrm{GL}(n, \mathbb{R})$, the most famous group is the projective linear group which was studied by Cartan, Fubini, Blaschke and etc in the 19th century. The corresponding geometry is called projective differential geometry. It requires to study higher order invariants. In Euclidean geometry, the fundamental theorem says that the first fundamental form and the second fundamental form decide surfaces locally. In affine geometry, we study hypersurfaces in affine space and their properties which are invariant under affine group given by $x \mapsto Ax + b$. In many cases we assume that $\det A = 1$, which may not be necessary.

2.2.1 Hypersurfaces in affine geometry

For a hypersurface $X : M \to \mathbb{R}^{n+1}$ in affine geometry, its first fundamental form is not invariant under affine transformation group. One needs to look at the following invariant, the first affine fundamental form defined by

$$\mathrm{I}_A = \frac{\mathrm{II}}{K^{n+1}},$$

where II denotes the second fundamental form of M in the Euclidean sense and K denotes the Gauss curvature of M. Note that I_A is not defined when $K = 0$. Then we assume $\mathrm{II} > 0$. This condition allows the first affine fundamental form nondegenerate. Remark that $K \neq 0$ is assumed in all the significant works we know.

The concept of affine normal can be defined completely by the linear structure and the volume element. Given a tangent plane to the hypersurface at one point, one can always move it along parallel lines. Then the centers of gravity trace a curve if the hypersurface is convex. The affine normal is simply the tangent line to this curve.

Once the affine normal n is defined, one can define the second fundamental form $\mathrm{II}_A := -\langle \mathrm{d}n, \mathrm{d}X \rangle$ by the affine normal in a usual way. More interestingly, one can define a cubic form III associated with the Fubini-Pick cubic form. Then $\mathrm{I}_A, \mathrm{II}_A$ and the cubic form III form the important invariants of affine geometry which can determine an affine hypersurface locally up to affine transformations. This is the fundamental theorem in affine geometry. Such subject was called the equivalence problem in Klein's Erlangen Program.

The affine mean curvature is defined to be the trace of II_A with respect to the metric I_A. An affine sphere satisfies $\mathrm{II}_A = c\mathrm{I}_A$, where c is a constant. So it seems like the umbilical surfaces in the Euclidean geometry. The Monge-Ampère equation is a second order equation but affine mean curvature can give a forth-order equation. The affine mean curvature becomes more complicated than affine sphere, as a fourth-order quantity.

On the other hand, just as the classical mean curvature flow of hypersurfaces in Euclidean space, one can deform a hypersurface in affine space along its affine mean curvature. The following is very interesting.

Question 2.1. *The affine mean curvature flow should deform a convex set to a set with constant affine mean curvature. Is this always an ellipsoid? Also, can we try to use the affine mean curvature flow to reprove John and Loewner's results on the ellipsoids?*

$K \neq 0$ is assumed in all the significant works in affine geometry we know. It is interesting to ask

Question 2.2. *How to handle affine geometry when the hypersurface is not convex (i.e.,* I_A *has singularities)?*

There is an intrinsic difficulty since Monge-Ampère equation has been no progress so far when the right side Changes signs. But it is a fundamental question.

So far one knows how to handle hypersurfaces in affine geometry effectively. Another question is

Question 2.3. *What happens when the codimension of* M *is greater than* 1*?*

There are some studies but none of them is significant enough in relation to the known results in Riemannian geometry.

At last of this subsection, I'd like to mention a simple application of affine geometry in numerical calculations ([47]). One can calculate affine minimums of a function by using Newton's method, namely choosing normal directions then rectifying iteratively. On the other hand, we can consider minimization in terms of the affine normal of the level set hypersurfaces of the function. The difficulty is how to calculate the affine normal effectively, especially in the case of higher dimensions.

Question 2.4. *How does one numerically calculate affine normal for a hypersurface* $M^n \subset \mathbb{R}^n$ *with* n *is large?*

2.2.2 Affine spheres

An affine hypersurface is called an affine sphere if all the affine normals converge to a point. When the affine normals converge to a point at the concave side of the hypersurface, it is called an elliptic affine sphere; When the affine normals converge to a point at the convex side, it is called a hyperbolic affine sphere; When the affine normals converge to ∞, i.e., parallel to each other, it is called an improper affine sphere.

An affine sphere satisfies $\mathrm{II}_A = c\mathrm{I}_A$, where c is a constant. To study geometric properties of affine hypersurfaces, the first step is to study affine spheres. While the classification of umbilical submanifolds in Euclidean geometry is trivial, the classification of umbilical affine hypersurfaces is very intricate and global.

Each of the three distinguish affine spheres associates to an interesting Monge-Ampère equation. The most important one is the improper affine hypersphere, which corresponds to Monge-Amprère equation $\det(\partial^2 u/\partial x_i \partial x_j) = 1$. So for person who is interested in Monge-Amprère equation, this is the important test to understand the equation.

In the case of dimension 2, by using a Legendre transformation, one can show that the study of convex improper affine spheres is equivalent to the study of the convex solutions of the equation $\det(\partial^2 u/\partial x_i \partial x_j) = 1$. It is solved by Jörgens in 1950s by using complex analysis that is powerful enough to 2-dimensional geometry. He showed that if u is convex and defined over the entire $\mathbb{R}^2$, then u is quadratic. This is the famous Jörgens theorem. It took a long time to solve the same question for higher dimensions. In Calabi's remarkable paper ([32]), he computed the Laplacian of the cubic form and made very important contribution. Calabi proved that when $n \leq 5$, u is quadratic. He also showed that any compact elliptic affine sphere is an ellipsoid. So an ellipsoid is linearly equivalent to a sphere and there are no differences between ellipsoids and spheres under affine transformations. The proof that every complete improper affine sphere is a paraboloid for all dimensions is due to Pogorelov and Cheng-Yau independently, using different gradient estimates.

Hyperbolic affine spheres are far more complicated. In 1950s, Calabi studied improper affine spheres and hyperbolic affine spheres when $n > 5$. It turns out that a hyperbolic affine sphere, after suitable Legendrian transformation, can be written as

$$\det(u_{ij}) = (-\frac{1}{u})^{n+2}. \tag{2.2}$$

We know that a hyperboloid is asymptotic to the light cone. The important question on hyperbolic affine spheres is

Question 2.5. *Given any convex cone, prove the existence of a hyperbolic affine sphere which is asymptotic to this cone.*

It is not unique in the sense that they are not only hyperboloids defined by quadric polynomials. The question is to solve the following equation

$$\begin{aligned} &\det(u_{ij}) = (-\frac{1}{u})^{n+2} \text{ on } \Omega \subseteq \mathbb{R}^{n+1}, \\ &u = 0 \text{ in } \partial\Omega, \end{aligned} \tag{2.3}$$

where u is defined on a convex domain $\Omega \subseteq \mathbb{R}^{n+1}$. This problem was solved by Cheng-Yau in 1970s. They developed gradient estimates in [49, 52]. That is useful for later development of studying Monge-Ampère equation.

When a hypersurface in Euclidean space is globally defined, there are two important metrics defined on the hypersurface. One is the affine metric, another is the induced Euclidean metric. The major question is whether the completeness of the induced Euclidean metric implies the completeness of the affine metric. The reason is that the equation of the affine sphere implies several inequalities. The first one is the Ricci curvature of the improper affine sphere with respect to the affine metric is nonnegative:

$$\mathrm{Ric}_{\mathrm{I}_A} \geq 0. \tag{2.4}$$

In fact, for the elliptic case, the Ricci curvature is bounded below by a positive constant. In the case of hyperbolic affine sphere, the Ricci curvature has a lower bound $\mathrm{Ric}_{\mathrm{I}_A} \geq -C$. Once we know that the metric is complete and the Ricci curvature $\mathrm{Ric}_{\mathrm{I}_A} \geq 0$, then we have the following inequality for the cubic form III:

$$\triangle_{\mathrm{I}_A}||\mathrm{III}||^2 \geq C_1||\mathrm{III}||^4 + \text{lower terms}, \tag{2.5}$$

where $C_1 > 0$. This is an extremely useful inequality for the affine sphere because it tells how to control the cubic form which controls everything of the affine sphere. The two consequences (2.4) and (2.5) are all in Calabi's paper [32]. Based on this, Calabi showed the result up to dimension 5. He didn't go further since he didn't have enough analytic means.

In order to prove the metric is complete, Cheng and Yau did the following thing. Move the tangent plane to the convex side if necessary, we may use the height function u for gradient estimate. u is non-negative and proper. Thus we get the affine gradient estimate

$$\sup_{u\leq c\leq R} \frac{|\nabla_{\mathrm{I}_A} u|^2}{(c-u)^2} \leq C,$$

where the constant C is independent of R. If the affine sphere is determined globally, then for any path γ on the affine sphere, we get

$$\log \frac{c-u}{c} \leq \int_\gamma |\nabla_{\mathrm{I}_A}| \log(c-u)||^2 \leq C \cdot \text{length of } \gamma \text{ w.r.t } \mathrm{I}_A.$$

As a result the length of the path γ goes to infinity so the improper affine sphere is affine complete.

The argument discussed here is more general and it allows us to prove the hyperbolic affine sphere metric is affine complete. Based on this, one can prove that there exists a unique hyperbolic affine sphere to (2.3) which is asymptotic to the given convex cone. This proved a conjecture of Calabi ([33]) is true. He also conjectured that any complete hyperbolic affine sphere is asymptotic to a convex cone. This was also proved by Cheng-Yau in [52].

2.2.3 Affine minimal hypersurfaces and the Bernstein problem

An affine minimal hypersurface is defined by the affine mean curvature is zero. Chern proposed a Bernstein problem for this ([55], [56]).

Let u be a convex function on $\mathbb{R}^n$. Assume that its graph grap(u) is an affine minimal hypersurface in $\mathbb{R}^{n+1}$. The question is whether u is quadric or not.

Trudinger and Wang showed that it is true in the 2 dimensional case ([191]). There are counter examples for the dimension $n \geq 10$. So it is interesting to figure out what happens for the dimensions in between. Maybe it is similar to the Bernstein problem in the Euclidean spaces, where the counter examples happen for the dimension $n \geq 7$ and the Bernstein problem is true for $n < 7$ and false for $n \geq 7$.

The Dirichlet problem of the affine minimal surface is interesting. Even for the curve σ in the Euclidean space, the fourth-order equation can be associated with two boundary conditions. Namely, for the surface Σ, not only $\partial\Sigma = \sigma$ on the boundary, but also the tangent planes $T\Sigma$ of Σ along σ could be given as $T\Sigma|_\sigma = \gamma$, i.e.,

$$\begin{aligned} \partial\Sigma &= \sigma, \\ T\Sigma|_\sigma &= \gamma. \end{aligned}$$

It would be very interesting to solve this Dirichlet problem by numerical method effectively.

2.3 Affine flat manifolds

We have already discussed the affine geometry of hypersurfaces in Euclidean space. It is natural to ask whether the real Monge-Ampère equation can solve some problems in intrinsic geometry.

A manifold M is called an affine flat manifold, if it is covered by coordinate chart system which is glued by affine transforms. Explicitly, $M = \cup \mathcal{O}_i$, where $(x_1, \cdots, x_n) \in \mathcal{O}_i$ is a coordinate chart. Suppose that $\mathcal{O}_j$ is another coordinate chart, $(y_1, \cdots, y_n) \in \mathcal{O}_j$ and $\mathcal{O}_i \cap \mathcal{O}_j \neq \emptyset$, then the coordinate transformation is given by the affine transformation $y = Ax + b$. It is equivalent to that there exists a torsion free connection with zero curvature on the tangent bundle. The torsion freeness is an important condition.

A very interesting problem is which manifold admits affine structure. It is not so easy even to the 2 dimensional case. Actually the problem to decide which bundle admits flat connection is already nontrivial.

2.3.1 Flat vector bundles

We say a vector bundle $V \to M$ is flat if there exists an affine connection ∇ such that the curvature of ∇ vanishes. There is not much known in this case except an important theorem of Milnor.

Theorem 2.6 (Milnor, 1957). *A rank 2 vector bundle V over a Riemannian surface $\Sigma_g (g \geq 1)$ is flat then*

$$|\chi(\Sigma_g)| \leq -\frac{1}{2}\chi(V). \tag{2.6}$$

This is a constraint for the Euler number of the vector bundle. Applying to $V = TM$ shows that only genus one Riemann surface supports a flat connection on its tangent bundle. Let's avoid non-orientable manifolds. The argument is very powerful but it only works for dimension 2. And Milnor didn't use torsion freeness which is not necessary for the existence of flat connections in the 2 dimensional case.

Question 2.7. *Is there any higher dimensional generalization of Milnor's theorem? For the higher dimensional case, does the flatness of the tangent bundle TM imply $\chi(M) = 0$?*

It is a famous problem proposed almost 100 years ago. Gauss-Bonnet formula says that if the Riemannian curvature is zero, then the Euler number must be zero. That is not true here because the Gauss-Bonnet formula used the Pfaffian, which depends on the metric and the skew-symmetry of the curvature tensor, and here we talk about affine connections, where there are no metrics involving and they are not Levi-Civita connections. So it is natural to ask whether the weaker Gauss-Bonnet theorem holds or not, i.e., whether the flatness of the tangent bundle TM implies $\chi(M) = 0$. The answer is negative. The counter examples were given by John Smillie ([174]). He constructed a flat 4-manifold N with $\chi(N) = 4$ and a flat 6-manifold Q with $\chi(Q) = 8$ given by

$$N := (\Sigma_3 \times \Sigma_3)\#\underbrace{(S^1 \times S^3)\#\cdots\#(S^1 \times S^3)}_{6-\text{ copies}},$$

$$Q := \{(\Sigma_3 \times \Sigma_3)\#\underbrace{(S^1 \times S^3)\#\cdots\#(S^1 \times S^3)}_{9-\text{ copies}}\}\#\Sigma_3,$$

where Σ_3 is a closed Riemann surface of genus 3. Products of N, Q with themselves give examples of flat manifolds with non-zero Euler numbers of any even dimension greater than 2. These examples are not torsion free. Torsion free condition is an important and powerful assumption, which makes deeper than what topology can offer since it is not a topological property.

2.3.2 Complete affine flat structure

One can ask whether a complete affine flat structure has zero Euler characteristic number. The completeness means geodesically completeness of the connection. In this case, they are in fact covered by $\mathbb{R}^n$ and thus $M = \mathbb{R}^n \backslash \Gamma$ for a discrete subgroup $\Gamma \subset \mathrm{GL}(n, \mathbb{R}) \ltimes \mathbb{R}^n$ and Γ is torsion free (without elements of finite order).

In 1970s, Milnor classified the fundamental group of a complete affine flat manifold. He obtained

Theorem 2.8 (Milnor [140]). *The fundamental group of a complete affine flat manifold is virtually polycyclic and torsion free. Conversely, for any virtually polycyclic group G, there exists a complete affine flat manifold M such that $\pi_1(M) = G$.*

Here, a group G is called virtually polycyclic, if there is a finite index subgroup H of G such that H is polycyclic. (G has an extension by cyclic group)

For any virtually polycyclic group G, we can find one complete affine flat manifold M such that $\pi_1(M) = G$. But how many of them are not clear. It is interesting to study the moduli of such manifolds. In Milnor's paper, it is not known whether the manifold can always be chosen to be compact. The compactness of an affine flat manifold does not imply completeness since there is no metric to help you to prove the convergence.

However, Benoist [12] gave an example of a torsion free, virtually polycyclic group which cannot be the fundamental group of a *compact*, complete affine flat manifold.

So it is interesting to arise the following question:

Question 2.9. *Which virtually polycyclic group could be the fundamental group of a compact complete affine flat manifold?*

Nevertheless, it is known, by work of Kostant–Sullivan [119], that for a compact, complete affine flat manifold M, $\chi(M) = 0$.

2.3.3 Compact affine manifolds

We recall the conjecture:

Conjecture 2.10. *A compact affine manifold has zero Euler characteristic number.*

Assuming completeness, this was solved by Kostant–Sullivan ([119]) in 1975.

In 1975, Hirsch–Thurston ([103]) proved that if $\pi_1(M)$ is built out of taking free products or forming finite extensions of solvable groups or groups of polynomial growth, then $\chi(M) = 0$.

In fact, they proved $\chi(M) = 0$ under a more general setting. We recall the holonomy representation of an affine flat manifold, $h : \pi_1(M) \to \Gamma \subset \mathrm{GL}(n)$, defined by the parallel transport of the affine connection, or equivalently, by a chain of coordinate patches along a closed loop representing the classes of $\pi_1(M)$. Their main theorem says that if h factors go through a group in the class $\mathcal{C}$, or if Γ preserves a content on $\mathbf{S}^{n-1}$, then $\chi(M) = 0$. We now briefly explain the notion of class $\mathcal{C}$ and content.

(1) An *amenable* group Γ is a group with a left-invariant linear function A on $B(T)$, the bounded functions on Γ, such that $A(f) > 0$ when $f > 0$ and $A(1) = 1$. The class of amenable groups is closed under the operations of taking quotients, subgroups, extensions of amenable groups by amenable groups, and direct limits. Note that abelian groups and finite groups are amenable.
(2) *Class* $\mathcal{C}$ is the smallest class of groups which contains all amenable groups, and which satisfies: if $G, H \in \mathcal{C}$, then the free product $G * H \in \mathcal{C}$, and if G has finite index in K then $K \in \mathcal{C}$. In particular $F \in \mathcal{C}$ if F is solvable, free or of sub-exponential growth.
(3) A *content* on a set V is a linear functional S on $C(V)$, the vector space of continuous functions on V such that $S(c) = c$ for all constant functions c. Note that if we require additionally that S is bounded, then by Riesz representation theorem, S is a measure. But there are other examples of contents.
(4) If a group Γ acts on V, it acts on the set of contents by $(\gamma \cdot S)(f) = S(f \circ \gamma^{-1})$. For an amenable group acting on a compact space, one can easily cook up a content preserved by the group from the function A.

It is known that any complete Riemannian flat manifold must be a flat bundle over a compact Riemannian flat manifold, due to Bieberbach. We remark that this statement is actually true for complete connected Riemannian manifolds with nonnegative sectional curvature, by Cheeger-Gromoll Soul theorem in Riemannian geometry.

Compact flat manifolds are sometimes called Bieberbach manifolds, which are classified in term of of group theory. Bieberbach theorem says that given any finite group Γ, stronger than polycyclic, if the group is torsion free and an extension of the integer lattice by finite group, then it is a fundamental group of a compact Riemannian flat manifold. This gives a necessary and sufficient condition for a torsion free group to be the fundamental group of a compact Riemannian flat manifold. This complete classification theorem reduces a geometry problem to a group representation problem, but it does not mean the problem has been solved, because how to determine which group could appear in this way turns out to be a nontrivial question in representation theory of finite groups.

The complete classification is known for $n = 3$. There is a list for $n = 4$, given by E. Calabi, which hasn't been published.

An interesting question is whether the fundamental group of a manifold $\pi_1(M)$ is *large*[1]. Based on Milnor's work, we would expect that the fundamental group of a compact, complete affine flat manifold cannot be large. This is not true if the completeness assumption is dropped. For example, let Σ_2 be a Riemann surface of genus two, the fundamental group of $\mathbf{S}^1 \times \Sigma_2$ can be realized as a discrete group in $O(2,1)$, thus preserving the open hyperboloid $x^2 > y^2 + z^2$. With this, one can prove that it admits an affine flat structure.

Complete affine flat manifolds are more or less classified. On the other hand, it is interesting to know whether compactness give rise to more constrains or just virtually polycyclic. But affine flat manifolds are far from being classified, even for 3-dimensional case.

In 1981, Smillie ([175]) showed that if M is compact and if $\pi_1(M)$ is built out of finite groups by taking free product, direct product, and finite extensions, then M does not have any affine flat structure. In particular, connected sums of Lens spaces do not have affine flat structure. Also, if $\pi_1(M)$ is built out of finite groups and infinite cyclic groups by taking free product and finite extensions, then M does not admit volume preserving affine flat structure ($\det A = 1$). Therefore, connected sums of $\mathbf{S}^1 \times \mathbf{S}^2$ and Lens spaces do not have volume preserving affine flat structure.

We now briefly explain Smillie's proof. He showed that there exists a map from $H^n(BG^\delta, V) \to H^n(M, V)$, where G^δ is the group but with the discrete topology and BG^δ is the classifying space of G^δ. There always exists a map from $M \to BG^\delta$ such that the bundle V over M is the pull back of the universal bundle over BG^δ. Smillie proved that the map $M \to BG^\delta$ factors go through $K(\Gamma, 1)$ where $\Gamma = \pi_1(M)$, and he did the study on $K(\Gamma, 1)$.

These results do not touch the hyperbolic group. On the other hand, large class of 3-dimensional manifolds are given by hyperbolic manifolds. We may ask:

Question 2.11. *Can one prove that hyperbolic 3-manifolds do not admit affine flat structures? The 2-dimensional case follows from Milnor's result.*

[1] A group Γ is called large if it contains a subgroup of finite index, $\Gamma' \subset \Gamma$, such that Γ' admits a surjective homomorphism onto the free group with two generators, see [90, P. 82].

There are no tools to solve this problem. Maybe we could flow the affine connection. But since here the curvature is zero, maybe we should find a better way to move in the class of affine structures. This is a fundamental question.

With a connection ∇, we have the notion of *parallel tensor fields*, i.e. a tensor T satisfying $\nabla T = 0$. More generally, we can consider *polynomial tensor fields*. A tensor field T is polynomial of degree less than k if $(\nabla \circ \cdots \circ \nabla)(T) = 0$, where we take k-times of ∇. Equivalently, the local expression of T is a polynomial of degree less than k in terms of affine coordinates.

Clearly, the existence of a parallel volume form is equivalent to that image of the linear holonomy representation, $\lambda : \pi_1(M) \to \mathrm{GL}(n, \mathbb{R})$, lies in $\mathrm{SL}(n, \mathbb{R})$. This is related to being complete:

Conjecture 2.12 (Markus conjecture, [134]). *A compact orientable affine manifold is complete if and only if it has a parallel volume form.*

There are many works concerning the Markus conjecture. For instance, Furness–Fedida ([78]) confirmed the conjecture in the case of pseudo–Riemannian with $\dim M \leq 3$, and Smillie's thesis ([173]) studied the case of abelian holonomy. Fried–Goldman–Hirsch ([75]) proved the Markus conjecture for affine manifolds with nilpotent holonomy. We now briefly discuss their work.

2.3.4 Nilpotent holonomy

It turns out that the Markus conjecture is quite difficult. Consider the following setting: Let G be a Lie group with a left-invariant affine structure, and $\Gamma \subset G$ be a discrete subgroup. Then the homogeneous space of right cosets $\Gamma\backslash G$ inherits an affine structure. In particular, if G is simply connected and nilpotent, $\Gamma\backslash G$ is called an *affine nilmanifold.*

This notion seems to be restricted, but there are some nontrivial examples. Let G be $\mathbb{R}^2$, and the left-invariant affine structure be defined by $\phi : G \to \mathrm{Aff}(\mathbb{R}^2)$

$$\phi(s,t)(z) = \begin{bmatrix} 1 & s \\ 0 & 1 \end{bmatrix} z + \begin{bmatrix} \frac{s^2}{2} + t \\ s \end{bmatrix}.$$

Equivalently, it is defined by

$$\begin{bmatrix} 1 & s & \frac{s^2}{2} + t \\ 0 & 1 & s \\ 0 & 0 & 1 \end{bmatrix} = \exp\left(\begin{bmatrix} 0 & s & 0 \\ 0 & 0 & s \\ 0 & 0 & 0 \end{bmatrix} + \begin{bmatrix} 0 & 0 & t \\ 0 & 0 & 0 \\ 0 & 0 & 0 \end{bmatrix} \right).$$

Given a discrete subgroup $\Gamma \subset G$, it defines an affine structure on $\Gamma\backslash G = \mathbf{T}^2$. The left-invariant vector fields are

$$\frac{\partial}{\partial t} \mapsto \frac{\partial}{\partial x}, \qquad \frac{\partial}{\partial s} \mapsto y(\frac{\partial}{\partial x} + \frac{\partial}{\partial y}),$$

and their integral curves are $y =$ constants and $y^2 - 2x =$ constants, respectively. This gives a non-standard affine structure on $\mathbf{T}^2$. Nagano–Yagi [145] proved that this is the only complete affine structure on $\mathbf{T}^2$ other than the standard one. The punch line here is that even on $\mathbf{T}^2$, the complete affine structure is not easy. Also, the situation of higher dimensional tori is still unknown.

Problem 2.13. *How to classify affine structures on* $\mathbf{T}^3$*?*

Here is the main results of Fried–Goldman–Hirsch [75, Theorem A]:

Theorem 2.14. *Let M be a compact affine manifold whose affine holonomy group [2] is nilpotent. Then the following statements are equivalent:*

(1) M is complete.
(2) The developing map is surjective.
(3) Its linear holonomy group is nilpotent.
(4) Its linear holonomy group preserves volume.
(5) M is a complete affine nilmanifold, $\Gamma\backslash G$.
(6) The alpine holonomy is indecomposable.
(7) M is a complete affine nilmanifold.
(8) M has a polynomial volume form.
(9) M is orientable, and the de Rham cohomology of M is the same as the cohomology of the complex of polynomial exterior forms.

2.3.5 Radiant affine manifold

A large class of affine manifolds are radiant manifolds. An affine manifold is called *radiant* if its affine holonomy group has a fixed point. We can take the fixed point to be the origin of the affine space, then the radiant vector field $R = \sum_{i=1}^n x^i \partial_i$ is invariant under the affine holonomy group. Thus it gives a vector field on M. In some sense radiant manifolds are at the opposite extreme from complete affine manifolds.

Here is an example: The manifold $\Sigma_{g\geq 2} \times \mathbf{S}^1$ has a radiant affine structure. Regard $\pi_1(\Sigma)$ as a discrete subgroup of $\mathrm{SO}(1,2)$, and $\mathbb{Z} = \pi_1(\mathbf{S}^1) \subset \mathrm{GL}(\mathbb{R}^3)$ be generated by $\alpha\mathbf{I}$ where $0 < \alpha < 1$. The quotient of a component of the interior of the light cone is $\Sigma \times \mathbf{S}^1$. One can check that the construction induces a radiant affine structure. There is another conjecture about the affine structure on such manifolds:

Conjecture 2.15. *The manifold $\Sigma_{g\geq 2} \times \mathbf{S}^1$ does not admit any complete affine structure.*

For a compact affine manifold M^n, we can define a map from its universal covering $\tilde{M}$ to the affine space $\mathbb{R}^n$ by patching the transition functions. This map is denoted by D, and is called the *developing map*. The affine manifold M is said to be *complete* if the developing map D is a diffeomorphism[3].

[2] The affine holonomy representation is the map $\alpha : \pi_1(M) \to \mathrm{Aff}(\mathbb{R}^n)$ induced by the developing map $D : \tilde{M} \to \mathbb{R}^n$. The linear holonomy representation, $\lambda : \pi_1(M) \to \mathrm{GL}(n, \mathbb{R})$, is obtained from α by forgetting the translation, or equivalently, defined by the parallel transport of ∇.

[3] Auslander–Markus [7] proved that an affine manifold is complete if and only if the connection ∇ is geodesically complete.

Sullivan–Thurston [182], also Smillie [173], constructed many radiant affine structures on $\Sigma_{g\geq 1} \times \mathbf{S}^1$ such that the developing map is not a covering map onto its image. They actually found the image of the developing map satisfies $D(\widetilde{\mathbf{T}^3}) = \mathbb{R}^3 \backslash \{\text{three axes}\}$ and $D(\widetilde{\Sigma \times \mathbf{S}^1}) = \mathbb{R}^3 \backslash \{0\}$. The general interest is to understand the developing map.

Question 2.16. *For an affine manifold M^n, can one show that the developing map $D : \tilde{M} \to \mathbb{R}^n$ is surjective if and only if M is complete? Also, can one find out certain reasonable condition under which the developing map D is injective?*

As a contrast, Schoen and Yau [165] studied the developing map in conformal geometry. Namely, instead of affine transforms, the model space is the sphere $\mathbf{S}^n$, and the transitions are Möbius transforms. They proved that for a conformally flat manifold M, if the map $f : \widetilde{M} \to \mathbf{S}^n$ is conformal, $\widetilde{M}$ is complete and the scalar curvature is nonnegative, then f is one to one. Hence for a compact, locally conformally flat manifold, if the Yamabe constant is positive, then the developing map is one to one. The argument is using the Green function for the conformal Laplacian. The points in the pre-image $f^{-1}(p \in \mathbf{S}^n)$ can be governed by the Green's function, then the argument requires the positive mass theorem.

Based on this, it would be helpful to study the kernel of certain affine invariant differential operator. Remember that the Monge–Ampère equation plays an important role in affine geometry. However, the kernel of

$$\det\left(\frac{\partial^2 u}{\partial x_i \partial x_j}\right) = \delta$$

is not unique. Also, there is no radial symmetry because affine geometry does not distinguish spheres from ellipsoids, while the conformal Green's function has the radial symmetry.

Instead of real Monge–Ampère operator, we can also consider the kernel of the complex Monge–Ampère operator. This is related to the Jacobian conjecture: Consider the map from $\mathbb{C}^2$ to itself given by two polynomials (f, g). If the Jacobian $\frac{\partial(f,g)}{\partial(z,w)} = 1$, then the map is invertible.

2.3.6 The Lorentzian case

Recall that for an affine flat manifold, the coordinate transformation is given by $x \mapsto y = Ax + c$ where $A \in \mathrm{GL}(n, \mathbb{R})$. In case when all $A's$ are in a subgroup $G \subset \mathrm{GL}(n)$, we say the manifold has a flat G-structure and G is the holonomy group of the connection. An interesting case is when $G = \mathrm{O}(p, q)$. Does the Gauss–Bonnet theorem implies $\chi(M) = 0$ for a Lorentzian manifold? A good example is the $\Sigma_g \times \mathbf{S}^1$ we discussed before in which the holonomy group is $\mathrm{O}(2, 1)$. We can ask when a compact manifold has a Lorentzian flat connection.

Question 2.17. *Can one generalize the above example to a circle bundle over a Riemann surface? Or more generally, an orbifold bundle with smooth total space, i.e. the Seifert fiber space. Do these spaces admit affine (or Lorentzian) flat structure?*

The example $\Sigma_g \times \mathbf{S}^1$ with a flat Lorentzian metric can be interpreted in general relativity. The simpler case is to consider a Riemann surface in a 3-dimensional Lorentzian manifold which satisfies the 3-D vacuum Einstein equation $R_{ij} - \frac{R}{2} g_{ij} = 0$. This equation implies $R_{ij} = 0$ which is equivalent to the vanishing of the full Riemannian curvature in 3-D, and the Lorentzian manifold is locally isometric to the 3-dimensional Minkowski space. To solve the Einstein equation, we need to impose constraint equations on the initial data Σ_g with induced metric $g_{ij}|_{\Sigma_g}$ and the second fundamental form $h_{ij}|_{\Sigma_g}$. The constraint equation is equivalent to the Gauss equation and Codazzi equation for a surface in $\mathbb{R}^{2,1}$. If we assume $\mathrm{tr}_g h$ is constant initially, a gauge choice will give us a CMC (constant mean curvature) foliation such that each leaf is a CMC surface. The constraint equation implies $h_{11} - h_{22} + 2\sqrt{-1}h_{12}$ is a holomorphic quadratic differential in $H^0(K^2)$. Thus the initial data consists of a conformal structure on Σ_g and an element in $H^0(K^2)$ which is isomorphic to $H^1(T^{1,0}M)^* = (T_{(\Sigma_g, g_{ij})}\mathfrak{M})^*$, where $\mathfrak{M}$ is the moduli space of Σ_g. This pair is an element in the cotangent space of $\mathfrak{M}$ whose complex dimension is $3g-3$. The $2+1$ Einstein equation forms a curve in the Teichmüller space and hits the boundary. If we vary the direction at a point, we should get the Thurston boundary but this has not been proved yet. For details, consult Moncrief's article [143].

Note that if we identify the universal cover of the initial data as the hyperbola in $\mathbb{R}^{2,1}$, the solution actually forms a family of CMCs. What is the higher dimensional generalization of this picture? Can we move a CMC hypersurface in $\mathbb{R}^{n,1}$? We can ask about CMC hypersurface in $\mathbb{R}^{n,1}$ and their deformation. Cheng–Yau [48] proved that a CMC hypersurfaces in $\mathbb{R}^{n,1}$ is complete, convex, and with non-negative Ricci curvature and asymptotic to a cone. Treibergs [190] has studied the boundary value problem of CMC hypersurface in $\mathbb{R}^{n,1}$ and classified this kind of convex cones. Later Z.-C. Han, L.-F. Tam, A. Treibergs and T. Wan ([100]) did much more understanding on CMC hypersurfaces in $\mathbb{R}^{2,1}$ by using Gauss map from a surface to a hyperbolic disk. Whether it is good enough to understand the dynamic of such Einstein flow is interesting. Einstein flow inside $\mathbb{R}^{n,1}$ is certain not a mean curvature flow since the mean curvature H is constant. It should be a new curvature flow in $\mathbb{R}^{n,1}$.

2.4 Holomorphic affine connections

For complex manifolds, there is a notion of holomorphic affine connection on the tangent bundle.

We briefly review some relevant notions: Denote the holomorphic tangent bundle by T, and the holomorphic k-form by Ω^k. A *holomorphic affine connection* on T is a connection with holomorphic Christoffel symbol, i.e., $\nabla \frac{\partial}{\partial z^\alpha} = (\Gamma^\mu_{\alpha\beta} dz^\beta)\frac{\partial}{\partial z^\mu}$, where $\Gamma^\mu_{\alpha\beta}$ are holomorphic functions. It turns out that the existence of a holomorphic affine connection already gives some constraints on the manifold M. For instance, it implies that $c_i(M) = 0$ for all $i \geq \dim_{\mathbb{C}} M/2$. If M is furthermore Kähler, $c_i(M) = 0$ for all i. This implies that if a Kähler manifold admits a holomorphic affine connection, then it is covered by a complex tori.

In general, the existence of holomorphic affine connection is determined by the following Čech 1-cocycle:

$$c_{VU} = \frac{\partial^2 v^\mu}{\partial u^\alpha \partial u^\beta} du^\alpha \otimes du^\beta \otimes \frac{\partial}{\partial v^\mu} \in H^0(U \cap V, \Omega^1 \otimes \Omega^1 \otimes T),$$

where $\{U, V, W, \cdots\}$ is an open cover of M with local coordinates $u^\alpha, v^\alpha, w^\alpha, \cdots$. It is a simple exercise for chain rule to check that $c_{VU} + c_{UV} = 0$ and $c_{VU} + c_{UW} + c_{WV} = 0$. With a holomorphic affine connection, define

$$b_U = \Gamma^\mu_{\alpha\beta}(U) du^\alpha \otimes du^\beta \otimes \frac{\partial}{\partial u^\mu}.$$

The transformation law for the Christoffel symbol would imply that $c_{VU} = b_U - b_V$. On the other hand, if the cocyle $[c_{VU}]$ is coexact, one can cook up a holomorphic affine connection. To sum up, $[c_{VU}] = 0$ is equivalent to the existence of holomorphic affine connections. Hence, $H^1(\Omega^1 \otimes \Omega^1 \otimes T) = 0$ implies that M admits a holomorphic affine connection.

The torsion and the difference of holomorphic affine connections are both elements in $H^0(\Omega^1 \otimes \Omega^1 \otimes T)$. $H^0(\Omega^1 \otimes \Omega^1 \otimes T) = 0$. This implies that M has a unique torsion free holomorphic affine connection.

The curvature of a holomorphic affine connection defines an element in $H^0(\Omega^1 \otimes T \otimes \Omega^2)$ and $H^0(\Omega^1 \otimes T \otimes \Omega^2) = 0$. This implies that every holomorphic affine connection is flat. For surfaces, $T \otimes \Omega^2 = \Omega^1$. Thus, we can look at $H^0(\Omega^1 \otimes \Omega^1)$ instead.

2.4.1 Holomorphic affine structures on complex surfaces

We now discuss the holomorphic affine structures on surfaces. We already know that $c_1(M) = c_2(M) = 0$. Moreover, a simple computation with the adjunction formula shows that M has no exceptional curve, i.e. a rational curve $\mathbb{P}^1$ whose self-intersection number is -1. Inoue–Kobayashi–Ochiai [113] studied both flat and non-flat holomorphic affine connections based on Kodaira's classification of surfaces. Let us focus on the *holomorphic affine structures*, i.e. flat holomorphic affine connections.

- Class I_0 = {rational and ruled surfaces}: This class does not admit a holomorphic affine structure.
- Class II_0 = {K3 surfaces}: This class does not admit a holomorphic affine structure.
- Class III_0 = {complex tori}: This class admits a holomorphic affine structure.
- Class IV_0 = {elliptic surfaces with $b_1 = \text{even}, p_g > 0$, and $c_1^2 = 0$}: This class does not admit a holomorphic affine structure.
- Class V_0 = {algebraic surfaces of general type, $p_g > 0$, and $c_1^2 > 0$}: This class does not admit a holomorphic affine structure.
- Class VI_0 = {elliptic surfaces with $b_1 = \text{odd}, p_g > 0$, and $c_1^2 = 0$}: This class admits a holomorphic affine structure.

- Class $\mathrm{VII}_0 = \{\text{surfaces with } b_1 = 1 \text{ and } p_g = 0\}$: Except the Hopf surfaces[4], those with $b_2 = 0$ admit holomorphic affine structures. Those with $b_2 \neq 0$ do not admit a holomorphic affine structures.

In VI_0 and elliptic surfaces of VII_0, surfaces admit holomorphic affine structures except for the Hopf surfaces. We now briefly explain the story for class VI_0 and VII_0.

2.4.2 Class VI_0

Kodaira proved that complex surfaces of class VI_0 are covered by $E \times \Delta$, where E is an elliptic curve, and Δ is either $\mathbb{C}$ or $\mathbb{H}$, where $\mathbb{H}$ denotes the upper half plane.

(1) For $\Delta = \mathbb{C}$, they are called the Kodaira surfaces. M is $\mathbb{C}^2/G$, where G is a discrete subgroup of $\mathrm{SL}(2,\mathbb{C})$. M has a holomorphic volume form induced by $\mathrm{d}z^1 \wedge \mathrm{d}z^2$, but M is non-Kähler. These manifolds might have some meaning in physics.

(2) For $\Delta = \mathbb{H}$, $M = (\mathbb{C}^* \times \mathbb{H})/\langle \alpha, \Gamma; \chi \rangle$, where Γ is a discrete subgroup of $\mathrm{SL}(2,\mathbb{R})$ with a compact quotient $\mathbb{H}/\Gamma$, $E = \mathbb{C}^*/\langle \alpha \rangle$, $0 \neq |\alpha| < 1$, and $\chi : \Gamma \to \mathbb{C}^*$ is a character of Γ. The group action of $g = (\alpha^n, \gamma \in \Gamma)$ on $\mathbb{C}^* \times \mathbb{H}$ is defined by

$$g(z,\zeta) = \big(\alpha^n \chi(\gamma)(c\zeta + d)z, \gamma(\zeta)\big)$$

where c and d come from

$$\gamma = \begin{bmatrix} a & b \\ c & d \end{bmatrix} \in \Gamma \subset \mathrm{SL}(2,\mathbb{R}).$$

The fibration structure is given by $\mathbb{C}^*/\langle \alpha \rangle \to M \to \mathbb{H}/\Gamma$. It is not obvious that M has a holomorphic affine structure. To see this, consider the isomorphism

$$\begin{aligned} \mathbb{C}^* \times \mathbb{H} &\longrightarrow \mathcal{D} = \{(z,w) \in \mathbb{C}^2; \mathrm{Im}(w\bar{z}) > 0\}, \\ (z,\zeta) &\mapsto (z, w = z\zeta). \end{aligned}$$

The group action of $g = (\alpha^n, \gamma \in \Gamma)$ on $\mathcal{D}$ is then given by

$$g(z,w) = \big(\alpha^n \chi(\gamma)(cw + dz), \alpha^n \chi(\gamma)(aw + bz)\big)$$

which is clearly affine.

It is interesting to think about the higher dimensional generalization of the second construction. Namely, construct a (holomorphic) affine structure for a manifold M which has a fibration structure $\mathbf{T}^k \to M \to \mathbb{H}^k/\Gamma$.

---

[4]The Hopf surfaces are quotients of $\mathbb{C}^2\backslash\{0\}$ by some discrete group action. The discussion of holomorphic affine structure depends on the discrete group. We skipped this part.

2.4.3 Class VII_0

In the class VII_0, besides the Hopf surfaces, examples with $b_2 = 0$ are constructed by Inoue [110], based on the work and ideas of Kodaira. There are 3 families: S_M, $S^{(+)}_{N,p,q,r;t}$, and $S^{(-)}_{N,p,q,r}$, and they are all quotients of $\mathbb{H} \times \mathbb{C}$ by some affine transformation G.

In fact, these are all possible examples with $b_2 = 0$. At first, if there exists a line bundle F such that $H^0(M, \Omega^1(F)) \neq 0$, then one can argue that M must be those examples. Then we only need to consider the case $H^0(M, \Omega^1(F)) = 0$ for all line bundle F, and prove that this cannot happen. Bogomolov gave a proof [22, 23], but it seems too complicated to understand. With the help of Hermitian–Yang–Mills connections ([192]) for stable bundles on non-Kähler manifolds ([130]), Li, Yau and Zheng ([131, 132]) gave another proof. Note that the flat affine connection might not preserve the Hermitian metric. However, the condition $H^0(M, \Omega^1(F)) = 0$ for all F implies stability. With this, we proved the manifold must be projectively flat, and then the proof ends up with a contradiction.

The classification for $b_2 \neq 0$ is still open. Inoue ([111, 112]) constructed some examples. Also consult the paper by Nakamura ([146]). It is believed that these are all examples, but the proof is still missing. In contrast to the case of $b_2 = 0$, these surfaces have finite number of curves. The construction starts with $(\mathbf{D} \times \mathbf{D})/G$. The quotient is not compact but of finite volume. Then do the blow-up over the cusps, and it ends up with a surface with finite curves.

Here is a proposal for the classification: Study the flat holomorphic affine connection on M outside those curves, $\cup C_i$. One should impose some suitable condition for the connection near C_i's. Then modify the argument of Li–Yau–Zheng ([132]).

2.5 Affine Kähler geometry and SYZ conjecture

Affine structure is a good invariant. Recall that it is the structure invariant under affine transformations when we move from one canonical coordinate chart to another. Sometimes we assume $\det A = 1$ if the manifold is orientable.

There is one class of structures introduced by Cheng and Yau in 1970s in order to understand the Monge-Ampère equation. They were interested in Kähler geometry. If the Kähler metric is given by $\omega = \sqrt{-1} \sum g_{i\bar{j}} \mathrm{d}z^i \wedge \mathrm{d}\bar{z}^j$ and it is Kähler-Einstein means that $R_{i\bar{j}} = c g_{i\bar{j}}$. It is Kähler means that $g_{i\bar{j}}$ is the Hessian of some potential which is positive definite, i.e., $g_{i\bar{j}} = \frac{\partial^2 u_\alpha}{\partial z_i \partial \bar{z}_j} > 0$. Locally this is true on a coordinate chart $\mathfrak{U}_\alpha$. u_α is called a Kähler potential defined on $\mathfrak{U}_\alpha$. Kähler potential has the following property:

$$\partial\bar{\partial} u_\alpha = \partial\bar{\partial} u_\beta,$$

when one moves from one chart $\mathfrak{U}_\alpha$ to another chart $\mathfrak{U}_\beta$, so Kähler metric is invariant. Therefore,

$$\partial\bar{\partial}(u_\alpha - u_\beta) = 0,$$

i.e., $u_\alpha - u_\beta$ is pluriharmonic. Locally $u_\alpha - u_\beta$ is written as $f+\bar{g}$ for two holomorphic functions f and g. Calabi proposed a way in 1974 to construct inhomogeneous Kähler-Einstein metrics in a tube domain.

If a holomorphic coordinate is given as $z_i = x_i + \sqrt{-1}y_i$, look for the metric invariant under the translations in the y-direction. Then the metric $g_{i\bar{j}}(x_i + \sqrt{-1}y_i) = g_{i\bar{j}}(x)$ is independent of the choice of the height. Since the metric in this setting is given by the Hessian form, the metric now is actually given by

$$\sum \frac{\partial^2 u(x)}{\partial x_i \partial x_j} \mathrm{d}x_i \mathrm{d}x_j,$$

i.e., the Kähler metric becomes a Hessian in x. Thus the equation of Kähler-Einstein metric becomes

$$\det(\frac{\partial^2 u}{\partial z_i \partial \bar{z}_j}) = ||F||^2 e^{cu},$$

where F is a nonzero holomorphic function. In fact, the Ricci curvature tensor is given by

$$R_{i\bar{j}} = -\partial\bar{\partial} \log \det(g_{i\bar{j}}). \tag{2.7}$$

Then $R_{i\bar{j}} = c\partial\bar{\partial}u = cg_{i\bar{j}}$. The expression (2.7) is important, in fact, it is the reason why Kähler geometry is easy to handle than other geometry. If $g_{i\bar{j}}$ only depends on x and $F = 1$, we obtain an equation of Monge-Ampère type

$$\det(\frac{\partial^2 u}{\partial x_i \partial x_j}) = e^{-cu}. \tag{2.8}$$

In 1974, Calabi initiated the study of this interesting equation. By reducing to real 2-dimensional domain, inhomogeneous Kähler-Einstein metrics were explicitly constructed for the first time. When $c = 0$, the equation becomes $\det(u_{ij}) = 1$. If u is globally defined, then u is quadratic. when $n = 2$, it is proven by Jörgens; $n \leq 5$ is due to Calabi; For arbitrary n, it is due to Pogorelov, and Cheng-Yau.

When $c > 0$, we have $R_{i\bar{j}} \geq g_{i\bar{j}}$ essentially, then the completeness implies compactness by Bonnet theorem. Therefore, u is not globally defined, i.e., there is no global examples and singularities may occur.

When $c < 0$, there are quite a lot homogeneous tube domains with Kähler-Einstein metrics of negative Ricci curvature. In fact, for any convex domain Ω, there exist global solutions. It is true in more general situations. The solutions are not necessarily obtained by reduction. Actually for a bounded, smooth, strictly pseudo-convex domain Ω in $\mathbb{C}^n$, Cheng-Yau ([50]) proved that there exists a unique complete Kähler-Einstein metric of scalar curvature -1. The regularity assumption of the boundary was removed later by Mok and Yau ([142]). This result is remarkable since this is the only known complete intrinsic Kähler-Einstein metric we know for an arbitrary domain; on the other hand, Bergman metric and Kobayashi metric may not be complete for bounded domains.

Question 2.18. *For any unbounded pseudo-convex domain, when does it admit a complete Kähler-Einstein metric?*

This is actually a technical problem. Any unbounded domain can be transformed to a bounded domain by Cayley transforms. But in general it would be nice to have a good description on it. The only condition you need is that there exists a volume form on Ω, denoted by $V dz^1 \wedge d\bar{z}^1 \wedge \cdots \wedge dz^n \wedge d\bar{z}^n$ in local coordinates, such that $\partial\bar{\partial} \log V > 0$. It is independent of the choice of the metric. Then $(\sqrt{-1}\partial\bar{\partial}\log V)^n \geq cV$ for some constant $c > 0$, i.e., $(\mathrm{Ric})^n \geq cV$. This is a completely intrinsic question on the volume form. More canonical question is what canonical bundle over the manifold has uniformly large first Chern form and good curvature, i.e., $c_1^n(K) \geq cV$. This is the condition one needs to prove. It is always bounded from below. So you get a complete Kähler-Einstein metric over the bounded domain. In order to prove the convergence of the metric, you need something could support over it. The canonical bundle would be the normal bundle of it so it wouldn't disappear. Local convexity is not a problem and the only problem is that it may disappear in the limit. This is roughly what Kähler-Einstein metric happens for a domain.

Cheng and Yau looked at this problem carefully in 1974 since they were working on the Calabi conjecture and the Monge-Ampère equation at that time. Later they were interested in the reduction for the sake of giving a better understanding on the complex Monge-Ampère equation. Around 1978 or 1979, they were able to solve the following equation

$$\det(\frac{\partial^2 u}{\partial x_i \partial x_j}) = F(x, u, \nabla u), \tag{2.9}$$

on the convex domain as far as F is positive in a right way: $\frac{\partial F}{\partial u} \geq 0$. It relates to the fact that (2.8) is solvable for c negative and not solvable for c positive. They knew how to solve the equation

$$\det(u_{ij}) = (-\frac{1}{u})^{n+2}$$

related to hyperbolic affine spheres. Using this, they studied the general equation (2.9) and solved the Dirichlet problem to get $u \in C^\infty(\Omega) \cap C^{1,1}(\bar{\Omega})$ for a given smooth boundary data. The boundary regularity was eventually solved by Caffarelli-Nirenberg-Spruck ([31]) later to obtain $u \in C^\infty(\bar{\Omega})$.

Cheng and Yau wrote a paper [51] to generalize this question to a manifold. Let M be an affine manifold with a coordinate chart system $\{\mathfrak{U}_\alpha\}$. Assume the metric is a Hessian metric locally, i.e.,

$$g_{ij} = \frac{\partial^2 u_\alpha}{\partial x_i \partial x_j} \tag{2.10}$$

in a coordinate chart $\mathfrak{U}_\alpha$. On each coordinate chart $\mathfrak{U}_\alpha$, there is a potential u_α. The coordinate transforms are given by linear transforms, which preserve this structure. Such an affine manifold which admits a coordinate chart so that the metric can be defined by a Hessian is studied in affine-Kähler geometry. It is just a reduction of Kähler geometry from complex coordinates to real coordinates. Suppose $\det A = 1$. There is a volume form $dV = dx^1 \wedge \cdots \wedge dx^n$ globally defined on the affine manifold. Then $\det(g_{ij}) = 1$. So global geometry does not support such manifold. In another words, if an affine manifold M admits the metric satisfying $\det(g_{ij}) = 1$ and compact, then $M \simeq T^n$. Although there is no global examples in such kind of geometry, it is still interesting because it relates to structures with some singularities.

In 1995, it motivated Stronger, Yau and Zaslow's study on Calabi-Yau manifolds. They proposed there is a mirror symmetry ([180]). A manifold M^3 is called a Calabi-Yau manifold, if M is compact Kähler and Ricci flat. They proposed that the Calabi-Yau manifold M^3 has a T^3 fibration over S^3, where T^3 is a special Lagrangian torus. Here S^3 is just one possibility and it is the major example.

$$\begin{array}{ccc} T^3 & \longrightarrow & M^3 \\ & & \downarrow \\ & & S^3 \end{array}$$

The fiber T^3 is a special Lagrangian submanifold of M^3. The symplectic form of M is given by the Kähler form $\omega = \sqrt{-1}\sum g_{i\bar{j}} dz_i \wedge d\bar{z}_j$. T^3 is Lagrangian if $\omega|_{T^3} = 0$. On the Calabi-Yau manifold M, there is a holomorphic volume form $\Omega = f dz^1 \wedge dz^2 \wedge dz^3$, where f is a holomorphic nonzero function. The Ricci curvature of M is zero implies $\nabla\Omega = 0$ with respect to the connection coming from the Kähler metric. A Lagrangian torus T^3 is called special, if $\Omega|_{T^3} = c\,\mathrm{dvol}_{T^3}$ for a unitary constant c. Special Lagrangian submanifolds have minimizing volume comparing with nearby cycles, which were studied firstly by Harvey and Lawson. This is just like the case in algebraic geometry where the fiber converges to some degenerate structure. In algebraic surface $M^2_{\mathbb{C}}$, there is an elliptic fibration map to $\mathbb{P}^1$, where the fiber is an elliptic curve:

$$E \longrightarrow M^2_{\mathbb{C}} \xrightarrow[\pi]{} P^1$$

Naturally the singularities of the singular curves are the finite number of points on P^1. The inverse images could be a bunch of singular curves, which are rational curves. Nearby there are nice elliptic curves. Kodaria classified all the possible configuration of singular fibers in such elliptic fibrations, in the spirit following by Italian geometers. The 2-dimensional case is much more well understood in this picture.

Based on the two dimensional case, Stronger, Yau and Zaslow proposed mirror symmetry. Now we talk about the picture of complex 3-dimension. For this construction, there will be a codimension 2 set, a graph G, sitting in S^3, where the torus will degenerate. When the fibration is good, i.e., outside of the graph G, $\pi : M^3 \to S^3$ is a submersion and the fiber is a torus T^3. When you converge to a singular point on G, the fiber is degenerated. Hopefully the degeneration is just like in the case of the Kodaria fibration, because we talk about the part transversal to G. What happens at the vertex in the graph G? What kind of monodromy group do we expect at the singular touches, called vertex of the graph? This is the major problem. Quite a lot of studies have been done from the point of view of quantum physics.

We expect this is true only for the so-called large radius limit. When we deal with a Calabi-Yau manifold, all the information we have great confidence is when the scale goes to some limit, which we call large complex limit.

Now let's see what the large radius limit looks like. A famous example is the quintic given by

$$M_\psi = \{z_0^5 + \cdots + z_4^5 + 5\psi z_0 \cdots z_4 = 0\} \subset \mathbb{CP}^4,$$

for a complex number ψ. This is a one-parameter family of quintic hypersurfaces sitting in $\mathbb{CP}^4$. Each of them is a Calabi-Yau threefold. Here $z_0, \cdots, z_4$ are homogeneous coordinates depending on the parameter ψ. When $\psi \to \infty$, this is called the large radius limit. The manifold converges to $\{z_0 \cdots z_4 = 0\} = \cup_{i=0}^4 \{z_i = 0\}$, corresponding to five hyperplanes in $\mathbb{CP}^4$. So each hyperplane is $\mathbb{CP}^3$, denoted by $\mathbb{CP}_i^3$ for $i = 1, \cdots, 5$. The five $\mathbb{CP}^3$ glue together. Then

$$\mathbb{CP}_1^3 \bigcup_{i \neq 1} \{\mathbb{CP}_i^3 \cap \mathbb{CP}_1^3\} = \mathbb{C}^* \times \mathbb{C}^* \times \mathbb{C}^*$$

It has a nice Ricci-flat metric given by $\sum \frac{|dz_i|^2}{|z_i|^2}$. Each of $\mathbb{C}^*$ has a circle. Here $S^1 \times S^1 \times S^1 = T^3$ is the torus we talked in the Lagrangian torus fibration $T^3 \to M^3 \to S^3$. When $|\psi| < \infty$, M_ψ is close to the hyperplane in the large radius limit and the torus survives after perturbation when ψ is close to ∞ but not ∞. So this is not hard at the nonsingular points for $\psi = \infty$. There are more difficulties at the singular points. The limiting Calabi-Yau manifold is given by five open sets gluing with $\mathbb{C}^* \times \mathbb{C}^* \times \mathbb{C}^*$, $\mathbb{C}^* \times \mathbb{C}^*$, $\mathbb{C}^*$ and a point. Each of them has a nice torus sitting on. This is a typical example for large radius limit. We'd like to see how to go back to the original nonsingular geometry.

At the limit, every torus has a linear structure. In general, we use the lattice $Z^3 \subset \mathbb{R}^3$. The picture of mirror symmetry is

$$\begin{array}{ccccc} T^3 \longrightarrow M^3 & & & & (\tilde{M}^3)^* \longleftarrow (T^3)^* \\ & \searrow & & \swarrow & \\ & & S^3 & & \end{array}$$

where $T^3 : Z^3 \subseteq \mathbb{R}^3$ is defined by the lattice of T^3 and the dual lattice is given by $(T^3)^* : (Z^3)^* \subseteq (\mathbb{R}^3)^*$, which is a linear functional defined on $\mathbb{R}^3$ taking values of integer at the lattice. In the case of linearity of the torus, we expect to the picture is as follows. Locally near large radius limit, $x_i + \sqrt{-1}y_i \mapsto x_i$. Then the Ricci-flat metric $\sum g_{i\bar{j}} dz^i \wedge d\bar{z}^j$ given by a potential, satisfies $\det(\partial\bar{\partial}u_\alpha) = 1$ locally. Correspondingly, in the limit, $\det(\frac{\partial^2 u_\alpha}{\partial x^i \partial x^j}) = 1$.

We expect that on S^3 with a graph G of codimension 2, there exists an affine structure on $S^3 \backslash G$, i.e., $S^3 \backslash G = \cup_\alpha \mathfrak{U}_\alpha$ is covered by coordinate charts glued by linear transformations. On each $\mathfrak{U}_\alpha$ there is a convex function u_α such that the affine-Kähler metric is given by $g_{ij} = \frac{\partial^2 u_\alpha}{\partial x^i \partial x^j} > 0$ and $\det(\frac{\partial^2 u_\alpha}{\partial x^i \partial x^j}) = 1$.

The function u_α is globally defined. $u_\alpha - u_\beta$ is linear, then $\frac{\partial^2(u_\alpha - u_\beta)}{\partial x^i \partial x^j} = 0$. The monodromy is given by linear functions. There is another monodromy. Each fiber is a torus. Keep moving the torus then come back. Then the monodromy sits in $\mathrm{SL}(3, \mathbb{Z})$. This relates to the linear transformation between the coordinate charts in order to make a well-defined bundle. The whole setting is more complicated.

Question 2.19. *Find graphs whose compliments admit affine structures. In fact, there is a finite number of such graphs in each dimension. There should be a finite number of such combinatorial torus structures whose compliments admit affine structures. This is the first challenging question. Classify them if possible.*

Given a graph, solve the equation $\det(\frac{\partial^2 u}{\partial x^i \partial x^j}) = 1$ with prescribed monodromy. This becomes non-linear Riemann-Hilbert correspondence like Riemann-Hilbert deal with linear equations, e.g. Picard-Fuchs equation. Here we do Monge-Ampère equation where u_α could have monodromy. We need to develop a method to solve this equation with given monodromy.

3 Einstein metrics

3.1 Einstein equation

3.1.1 Einstein equation

In this chapter we discuss the Ricci curvature and Einstein metrics. The Einstein equation is the Euler-Lagrange equation of the Hilbert action $\int R\sqrt{g}$, where R is the scalar curvature.

From the variation

$$\delta \int R\sqrt{g} = \int (R_{ij} - \frac{R}{2} g_{ij})\delta g^{ij},$$

we derive the Einstein equation

$$R_{ij} - \frac{R}{2} g_{ij} = 0. \tag{3.1}$$

From the differential equation point of view, it is important that the Einstein tensor $E_{ij} = R_{ij} - \frac{R}{2} g_{ij}$ satisfies the conservation law. It follows from the Bianchi identity that $\sum_j E_{ij,j} = 0$ for $i = 0, 1, 2, 3$.

We can also consider the Einstein equation coupled with matter fields:

$$R_{ij} - \frac{R}{2} g_{ij} = T_{ij} \tag{3.2}$$

In fact, the term $-\frac{R}{2} g_{ij}$ in (3.2) was missing in Einstein's original formulation and this is not consistent with the conservation of the energy-momentum tensor T.

Where does the matter tensor term come from? Consider $\int R\sqrt{g} + \int L(\phi, g)$, where the Lagrangian $L(\phi, g)$ depends on a matter field ϕ and the metric g. Then we can derive a coupled system between the gravity and the matter field ϕ by the variation principle:

$$\delta_g(\int R\sqrt{g} + \int L(\phi, g)) = 0,$$
$$\delta_\phi(\int R\sqrt{g} + \int L(\phi, g)) = 0.$$

Thus the first equation gives (3.2), where $T_{ij} = -\frac{\delta L}{\delta g_{ij}}$ and the second equation becomes an equation for the matter field ϕ. There are various choices of L. For example, one choice is $L(\phi, g) = \int |\nabla\phi|^2$ for a scalar field ϕ, where the gradient is with respect to the Lorentzian metric. Then the second equation becomes the wave equation $\Box\phi = 0$. Another choice is to take the curvature F of a gauge theory and let $L(\phi, g) = \int_M |F|^2$, then the Einstein equation is coupled with the Maxwell equation or the Yang-Mills equation.

With cosmological constant, the full Einstein equation is

$$R_{ij} - \frac{1}{2}Rg_{ij} + \Lambda g_{ij} = 8\pi G T_{ij}, \tag{3.3}$$

where G is the Newton constant and Λ is the cosmological constant. It comes from the action

$$\frac{1}{16\pi G}\int_M (R - 2\Lambda)\sqrt{g} + \int_M L(\phi, g). \tag{3.4}$$

In the Euclidean signature case with $T_{ij} = 0$, the Einstein equation (3.3) gives rise to an Einstein metric with $R_{ij} = cg_{ij}$ for a constant c.

The Einstein equation for spacetime with a fixed boundary was later proposed. We apply variational principle on the following functional

$$\mathfrak{L} = \frac{1}{16\pi G}\int_M (R - 2\Lambda)\sqrt{g} + \int_M L(\phi, g) + \frac{1}{8\pi G}\int_{\partial M} \mathrm{Tr}K, \tag{3.5}$$

where K is the second fundamental form of the boundary ∂M. The second term may be dropped if we focus on gravity. We assume that the metric g_{ij} is fixed on the boundary, i.e., $\delta g_{ij} = 0$ near ∂M.

Applying variational principle to the above functional gives equation (3.3) in the interior because we can take the variational deformation to be with compact support.

In general, we can only fix the induced metric on the boundary and should allow the metric to change in the normal direction. In this case, the variation of the functional (3.4) contains a boundary term that is cancelled by $\frac{1}{8\pi G}\int_{\partial M} \mathrm{Tr}K$. This is why this term is added to the functional (3.5), and this turns out to be important.

From the point of view of Lorentzian geometry or hyperbolic equations, one can argue whether the boundary condition should be fixed. But from the point of view of quantum gravity, it is quite natural. It was called the thin Sandwich problem by some physicists.

Consider a configuration ϕ' at time t_1 and suppose it moves to another configuration ϕ'' at time t_2. The Feynman integral on the space B of all paths from ϕ' to ϕ'' is given by

$$\int_B D[\phi]\exp(-i\mathfrak{L}[\phi]). \tag{3.6}$$

This describes the probability of going from one configuration to another, where $D[\phi]$ is a certain measure on the space of all fields ϕ and $\mathfrak{L}[\phi]$ is the action for the quantum field ϕ on a Lorentzian manifold M. The conservation of probability is stated as follows:

$$\langle \phi''', t_3 | \phi', t_1 \rangle = \sum_{\phi''} \langle \phi''', t_3 | \phi'', t_2 \rangle \langle \phi'', t_2 | \phi', t_1 \rangle.$$

This has been important since Feynman defined quantum mechanical reformulation coupling with nonlinear fields. It is important to look at the critical point of $\mathfrak{L}$. A field ϕ that satisfies $\delta\mathfrak{L}(\phi) = 0$ is called a stationary field.

M. Kac added a minus sign in $\mathfrak{L}$ to treat the Euclidean actions case. Thus he formulated the Euclidean version of the Feynman path integral. This is where the Euclidean motion and Feynman–Kac formula come in.

Similarly, in gravitation the Euclidean version was introduced by Hawking and Gibbons ([83, 101]). They dropped the i in (3.6) and consider an expression of the form

$$Z \sim \int_{B'} D(g) \exp(-I(g)),$$

where $I(g)$ is the Euclidean gravitational action given by

$$I(g) = -\frac{1}{16\pi} \int_M (R - 2\Lambda) - \frac{1}{8\pi} \int_{\partial M} \mathrm{Tr} K + C$$

on the 4-dimensional manifold M with the boundary ∂M. Here the constant C is not essential, and is only added to guarantee that the action is zero for the flat Euclidean metric.

Now we want to understand $\delta I(g) = 0$ as before. The first important class of metrics are those *asymptotically flat* metrics. Thus the gravity vanishes at infinity. Hawking called this class of metrics *gravitational instantons*. The classification of them turns out to be a difficult problem. The case when $\Lambda = 0$ and $T = 0$ is the most important one to begin with. We are simply looking for Ricci flat metrics. No such nontrivial metric is known to exist until the solution of the Calabi conjecture in 1976. On a $K3$ surface, there exist metrics which are Ricci-flat but not flat.

The $K3$ surfaces are complex surfaces which can be deformed to standard quartic surfaces $x_0^4 + x_1^4 + x_2^4 + x_3^4 = 0$ in $\mathbb{CP}^3$. I proved that a quartic has a Ricci flat Kähler metric. The construction does not involve the explicit form of the metric. After such a metric was proved to exist, many experts including Hawking tried to explicitly construct the metric. But up to now this is still unsolved. It is one of the most important open problems, i.e.,

Question 3.1. *Find the explicit form of a Ricci-flat metric on a quartic surface in a closed form.*

It is likely that perturbation theory in string theory may help. This is nevertheless still a challenging problem.

3.2 Wick rotation

Though the Einstein equation $R_{ij} - \frac{R}{2} g_{ij} = T_{ij}$ is an equation for a Lorentzian metric, there is the Wick rotation of Hawking-Gibbons [84] (see also [154]) that can turn a Lorentzian metric into a Riemannian one. It is not a well-defined process even formally. For example, we don't know how to do this for the canonical bundles. Only some special cases are considered, but they are already very interesting.

Gibbons and Hawking obtained a continuation from the space of Lorentzian Ricci flat metrics to the space of Riemanian Ricci flat metrics. This is the so called Wick rotation, and is one of the important constructions of Ricci flat Einstein metrics. This method is successful. But we have to take care of what happens at the zeros of the Killing field.

The Killing field $K = \frac{\partial}{\partial t} = \sum k^a \frac{\partial}{\partial x^a}$ ($a = 0, 1, 2, 3$ corresponds the index of an arbitrary space-time coordinate system) could have zeros, and the zeros could have limits. A zero of the Killing field corresponds to a fixed point of the associated group action by G. The group G induces an action on the tangent bundle of M. At a fixed point p of G, the differential $T_pM \to T_pM$ is generated by the anti-symmetric tensor $\nabla_a k_b$. This tensor can have rank 2 or 4. The rank is the codimension of the fixed point set. If $\nabla_a k_b$ has rank 2, then the fixed point set is called a "bolt". If it has rank 4, then the fixed point set is called a "nut". When $\nabla_a k_b$ has rank two, the group action in an orthonormal frame at p can be expressed as

$$\begin{pmatrix} 1 & 0 & 0 & 0 \\ 0 & 1 & 0 & 0 \\ 0 & 0 & \cos kt & \sin kt \\ 0 & 0 & -\sin kt & \cos kt \end{pmatrix},$$

where k is surface gravity of the bolt. Identifying t with $t + \frac{2\pi}{k}$, the bolt closes up under this action. It is actually a 2-dimensional compact submanifold. You can try to generalize all of these to the higher dimensional case. This two dimensional manifold (bolt) may be assigned an Euler number, $\hat{\chi}$. The Euler number of a spherical bolt is 2 and in general, $\hat{\chi} = 2(1 - g)$.

The nut is a point and in the case, the group action is of the form

$$\begin{pmatrix} \cos \alpha t & -\sin \alpha t & 0 & 0 \\ \sin \alpha t & \cos \alpha t & 0 & 0 \\ 0 & 0 & \cos \beta t & -\sin \beta t \\ 0 & 0 & \sin \beta t & \cos \beta t \end{pmatrix}$$

in an orthonormal frame at p where $\alpha\beta \neq 0$. It is called a nut if $\alpha\beta > 0$ and an "antinut" if $\alpha\beta < 0$. If $\alpha/\beta = p/q$ is rational number for a pair of coprime integers p, q, then this is called a nut of type (p, q). If α/β is irrational, then you can generate two linear independent Killing vector fields.

$\nabla_a k_b$ can be split into its self-dual K^+_{ab} and anti-self-dual pieces K^-_{ab}. A nut is self-dual if $K^-_{ab} = 0$, then $p = q = 1$. It is anti-self-dual if $K^+_{ab} = 0$, then $p = -q = 1$.

By the fixed point formula for Killing fields, we have

$$\chi(M) = \text{number of nuts} + \text{number of antinuts} + \sum_{\text{bolts}} \hat{\chi},$$

and

$$\tau(M) = - \sum_{\text{nuts, antinuts}} \cot(p\theta)\cot(q\theta) + \sum_{\text{bolts}} Y/\sin^2\theta + \eta(0,\theta),$$

where $\eta(0,\theta)$ is the η-invariant which is determined by boundary information. For example, for an asymptotically flat metric, $\eta(0,\theta) = 0$. For an asymptotically locally flat metric such as $\mathbb{R}^4/\mathbb{Z}_k$, $\eta(0,\theta) = k\csc^2\theta - 1$.

Question 3.2. *Generalize the Wick rotation to study under what conditions one can change Lorentzian metrics to Riemannian metrics.*

We'd like to find a way to use this procedure more systematically. Hawking and other people used complex analytic continuation, which is a rather mysterious procedure, particularly on singularities of the manifold.

The Wick rotation sometimes kills singularities so one gets nonsingular Riemannian manifolds from singular Lorentzian manifolds. We ask whether the procedure can be reversed.

Question 3.3. *Given a Riemannian Einstein manifold with zero Ricci curvature, find a Killing field in order to reverse the procedure of the Wick rotation to get a Ricci-flat Lorentzian manifold.*

It is an interesting and rather nontrivial question. We do know a great deal of Riemannian Einstein manifolds. The important question is to understand this procedure in more general settings. One may complexify the metrics to get complex Euclidean metrics and Lorentzian metrics. It is also very interesting to study the Wick rotation in the following cases:

(1) with cosmology constant.
(2) higher dimensional.
(3) coupled with other matter fields such as Yang-Mills.

In the following we discuss some examples of Lorentzian metrics on which the Wick rotation works.

3.2.1 Kerr metric

The Wick rotation turns Lorentzian Ricci flat metrics to Riemannian Ricci flat metrics. The most important example is the following Euclidean Kerr metric ([83, 101]). In Boyer-Lindquist coordinates, the metric is given by

$$\begin{aligned} ds^2 &= (r^2 - a^2\cos^2\theta)\frac{dr^2}{\Delta} + d\theta^2 \\ &+ \frac{1}{r^2 - a^2\cos^2\theta}[\Delta(dt + a\sin^2\theta d\phi)^2 + \sin^2((r^2 - a^2)d\phi - adt)^2], \end{aligned}$$

where $\Delta = r^2 - 2mr - a^2$, m is the mass of the black hole and a is the angular momentum. We are interested in the region where $\infty \geq r \geq m + (a^2+m^2)^{\frac{1}{2}}$, which is used to describe the rotational black holes. The Killing field $\frac{\partial}{\partial t}$ has a spherical bolt at $r = m + (a^2 + m^2)^{\frac{1}{2}}$. θ and ϕ are the standard spherical coordinates on a 2-sphere. The Euclidean Kerr metric is obtained from the Lorentzian Kerr metric by changing t to it and identifying

$$(r, t, \theta, \phi) \to (r, t, \theta, \phi + 2\pi),$$
$$(r, t, \theta, \phi) \to (r, t + \frac{2\pi}{k}, \theta, \phi + 2\pi\frac{\Omega}{k}),$$

where $k = \frac{(r^2+a^2)^{\frac{1}{2}}}{2m(m+\sqrt{m^2+a^2})}$ is the surface gravity of the bolt and $\Omega = \frac{1}{2}a\frac{1}{m^2+m\sqrt{m^2+a^2}}$. Thus the manifold has the topology of $\mathbb{R}^2 \times S^2$. The singularity turns out to be very similar to what happens in the polar coordinate system. Changing t to it induces a conic singularity. The identification procedure kills the conic singularity and makes a nice smooth metric. The Kerr metric is one of the most important and remarkable complete Ricci-flat metric in general relativity and geometry. This is essentially the only complete Ricci-flat Riemannian manifold which is not constructed from Kähler metrics.

A rather important conclusion one wants to draw for this metric is the uniqueness, which is not known. The uniqueness means that if a 4-dimensional Ricci flat manifold M is asymptotic to the Euclidean Kerr metric $\mathbb{R}^2 \times S^2$ at infinity, then M is exactly the Euclidean Kerr metric. This means that black hole is unique in the Euclidean case too. The uniqueness of black hole is known for Lorentzian case at least under certain assumptions. The "no-hair theorem", due to Carter, Hawking, Israel and Robinson, etc., states that stationary Lorentzian black holes are determined by mass, angular momentum and electric charge q. It is interesting to prove the Euclidean version.

(Remark: Walter Simon ([170]) proved that Kerr instantons are unique among complete, Ricci-flat, asymptotically flat metrics on $\mathbb{R}^2 \times S^2$ which have a $U(1)$-group of isometrics with only isolated fixed points and orbits of bounded length at infinity.)

Relating to this is another uniqueness result we would like to prove:

Question 3.4. *Does S^4 have a unique Einstein metric?*

This is not true on higher dimensional spheres and there are exotic Einstein metrics.

The complex version is

Question 3.5. *Does a $K3$ surface have a unique Ricci flat metric?*

Before I proved the Calabi's conjecture, many arguments were built to disprove it. Here is one of them. It was observed by Green and proved by Hitchin ([104]) that given a 4-dimensional compact Ricci-flat manifold, if it satisfies $3\chi = 2|\tau|$, then the manifold must be isometric to a $K3$ surface, or the quotient of a $K3$ surface modulo $\mathbb{Z}_2$ (Enriques surface), or $\mathbb{Z}_2 \times \mathbb{Z}_2$ (the quotient surface by an anti-holomorphic free involution). This is a very strong uniqueness result which only depends on the homotopy type of the manifold. It would be interesting even to drop the assumption $3\chi = 2|\tau|$ and establish a stronger uniqueness theorem. We don't know whether it is true or not, but either answer will be a remarkable result.

3.2.2 Taub–Nut metric

There is also a well-known metric, so-called Taub–Nut metric, which is written as

$$\mathrm{d}s^2 = \frac{1}{V}(\mathrm{d}t + \vec{\omega} \cdot \mathrm{d}\vec{x})^2 + V \mathrm{d}\vec{x} \cdot \mathrm{d}\vec{x}$$

where $\vec{x} \in \mathbb{R}^3$ is the position vector. This is already the Euclidean version. We impose the condition

$$\vec{\nabla} V = \pm \vec{\nabla} \times \vec{\omega}, \tag{3.7}$$

which implies $\triangle V = 0$. Equation (3.7) turns out to be the condition for Ricci-flatness, where $\pm$ is chosen depending on whether the metric is self-dual or anti-self-dual. If one assumes that V is asymptotic to 1 at ∞, then the metric is asymptotically Euclidean. Since $\triangle V = 0$, Liouville theorem will imply $V \equiv 1$ if V is smooth everywhere. So V must have singularities and we assume

$$V = 2\lambda \sum_{i=1}^{k} \frac{1}{|\vec{x} - \vec{x}_i|},$$

where $\vec{x}_i$ are arbitrary three vectors. $\partial/\partial t$ is the Killing field and has fixed points. The locations $\vec{x} = \vec{x}_i$ are nuts with respect to the Killing field $\partial/\partial t$. The Killing field can be written as $K^a \frac{\partial}{\partial x^a}$ in general local coordinates x^a. Then the infinitesimal action $\nabla_a K_b$ at the fixed points of the Killing field is skew symmetric.

The eigenvalues of $\nabla_a K_b$ are $\pm\frac{1}{4\lambda}$. $\vec{\omega}$ has string singularities $\vec{x}_1, \vec{x}_2, \cdots$. We can construct a pair of coordinate patches $(t_n, \vec{x})$ $(t_{n+1}, \vec{x})$ near the nuts $\vec{x} = \vec{x}_n$. Denote the azimuthal angle around the line from $\vec{x}_n$ to $\vec{x}_{n+1}$ by ϕ_n, make the identifications

$$t_{n+1} = t_n + 4|\lambda|\phi_n,$$

then the singularities can be resolved ([84]).

After these identifications, the manifold constructed in this way becomes nonsingular and asymptotic to a cone over $S^3/\mathbb{Z}_k$, where k is the number of the singularities of $\vec{\omega}$. Thus the Euler number is $\chi = k$ and the signature is $\tau = \pm(k-1)$.

In general we need to impose boundary conditions for noncompact manifolds. In most cases in general relativity, we assume that the metric is asymptotic to the flat metric. There are the following different conditions:

(1) A space is *asymptotically Euclidean*(AE) if outside a compact set the metric approaches the flat metric on $\mathbb{R}^4$. One may regard the boundary of such spaces as an S^3 at infinity.

(2) A space is *asymptotically locally Euclidean* (ALE) if outside a compact set the metric approaches the flat metric on $\mathbb{R}^4/\Gamma$, where Γ is a discrete subgroup of SO(4) which acts freely on S^3 at infinity. Thus the boundary of such spaces

is S^3/Γ. If Γ is the identity, this class degenerates to AE. As an example, Γ may be $\mathbb{Z}_2$ (as in the Eguchi–Hanson instanton).

(3) A space is *asymptotically flat*(AF) if outside of a compact set the metric approaches the flat metric $\mathbb{R}^3 \times S^1$. One may regard the boundary at infinity of such objects as $S^2 \times S^1$.

(4) A space is *asymptotically locally flat* (ALF) if outside a compact set the metric approaches the metric $\mathrm{d}s^2 = \mathrm{d}r^2 + r^2(\sigma_1^2 + \sigma_2^2) + \sigma_3^2$, where σ_i are the left-invariant 1-forms on S^3/Γ where Γ is an isometry of this $U(1) \otimes \mathrm{SU}(2)$ invariant metric.)

Consider the action $I(g) = \int_M (R - 2\Lambda) + \int_{\partial M} \mathrm{Tr} K$ with boundary term. In 1970's, Hawking proposed the so-called Positive action conjecture, which states that any asymptotically Euclidean 4-manifold with non-negative scalar curvature has positive action $I(g)$. The action $I(g) = 0$ if and only if the metric g is Euclidean. It was proved by Gibbons-Hawking-Perry ([85]) and Schoen–Yau ([164]). The assumption is important because the above Taub–Nut example, which is only asymptotically locally Euclidean. In fact, one can construct asymptotically locally flat 4-manifolds with $R = 0$ whose action is negative ([125]).

The Taub–Nut metric itself is Kähler–Einstein, which is given by

$$\mathrm{d}s^2 = \frac{1}{V}(\mathrm{d}t + \vec{\omega} \cdot \mathrm{d}\vec{x})^2 + V \mathrm{d}\vec{x} \cdot \mathrm{d}\vec{x},$$

where

$$V = 1 + 2N \sum_{i=1}^{k} \frac{1}{|\vec{x} - \vec{x}_i|}. \tag{3.8}$$

LeBrun ([126]) was able to write down the complex coordinates by the argument of holonomy groups. The holonomy group of this metric is SU(2) due to the presence of parallell spinors. An important example, called the Eguchi-Hanson metric, was discovered before ([67, 68]).

If we choose $k = 2$ in (3.8) of the Taub–Nut metric, we get the Eguchi–Hanson metric. It was first discovered in a different coordinate system which was given by

$$\mathrm{d}s^2 = \frac{1}{f}\mathrm{d}r^2 + \frac{1}{4}r^2(\sigma_1^2 + \sigma_2^2 + f\sigma_3^2),$$

where

$$f(r) = 1 - \frac{a^4}{r^4},$$

r is a radial coordinate and σ_i are the standard 1-forms on S^3 given by

$$\sigma_1^2 + \sigma_2^2 = \mathrm{d}\theta^2 + \sin^2\theta\phi^2,$$
$$\sigma_3 = \mathrm{d}\psi + \cos\theta \mathrm{d}\phi.$$

Here θ, ϕ and ψ are Euler coordinates on S^3. One can then write down the complex coordinates.

3.2.3 Kerr–Taub–Nut metric

There is the third class of examples called the Kerr–Taub–Nut metrics, which have features of both. These form a 3-parameter family of Ricci-flat metrics.

Remark 3.6. *If a Lorentzian–Einstein manifold has a Killing vector field which is hypersurface orthogonal, the Wick rotation turns it into a Ricci flat Riemannian manifold. In particular, we get many examples from the Kerr, Taub–NUT, and other space-times. For these examples, an interesting phenomenon is that the singularity of the space-time disappears after the Wick rotation. It is worth understanding more about the relation between this phenomenon and the curvature degeneracy. The curvature degeneracy can be studied in terms of the eigenvalues of the Weyl tensor, i.e. the Petrov classification [159]. For instance, the Kerr metric is of type D. We will discuss more about the Petrov classification later.*

3.3 Reduction by symmetries

Another approach to construct Einstein metrics is the reduction by symmetry. For $G \subset \mathrm{Isom}(M)$, consider the orbit space M/G. It is in general an orbifold. We can as well consider $\delta_g \int s_g^2$ on M/G, but we should be careful about the singularity of M/G. Presumably, an extremal metric on M/G together with a canonical metric on G gives an Einstein metric on M.

Let $\pi : M \to B$ be a Riemannian submersion with totally geodesic fibers G for simplicity. We would like to find conditions so that the manifold M is Einstein. Decompose the tangent space $T_{\pi^{-1}(p)}M$ of M at $p \in B$ into horizontal component $\mathcal{H}$ and vertical component $\mathcal{V} = \ker(\pi_*)$. On the horizontal distribution $\mathcal{H}$, one needs to find a connection so that π_* is an isometry when restricting to horizontal distributions. The fibers are isomorphic to one another. Major examples are principal G-bundles. Consider a principal G-bundle $\pi : M \to B$. Assume that π is a Riemannian submersion with totally geodesic fibers isometric to the standard metric on G. The following three data allow us to construct the metric on the total space: a left invariant metric on the fiber space (if G is compact, one can take bi-invariant metric, but not necessary), a connection A for the fiber bundle, and a metric on the base B.

To construct an Einstein metric on the total space M, we require that

(1) The connection is a Yang-Mills connection;
(2) The scalar curvature s of the base B is constant;
(3) $\mathrm{Ric}_B(X,Y) - 2(A_X, A_Y) = \frac{s+|A|^2}{\dim G}(X,Y)$ for any horizontal vector fields X and Y,

where $A_X(Z) = \frac{1}{2}\mathcal{V}([X,Z])$ for $Z \in \mathcal{H}$ and $\frac{s+|A|^2}{\dim G}$ is a constant.

There is the so-called Kaluza–Klein ansatz for the Einstein field equation in spacetime. On the circle bundle $S^1 \to M^5 \to N^4$, the metric satisfies the Einstein equation on the 5-dimensional manifold M. Physicists wrote down the following block matrix

$$\begin{pmatrix} \text{scalar field} & \text{connections} \\ \text{connections} & \text{metrics on the base} \end{pmatrix}.$$

The 5-dimensional Lorentizian metric on M^5 comes from the metric on the base, the connection, and the scalar field. This is the Kaluza–Klein model.

If we go up one higher dimension, we get a 4-dimensional field coupled with a Maxwell field and a scalar field. This is very remarkable. I think that Einstein was very pleased at this theory except that the scalar field was never found.

The reduction we considered here is similar except that Yang-Mills theory is non-abelian. Even with this reduction, there is still no general method to construct Yang-Mills fields except for Taubes' construction of self-dual and anti-self-dual connections on four manifolds. One can also construct Yang-Mills connections over spheres or other special manifolds.

The simplest example for such a reduction construction is $\mathbf{S}(K_M)$, the unit circle bundle of the canonical bundle of a Kähler manifold M. It admits a free $\mathbf{S}^1$-action. If the underlying manifold M admits a Kähler–Einstein metric with scalar curvature $s > 0$, we can then construct an Einstein metric on $\mathbf{S}(K_M)$. This is the inverse procedure of the Kaluza–Klein ansatz.

Wang–Ziller [198] generalized this construction to principal torus bundles. They proved the following theorem: Let (M_i, g_i) be Kähler–Einstein manifolds with $c_1(M_i) > 0$ and $c_1(M_i) = q_i\alpha_i$ with α_i indivisible. Suppose P is the total space of a principal torus bundle over $B = M_1 \times M_2 \times \cdots \times M_m$ whose characteristic classes in $H^2(B, \mathbb{Z})$ are integral linear combinations of $\alpha_1, \alpha_2, \cdots, \alpha_m$. If $\pi_1(P)$ is finite, then P carries an Einstein metric with $s > 0$.

For a fixed B, suppose that there is a sequence of P which are pairwise inequivalent as bundles over B. They showed that if the Einstein metric on P is normalized to be of unit volume, then there exists a subsequence whose scalar curvature converges to zero.

They examined the total space $M^{p,q}_{k,l}$ of circle bundles over $\mathbb{CP}^p \times \mathbb{CP}^q$ with $c_1 = k\alpha_1 + l\alpha_2$, where α_1 and α_2 are the hyperplane sections of $\mathbb{CP}^p$ and $\mathbb{CP}^q$, respectively. A standard topological argument shows that $M^{p,q}_{k,l}$ is simply-connected if and only if $(k, l) = 1$. If $(k, l) = 1$, then

(1) For $2 \leq p < q$, $M^{p,q}_{k,l}$ is diffeomorphic to $M^{p,q}_{k',l'}$ if and only if $|k| = |k'|$ and $|l| = |l'|$.
(2) All the $M^{p,p}_{k,l}$'s have the same cohomology ring. For $p \geq 4$, $M^{p,p}_{k,l}$ is diffeomorphic to $M^{p,p}_{k',l'}$ if and only if $(|k|, |l|)$ is $(|k'|, |l'|)$ or $(|l'|, |k'|)$.
(3) For $1 = p \leq q$,
    (a) $M^{1,1}_{k,l}$ is $\mathbf{S}^2 \times \mathbf{S}^3$ for all (k, l).
    (b) $M^{1,q>1}_{k,\pm 1}$ is $\mathbf{S}^2 \times \mathbf{S}^{2q+1}$ if q is odd, or if q and k are even. It is the nontrivial $\mathbf{S}^{2q+1}$-bundle over $\mathbf{S}^2$ if q is even and k is odd.
    (c) $M^{1,q>1}_{k,\pm 2}$ is one of the three nontrivial $\mathbb{RP}^{2q+1}$-bundles over $\mathbf{S}^2$.
    (d) In general, fix $(q, |l|)$, there are only finitely many diffeomorphism types of $M^{1,q}_{k,l}$.

In particular, their results imply that $\mathbf{S}^2 \times \mathbf{S}^{2q+1}$, the nontrivial $\mathbf{S}^{2q+1}$-bundled over $\mathbf{S}^2$, and some nontrivial $\mathbb{RP}^{2q+1}$-bundles over $\mathbf{S}^2$ admit infinitely many non-isometric Einstein metrics with positive scalar curvature. That is to say, the moduli space of Einstein metrics for these manifolds is disconnected, and has infinitely many components. This is an interesting phenomenon. The construction comes from complex bundles over Kähler–Einstein manifolds. However, such a behavior of moduli space does not seem to happen in algebraic geometry. We ask:

Question 3.7. *Can one collapse the moduli space of the Einstein metrics in a reasonable way so there are only a finite number of components?*

Note that inequivalent Kähler-Einstein metrics can be deformed to each other as the complex moduli space is continuous.

Notice that this method can only construct Einstein metrics with positive scalar curvatures since manifolds with negative scalar curvatures have no continuous symmetries.

Question 3.8. *Find a way to construct Einstein metrics with negative scalar curvatures.*

Question 3.9. (*Alekseevsky's conjecture* [*18*, 7.57]) [5] *A homogeneous Einstein manifold with negative scalar curvature is conjectured to be diffeomorphic to a Euclidean space. The homogeneous manifold is of the form G/K where K is a maximal compact subgroup of G, which is unique up to conjugations. As a consequence of this more detailed description, the classification problem is reduced to looking at left invariant metrics on solvable Lie groups.*

3.3.1 Sketch of the construction of Wang–Ziller

We need a connection θ on $P \to B$ to construct an Einstein metric. A connection is equivalent to a G-invariant horizontal distribution $\mathcal{H} \overset{\pi_*}{\cong} TB$. The vertical bundle is canonically defined by $\mathcal{V} = \ker(\pi_*)$, which is isomorphic to the Lie algebra $\mathfrak{g}$. With the connection θ, a metric on B and a metric on G, we can define a metric on P by

$$\langle X, Y\rangle_P = \langle \pi_* X, \pi_* Y\rangle_B + \langle \theta(X), \theta(Y)\rangle_G.$$

A straightforward computation shows that the metric is Einstein with constant E if and only if

(1) The connection θ is Yang–Mills with respect to $\langle\ ,\ \rangle_B$ and $\langle\ ,\ \rangle_G$.
(2) $\mathrm{Ric}_G(\theta(X), \theta(Y)) + \frac{1}{4}\sum_{i,j}\langle \Omega(U_i, U_j), \theta(X)\rangle_G \langle \Omega(U_i, U_j), \theta(Y)\rangle_G = E\langle \theta(X), \theta(Y)\rangle_G$.
(3) $\mathrm{Ric}_B(\pi_* X, \pi_* Y) - \frac{1}{2}\sum_i \langle \Omega(X, U_i), \Omega(X, U_i)\rangle_G = E\langle \pi_* X, \pi_* Y\rangle_B$.

where Ω is the curvature of θ, and $\{U_i\}$ is an orthonormal basis for $\mathcal{H}$ or TB.

In general, it is difficult to solve the system of equations. Wang–Ziller [198] studied the case when $G = \mathbf{T}^r$. The curvature is $\Omega = d\theta = \pi^*\eta$ for a closed 2-form η on B. The Einstein conditions become

[5] The author thanks McKenzie Wang for his explanation and the references he provided.

(1) η is harmonic with respect to $\langle\ ,\ \rangle_B$.
(2) $\frac{1}{4}\sum_{i,j}\langle\eta(U_i,U_j),\theta(X)\rangle_G\langle\eta(U_i,U_j),\theta(Y)\rangle_G = E\langle X,Y\rangle_G$.
(3) $\mathrm{Ric}_B(X,Y) - \frac{1}{2}\sum_i\langle\eta(X,U_i),\eta(X,U_i)\rangle_G = E\langle X,Y\rangle_B$.

They first consider the principal $\mathbf{S}^1$-bundle case and assume that (M_i, g_i) are Kähler–Einstein manifolds with $c_1(M_i) > 0$, and $c_1(M_i) = q_i\alpha_i$ where α_i is indivisible. Let P be the principal $\mathbf{S}^1$-bundle whose Euler class $e(P) = b_i\pi^*\alpha_i \neq 0$, $b_i \in \mathbb{Z}$. The metric on B is chosen to be x_ig_i, and the metric on $G = \mathbf{S}^1$ is the standard one with total length $2\pi\rho$. The curvature form η is the unique harmonic 2-form representing $e(P)$. It gives a unique connection θ on P, up to gauge equivalence. The Einstein conditions become

$$\rho^2\sum_{i=1}^m n_i\frac{b_i^2}{x_i^2} = 4E;$$

$$\frac{q_i}{x_i} - \frac{1}{2}\rho^2\Big(\frac{b_i}{x_i}\Big)^2 = E, \qquad \text{for } i = 1,2,\cdots,m$$

where $n_i = 2||\omega_i||^2_{g_i}$ and q_i is the divisibility of $c_1(M_i)$. They prove this system of algebraic equations in x_i and ρ always has a solution under the above assumptions. Therefore, P has an Einstein metric.

They also consider the case when $G = \mathbf{T}^r$. If P has first Chern classes $\beta_j = \sum_{i=1}^m b_{ji}\pi^*\alpha_i$, $j = 1,2,\cdots,r$ with $[b_{ji}] \in M_{r\times m}(\mathbb{Z})$ of maximal rank, then P has an Einstein metric of positive scalar curvature. The algebraic system becomes more complicated, but it still has a solution.

3.3.2 Sasaki–Einstein metrics

I should mention that the same method can be used to construct the so-called Sasaki–Einstein metrics. Let $M \hookrightarrow \mathbb{CP}^n$ be a Fano manifold. Suppose there exists a Kähler–Einstein metric on M with $R_{ij} = g_{ij}$. According to my conjecture, there always exists a Kähler–Einstein metric on any Fano manifold M, when M is stable in the sense of algebraic geometry. Then the unit canonical bundle $S(K_M)$ over M admits a S^1-invariant Sasakian–Einstein metric. Thus the cone $C(S(K_M))$ over $S(K_M)$ has a Ricci-flat Kähler–Einstein metric, i.e., Calabi–Yau metric, with conic singularity. This procedure deforms the Einstein metric which is invariant under circle action. In this class of manifolds, one can define quivers and consider conformal field theory which gives rise to many interesting invariants of $C(S(K_M))$. There are many interesting questions one can ask. The work was done by J. Sparks, etc. One can consult the article [177] by Sparks for more about Sasaki–Einstein manifolds. It seems to be natural to generalize his work to circle bundles over product manifolds.

3.3.3 SU(2) action

There is a relevant construction due to Lawson and myself [124]: A manifold with a SU(2)-action admits an invariant metric with positive scalar curvature. Our construction is also a principal bundle construction. It is more complicated because of the fixed point set. We ask:

Question 3.10. *Does a manifold with a* SU(2)*-action always admit an Einstein metric if the orbit space satisfies some nice condition such as the existence of an Einstein metric?*

One way to understand this is to take $G \times G/G$, where the action is the diagonal action. If there is a non-abelian group action, the manifold must be rather round. I proposed that the manifold must admit a metric with positive curvature. We can make the fibers very tiny to get bunches of small spheres around. The fibers have lots of positive curvature. If there is no fixed point, one can use the argument here to prove the metric of the total space has positive scalar curvature. Additional work is needed if there are fixed points. My work with Lawson ([124]) proved this is indeed possible even with fixed points. There are various obstructions to the existence of metrics with positive scalar curvature. For example, the $\hat{A}$-genus $\hat{A}(M)$ of the manifold and the topological α-invariant $\alpha(M)$ must vanish. This gives rise to an interesting topological theorem for obstructions of SU(2)-actions. I believe they can be generalized to allow more general SU(2) group actions. Instead of conditions on the scalar curvature, one may also try the Ricci curvature.

3.4 Inhomogeneous Einstein metrics

By the principal bundle construction, Wang–Ziller [198] showed that the moduli space of Einstein metrics can have infinitely many components.

Jensen [114] used such a construction to build a nontrivial Einstein metric on $\mathbf{S}^{4m+3}$. Bourguignon–Karcher [27] described a further Einstein metric on $\mathbf{S}^{15}$. They are all homogeneous and with positive scalar curvature. Ziller [207] proved that these are all homogeneous Einstein metrics on $\mathbf{S}^n$.

On the other hand, there could be inhomogeneous Einstein metrics on $\mathbf{S}^n$. Böhm [24] constructed infinitely many cohomogeneity one Einstein metrics on $\mathbf{S}^5$, $\mathbf{S}^6$, $\mathbf{S}^7$, $\mathbf{S}^8$, $\mathbf{S}^9$, and other low dimensional spaces such as $\mathbf{S}^3 \times \mathbf{S}^2$, $\mathbf{S}^3 \times \mathbf{S}^3$, $\mathbf{S}^3 \times \mathbb{CP}^2$, $\mathbf{S}^4 \times \mathbb{RP}^2 \times \mathbb{RP}^2$, $\mathbf{S}^3 \times \mathbf{S}^3 \times \mathbf{S}^3$, $\mathbf{S}^7 \times \mathbf{S}^2$. These infinite sequences of Einstein metrics converge in some weak sense. The construction does not work for dimensions greater than 9.

The first example of inhomogeneous Einstein metric with positive scalar curvature is the Page metric [151] on $\mathbb{CP}^2 \# \overline{\mathbb{CP}}^2$. Böhm's construction produces an Einstein metric on $\mathbb{HP}^2 \# \overline{\mathbb{HP}}^2$. It is believed that $\mathbb{HP}^{n+1} \# \overline{\mathbb{HP}}^{n+1}$ admits an Einstein metric and Page–Pope [152] produced numerical solutions. It is not clear for $\mathbf{Ca}\mathbb{P}^2 \# \overline{\mathbf{Ca}\mathbb{P}}^2$ though.

We now explain the key ingredient of Böhm's construction. Let G be a compact Lie group acting by isometry on $(\hat{M}, \hat{g})$ with cohomogeneity one[6]. We assume that there exists a singular orbit Q with $\dim Q = d_Q > 0$. Let P denote a principal orbit which is a distance tube around Q. We have the fibration $H/K \to P = G/K \to Q = G/H$ whose fiber H/K is a d_S-dimensional sphere. Let M be

[6] It means that the principal orbit has codimension one, or equivalently, the orbit space is of dimension one.

a tubular neighborhood of Q. The metric $\hat{g}|_{M\setminus Q}$ can be described as $\mathrm{d}t^2 + g(t)$, where $g(t)$ is one parameter family of G-invariant metric on P.

We assume the space of G-invariant metric on P is two dimensional. Thus

$$\hat{g}|_{M\setminus Q} = \mathrm{d}t^2 + f^2(t)g^S + h^2(t)g^Q,$$

where g^S is the standard metric on the sphere H/K and g^Q is the G-invariant metric on $Q = G/H$. The Einstein equation for $\hat{g}|_{M\setminus Q}$ with Einstein constant λ reads

$$\begin{aligned}
&d_S\frac{f''}{f} + d_Q\frac{h''}{h} = -\lambda, && (1)\\
&\frac{f''}{f} + (d_S-1)\frac{(f')^2}{f^2} + d_Q\frac{f'h'}{fh} - \frac{1}{f^2}(d_S-1) - \frac{d_Q}{d_S}||\mathcal{A}||^2\frac{f^2}{h^4} = -\lambda, && (2)\\
&\frac{h''}{h} + (d_Q-1)\frac{(h')^2}{h^2} + d_S\frac{f'h'}{fh} - \frac{1}{h^2}\mathrm{Ric}^Q + 2||\mathcal{A}||^2\frac{f^2}{h^4} = -\lambda, && (3)\\
&\mathrm{Ric}^{M\setminus Q}(X,N) = 0. && (4)
\end{aligned} \tag{3.9}$$

The quantities are defined as follows: Ric^Q is the Einstein constant of (Q, g^Q). Let D^1 be the Levi–Civita connection on P with respect to $g^1 = g^S + g^Q$, then $||\mathcal{A}||^2 = \sum_{i=1}^{d_Q} g^S((D^1_{H_1}H_i)|_{\mathrm{Vert}}, (D^1_{H_1}H_i)|_{\mathrm{Vert}})$ where $\{H_i\}$ form a horizontal orthonormal basis with respect to the Riemannian submersion $(P, g^1) \to (Q, g^Q)$ (we can choose any H_k in place of H_1 and the value is the same). In (4), X is tangent to the principal orbits, and N is normal to the principal orbits.

These equations have been studied by Eschenburg–Wang [72]. They proved that if $(f(t), h(t))$ solves (2) and (3), and $\hat{g}|_{M\setminus Q} = \mathrm{d}t^2 + f^2(t)g^S + h^2(t)g^Q$ can be extended to a smooth metric g on M, then g is Einstein. In this regard, we focus on (2) and (3) with the initial value $(f(0), f'(0), h(0), h'(0)) = (0, 1, \bar{h}, 0)$ for $\bar{h} > 0$.

To solve (2) and (3), consider

$$y'(t) = A(y(t)) + \frac{1}{t}B(y(t)) \tag{3.10}$$

where $A, B : U^j_a \to \mathbb{R}^j$ are analytic functions on an open neighborhood U^j_a of $a \in \mathbb{R}^j$. It has a unique solution $y(t)$ with $y(0) = a$ if $B(a) = 0$ and $\mathbf{I}_j - \frac{1}{n}DB_a$ is invertible for all $n \in \mathbb{N}$. Furthermore, the solution $y(t)$ depends continuously on $a \in \{b \in U^j_a | B(b) = 0\}$.

Let $y(t) = (f(t), f'(t), h(t), h'(t))$. Up to a scaling of $1/f'$, we can apply the above result to equations (2) and (3). Thus, for any $\bar{h} > 0$, there exists a unique solution $c_{\bar{h}}(t) = (f(t), f'(t), h(t), h'(t))$ of (2) and (3) with $c_{\bar{h}}(0) = (0, 1, \bar{h}, 0)$. Moreover, Eschenburg–Wang showed that $\hat{g}|_{M\setminus Q} = \mathrm{d}t^2 + f^2(t)g^S + h^2(t)g^Q$ can be extended to a smooth Einstein metric on M.

So far we have obtained Einstein metrics on the open manifold M. If there exist positive constants $\bar{h}_1, \bar{h}_2$ and t_0 such that $\lim_{t\to t_0} c_{\bar{h}_1}(t) = (0, -1, \bar{h}_2, 0)$, then $c_{\bar{h}_1}(t)$ defines a smooth metric $\hat{g}(\bar{h}_1, \bar{h}_2)$ on the double of M, $M \cup_{\partial M} \bar{M}$.

If $\bar{h}_1 = \bar{h}_2$, the solution is called *symmetric*. A symmetric metric $\hat{g}(\bar{h}, \bar{h})$ has better geometry: the geodesic reflection along normal geodesics of the middle principal orbit is an isometry. Böhm proved that if $\mathrm{Ric}^Q > 0$, $||\mathcal{A}|| = 0$ and $d_P = \dim P = d_S + d_Q \leq 8$, then there exist infinitely many symmetric solutions.

Böhm's construction generalizes the principal bundle construction to allow singular orbits. Lawson and myself have a similar construction in [124]. We proved that any manifold with a nontrivial SU(2)-action admits a metric of positive scalar curvature. For SU(2)-actions, the principal orbits are $\mathbf{S}^3$ or $\mathbf{S}^2$. We shrink the size of the fiber to gain positive curvature. Near the singular orbit, we construct a local model. These two pieces can be glued together to obtain a metric of positive scalar curvature.

Conjecture 3.11. *If M admit a nontrivial G-action, then it carries a metric with certain positivity. For instance, if $k = \dim G$, then $\lambda_1 + \cdots + \lambda_k \geq 0$ where λ_j are eigenvalues of the Ricci curvature.*

3.5 Kähler–Einstein metrics

Another general construction for the Einstein metric is under the Kähler assumption. A Kähler manifold is a complex manifold (M, J) with a Riemannian metric $g = g_{i\bar{j}} \mathrm{d}z^i \otimes \mathrm{d}\bar{z}^j$ such that $\omega = \sqrt{-1} g_{i\bar{j}} \mathrm{d}z^i \wedge \mathrm{d}\bar{z}^j$ is d-closed. The Ricci curvature is given by $R_{i\bar{j}} = -\partial_j \partial_{\bar{j}} \log \det(g_{k\bar{l}})$. We associate to it the Ricci form $\mathrm{Ric}(\omega) = \sqrt{-1} R_{i\bar{j}} \mathrm{d}z^i \wedge \mathrm{d}\bar{z}^j$, then $\mathrm{Ric}(\omega)/2\pi \in H^2(M, \mathbb{Z})$ gives $c_1(M)$.

The advantage of Kähler geometry is the $\partial\bar{\partial}$-lemma. In particular two Kähler metrics representing the same class differ by the $\partial\bar{\partial}$ of a function,

$$\omega = \omega_0 + \partial\bar{\partial}\phi \ .$$

To construct a Kähler–Einstein metric, we consider

$$\det(g^0_{i\bar{j}} + \partial_i \partial_{\bar{j}} \phi) = c e^{-\alpha\phi + F} \det(g^0_{i\bar{j}}) \ . \tag{3.11}$$

The constant α and the function F are chosen in the following way. If TM is negative[7], take $\alpha = -1$ and ω^0 in the class $-2\pi c_1(M)$. If TM is zero, take $\alpha = 0$. If TM is positive, take $\alpha = +1$ and ω^0 in the class of $2\pi c_1(M)$. With this choice of α, $R^0_{i\bar{j}}$ and $\alpha g^0_{i\bar{j}}$ are in the same class, the $\partial\bar{\partial}$-lemma implies that

$$R^0_{i\bar{j}} - \alpha g^0_{i\bar{j}} = \partial_i \partial_{\bar{j}} F$$

for some function F which is the one chosen in (3.11).

If $g_{i\bar{j}} = g^0_{i\bar{j}} + \partial_i \partial_{\bar{j}} \phi$ solves (3.11) and is everywhere positive, then it gives a Kähler–Einstein metric. To see it, take $-\partial\bar{\partial} \log$ on (3.11);

$$\begin{aligned} R_{i\bar{j}} &= \alpha \partial_i \partial_{\bar{j}} \phi - \partial_i \partial_{\bar{j}} F + R^0_{i\bar{j}} \\ &= \alpha(g_{i\bar{j}} - g^0_{i\bar{j}}) - \partial_i \partial_{\bar{j}} F + R^0_{i\bar{j}} = \alpha g_{i\bar{j}} \ . \end{aligned}$$

For $\alpha \leq 0$ ($c_1(M) = 0$ or $c_1(M) \leq 0$), the equation (3.11) always has a solution. For $\alpha > 0$ ($c_1(M) > 0$), there are obstructions from stability condition.

[7]It means that $c_1(M)$ can be represented by a everywhere negative $(1, 1)$-form.

3.5.1 Positive tangent bundle

Although (3.11) does not need to have a solution for $c_1(M) > 0$, we can still do the following: Assume $c_1(M) \geq 0$, i.e. $c_1(M)$ can be represented by an everywhere non-negative $(1,1)$-form $\tilde{\omega}/2\pi$. By the $\partial\bar{\partial}$-lemma,

$$R^0_{i\bar{j}} = -\partial_i\partial_{\bar{j}} \log\det(\omega_0) = \tilde{\omega}_{i\bar{j}} + \partial_i\partial_{\bar{j}} f$$

for some function f. Consider the equation

$$\det(\omega_0 + \partial\bar{\partial}\phi) = e^f \det(\omega_0) \tag{3.12}$$

My solution [206] to the Calabi conjecture says that (3.12) always has a solution with $\omega_0 + \partial\bar{\partial}\phi$ everywhere positive. Taking $-\partial\bar{\partial}\log$ on (3.12) gives the Ricci curvature of $\omega_0 + \partial\bar{\partial}\phi$,

$$R_{i\bar{j}} = -\partial_i\partial_{\bar{j}} f + R^0_{i\bar{j}} = \tilde{\omega}_{i\bar{j}} > 0 \ .$$

The punch line here is that if $c_1(M) \geq 0$, any class of the Kähler cone can be represented by a metric whose Ricci form is everywhere non-negative.

By combining with the splitting theorem of Cheeger–Gromoll [53], we obtain the structure theorem for Kähler manifold M with $c_1(M) \geq 0$: M is finitely covered by $N \times \mathbf{T}^k$ where N is another compact complex manifolds and $k = b^1(M)$.

In fact, Calabi constructed the map $M \to \mathbf{T}^k$: Let $\{\alpha_1, \alpha_2, \ldots, \alpha_k\}$ be a basis for $H^1(M,\mathbb{Z}) \subset H^1(M,\mathbb{R})$. Since $R_{i\bar{j}} \geq 0$, the Böchner [21] argument shows that α_j can be taken to be parallel 1-forms. Then take the Abel–Jacobi map to the Albanese torus:

$$\begin{aligned} M &\to \mathbf{T}^k = H^1(M,\mathbb{R})/H^1(M,\mathbb{Z}) \\ x &\mapsto (\int_{x_0}^{x} \alpha_1, \ldots, \int_{x_0}^{x} \alpha_k) \ . \end{aligned}$$

The map is independent of the path of integration because the integration lattice has been mod out. Later on, Matsushima and Lichnerowicz generalized the Calabi construction under a weaker assumption. However, these arguments only use $H^1(M)$ and never touch $\pi_1(M)$. The argument of Cheeger–Gromoll handles $\pi_1(M)$ and gives a stronger structure theorem.

Before proving the Calabi conjecture, I tried to use the structure theorem to construct counterexamples. I constructed a Kähler surface M which is numerically non-negative[8] and which cannot be covered by $N \times \mathbf{T}^k$. However, numerically non-negative does not imply $c_1(M) \geq 0$.

3.5.2 Degenerate Kähler metric

For any class in the boundary of the Kähler cone, we would like to know when it can be represented by an everywhere non-negative kähler form.

[8]A kähler manifold M is said to be numerically non-negative if $c_1(M) + \varepsilon\omega > 0$.

We can separate it into two cases. The first one is the non-collapsing case: $[\omega]^n \neq 0$. In this case, we hope that there exists a Ricci-flat quasi-metric which degenerates along some subvariety. The second case is the collapsing case: $[\omega]^n = 0$. In this case, we hope that there is a fibration $F \to M \to N$ such that N is non-collapsing. My former student Tosatti studied this in his thesis [189], and obtained some nice results. The strategy is to understand the boundary of the Kähler cone through the canonical metric in the interior of the Kähler cone.

Here is the main question we are interested in:

Question 3.12. *If $\omega \geq 0$, can one construct a Ricci-flat or Kähler–Einstein quasi-metric from ω?*

I have some general results in [206]. However, in order to say more, we need to understand the degeneracy of $c_j(M, \omega)$ on the singularities of the metric.

The orbifold singularity is a special case we do know how to handle. Let G be a finite subgroup of $\mathrm{SU}(n)$ acting on the complex ball $\mathbf{B}^n$ with the origin as the isolated fixed point. Then the origin is the orbifold singularity of $\mathbf{B}^n/G$. We can resolve the singularity to get a smooth variety D, and let $S_0 \subset D$ be the preimage of the origin. The standard Euclidean metric on $\mathbf{B}^n$ gives a quasi-metric on D which degenerates along S_0.

Let's say, we have a Kähler manifold M with a divisor S such that S has a neighborhood modeled on $S_0 \subset D$. If $c_1(M) < 0$, we can solve the Kähler–Einstein metric which is asymptotic to the induced quasi-metric on (D, S_0) near S.

This special case already has some interesting application. Let M be a Kähler surface, which is minimal and of general type. Thus $c_1(M) \leq 0$. For $m >> 1$, the orthonomul sections of K^m define a map $M \to \mathbb{CP}^N$. The image can have singularities which are ordinary double points. Let $S \subset M$ be the preimage of these singularities. The above construction gives a Kähler–Einstein quasi-metric $\mathrm{d}s^2$ which degenerates along S. Because of the Kähler–Einstein condition, $3c_2(\mathrm{d}s^2) - c_1^2(\mathrm{d}s^2) \geq 0$ everywhere. Since $\mathrm{d}s^2$ is not a smooth metric, these curvature forms do not give topological Chern classes. However, one can find

$$3c_2(M) - c_1^2(M) - (\text{contribution from } S) = 3c_2(\mathrm{d}s^2) - c_1^2(\mathrm{d}s^2) \geq 0 \,.$$

In particular, this improves the Miyaoka–Yau inequality

$$3c_2(M) - c_1^2(M) \geq (\text{contribution from } S) \,, \tag{3.13}$$

and the contribution from S can be evaluated explicitly.

I expect that one can generalize this to higher dimensions', at least for 3-folds. There are more singularity types one need to handle. For 3-folds, one have to handle canonical singularities.

Remark 3.13. *There are not many methods to construct Einstein metrics. So far, we have discussed three constructions: Wick rotation from a Lorentzian Einstein manifold, reduction with symmetry, and Kähler-Einstein metrics. We can also* *consider the mixed case. For example, M admits an $\mathbf{S}^1$-action and $M/\mathbf{S}^1$ is a Kähler–Einstein orbifold. This is basically the Sasaki–Einstein metric, and there are lots of interesting examples.*

3.6 Einstein metrics in dimension 4

In dimension 3, Einstein metrics have constant sectional curvature, and the uniformization theorem applies. There is no known example of Einstein metrics on a smooth 4-manifold with opposite signs of scalar curvature. On the other hand, there do exist such examples on a topological 4-manifold, and they support different smooth structures. The examples are constructed by Catanese–LeBrun [37], and independently by Kotschick [120].

We now briefly explain the example of Kotschick. Let M_1 be the blow-up of $\mathbb{CP}^2$ at 8 points, $\mathbb{CP}^2\#8\overline{\mathbb{CP}^2}$. By the work of Tian–Yau [188], M_1 admits a Kähler–Einstein metric of positive scalar curvature. Take M_2 to be the simply-connected numerical Godeaux surface constructed by Craighero–Gattazzo [61]. Its canonical bundle is ample. By my theorem [206] it admits a Kähler–Einstein metric of negative scalar curvature.

The surface M_2 is a desingularization of a quintic in $\mathbb{CP}^3$. It has $q = p_g = 0$, $P_2 \neq 0$, then $h^1(M_2) = 0 = h^{2,0}(M_2)$. Furthermore, $H^{1,1}(M_2)$ is the same as that of M_1, with equivalent intersection form. Freedman's theorem [74] implies that M_2 and M_1 are homeomorphic, and h-cobordant[9]. They are not diffeomorphic by Donaldson theory or Seiberg–Witten theory.

We can use M_1 and M_2 to construct examples in higher dimensions. Consider $M_1 \times M_1$ and $M_2 \times M_2$. They are still h-cobordant. The h-cobordism theorem in higher dimensions says that they are diffeomorphic. The product metrics are still Kähler–Einstein metrics with opposite sign of scalar curvature.

Question 3.14. *Based on the above example, we ask:*

(1) *How about dimension 6? In particular, do there exist diffeomorphic complex manifolds X^3 and Y^3 with $K_X > 0$ and $K_Y < 0$? In complex dimension two, this cannot happen. The Seiberg–Witten theory implies that the sign of the canonical bundle is a differentiable invariant.*

(2) *Does there exist a smooth 4-manifold with Einstein metrics of opposite sign of scalar curvature? The above remark implies that such an example cannot be a complex surface.*

(3) *Does the moduli space of Einstein metrics have finitely many components? If so, the number is a topological invariant.*

(4) *If non-empty, is the moduli space of Kähler–Einstein metrics always connected? It is true for all known examples in complex dimension two. The moduli space may be reducible, but is always connected. The Horikawa surface [107] is a good example.*

With a sequence of Einstein metrics, we can normalize them to be unit volume, and consider the weak limit under Gromov–Hausdorff convergence. The

[9] h-cobordant to X means that it is cobordant to X, and the cobordism is homotopic to a cylinder on X.

formula

$$\chi(M) = \frac{1}{8\pi^2}\int \frac{s^2}{24} + |W_+|^2 + |W_-|^2$$

gives integral curvature bound. If the curvature of the Gromov-Hausdorff limit blows up at some points, it bubbles out ALE spaces. Especially for positive scalar curvature $s > 0$, there is much known about the limit.

3.7 Obstructions to Einstein metrics

We now explain some obstructions to the existence of an Einstein metric in dimension 4. When M is a 4-dimensional compact oriented Riemannian manifold, the generalized Gauss-Bonnet theorem provides a formula for the Euler characteristic $\chi(M)$ as follows.

$$\chi(M) = \frac{1}{8\pi^2}\int_M \left(\frac{s^2}{24} + |W_+|^2 + |W_-|^2 - \frac{|\mathring{R}|^2}{2}\right)\mu_g, \tag{3.14}$$

where W is the Weyl tensor, $\mathring{R}$ is the traceless Ricci tensor and s is the scalar curvature. Here W^+ and W^- are irreducible components of the Weyl tensor W under the action of the special orthogonal group. $W = W^+ + W^-$ and $*W^+ = W^+*$, $*W^- = -W^-*$. The signature $\tau(M)$ of a compact oriented 4-manifold M is given by

$$\tau(M) = \frac{1}{12\pi^2}\int_M (|W^+|^2 - |W^-|^2)\mu_g.$$

We now explain the major obstruction to the Einstein metric in dimension 4.

3.7.1 Euler characteristic and signature

The Einstein equation $R_{ij} = \frac{s}{4}g_{ij}$ is exactly $\mathring{R} = 0$. If (M, g) is a 4-dimensional Einstein manifold, then $\chi(M)$ is positive and can be zero only if (M, g) is flat. This theorem is due to Berger [15].

Let M be a compact oriented Einstein 4-manifold. Thorpe showed that the Euler characteristic $\chi(M)$ and the signature $\tau(M)$ satisfies the inequality ([186])

$$\chi(M) \geq \frac{3}{2}|\tau(M)|.$$

Hitchin ([104]) further proved that when the equality holds, the manifold M is either flat or its universal cover is a $K3$ surface, which means M is either a $K3$ surface, or $K3/\mathbb{Z}_2$, or $K3/\mathbb{Z}_2 \times \mathbb{Z}_2$.

3.7.1.1 *Some remarks on Hitchin–Thorpe inequality.* In the formula (3.14) of χ, only the $\mathring{R}$ with a minus sign, the Euler characteristic of a compact Einstein 4-manifold is nonnegative, and it is zero if and only if $R = 0 = W^+ = W^-$. It is observed by Berger [15], Thorpe [186], Gray [87], Hitchin [104], etc.

The same argument shows that $2\chi \geq 3|\tau|$. The equality holds if and only if $R = 0$ and self-dual, $W^- = 0$ (or anti-self-dual, $W^+ = 0$; this depends on the orientation). Moreover, by analyzing the holonomy group, we conclude that a compact Einstein 4-manifold with $2\chi = 3|\tau|$ must be Kähler. As a corollary, a simply-connected Einstein manifold homotopically equivalent to a $K3$ surface is Kähler and biholomorphic to a $K3$ surface.

There are some smooth 4-manifolds which are homotopically equivalent to $K3$ surfaces, but not diffeomorphic. They are first constructed by Kodaira [118]: $K3$ admits an elliptic fibration over $\mathbb{P}^1$ with some singular fibers. The fake $K3$'s are constructed by logarithmic transform. Namely, remove the tubular neighborhood of a singular fiber, and glue back some local model. The resulting manifold is still complex, and homotopically equivalent to $K3$, but $c_1 \neq 0$. The Donaldson theory [76, 77] shows that they are not diffeomorphic to $K3$. Therefore, these fake $K3$'s do not admit any Einstein metric.

On the other hand, we ask:

Question 3.15. *Is there any interesting example of Ricci flat manifold with $2\chi > 3|\tau|$? Does $\mathbf{S}^4$ admit a Ricci flat metric?*

There are also some special surfaces with $2\chi \geq 3|\tau|$ such as the Enriques surfaces $K3/\mathbb{Z}_2$, or $K3/\mathbb{Z}_2 \times \mathbb{Z}_2$. The group $\mathbb{Z}_2 \times \mathbb{Z}_2$ acts freely. The first factor acts holomorphically, and the second one acts anti-holomorphically. There are more non-compact examples.

3.7.1.2 *Boundary contribution of characteristics.* If the manifold M has boundary, the Atiyah–Patodi–Singer index formula [4] says that

$$\begin{aligned}
\chi(M) &= \frac{1}{128\pi^2}\int_M \varepsilon_{abef}\varepsilon^{abgh}R^{efgh}R_{cdgh}\sqrt{g}\mathrm{d}^4x \\
&\quad + \frac{1}{128\pi^2}\int_{\partial M}(R_{abcd}K^{ac}n^b n^d + 64\det(K^a_b))\sqrt{h}\mathrm{d}^3x, \\
\tau(M) &= \frac{1}{96\pi^2}\int_M \varepsilon_{abcd}R^{abef}R^{cd}{}_{ef}\sqrt{g}\mathrm{d}^4x \\
&\quad - \frac{1}{48\pi^2}\int_{\partial M} R_{abcd}\varepsilon^{cdef}n^a n_e K^b_f\sqrt{h}\mathrm{d}^3x - \eta(0)
\end{aligned}$$

where h is the induced metric on ∂M, n is the unit outward normal, and K is the second fundamental form. The formula for $\tau(M)$ involves a non-local term $\eta(0)$. The η function is defined by $\eta(s) = \sum_j \mathrm{sign}(\lambda_j)|\lambda_j|^{-s}$ where λ_j are the eigenvalues of $(-1)^p(*\mathrm{d} - \mathrm{d}*)$ on $2p$-forms of ∂M. Due to [4], $\eta(s)$ converges absolutely for $\mathrm{Re}(s) >> 1$, and admits an analytic continuation to $\mathrm{Re}(s) > -1/2$.

The second line in the formula for χ and τ will be referred as the *boundary contribution*. It can be calculated explicitly for certain boundary behaviors. We list some results:

boundary contribution	χ	τ
AE	1	0
AF	0	0
ALE, $\Gamma = \mathbb{Z}_k$	$\frac{1}{k}$	$\frac{(k-2)(k-1)}{3k}$
$\Gamma = \mathbf{D}_k^*$	$\frac{1}{4k}$	$\frac{2k^2+1}{6}$
$\Gamma = \mathbf{T}^*$	$\frac{1}{24}$	$\frac{49}{36}$
$\Gamma = \mathbf{O}^*$	$\frac{1}{48}$	$\frac{121}{72}$
$\Gamma = \mathbf{I}^*$	$\frac{1}{120}$	$\frac{361}{180}$

In the table, AE stands for asymptotically Euclidean. A space is called AE if outside some compact set, the metric approaches the flat metric on $\mathbb{R}^4$, $dr^2 + r^2 g_{\text{std}}(\mathbf{S}^3)$. ALE stands for asymptotically locally Euclidean. Its geometry is the same as AE locally. Its topology at infinity is allowed to be $\mathbf{S}^3/\Gamma$ for some discrete subgroup $\Gamma \subset \text{SU}(2)$ acting freely on $\mathbf{S}^3$. The possible choices for such Γ are the cyclic group $\mathbb{Z}_k$, the binary dihedral group of order $4k$ $\mathbf{D}_k^*$, the binary tetrahedral group $\mathbf{T}^*$, the binary octahedral group $\mathbf{O}^*$, and the binary icosahedral group $\mathbf{I}^*$. AF stands for asymptotically flat. A space is called AF if outside some compact set, the metric approaches the flat metric on $\mathbb{R}^3 \times \mathbf{S}^1$, $dr^2 + r^2 g_{\text{std}}(\mathbf{S}^2) + (\text{Hopf form})^2$. The topology at infinity is $\mathbf{S}^2 \times \mathbf{S}^1$. One can consult [154] and the references therein for more explanations. The ALE gravitational instantons are constructed by Hitchin [105] for $\mathbb{Z}_k$, and Kronheimer [123] for all Γ.

For compact complex surfaces, $c_2 = \chi$ and $c_1^2 = 2\chi + 3\tau$. We ask:

Question 3.16. *What is the boundary contribution in the Chern numbers for complete Kähler surfaces, especially for Kähler–Einstein metrics?*

There are some known results for locally Hermitian symmetric space of finite volume. We consider the Borel–Baily compactification of $M \hookrightarrow \bar{M}$. The L^2-index can be computed, and depends on the boundary behavior. See the thesis of Stern [178].

3.7.2 Gromov's simplicial volume

The second class of obstruction of Einstein metrics is known as Gromov's simplicial volume.

The *Gromov norm* [90] is a topological invariant for studying negatively curved manifolds. It is defined by

$$||M|| = \inf \left\{ \sum \alpha_i \,\middle|\, [M] = \sum \alpha_i [C_i] \text{ where } \alpha_i \in \mathbb{Q} \right\}$$

where C_i are simplices. Namely, it is the infimum of the coefficients taken over all expressions of the fundamental class. The Gromov norm has the following significant properties:

(1) $||M|| = 0$ if and only if $\pi_1(M)$ is finite.
(2) If M is hyperbolic, then $||M|| = c_n \text{vol}(M)$ where c_n is a universal coefficient depending only on the dimension.

(3) Given any map $f : M_1 \to M_2$, $||M_1|| \geq \deg(f)||M_2||$. The equality holds if and only if f is homotopic to an isometry.
(4) $||M_1 \times M_2|| \geq \tilde{c}||M_1||\,||M_2||$.
(5) If $R_{ij} \geq -g_{ij}$, property 2 becomes $c_n \mathrm{vol}(M) \geq ||M||$.

If the scalar curvature of a four-dimensional Einstein metric is negative, we can normalize to $R_{ij} = -g_{ij}$. Then by (3.14),

$$\chi(M) \geq c\mathrm{vol}(M, g)$$

for some constant $c > 0$. Therefore,

$$\chi(M) \geq c\mathrm{vol}(M, g) \geq c'||M||.$$

Note that both $\chi(M)$ and $||M||$ are topological invariants. The constant c' can be calculated, and is equal to $\frac{1}{2592\pi^2}$.

Here is an application of the inequality. Consider the manifold with boundary $\Sigma_g \times \Sigma_g \backslash \{B_1 \cup B_2 \cup \cdots \cup B_p\}$, where Σ_g denotes a Riemann surface of genus g. Let M be the double of it. At first, $\chi(M) = 8(g-1)^2 - 2p$ and $\tau(M) = 0$. The manifold M admits a degree 2 map over $\Sigma_g \times \Sigma_g$, and then $||M|| \geq 2||\Sigma_g \times \Sigma_g|| \geq 32(g-1)^2$. For a suitable choice of g and p, it has $2592\pi^2\chi(M) \leq ||M||$. Thus, it has no Einstein metric with $s < 0$. By examining $\pi_1(M)$, it has no Einstein metric with $s \geq 0$. Therefore, M cannot admit any Einstein metric.

There is a similar notion in complex geometry, the *Kobayashi measure.* For a complex manifold M, its Kobayashi volume [117] is defined to be

$$\mu(M) = \inf\Big\{\sum_j \mathrm{vol}(U_j)\big|f_j(U_j) \text{ covers } M\Big\},$$

where U_j are subsets of the complex ball, f_j are holomorphic maps and the volume is given by the Bergmann metric. The infinitesimal version is found by Eisenman [69], as defined by

$$E(p, e_1 \wedge \cdots \wedge e_n) = \inf\Big\{||v_1 \wedge \cdots \wedge v_n||\big|f_*(v_1 \wedge \cdots \wedge v_n) = e_1 \wedge \cdots \wedge e_n\Big\}$$

where e_j is a basis for T_pM, v_j is a basis for T_0B, and f is a holomorphic map $B \to M$ with $f(0) = p$.

Question 3.17. *Is $\mu(M)$ greater than or equal to $||M||$? If we allow meromorphic maps, the volume is a birational invariant. What is the property of this quantity?*

3.7.3 Seiberg–Witten invariants

Another class of obstructions comes from the Seiberg–Witten invariants. The solution of the Seiberg–Witten equation [167, 200] consists of a monopole Φ and a connection A. There is a topological condition that guarantees the existence of the solution, and the solution imposes geometric constraints.

For a 4-manifold M, the second Stiefel–Whitney class $w_2(M)$ lies in the image of the natural homomorphism $H^2(M,\mathbb{Z}) \to H^2(M,\mathbb{Z}/2)$. For a metric g on M and a line bundle L with $c_1(L) = w_2(M) \pmod 2$, there are two $U(2)$-bundles

$$\mathbf{V}_\pm = \mathbf{S}_\pm \otimes L^{\frac{1}{2}}$$

where $\mathbf{S}_\pm$ are the locally defined left- and right-handed spinor bundles of (M,g). Any unitary connection A on L induces a connection

$$\nabla_A : \Gamma(\mathbf{V}_+) \to \Gamma(\Lambda^1 \otimes \mathbf{V}_+).$$

Composing the connection with the Clifford multiplication

$$\mathrm{cl} : \Lambda^1 \otimes \mathbf{V}_+ \to \mathbf{V}_-$$

defines a Dirac operator

$$D_A = \mathrm{cl} \circ \nabla_A : \Gamma(\mathbf{V}_+) \to \Gamma(\mathbf{V}_-).$$

A monopole $\Phi \in \Gamma(\mathbf{V}_+)$ and a connection A always obey the Weitzenböck formula:

$$\langle \Phi, D_A^* D_A \Phi \rangle = \frac{1}{2} \mathrm{d}^*\mathrm{d}|\Phi|^2 + |\nabla_A \Phi|^2 + \frac{s}{4}|\Phi|^2 + 2\langle -iF_A^+, \sigma(\Phi)\rangle \tag{3.15}$$

where $\sigma : \mathbf{V}_+ \to \Lambda_+^2$ is a natural quadratic map with $|\sigma(\Phi)| = \frac{1}{2\sqrt{2}}|\Phi|^2$.

For the Spin structure $\mathbf{S}_\pm$, A is flat and the term $\langle -iF_A^+, \sigma(\Phi)\rangle$ vanishes. Lichnerowicz [128] used the formula (3.15) to show that manifolds with positive scalar curvature have no harmonic spinor. The Atiyah–Singer [5] index formula says that $\hat{A}$-genus is $\mathrm{index}(\not{D}) = \dim(\ker \not{D}) - \dim(\mathrm{coker}\, \not{D})$. Thus, any M^{4k} with positive scalar curvature has zero $\hat{A}$-genus. For dimension $8k+1$ and $8k+2$, Atiyah–Singer [5] proved that $\alpha(M) = \dim(\ker \not{D}) \pmod 2$ is a topological invariant. By the same token, any $8k+1$ or $8k+2$-dimensional manifold with positive scalar curvature has zero α-invariant.

The Seiberg–Witten equation reads

$$\begin{aligned} D_A\Phi &= 0 \\ F_A^+ &= i\sigma(\Phi). \end{aligned} \tag{3.16}$$

After gauge fixing $\mathrm{d}^*(A - A_0) = 0$, the Seiberg–Witten equation is a first order elliptic system. Plugging (3.16) into (3.15) gives

$$0 = 2\mathrm{d}^*\mathrm{d}|\Phi|^2 + |\nabla_A\Phi|^2 + s|\Phi|^2 + |\Phi|^4.$$

Take the integration

$$\int |\Phi|^4 \le \int -s|\Phi|^2 \le \Big(\int s^2\Big)^{\frac{1}{2}} (|\Phi|^4)^{\frac{1}{2}}.$$

Hence, a solution always obeys $\int |\Phi|^4 \le \int s^2$. After some elliptic estimate, the L^4-norm bound implies that the solution space is compact. If $b_+^2 \ge 2$, one can avoid degeneracy of the solution space, and the number of solutions to (3.16) is a topological invariant. In particular, if the invariant is nonzero, the Seiberg–Witten equation (3.16) has a solution.

With $F_A^+ = i\sigma(\Phi)$, we conclude that

$$\int s_g^2 \geq \int |\Phi|^2 = 8\int |F_A^+|^2 = 32\pi^2[c_1^+(L)]^2, \tag{3.17}$$

if L has a solution for (3.16). In the above derivation, the inequality is strict unless $\nabla_A\Phi \equiv 0$ and s is constant. The existence of parallel spinor on $\mathbf{S}_+ \otimes L^{\frac{1}{2}}$ implies that g has special holonomy. Therefore, if $[c_1^+(L)] \neq 0$, $\int s_g^2 = 32\pi^2[c_1^+(L)]^2$ can only occur when g is a Kähler metric of constant scalar curvature.

If L has a solution for (3.16) with respect to any metric g on M, $[c_1(L)] \in H^2(M,\mathbb{Z})/\mathrm{Tor}$ is called a *monopole class*. After some more arguments, (3.17) also leads to an a priori estimate on $[c_1^-(L)]^2$. Therefore, the monopole class is compact in $H^2(M,\mathbb{Z})/\mathrm{Tor}$, and is finite. Let $\mathfrak{C} \subset H^2(M,\mathbb{Z})/\mathrm{Tor}$ be the set of all nonzero monopole classes.

Consider the open Grassmannian $\mathfrak{G} = \mathrm{Gr}^+_{b^2_+}(H^2(M,\mathbb{R}))$ of all maximal subspaces of $H^2(M,\mathbb{R})$ for which the restriction of the intersection pairing is positive definite. Each element $\mathcal{H} \in \mathfrak{G}$ determines an orthogonal decomposition $H^2(M,\mathbb{R}) = \mathcal{H} \oplus \mathcal{H}^\perp$. Given a monopole class a, denote its $\mathcal{H}$-component by a^+. Consider the quantity

$$\alpha(M) = \inf_{\mathcal{H}\in\mathfrak{G}}\left(\max_{a\in\mathfrak{C}}\sqrt{(a^+)^2}\right).$$

Note that $\alpha(M)$ is a (differential) topological quantity. By (3.17), *any* metric g on M obeys

$$\int s_g^2 \geq 32\pi^2\alpha^2(M), \tag{3.18}$$

and the equality holds only if g is a Kähler metric of constant scalar curvature.

On a complex surface, the solutions to the Seiberg–Witten equation can be described in terms of algebraic geometry, and the value of α can be calculated. The value of $\alpha^2(M)$ for any complex surface M with $b_+^2 > 1$ is equal to $c_1^2(X)$, where X is the minimal model of M. There are also other results. For example, if X_1, X_2 and X_3 are minimal, simply connected complex surfaces with $b_+^2 = 3 (\mathrm{mod}\ 4)$, then $\alpha^2(M) = c_1^2(X_1) + c_1^2(X_2) + c_1^2(X_3)$ where M the connected sum of X_1, X_2, and X_3.

The Gauss–Bonnet formula (3.14) with (3.18) improves the Hitchin–Thorpe inequality. Let M be an Einstein manifold with $b_+^2 > 1$, then

(1) $2\chi(M) - 3\tau(M) \geq \frac{1}{3}\alpha^2(M)$. Moreover, if $\alpha(M) \neq 0$, the equality holds if and only if M is $\mathbb{CH}^2/\Gamma$ with the standard Kähler–Einstein metric.
(2) $2\chi(M) + 3\tau(M) \geq \frac{2}{3}\alpha^2(M)$. The equality holds only if both sides vanish, in which case M is $K3$ or $\mathbf{T}^4$ with hyper-Kähler metric.

3.7.3.1 *Characterization of complex hyperbolic surfaces.* In many aspects, a complex hyperbolic space $\mathbb{CH}^2/\Gamma$ is quite rigid. We ask:

Question 3.18. *Does the Ricci flow behave nicely on* $\mathbb{CH}^2/\Gamma$*? Is the Ricci soliton of the complex hyperbolic space unique?*

The topological characterization of $\mathbb{CH}^2/\Gamma$ is also interesting. Thurston's hyperbolization theorem says that an atoroidal Haken manifold is hyperbolic. For the characterization of complex hyperbolic surface, we need to find some additional conditions. For instance,

(1) There is the characterization [205] if M^2 has a complex structure
(2) Instead of an integrable complex structure, what happens if M^2 admits an almost complex structure?
(3) If $\pi_1(M)$ admits a faithful representation in $\mathrm{SU}(2,1)$, is M complex hyperbolic?

The smooth Poincaré conjecture in dimension four is still open. The above discussion can help to understand the possible differentiable structure on $(\mathbb{CH}^2/\Gamma)\#$ (exotic spheres). It might lead to some interesting results for 4-manifolds.

3.7.4 Higher dimensional case

We just discussed the obstructions to Einstein metrics in dimension 4. But none of these three obstructions can be generalized to dimension ≥ 5. Up to now, we know almost nothing about the obstructions to Einstein metrics on higher dimensional manifolds.

Question 3.19. *Does a compact oriented 6-manifold M with $\chi(M) > 0$ admit a metric of positive sectional curvature?*

The statement is true in dimension 4. Geroch found an example on dimension 6 case for which the Gauss-Bonnet integrand could be negative ([82]). But the statement about the Euler characteristic still could be true. Dimension 6 case is already very complicated.

Question 3.20. *Does M^n $(n \geq 5)$ always admit an Einstein metric?*

There is no counter examples so far. This is a major question about Einstein metrics. There are many rigidity results in 4-dimension but not for dimension grater or equal to 5.

The following question is related to previous one:

Question 3.21. *If M^{2n} admits an almost complex structure, does it admit an integrable complex structure for $n > 2$?*

When the dimension is large enough, there is probably no obstruction to the existence of integrable complex structures. Constructing Einstein metrics are different between four dimensions and higher dimensions. We can construct infinitely many exotic Einstein structures on manifolds of dimension greater or equal to 5. Riemann-Roch formula, formulae for $\chi(M)$ and $\tau(M)$ here, give many restrictions in dimension 4 but not for higher dimensions.

3.8 Hermitian Einstein metrics

There are some topological obstructions to the existence of Einstein metrics of nonnegative scalar curvature in dimension 4. For instance, $\pi_1(M)$ consists of a semi-direct product of a Bieberbach group and a finite group. If M is spin, then $\hat{A}(M) = 0$ or certain mod 2 index is zero. However, these obstructions come from the nonnegativity of scalar curvature and not from the Einstein condition.

It seems to be very flexible if we simply consider the Einstein condition. We can consider Einstein metrics with other structures. On a Kähler manifold, the Kähler–Einstein metric is much more rigid. It is basically unique, and depends only on the complex structure. We can consider weaker assumptions than Kähler. For example,

(1) Suppose that M^{2n} has an almost complex structure J, and $\omega(X,Y) = g(JX,Y)$ is a symplectic form. Namely, ω is closed, and $\omega^n > 0$.
(2) Suppose that M^{2n} has an integrable complex structure J, and g is invariant under J. Namely, $g(JX,JY) = g(X,Y)$.

We refer the Einstein metric with the first condition as *symplectic–Einstein*, and the second one as *Hermitian–Einstein*.

Question 3.22. *Are there only finitely many symplectic/Hermitian–Einstein metrics on a manifold M? (for a fixed ω or J, respectively)*

In the Kähler case, $c_1(M) \leq 0$ is completely solved. For $c_1(M) > 0$, the existence of Kähler–Einstein metric implies that TM is stable, and M is Chow stable. The remaining issue is to prove the existence of Kähler–Einstein metric with stability assumption. There are many works on it, especially by Donaldson [64].

Question 3.23. *Does the existence of a Hermitian–Einstein metric imply some stability condition in algebraic/complex geometry? If so, is the condition sufficient to find a Hermitian–Einstein metric?*

Hermitian Einstein metrics on complex surfaces are recently classified by LeBrun [127]. Here is his classification:

(1) The metric is Kähler–Einstein. The construction is given by
   (a) If $c_1(M) = 0$, it is constructed by myself [205, 206].
   (b) If $c_1(M) < 0$ or K_M is ample, it is done by Aubin [6] and myself [205, 206].
   (c) If $c_1(M) > 0$, the Fano case is more subtle, and the construction is not completely finished yet. We need to solve the complex Monge–Ampère equation $\det(g_{i\bar{j}} + \partial_{z^i}\partial_{\bar{z}^j}u) = e^{-cu+F}\det(g_{i\bar{j}})$. All the argument in [206] still work, except that the C^0 estimate is missing. In the Fano case, we need to consider

$$\alpha(M) = \sup\Big\{\alpha \Big| \int e^{-\alpha u} < \infty, \text{ for all positive } u \\ \text{with } (g_{i\bar{j}} + \partial_{z^i}\partial_{\bar{z}^j}u)\mathrm{d}z^i\mathrm{d}\bar{z}^j \text{ positive definite}\Big\}$$

If $\alpha(M)$ is big enough, one can follow the argument in [206]. If M is $\mathbb{CP}^2$ blow-up along finite points, one can arrange the points in a symmetric position to achieve this goal. This is studied by Tian [187].

(2) The Page metric on $\mathbb{CP}^2\#\overline{\mathbb{CP}^2}$.

(3) The conformally Kähler metric on $\mathbb{CP}^2\#2\overline{\mathbb{CP}^2}$ constructed by Chen–LeBrun–Weber [46]. The manifold $\mathbb{CP}^2\#2\overline{\mathbb{CP}^2}$ is $\mathbb{CP}^1\times\mathbb{CP}^1$ blow-up at a point. Their construction starts with a $\mathbf{S}^1\times\mathbf{S}^1$-invariant metric on $\mathbb{CP}^1\times\mathbb{CP}^1$.

In the following we first explain the constructions of the Page metric and the Chen-LeBrun-Weber metric.

3.8.1 The Page metric

Page [151] considered the Wick rotation of the Kerr–de Sitter space, which obeys the Einstein equation with a cosmology term $\Lambda > 0$. It gives a *compact*, Einstein manifold with $R_{ij} = \Lambda g_{ij}$. The Wick rotation in this case is more complicated than the previous cases.

Topologically, the manifold is $\mathbb{CP}^2$ blow-up at one point, i.e. $\mathbb{CP}^2\#\overline{\mathbb{CP}^2}$. The metric is not Kähler, but conformally Kähler. There exists an integrable complex structure J and the metric is Hermitian with respect to J. Though $\nabla J \neq 0$, there exists a smooth function f such that $e^f g$ is Kähler.

3.8.2 Characterization of the Page metric as a Bach flat metric

The Page metric can be produced by another method. The 4-dimensional curvature tensor as an endomorphism on $\Lambda^2_+ \oplus \Lambda^2_-$ splits as

$$\begin{bmatrix} W_+ + \frac{s}{12} & \overset{\circ}{R} \\ (\overset{\circ}{R})^T & W_- + \frac{s}{12} \end{bmatrix}$$

where $\overset{\circ}{R}_{ij} = R_{ij} - \frac{s}{4}g_{ij}$ is the traceless Ricci tensor. Consider the variational equation

$$\delta_g \int |W|_g^2 \mathrm{d}\mu_g = \int (\delta g_{ij}) B_{ij} \mathrm{d}\mu_g = 0.$$

A direct computation shows that

$$B_{ij} = (\nabla^k \nabla^l + \frac{1}{2} R^{kl}) W_{ikjl}.$$

This is called the Bach tensor. Since the functional $\int |W|_g^2 \mathrm{d}\mu_g$ is invariant under self-diffeomorphism, by setting $\delta g_{ij} = V_{i;j} + V_{j;i}$, we derive that the Bach tensor is divergence free

$$\nabla^i B_{ij} = 0.$$

The first Bianchi identity leads to the following identity:

$$(\nabla^i \nabla^j + \frac{1}{2} R^{ij})(*W)_{kijl} = 0.$$

Hence, the Bach tensor is

$$B_{ij} = (2\nabla^k\nabla^l + R^{kl})(W_+)_{ikjl}$$

where $W_+ = \frac{W+*W}{2}$. In fact, this is the variational equation for $2\int |W_+|^2$ which is equal to $\int |W|^2 + 12\pi^2\tau(M)$.

An important observation of Derdziński [62] says that if g is a Bach flat metric with nowhere zero scalar curvature s, then g/s^2 is an Einstein metric. The Page metric on $\mathbb{CP}^2\#\overline{\mathbb{CP}^2}$ is of this form.

3.8.3 Extremal metric on Kähler surfaces

Calabi [35, 36] considered the variational equation $\delta_g \int s_g^2 d\mu_g$ for a fixed Kähler class. A critical point of this functional is called an extremal metric. We can also consider other functionals, such as $\delta_g \int |R_{ij}|^2$. It turns out that these equations are equivalent to each other. It follows from the topological invariance of the following curvature combinations:

$$\chi(M) = \frac{1}{8\pi^2}\int \frac{s^2}{24} + |W_+|^2 + |W_-|^2 - \frac{|\mathring{R}|^2}{2}$$

$$\tau(M) = \frac{1}{12\pi^2}\int |W_+|^2 - |W_-|^2.$$

And the fact that $|W_+|^2 = s^2/24$ for a Kähler metric on a complex surface. This implies that if the extremal Kähler metric of $\int s^2$ is also extremal over variation of Kähler classes, then it is Bach flat.

There are two possibilities for the extremal metric:

(1) If s is a constant, the first Chern form $c_1(M)$ is harmonic. If we further know that $\lambda[\omega] = [c_1(M)]$ for some constant λ, the harmonicity of ω implies $\lambda\omega = c_1(M)$. Thus, the extremal metric is Kähler–Einstein.

(2) If s is not a constant but happened to be nowhere zero, the Bach flatness of g implies that g/s^2 is Einstein. It is surely conformally Kähler.

The first case relies on the Kähler condition. We ask:

Question 3.24. *What about non-Kähler Hermitian metric? Is there a similar condition such that $c_1(M)$ is harmonic?*

3.8.4 The construction of Chen–LeBrun–Weber

Chen–LeBrun–Weber [46] constructed a conformally Kähler, Einstein metric on $\mathbb{CP}^2\#2\overline{\mathbb{CP}^2}$ by finding the extremal metric. Due to Derdziński [62], g/s^2 is Einstein for a Bach flat metric g with nonzero scalar curvature.

If ω is an extremal Kähler metric (for a fixed Kähler class), then $\xi = \nabla_\omega^{1,0}s = -ig^{\bar{a}b}\frac{\partial s}{\partial \bar{z}^a}\frac{\partial}{\partial z^b}$ is a holomorphic vector field, and $\frac{\partial^2 s}{\partial z^a \partial z^b} = 0$. The Calabi functional is defined by

$$\mathcal{C}(g) = s_0^2\int d\mu + \int (s-s_0)^2 d\mu = 32\pi^2\frac{(c_1(M)\cdot[\omega])^2}{[\omega]^2} - \mathcal{F}(\xi,[\omega]).$$

Here, $s_0 = \frac{\int s d\mu}{\int d\mu}$ is the average value of the scalar curvature, and $\mathcal{F}(\xi, [\omega])$ is the Futaki invariant for the holomorphic vector filed ξ [79, 81].

The crucial fact is that ξ can be determined up to conjugation without knowing the existence of the extremal metric [80]. Granted what was said, they considered the functional

$$\mathcal{C}_{\text{virtual}}([\omega]) = \frac{(c_1(M) \cdot [\omega])^2}{[\omega]^2} - \frac{1}{32\pi^2}\mathcal{F}(\xi, [\omega])$$

on the Kähler cone.

The manifold $\mathbb{CP}^2\#2\overline{\mathbb{CP}^2}$ is $\mathbb{CP}^1 \times \mathbb{CP}^1$ blow-up at a point. The holomorphic vector field ξ is a multiple of the generator of the $\mathbb{C}^\times$-action induced by

$$\begin{array}{ccc} \mathbb{CP}^1 \times \mathbb{CP}^1 & \to & \mathbb{CP}^1 \times \mathbb{CP}^1 \\ ([u_0 : u_1], [v_0 : v_1]) & \mapsto & ([u_0 : \zeta u_1], [v_0 : \zeta v_1]). \end{array}$$

Because of symmetry, they only need to look at $\mathcal{C}_{\text{virtual}}$ for the Kähler classes $(1 + x)(F_1 + F_2) - xE$, where $F_j, j = 1, 2$ is the fundamental class of each $\mathbb{CP}^1$ and E is the exception divisor. The function $\mathfrak{C}_{\text{virtual}}$ of x can be written down explicitly. It is the ratio of two degree six polynomials in x. It has only one critical point at some $x_0 > 0$. They did the local analysis to show that there is no bubbling and the extremal metric does exist.

LeBrun proved [127] that besides Kähler–Einstein metrics, there are only two classes of Hermitian–Einstein metrics on complex surfaces. the Page metric ([151]) on $\mathbb{CP}^2\#\overline{\mathbb{CP}^2}$ and the Chen–LeBrun–Weber metric on $\mathbb{CP}^2\#2\overline{\mathbb{CP}^2}$ ([46]). These two metrics are conformally Kähler, but not Kähler. Meanwhile, it is known that these two surfaces have no Kähler–Einstein metric. Therefore, the stability condition for Hermitian–Einstein metric must be different.

Because of the $\partial\bar{\partial}$-lemma, the deformation of Kähler metrics within the same class is given by a scalar function. It is important to find a suitable way to deform a Hermitian metric, and try to make the Hermitian–Einstein condition into a Monge–Ampère type equation. Ricci flow is another approach for searching Einstein metrics. The first step is to find out a condition on J or ∇J such that $g(JX, JY) - g(X, Y)$ vanishes along the flow.

To study the symplectic–Einstein metric, the symplectic cohomology established by Tseng and myself [185] could be useful.

We do not know much when $\dim_{\mathbb{C}} M \geq 3$.

Question 3.25. *Is there any interesting example of conformally Kähler-Einstein metric on a Fano 3-fold?*

Based on the result of Chen–LeBrun–Weber, I expect that there do exist some nice examples. In the Kähler case, we reduce the study on the Kähler potential u. For the conformally Kähler case, the metric is $e^f(g_{i\bar{j}} + \partial_{z^i}\partial_{\bar{z}^j}u)dz^i d\bar{z}^j$. The conformally Kähler–Einstein metric seems to be a reasonable proposal to pursue.

3.9 Other constructions of Einstein metrics

Roček ([161]) proposed a special construction of Einstein metrics, but there are not many interesting examples. The manifold M has two commuting integrable complex structures $[J^+, J^-] = 0$, and the metric g is Hermitian with respect to both of them. The corresponding symplectic form obeys $\partial^{\pm}\bar{\partial}^{\pm}\omega^{\pm} = 0$. The Hermitian connection satisfies $\Gamma^{\pm} = \Gamma^0 \pm T$ where Γ^0 is the Levi-Civita connection of g and T is the torsion tensor. After lowering indices, T becomes totally antisymmetric and can be regarded as a closed 3-form. It turns out that T can be written as $\pm\frac{1}{2}J^{\pm}\mathrm{d}\omega^{\pm}$.

These conditions imply that there exists a potential function K such that

$$g = \partial_i\partial_{\bar{j}}K\mathrm{d}u^i\mathrm{d}\bar{u}^j - \partial_a\partial_{\bar{b}}K\mathrm{d}v^a\mathrm{d}\bar{v}^b$$

where u, v are complex coordinates of $J^{\pm}$:

$$J^{\pm}\mathrm{d}u = i\mathrm{d}u \qquad J^{\pm}\mathrm{d}v = \pm i\mathrm{d}v \ .$$

The submanifolds defined by $u =$ constant or $v =$ constant are Kähler submanifold. The Ricci flat condition becomes

$$\det(\partial_i\partial_{\bar{j}}K) - \det(-\partial_a\partial_{\bar{b}}K) = 0 \ . \tag{3.19}$$

This is analogous to the Monge–Ampère equation in the usual Calabi–Yau case.

He provided an example: $\mathbf{S}^3 \times \mathbf{S}^1$ with the potential

$$K = -\int^{\frac{|v|^2}{|u|^2}} \frac{\log(1+x)}{x}\mathrm{d}x + \log u \log \bar{u} \ ,$$

and the metric

$$g = \frac{\mathrm{d}u\mathrm{d}\bar{u} + \mathrm{d}v\mathrm{d}\bar{v}}{|u|^2 + |v|^2} \ .$$

These conditions seem to be very restrictive. It would be nice to find more nontrivial examples.

3.10 Constructing Calabi–Yau metrics on $K3$

A compact, simply-connected, Ricci flat 4-manifold is a $K3$ surface. It satisfies $2\chi = 3|\tau|$, where $\chi = \sum_{j=0}^{4}(-1)^j b_j$ is the Euler characteristic and $\tau = b_2^+ - b_2^-$ is the Hirzebruch signature.

3.10.1 Kummer construction

There are two approaches to explicitly construct the Calabi–Yau metric on $K3$.

The first approach starts with $\mathbf{T}^2_{\mathbb{C}} = \mathbb{C}^2/\mathbb{Z}^4$. Take the quotient $P = \mathbf{T}^2_{\mathbb{C}}/\mathbb{Z}_2$, where the action is defined by $(z, w) \to (-z, -w)$. The action has 16 fixed points, and they become double point singularities of P. $K3$ is obtained by resolve these 16 singularities.

The local model of the singularity is $\mathbb{C}^2/\mathbb{Z}_2$. It embeds into $\mathbb{P}^3$ by $[1, z^2, zw, w^2]$. In terms of the homogeneous coordinate of $\mathbb{P}^3$, the image is defined by $x_1x_3 = x_2^2$. Blow up $\mathbb{P}^3$ at $[1, 0, 0, 0]$. The proper transform of $\mathbb{C}^2/\mathbb{Z}_2$ resolves the singularity. It is not hard to see that the procedure replaces the origin by a rational curve $\mathbb{P}^1$, and $\widehat{\mathbb{C}^2/\mathbb{Z}_2}$ is the total space of $T^*\mathbb{P}^1 = K_{\mathbb{P}^1} \to \mathbb{P}^1$.

However, $T^*\mathbb{P}^1$ has an AF Ricci flat metric. It is usually referred as the Eguchi–Hanson metric [67]. Calabi [34] also discovered the metric in a more general context.

3.10.2 Eguchi–Hanson metric

We first explain Calabi's construction. Let M^n be a projective Kähler–Einstein manifold, say $R_{i\bar{j}} = g_{i\bar{j}}$. Let (z, s) be coordinates on the total space of the canonical line bundle $K_M \to M$, where s is the coordinate for the fiber. The $(n+1, 0)$-form $\Omega = \mathrm{d}z^1 \wedge \cdots \wedge \mathrm{d}z^n \wedge \mathrm{d}s$ is a well-defined holomorphic volume form for K_M. If we can find a Kähler form ω for K_M such that

$$\omega^{n+1} = \Omega \wedge \bar{\Omega}, \qquad (\star)$$

the metric Ω is Ricci flat.

The initial metric $g_{i\bar{j}}$ on M naturally induces a Hermitian metric on K_M by $\det(g_{i\bar{j}})$. Assume $\omega = \sqrt{-1}\partial\bar{\partial}(F(\log ||s||^2))$, where F is some unknown function. Since $g_{i\bar{j}}$ is Kähler–Einstein, the equation $(\star)$ becomes a second order ODE of F, and the solution can be written down explicitly.

The method of Eguchi–Hanson is more straightforward. They solve the complex Monge–Ampère equation on $\mathbb{C}^2$, $\det(u_{i\bar{j}}) = 1$. The Kähler form is $\sqrt{-1}\sum_{i,j} u_{i\bar{j}}\mathrm{d}z_i \wedge \mathrm{d}\bar{z}_j$. They assume the function u is of the following form $F(|z_1|^2 + |z_2|^2)$. It turns out that there are two solutions. The first one is $F(t) = t$, and it gives the standard metric of $\mathbb{C}^2$. The second one blows up at the origin, and is asymptotic to t for $t \to \infty$. If we blow up at the origin, the second solution defines a smooth metric on $\hat{\mathbb{C}}^2$, and is invariant under the $\mathbb{Z}_2$-action. Since $\widehat{\mathbb{C}^2/\mathbb{Z}_2} = T^*\mathbb{P}^1$, this finishes the construction.

3.10.3 The expansion of the $K3$ metric

Let $\check{P}$ be a $K3$ surface given by the Kummer construction. It contains 16 exceptional $\mathbb{P}^1$'s. Fix the complex structure of $\widehat{P}$. Suppose that ω_ϵ is the Calabi–Yau metric of $K3$ with $\omega_\epsilon(\mathbb{P}^1) = \epsilon << 1$. It is known that

(1) As $\epsilon \to 0$, ω_ϵ converges to the flat metric of $\mathbf{T}^2_{\mathbb{C}}/\mathbb{Z}_2$.
(2) The linear term is the Eguchi–Hanson metric, i.e.

$$\omega_\epsilon = (\text{flat}) + \epsilon(\text{Eguchi–Hanson}) + o(\epsilon).$$

Question 3.26. *Suppose such an expansion exists,* ω_ϵ=*(flat)*+ϵ*(Eguchi–Hanson)*+ $\epsilon^2 g_2 + \epsilon^3 g_3 + \cdots$. *What are the higher order terms in the expansion?*

3.10.4 Elliptic fibration

A $K3$ surface admits an elliptic fibration $E \to K3 \to \mathbb{P}^1$. The number of singular fibers is at least 12, and at most 24. Greene, Shapere, Vafa and myself construct a Ricci flat metric based on an elliptic fibration. We call the metric the *semi-flat* metric. On smooth fibers, the induced metric is flat. On singular fibers, the metric is not smooth, but the singular behavior is in a controlled manner. The semi-flat metric can be used to perform the T-duality. We can do the fiberwise dual to obtain the mirror $E^* \to K3^* \to \mathbb{P}^1$, and the semi-flat metric is dual to the semi-flat metric on $K3^*$. This is related to the SYZ [180] construction of mirror Calabi–Yau manifolds.

We expect the the semi-flat metric can be used to construct the Calabi–Yau metric by: $\mathrm{d}s^2_{\mathrm{CY}} = \text{semi-flat} + \text{correction term}$. The correction term is more subtle. It is believed to come from Lagrangian cycles in the following way: Fix a fiber $\mathbf{T}^2$, and suppose that L is a flat line bundle over it. Consider maps $\mathbf{D} \to K3$ such that $\partial\mathbf{D} \mapsto \mathbf{T}^2$, $\omega|_{\mathbf{D}} = 0$ and $\Omega|_{\mathbf{D}} = e^{i\phi}\mathrm{vol}_{\mathbf{D}}$. It has two important quantities. The first one is the area $A = \mathrm{Area}(\mathbf{D})$; the second one is the holonomy $e^{i\theta}$ of L along $\partial\mathbf{D} \subset \mathbf{T}^2$. The Lagrangian cycle contributes to the correction term by $Ae^{i\theta}$.

On the other hand, there is a smooth Calabi–Yau metric on the local model of a singular fiber. Ooguri–Vafa [150] make an ansatz of the Taub–NUT metric to construct a Ricci flat metric on the local model. Unlike the semi-flat metric, the Ooguri–Vafa metric is smooth. They can also see holomorphic/Lagrangian disks in their local model. We ask:

Question 3.27. *Can one obtain a smooth Calabi–Yau metric for K3 by first gluing the semi-flat metric and the Ooguri–Vafa metric together and then performing singular perturbation?*

If so, $\mathrm{d}s^2_{\mathrm{CY}} = \text{semi-flat} + \text{Ooguri–Vafa} + \text{correction}'$. The correction$'$ term is more complicated. It involves the Lagrangian cycles going around many singular fibers. It is related to the wall crossing formula, and the work of Kontsevich–Soibelman [121, 122]. There are a lot to be done for this proposal. Even the Kähler class and the perturbation parameter are not chosen yet.

4 Manifolds with positive curvature

4.1 Manifolds with positive sectional curvature

One of the most well-known problems is Hopf's "does $\mathbf{S}^2 \times \mathbf{S}^2$ admit a complete metric of positive sectional curvature?"In general, if M_1 and M_2 are simply connected, it is not known if there exists any metric of positive sectional curvature on $M_1 \times M_2$. Without assuming simply connectedness, an application of Synge's theorem implies that $\mathbb{RP}^n \times \mathbb{RP}^n$ does not admit any metric of positive sectional curvature.

There are indeed very few examples of Riemannian manifolds with positive sectional curvature known. Here is a list:

(1) Compact rank one symmetric spaces.
(2) The homogeneous spaces are classified by Berger [16], Wallach [199], and Berard-Bergery [14]. Berger [16] classified even dimensional normal homogeneous spaces (i.e., the metric on G/H is induced by a bi-invariant metric on G). Wallach [199] classified even dimensional homogeneous spaces. Berard-Bergery [14] classified odd dimensional homogeneous spaces, and he proved that the examples by Berger and Wallach, together with the Aloff–Wallach manifolds [1] constitute the complete list. Note that besides in dimension 6, 7, 12, 13, and 24, all the examples are the compact rank one symmetric spaces.
(3) The only known non-homogeneous examples are biquotients, and have been constructed by Eschenburg [70, 71] in dimension 6 and 7, and then by Bazaikin [9] in dimension 13.

Most of these constructions are based on Cheeger's biquotient method [39]. Suppose that $\pi : M \to N$ is a Riemannian submersion. By O'Neill's formula, we have

$$K_N(v, w) = K_M(\tilde{v}, \tilde{w}) + \frac{3}{4}|[\tilde{v}, \tilde{w}]^{\mathrm{Vert}}|^2 \geq K_M(\tilde{v}, \tilde{w})$$

where $\tilde{v}$ and $\tilde{w}$ are the horizontal lifts of v and w. Therefore if $K_M \geq 0$ then K_N is even more positive. Cheeger considered a compact isometry group G of a Riemannian manifold M with $K_M \geq 0$. Take the bi-invariant metric on G and consider the Riemannian product $G \times M$. G acts on $G \times M$ by isometry through $g(p, m) = (pg^{-1}, gm)$, the quotient space $G \times M/G$ is equipped with another Riemannian metric. By O'Neill's formula, the sectional curvature of this quotient metric is greater than the original K_M. For example, if we apply this method to $\mathbf{S}^2 \times \mathbf{S}^2$, we can make the curvature more non-negative, but still not strictly positive.

Question 4.1. *If a manifold has finite fundamental group and has non-negative sectional curvature, does it admit a metric of positive sectional curvature?*

Regarding Hopf's conjecture, one can try to perturb the product metric on $\mathbf{S}^2 \times \mathbf{S}^2$. It was shown that up to first order it does not produce a metric of positive sectional curvature. The problem is to deal with curvature near totally geodesic subspaces such as $\mathbf{S}^1 \times \mathbf{S}^1 \subset \mathbf{S}^2 \times \mathbf{S}^2$. Another possible approach is to express $\mathbf{S}^2 \times \mathbf{S}^2$ as a double coset space $K_1 \backslash G / K_2$. There is some evidence that perturbing the double coset structure may increase the curvature.

Milnor showed that there are 28 differentiable structures $\mathbf{S}^7$. Gromoll–Meyer [89] constructed a metric of non-negative sectional curvature on one of Milnor's exotic 7-spheres which is represented as a $\mathbf{S}^3$-bundle over $\mathbf{S}^4$. The construction also applies to other exotic 7-spheres. Recently, Petersen and Wilhelm [157] claimed the metric can be made strictly positive. In general, it is very difficult to check the the positivity of the sectional curvature as one needs to consider all 2-planes in the tangent space. As these spaces are all $\mathbf{S}^3$-bundle over $\mathbf{S}^4$, the construction of these metrics may give some hint on how to approach the original Hopf conjecture on $\mathbf{S}^2 \times \mathbf{S}^2$.

For the unit tangent bundle of $\mathbf{S}^4$, it is known by [156] that there exists metrics with non-negative sectional curvature and strictly positive at some point. Such a metric also admits a free $\mathbf{S}^1$-action. The cohomology of the quotient $H^*(M/\mathbf{S}^1)$ is the same as $H^*(\mathbb{CP}^3)$, but with different ring structure.

Question 4.2. *If a manifold M has positive sectional curvature, does it admit a faithful $\mathbf{S}^1$ action? If this being the case, is the action isometric? This may be used to solve Hopf's conjecture as the existence of an isometric $\mathbf{S}^1$ action will reduce the problem to 3-manifolds of positive sectional curvature which are well-understood.*

So far all non-standard examples of manifolds of positive sectional curvature have dimension less than or equal to 24. I conjecture that

Question 4.3. *For dimension $n \geq 25$, all compact manifolds with positive sectional curvature are diffeomorphic to rank one symmetric spaces.*

There are some progresses in the case when the manifold is of cohomogeneity one, i.e. there exists a subgroup G of the isometry group of M such that the principal orbit has codimension one. This will essential reduce the problem to an ODE. For example, Verdiani [196, 197] showed that the conjecture is true for any even dimensional manifold of cohomogeneity one. For an odd dimensional manifold of cohomogeneity one, Grove–Wilking–Ziller [96] confirm the question except for two families of manifolds. They proved that the manifold must be one of the following:

(1) The standard spheres.
(2) E_p^7 (Eschenburg [70]) , B_p^{13} (Bazaikin [9]), or B^7 (Berger [16]).
(3) P_k, Q_k, or R.

It is not clear that whether the third class all admit a metric with positive sectional curvature, but [96] still says something about their geometry. The manifolds P_k and Q_k are 3-Sasakian manifolds (i.e., $M \times \mathbb{R}_+$ is hyper-Kähler.) In fact, up to taking the universal covering, they are the $SO(3)$ orbifold principal bundles of the vector bundle of self-dual 2-forms over some self-dual Einstein orbifolds, which were constructed by Hitchin. See [96] and the references therein for more about these manifolds.

Besides, Wilking [203] showed that if a simply connected compact manifold M has positive sectional curvature and if the dimension of the isometry group is greater than or equal to $2\dim M - 6$, then M is tangentially homotopically equivalent to a rank one symmetric space or isomorphic to a homogeneous space of positive sectional curvature. Also suppose that k is the cohomogeneity (i.e. codimension of the principal orbit) and if $\dim M \geq 18(k+1)^2$, then M is tangentially homotopically equivalent to a rank one symmetric space.

Recall that a homotopy equivalence $f : M_1 \longrightarrow M_2$ is a tangential homotopy equivalence if $f^{-1*}(TM_2) \bigoplus E \cong TM_1 \bigoplus E'$ are isomorphic as bundles, for some trivial bundles E and E'.

In [203], Wilking constructed a chain of totally geodesic manifolds $M = M_0 \subset M_1 \subset \cdots$, where each M_i admits certain isometric action and all inclusions are totally geodesic. The connectedness lemma is the key to to show that M_∞ has periodic cohomology ring, and to recover the homotopy type of M from M_∞.

Theorem 4.4 (connectedness lemma, [202]). *Let M^n be a compact Riemannian manifold with positive sectional curvature.*

(1) *Suppose that $N^{n-k} \subset M^n$ is a compact totally geodesic submanifold. Then the inclusion $\iota : N \hookrightarrow M$ is $(n-2k+1)$-connected (i.e., $\iota_* : \pi_j(N) \to \pi_j(M)$ is an isomorphism for $j < n-2k+1$, and is surjective for $j = n-2k+1$).*
(2) *Suppose that $N_1^{n-k_1}$ and $N_2^{n-k_2}$ are two compact totally geodesic submanifolds of M, with $k_1 \leq k_2$ and $k_1 + k_2 \leq n$. Then the intersection $N_1 \cap N_2$ is also a totally geodesic submanifold, and the inclusion map $\iota : N_1 \cap N_2 \hookrightarrow N_2$ is $(n - k_1 - k_2)$-connected.*

The proof of the connectedness lemma is to consider the relative loop space $\Omega(M, \iota(N))$, which are the space of paths in M with endpoints in $\iota(N)$. Then study $\Omega(M, \iota(N))$ via Morse theory, while critical points are geodesics, and compute the Morse index of them. Note that in complex geometry, the Morse theory argument can also be used to prove the Lefschetz hyperplane theorem [3, 26].

Instead of paths, one can consider surfaces, for example, minimal immersed $\mathbf{S}^2$. Maybe with some suitable assumption on the principal curvature, one can obtain a better version of the connectedness lemma.

One can probably answer the question by a geometric flow which preserves positive sectional curvature. Note that Ricci flow does not work here. The steady state of the flow should be Einstein metric as well. An Einstein metric with positive sectional curvature must be a rank one locally symmetric space.

Here is a list of other problems on manifolds with positive sectional curvature:

(1) Chern [55, P.167] conjectured that if G is an abelian subgroup of the fundamental group of a manifold M with positive sectional curvature, then G must be isomorphic to $\mathbb{Z}/k$. Preissman [160] showed that the statement is true for manifolds of negative sectional curvature. Wolf [201] showed that the statement is true if M is a finite quotient of the standard sphere. Chern's conjecture turned out to be false, see [168] for the counterexample. We can ask if every finite group appears as the first fundamental group of a manifold of positive sectional curvature (assume M is not a finite quotient of the standard sphere). Perhaps the rank of abelian subgroup of the fundamental group of a manifold of positive sectional curvature is bounded by a constant depending only on the dimension.
(2) Suppose M is a manifold of non-negative sectional curvature. Consider the rational homotopy groups $\pi_i(M^n) \otimes \mathbb{Q}$. Is there a constant $C(n)$ such that $\pi_i(M^n) \otimes \mathbb{Q} = 0$ for $i > C(n)$? I tried to look at harmonic maps $\mathbf{S}^i \to M$. Since the curvature of M is positive, there are some difficulties.
(3) Let $\Omega(M)$ be the loop space of M. Based on the result of [17], Bott conjectured that the Betti number $b_i(\Omega(M))$ grows polynomially if M has positive sectional curvature. By [98], this would imply a conjecture of Gromov [90] that $\sum b_i(M^n) \leq \sum b_i(\mathbf{T}^n) = 2^n$ where $\mathbf{T}^n$ is the n-dimensional torus. In

fact, based on the work of Sullivan [181], this problem is equivalent to the previous one. One can see [95] for more discussion on the consequences of this problem.

One may consider the double loop space $\Omega^2(M)$ which corresponds to maps $\mathbf{S}^2 \to M$. In this case, manifold of positive isotropic curvature naturally arises in the second variation of minimal spheres. This is related to the work of Sacks–Uhlenbeck and Siu–Yau [137, 162, 171].

(4) (Petersen and Wilhelm) [10] If a compact manifold M with positive sectional curvature has a Riemannian fibration over a compact manifold N, then the dimension of the fiber is less than the dimension of N.

Weinstein and Berard Bergery solved the case where the fibers are totally geodesic back in the 1960-70's. Wilking [204] had a paper on "Dual Foliations "in nonnegative curvature that also addressed this problem. But to date no one has found a general solution.

4.2 Manifolds with positive curvature operator

The Riemannian curvature tensor R_{ijkl} can be regarded as a self-adjoint a map $\Omega : \Lambda^2(M) \to \Lambda^2(M)$. A natural condition is that all the eigenvalues of Ω is positive; this is called *positive curvature operator.* In this case, the Bochner [21] argument implies all Betti numbers $b_i(M) = 0$ for $0 < i < \dim(M)$. Thus M is a homology sphere. In fact, with some more efforts, one can show M is a homology sphere. This implies M is a homeomorphic sphere in dimension > 4 by the Poincaré conjecture provided that M is simply connected. It was finally proved to be diffeomorphic to a sphere by Böhm–Wilking [25] using the Ricci flow. By Hamilton's theorem [99], positive curvature operator is preserved under the Ricci flow. A theorem of Berger (by maximum principle) says that an n-dimensional compact Einstein manifold whose sectional curvature K is pinched between $\frac{n-2}{n-1} < K \leq 1$ is a quotient of a sphere. [Maybe there is a mistake here. I think "non-negative sectional curvature" is not strong enough to get the result. I think here should be "suitably pinched positive curvature" (Remark by Entao Zhao)] Since steady states of (normalized) Ricci flow are Einstein manifolds, long time existence and convergence of Ricci flow imply that the manifold is a quotient of a sphere. In addition, if Ω is non-negative, Hamilton's strong maximum principle implies that M is diffeomorphic to $\mathbf{S}^n$ or a symmetric space. The Ricci flow argument does not apply to manifolds of non-negative sectional curvature as this condition is not preserved under the Ricci flow, see [147].

4.2.1 Quarter-pinched manifolds

There are recent results by Ricci flow for quarter-pinched manifolds. For a simply connected Riemannian manifold, it is known by the work of Rauch, Klingenberg and Berger that if the sectional curvature K is pinched $1 \geq K > \frac{1}{4}$, then the manifold is homeomorphic to the standard sphere. The differentiable case was

[10] The author thanks Peter Petersen for his explanation and the references he provided.

recently settled by Brendle–Schoen [29] under even weaker assumption that at each point $K > 0$ and $1 \geq K_{\min}/K_{\max} > \frac{1}{4}$. Also $1 \geq K_{\min}/K_{\max} \geq \frac{1}{4}$ implies diffeomorphism to the standard sphere or a symmetric space of rank one. The result was further improved by Petersen–Tao [155] to allow $\geq \frac{1}{4} - \delta_n$ for a dimension dependent constant δ_n.

It is important to study which algebraic conditions of the curvature is preserved under the Ricci flow. Böhm–Wilking [25] made the breakthrough to show that positive curvature operator is preserved. However, quarter-pinching is not preserved. Instead, Brendle–Schoen consider positive isotropic curvature conditions on $M \times \mathbb{R}$ and $M \times \mathbb{R}^2$ which are preserved along the Ricci flow. In the proof, they assume the manifold is compact to start with. However, the pinching condition $1 \geq K_{\min}/K_{\max} > \frac{1}{4}$ should imply compactness. In [29], assuming bounded curvature, they achieved the compactness under the stronger pinching condition $0 < K_{\max} < (4 - \epsilon)K_{\min}$, where $\epsilon > 0$ can be an arbitrary real number.

4.3 Kähler manifolds with positive bisectional curvature

For Kähler manifolds, we can consider positive bisectional curvature. Let (M, h) be a Kähler manifold with Kähler metric h. Remember that the bisectional curvature of any two complex direction $u_\alpha, u_\beta \in T^{1,0}M$ is defined to be

$$R_{\alpha\bar{\alpha}\beta\bar{\beta}} = h(R(u_\alpha, \bar{u}_\alpha)u_\beta, \bar{u}_\beta).$$

If we set $u_\alpha = (e_\alpha - iJe_\alpha)/\sqrt{2}$ and $u_\beta = (e_\beta - iJe_\beta)/\sqrt{2}$ for $e_\alpha, e_\beta \in T_{\mathbb{R}}M$, it is an easy exercise to show that

$$R_{\alpha\bar{\alpha}\beta\bar{\beta}} = R(e_\alpha, e_\beta, e_\beta, e_\alpha) + R(e_\alpha, Je_\beta, Je_\beta, e_\alpha).$$

In terms of local coordinates, if $h = h_{\alpha\bar{\beta}} dz^\alpha d\bar{z}^\beta$,

$$R_{\alpha\bar{\mu}\beta\bar{\nu}} = -\partial_\beta \partial_{\bar{\nu}} h_{\alpha\bar{\mu}} + h^{\sigma\bar{\eta}} \partial_\beta h_{\sigma\bar{\mu}} \partial_{\bar{\nu}} h_{\alpha\bar{\eta}}.$$

4.3.1 The work of Siu–Yau

The Frankel conjecture asserts that compact Kähler manifolds with positive bisectional curvature must be bi-holomorphic to $\mathbb{CP}^n$. It was proved by Siu–Yau [171]. There are three different approaches to this kind of problems. The other two are due to Mori's [144] and Taubes's [184] works. All three approaches involve construction of rational curves of low degree.

We first explain the work of Siu–Yau. By a theorem of Bishop–Goldberg [19], it is known that $H_2(M, \mathbb{Z}) = \mathbb{Z}$. It follows from Hirzebruch-Riemann-Roch theorem and Bonnet-Myers theorem that the curvature positivity implies the triviality of the fundamental group $\pi_1(M)$. Thus $\pi_2(M) = H_2(M, \mathbb{Z}) = \mathbb{Z}$. Then we can take a generator $[c] \in \pi_2(M)$ and consider a representative $f : \mathbf{S}^2 \to M$. We can minimize the energy $E(f) = \frac{1}{2}\int_{\mathbf{S}^2} |\nabla f|^2$ among $[f] = [c] \in \pi_2(M)$. An important step is to understand in details of the Sacks–Uhlenbeck's bubbling process [162] in detail. They consider a perturbation of the energy $E_\alpha(f) = \int_{\mathbf{S}^2}(|\nabla f|^2 + 1)^\alpha$ for

$\alpha > 1$ which has nice compactness property and there exists smooth minimizers f_α. In the limit $\alpha \to 1$, the original map breaks into several pieces and each one is a minimizer of the $E(f)$.

It follows from the second variation formula that each minimizer from S^2 to M is either holomorphic or anti-holomorphic. We use the fact that the f^*TM on $\mathbf{S}^2$ can be split into a direct sum of line bundles, each one is a holomorphic line bundle as the base manifold is two-dimensional. The second variation formula can be expressed in terms of the associated $\bar{\partial}$-operator. The suitable curvature condition for this argument is positive isotropic curvature which is implied by positive bisectional curvature. At the end, we use deformation theory to show that there is only one holomorphic map from $\mathbf{S}^2$ to M in the sum, which represents a generator of $\pi_2(M)$. By the result of Kobayashi-Ochiai, M is biholomorphic to $\mathbb{CP}^n$.

4.3.2 The work of Taubes

Taubes [184] used Seiberg–Witten theory to show that if a symplectic manifold (X, ω) is diffeomorphic to $\mathbb{CP}^2$, then it is symplectomorphic to $\mathbb{CP}^2$. Here is a brief explanation of his proof. Consider the Seiberg–Witten equation on $H \oplus HK^{-1}$, where $c_1(H)$ is the generator for $H_2(X, \mathbb{Z})$ and $c_1(H) \cdot [\omega] > 0$. Since the standard metric on $\mathbb{CP}^2$ has positive scalar curvature and $b_2^- = 0$, the unperturbed Seiberg–Witten equation has no solution for a generic choice of metric on X. Taubes perturbed the Seiberg–Witten equation by $r\omega$. At first, there is a wall-crossing phenomenon happening at some $r_0 > 0$, and it implies that the perturbed equation has a unique solution for any $r > r_0$. Then, he proved that as r goes to infinity, the zero locus of the solution would converge to an embedded pseudo-holomorphic $\mathbf{S}^2$, whose homology class is the Poincaré dual of $c_1(H)$.

The tubular neighborhood of this curves is standard by the symplectic neighborhood theorem. The construction of the symplectomorphism on the complement is due to Gromov's work [91] on the symplectic structure of $\mathbf{S}^2 \times \mathbf{S}^2$ and $\mathbb{R}^4$. Basically it is as same as Hirzebruch-Kodaira's idea [108] on using rational curves of low degree and deformation theory.

It is not known if this argument can be generalized to higher dimensions.

4.3.3 The work of Mori

Mori's argument [144] uses p-adic methods and shows that if the tangent bundle of M is positive then M is biholomorphic to $\mathbb{CP}^n$. Here the positivity of a bundle is defined as follows. Let $V \to M$ be a complex vector bundle. Consider the projectification of V, $\mathbb{P}(V)$. The positivity of the bundle V means that the hyperplane line bundle $H \to \mathbb{P}(V)$ is positive, i.e., there exists a Hermitian metric $\|\cdot\|$ on H such that the curvature is positive. Notice that the Hermitian metric induces a Finsler metric on the original bundle $V \to M$. It is interesting to understand what $TM > 0$ means in terms of Hermitian metrics.

Mori's argument actually does not need the positivity of the tangent bundle. It just needs the first Chern class $c_1(H) > 0$, which says that there exists an algebraic curve C such that $c_1(M) \cdot C > 0$. Then deform this curve C and find a neighborhood to get a rational curve. This deformation technique was created by Mori, used in the beginning algebraic geometry of characteristic p.

In view of Mori's theorem, we ask:

Question 4.5. *Suppose TM has a Hermitian metric (without assuming it is Kähler) with positive bisectional curvature, can one produce rational curves on M by an analytic method?*

Note that Mori's argument still works, but it is not clear for Siu–Yau's argument. As a remark, admitting a Kähler metric with positive bisectional curvature implies the positivity of the tangent bundle.

For Kähler manifolds, we consider non-negative bi-sectional curvature in Siu–Yau's [171] solution of Frankel's conjecture. It is known (Bando [8] for $\dim(M) = 3$, and Mok [141] for general dimensions) that Ricci flow preserves non-negative bi-sectional curvature. However, so far there is still no genuine Ricci flow proof of Frankel's conjecture (all works in this direction assume Siu–Yau's theorem and then prove convergence of Ricci flow). It is also interesting to look at Hermitian metric with positive bi-sectional curvature.

Question 4.6. *On a symplectic manifold (M, ω), we can consider an almost complex structure J which may not be integrable, and the compatible Riemannian metric g, $g(X,Y) = \omega(X, JY)$. Suppose this g has positive sectional curvature, does it imply M is diffeomorphic to $\mathbb{CP}^n$? Moreover, how about symplectomorphic?*

I have another conjecture regarding the bisectional curvature:

Conjecture 4.7. *For a complete Kähler manifold M^n with positive bisectional curvature, $\kappa_{\mathrm{bs}} > 0$, then M is biholomorphic to $\mathbb{C}^n$.*

There are some progress on this conjecture, by Shi [169], Chen–Zhu [44, 45], Chen–Tang–Zhu [42], and Chau–Tam [38]. Basically, they confirm the conjecture under the assumption of $\mathrm{vol}(B(r)) \geq cr^{2n}$ as $r >> 1$. The volume growth condition is used to rule out the cigar model.

4.4 Manifolds with positive isotropic curvature

For the Riemannian case, M. Micallef and J.D. Moore used the same idea to study manifolds with the so-called positive isotropic curvature ([137]), which is an analogue of positive bisectional curvature. We say a manifold (M, g) with dimension $n \geq 4$ has positive isotropic curvature if

$$\begin{aligned}&R(e_1, e_3, e_1, e_3) + R(e_1, e_4, e_1, e_4) + R(e_2, e_3, e_2, e_3) + R(e_2, e_4, e_2, e_4)\\&\quad -2R(e_1, e_2, e_3, e_4) > 0\end{aligned}$$

for all points $p \in M$ and all orthonormal four-frames $\{e_1, e_2, e_3, e_4\} \subset T_pM$.

Under the assumption of positive isotropic curvature, Micallef and Moore [137] apply Morse theory on the α-energy of Sacks and Uhlenbeck [162] and calculate the Morse index of harmonic maps $f : \mathbf{S}^2 \to M$ for M with $\pi_1(M) = 0$. They show that $\pi_i(M) = 0$ for $i < \dim(M)$. This implies that M is a homotopy sphere.

Hence, it follows that M is homeomorphic to $\mathbf{S}^n$ from Poincaré conjecture solved by Freedman [74] and Smale [172].

I hope that this argument can be generalized to reproduce Mori's result.

Question 4.8. *Let M be a compact Kähler manifold with positive Ricci curvature. Then there exists a rational curve. It is true by Mori's argument. We'd like to prove it by more geometric argument.*

Recently, Chen, Tang and Zhu [43] prove that in dimension 4, any compact manifold M with positive isotropic curvature has a finite covering that is diffeomorphic to connected sums of $\mathbf{S}^4$ and $\mathbf{S}^3 \times \mathbf{S}^1$. Is this true in all dimensions?

4.5 Manifolds with non-negative curvature

4.5.1 Non-compact manifold with non-negative sectional curvature

In the non-compact case, Cheeger–Gromoll–Meyer [40, 88] proved that a complete non-compact manifold with positive sectional curvature must be diffeomorphic to $\mathbb{R}^n$. The case for $n = 2$ was due to Cohn-Vossen, and Cheeger–Gromoll–Meyer generalized his argument to higher dimensions.

In the case of non-negative sectional curvature, Cheeger–Gromoll [41] proved the Soul theorem. It says that the manifold is homeomorphic to a vector bundle over a compact totally geodesic submanifold with non-negative sectional curvature. The earlier work of Gromoll–Meyer [88] only asserts they are homopotic equivalent. Perelman [153] was able to prove the manifold is diffeomorphic to $\mathbb{R}^n$ if the sectional curvature (in all direction) of the soul is positive at some point.

An interesting question is the converse statement.

Question 4.9. *Given a compact, simply-connected manifold M with non-negative sectional curvature and a vector bundle $V \to M$, does there exist a metric of non-negative sectional curvature on the total space?*

There are counter-examples to this statement if M is not simply-connected by the work of Belegradek–Kapovitch [10]. Suppose that C is a closed manifold and T is a torus. They showed that if a vector bundle V over $C \times T$ supports a metric of non-negative sectional curvature on the total space then V is "virtually" a pull-back bundle from C. They also showed that if M is flat and the total space of a vector bundle over M has non-negative sectional curvature, then V is a flat bundle. Thus the existence of metrics of non-negative sectional curvature on the total space of a vector bundle does have some obstructions [10, 11] in terms of the bundle structure when the base manifold is not simply-connected.

Recall a theorem of Novikov [148] asserts the following: Suppose V is a vector bundle over $\mathbf{S}^m$ and the total space of V is homeomorphic to $\mathbf{S}^m \times \mathbb{R}^n$, with $n \geq m + 2$ and $m \geq 5$, then V is a trivial bundle. This is the key ingredient for proving the famous theorem of Novikov on the homeomorphic invariance of the Pontryagin classes. Note that the definition of the Pontryagin classes only guarantees that they are diffeomorphic invariant. A good reference is Farrell's survey article [73]. The following question is interesting.

Question 4.10. *Can one prove Novikov's theorem by geometric method such as geometric flows?*

Sullivan–Teleman [183] claimed to have a geometric proof which still seemed to rely on the original argument of Novikov.

4.5.2 Analogue in complex geometry

We can consider similar questions in complex geometry.

Question 4.11. *Suppose that M is a compact Kähler manifold with non-negative bisectional curvature, and V is a holomorphic vector bundle over M. Suppose that there is a complete Kähler metric on the total space with non-negative bisectional curvature, what can be said about V? For example, is V stable in the sense of algebraic geometry (see [65, 66, 192] for stable bundles), or can one show that the enough twisting of V by an ample line bundle L, i.e., $V \otimes L^k$ for $k >> 1$ has a Hermitian metric with positive curvature?*

We can also ask a complex version of the Soul theorem.

Question 4.12. *Suppose N is a complete Kähler manifold with non-negative bisectional curvature. Is N biholomorphic to a holomorphic vector bundle V over a compact totally geodesic complex submanifold $M \subset N$?*

The case of strictly positive bisectional curvature is a conjecture of mine which has been resolved by Chau–Tam [38], Chen–Zhu [43, 45] under the assumption of Euclidean volume growth (this essentially excludes the bundle case, and leads to $\mathbb{C}^m$). Their method is again Ricci flow. A natural question is:

Question 4.13. *Can we prove the real or complex version of the Soul theorem by the Ricci flow? Note that Cheeger–Gromoll proved the Soul theorem by analyzing the Busemann function, and their argument envoked the Toponogov theorem. Maybe the Ricci flow can give a more direct proof.*

Since positive sectional curvature is not preserved by the Ricci flow, we may need to impose other curvature condition such as $M \times \mathbb{R}^k$ satisfies some curvature condition as considered by Brendle–Schoen [29]. In fact, even assuming non-negative curvature operator in the non-compact case is already interesting. For the non-compact complex case, we can ask if positive bisectional curvature can be replaced by other algebraic geometric conditions? For example, we can consider a quasi-projective variety and consider positive curvature conditions on the log-tangent bundle. In this setting, we need to look for affine lines (as opposed to rational curves in the compact case), i.e., maps from $\mathbb{C} \to M$. Can we still use analytic method such as the second variation formula on a non-compact minimal surface? Micallef's thesis [135] considered the case when M is of dimension 4. The converse statement is the following:

Question 4.14. *When does the total space of a holomorphic vector bundle over M admit a complete Kähler/Hermitian metric of positive bisectional curvature? This is not studied even in the case when the base manifold is a Hermitian symmetric space. Also, we might need certain stability condition on the vector bundle.*

The above discussions is based on the following philosophy: The geometry of the total space of a vector bundle should be closely related to the vector bundle structure.

For example, if the total space of V admits an affine linear structure, does it imply the structure group of V (as a vector bundle) can be reduced to the affine linear group? What about flat structure or projective flat structure, maybe assuming that the base manifold is also affine? In general, we would like to allow singularity on the base manifold such as orbifold singularity.

There is the plumbing construction that is adapted to vector bundle structures which is explained in the last chapter of Browder's book [30]. This will allow us to glue the total space of vector bundles along the base manifold and produce new simply-connected manifolds. It is possible that the plumbing construction preserve positives Ricci curvature as this is more related to differentiable structure (as opposed to connected sum construction). Positive isotropic curvature is preserved under connected sum by the work of Micaleff–Wang [138].

In [149], it was shown that if there exists a non-negative sectional curvature metric on the total space of a vector bundle E over M such that the induced metric on the zero section is flat, then E admits a flat connection.

A classification problem in algebraic geometry states the following. Suppose V_1 and V_2 are algebraic vector bundles over an algebraic manifold M, and the projectified bundles $P(V_1)$ and $P(V_2)$ are fiberwise biholomorphic. When does there exist a vector bundle isomorphism between V_1 and V_2? This is determined by the kernel of the map $H^1(M, \mathrm{GL}(n, \mathbb{C})) \to H^1(M, \mathrm{PGL}(n, \mathbb{C}))$. The quotient is called the Brauer group.

Another case is to consider a compactification of the vector bundle such as $V \hookrightarrow \bar{V} = P(V \oplus O)$. V is a quasi-projective variety and $D = \bar{V} \backslash V$ is a divisor in $\bar{V}$. Suppose the total space $(\bar{V}_1, D_1)$ is birationally equivalent to $(\bar{V}_2, D_2)$ such that the birational map is nonsingular or bihomorphic in the interiors $(\bar{V}_1, D_1) \to (\bar{V}_2, D_2)$. We ask if V_1 and V_2 are isomorphic as bundles. The assumptions seem very weak, but there are cancelation theorems in algebraic geometry which asserts that under suitable conditions biholomorphism of $M \times \mathbb{C}$ and $N \times \mathbb{C}$ forces M and N to be biholomorphic as well. It is interesting to find out the suitable assumption for the statement to be true. The difficulty is caused by the huge biholomophism group of an open variety such as $\mathbb{C}^n$. Similarly, the affine linear group causes big problem in solving the complex Monge-Ampère equation: $\det(\frac{\partial^2 u}{\partial z_i \partial \bar{z}_j}) = 1$.

4.6 Manifolds with positive scalar curvature

Before the work of Micallef–Wang, Schoen and Yau [163] proved that positive scalar curvature is preserved under connected sum construction, i.e., if both M and N have positive scalar curvature, so has $M \sharp N$. In general if M has positive scalar curvature and $\mathbf{S}^k \to M$ is of codimension ≥ 3, then we can perform surgery. In the simplest case when the unit normal bundle over $\mathbf{S}^k$ is trivial, we simply replace $\mathbf{S}^k \times \mathbf{D}^{n-k}$ by $\mathbf{D}^{k+1} \times \mathbf{S}^{n-k-1}$ by gluing along their common boundaries $\mathbf{S}^k \times \mathbf{S}^{n-k-1}$. This can be used to kill nontrivial elements in the homotopy group.

Thus if M has positive scalar curvature, so has $(M\backslash(\mathbf{S}^k \times \mathbf{D}^{n-k})) \cup (\mathbf{D}^{k+1} \times \mathbf{S}^{n-k-1})$. This procedure is quite general for surgery with respect to a compact embedded submanifold of codimension greater than or equal to 3 and with a nontrivial normal bundle. Gromov-Lawson ([92, 93]) have a different argument, but there remains some difficulties. They can perform codimension ≥ 3 surgery in the category of spin manifolds of positive scalar curvature.

For M being simply-connected and $\dim M \geq 5$, if M is spin cobordant to N, then one of them has positive scalar curvature implies that the other one also has. The positive scalar curvature condition would imply that there is no harmonic spinors by the Lichnerowicz [128] formula. In dimension $4k$, this leads to the vanishing of the index of Dirac operator, which is the same as the $\hat{A}$ genus by the Atiyah–Singer index theorem. In dimension $8k+1$ and $8k+2$, Hitchin [106] found a mod 2 index of the Dirac operator, which is also a topological invariant (called the α invariant), which generalize $\hat{A}$. With the same vanishing argument of Lichnerowicz, the α invariant is zero for positive scalar curvature. As a remark, by relating the α invariant with the work of Kervaire–Milnor, Hitchin [106] proved that: for every dimension $8k+1$ and $8k+2$ with $k \geq 1$, there exists exotic spheres without any metric with positive scalar curvature. For simply-connected spin manifolds, using the calculation of the spin cobordism group by Anderson–Brown–Peterson [2], Stolz [179] found that for $\dim M \geq 5$, admitting positive scalar curvature is equivalent to the vanishing of the α invariant (i.e., $M \equiv 0$ in the spin cobordism group). This is also well understood if π_1 is finite. It is far more complicated for manifolds with infinite fundamental group.

Recall that for manifolds of positive isotropic curvature, it is conjectured that they should be diffeomorphic to connected sums of quotients of $\mathbf{S}^n$, $\mathbf{S}^{n-1} \times \mathbf{S}^1, \ldots$ and so on. This conjecture in dimension 4 has been confirmed by Chen–Tang–Zhu [43].

It is a good exercise to check if conditions such as $s > c|W|$ is preserved under connected sum construction for some constant $c > 0$, where s is the scalar curvature and W is the Weyl curvature.

Suppose $M = \partial N$ and N is an Einstein manifold that is asymptotically hyperbolic in the sense that near the conformal boundary M at infinity, the metric behaves like $\frac{\mathrm{d}y^2+\mathrm{d}s^2}{y^2}$ where $\mathrm{d}s^2$ is the metric on M. In this case, the conformal structure of $\mathrm{d}s^2$ is uniquely determined (independent of the defining equation). Conversely, we would like to know if a positive scalar curved manifold M bounds such an N. This is is related to the AdS/CFT correspondence. For example, suppose we are given two such manifolds, can we perform surgery to glue them together? Along the boundaries with positive scalar curvature, we can use Schoen–Yau's [163] construction, and then use singular perturbation to perturb the interior metric to be Einstein. This will be an important direction.

In general, we would like to use our knowledge in topology and analysis to help constructing metrics. For example, we ask which exotic sphere admit positive sectional curvature. We already know examples in 7 dimension, if the exotic sphere is an $\mathbf{S}^3$ bundle over $\mathbf{S}^4$, then there is non-negative sectional curvature. Suppose an exotic sphere $\mathbf{S}^n$ admits a fibration $M \to \mathbf{S}^n \to N$, is it still true?

I conjecture that an exotic sphere $\mathbf{S}^n$ admits a metric of positive Ricci curvature if and only if $\mathbf{S}^n = \partial M$ for an M with trivial tangent bundle. There are known constructions of the "if" direction. Also, an exotic sphere admits positive scalar curvature if and only if $\mathbf{S}^n = \partial M$ for a spin manifold M.

4.7 Riemannian manifolds with positive or negative tangent bundle

If a manifold is Kähler and its tangent bundle $T > 0$, then M is holomorphic to $\mathbb{CP}^n$. If $T < 0$, then what can be said about M? Recall that Bloch-Gieseker proved [20] that for a rank k vector bundle $V^k \to M$ over M, if V is positive, then the top Chern class $c_k(V)$ is positive. This is rather interesting from the differential geometry point of view. Go back to tangent bundles. If the tangent bundle of a manifold is positive, then the Euler number $\chi(M)$ of the manifold is positive. It is a well-known conjecture in Riemannian Geometry that the sectional curvature is positive implies $\chi(M)$ is positive. This conjecture was based on the generalized Gauss-Bonnet Theorem. It is true in four dimensions. But the Gauss–Bonnet integrand can be non-positive. Nevertheless, there is no counterexample about the Euler characteristic so far. If the sectional curvature is negative, is $(-1)^{n+1}\chi(M) > 0$ hold? These are classical questions probably announced by Chern. It is nice to see if it is true at the Kähler case for the negative Chern class. The first one is not exciting since we know a Kähler manifold with positive Ricci curvature is biholomorphic to $\mathbb{CP}^n$. For the negative curvature case, it is nice to give an independent argument with Mori and Bloch-Gieseker, who used algebraic geometry over characteristic p. It would also be nice to understand the same question for complete noncompact manifolds. There we don't know any algebraic geometry arguments.

Manifolds with negative tangent bundle is a very important class of manifolds. They are far more complicated than manifolds with negative sectional curvature. Although there are some analogues, they are topological dramatically different. For example, manifolds with negative sectional curvature are not simply connected. But manifolds with negative tangent bundle can be simply connected.

4.8 Other related topics

It is interesting to study finite group (free) actions on $\mathbf{S}^n$. The so-called Vincent problem [195], solved by Wolf [201], who classified all finite linear actions (i.e., $\Gamma \subset \mathrm{SO}(n+1)$ on $\mathbf{S}^n$). This is the case of positive sectional curvature. It will be interesting to see if there is any exotic action in the case of positive isotropic curvature.

In general, suppose that $\mathbb{Z}/n$ acts on $\mathbf{S}^3$ by diffeomorphism, we ask if the action is conjugate to a linear action. When the action is fixed point free, this is implied by the Poincaré conjecture. When the fixed points are present, there is the Smith conjecture which asserts that the fixed point set must be unknotted curves. The Smith conjecture was resolved [176] by assembling the works of Thurston, Meeks–Yau, and Gordon.

The Smith conjecture is related to the orbifold version of the geometrization conjecture. In general, we can replace a manifold by an orbifold and ask the same question regarding positive sectional curvature. For example, the orbifold version of Poincaré conjecture is not resolved yet. Another question is that whether there exists some interesting examples of orbifolds with positive sectional curvature. Especially, for dim > 24, do the compact rank one symmetric spaces admit exotic actions? If so, are there infinite many such examples?

References

[1] A. Aloff and N.R. Wallach, *An infinite family of distinct 7-manifolds admitting postively curved Riemannian structures,* Bull. Amer. Math. Soc. **81** (1975), 93–97.

[2] D.W. Anderson, E.H. Jr. Brown and F.P. Peterson, *The structure of the Spin cobordism ring,* Ann. of Math. (2) **86** (1967), 271–298.

[3] A. Andreotti and T. Frankel, *The Lefschetz theorem on hyperplane sections,* Ann. of Math. (2) **69** (1959), 717–717.

[4] M.F. Atiyah, V.K. Patodi and I.M. Singer, *Spectral asymmetry and Riemannian geometry I,* Math. Proc. Cambridge Philos. Soc. **77** (1975), 43–69. *II.* ibid. **78** (1975), no. 3, 405–432. *III.* ibid. **79** (1976), no. 1, 71–99.

[5] M.F. Atiyah and I.M. Singer, *The index of elliptic operators. I and II,* Ann. of Math. (2) **87** (1968), 484–530. ibid. **87** (1968). 546–604.

[6] T. Aubin, *Équations du type Monge-Ampère sur les variétés Kähleriennes compactes,* C. R. Acad. Sci. Paris Sér. A-B **283** (1976), no. 3, Aiii, A119–A121.

[7] L. Auslander and L. Markus, *Holonomy of flat affinely connected manifolds,* Ann. of Math. (2) **62** (1955), 139–151.

[8] S. Bando, *On the classification of three-dimensional compact Kähler manifolds of nonnegative bisectional curvature,* J. Differential Geom. **19** (1984), no. 2, 283–297.

[9] Ya.V. Bazaikin, *On a family of* 13*-dimensional closed Riemannian manifolds of positive curvature,* Siberian Math. J. **37** (1996), no. 6, 1068–1085.

[10] I. Belegradek and V. Kapovitch, *Topological obstructions to nonnegative curvature,* Math. Ann. **320** (2001), no. 1, 167–190.

[11] I. Belegradek and V. Kapovitch, *Obstructions to nonnegative curvature and rational homotopy theory,* J. Amer. Math. Soc. **16** (2003), no. 2, 259–284.

[12] Y. Benoist, *Une nilvariete non affine,* J. Differential Geom. **41** (1995), no. 1, 21–52.

[13] Y. Benoist, *Convexes divisibles. I,* "Algebraic groups and arithmetic." 339–374, Tata Inst. Fund. Res., Mumbai, 2004.

[14] L. Berard-Bergery, *Les variétés riemanniennes homogènes simplement connexes de dimension impaire à courbure strictement positive,* J. Math. Pures Appl. (9) **55** (1976), no. 1, 47–67.

[15] M. Berger, *Sur quelques variétés d'Einstein compactes,* Annali di Math. Pura e Appl. **53** (1961), 89–96.

[16] M. Berger, *Les variétés riemanniennes homogènes normales simplement connexes à courbure strictement positive*, Ann. Scuola Norm. Sup. Pisa (3) **15** (1961), 179–246.

[17] M. Berger and R. Bott, *Sur les variétés à courbure strictement positive,* Topology **1** (1962), 301–311.

[18] A.L. Besse, *Einstein manifolds,* Reprint of the 1987 edition. Classics in Mathematics. Springer-Verlag, Berlin, 2008.

[19] R.L. Bishop and S.I. Goldberg, *On the second cohomology group of a Kaehler manifold of positive curvature*, Proc. Amer. Math. Soc. **16** (1965), 119–122.

[20] S. Bloch and D. Gieseker, *The positivity of the Chern classes of an ample vector bundle,* Invent. Math. **12** (1971), 112–117.

[21] S. Bochner, *Curvature and Betti numbers. I.* and *II*, Ann. of Math. (2) **49** (1948), 379–390. ibid. **50** (1949), 77–93.

[22] F.A. Bogomolov, *Classification of surfaces of class* VII_0 *with* $b_2 = 0$, Math. USSR-Izv. **10** (1976), 255–269.

[23] F.A. Bogomolov, *Surfaces of class* VII_0 *and affine geometry*, Math. USSR-Izv. **21** (1983), no. 1, 31–73.

[24] C. Böhm, *Inhomogeneous Einstein metrics on low-dimensional spheres and other low-dimensional spaces,* Invent. Math. **134** (1998), no. 1, 145–176.

[25] C. Böhm and B. Wilking, *Manifolds with positive curvature operators are space forms,* Ann. of Math. (2) **167** (2008), no. 3, 1079–1097.

[26] R. Bott, *On a theorem of Lefschetz*, Michigan Math. J. **6** (1959), 211–216.

[27] J.-P. Bourguignon and H. Karcher, *Curvature operators: pinching estimates and geometric examples*, Ann. Sci. École Norm. Sup. (4) **11** (1978), no. 1, 71–92.

[28] S. Brendle and R.M. Schoen, *Classification of manifolds with weakly* 1/4-*pinched curvatures*, Acta Math. **200** (2008), no. 1, 1–13.

[29] S. Brendle and R.M. Schoen, *Manifolds with* 1/4-*pinched curvature are space forms,* J. Amer. Math. Soc. **22** (2009), no. 1, 287–307.

[30] W. Browder, *Surgery on simply-connected manifolds,* Ergebnisse der Mathematik und ihrer Grenzgebiete, Band 65. Springer-Verlag, New York-Heidelberg, 1972. ix+132 pp.

[31] L. Caffarelli, L. Nirenberg and J. Spruck, *The Dirichlet problem for nonlinear second-order elliptic equations. I. Monge-Ampère equation,* Comm. Pure Appl. Math. **37**(1984), 369–402.

[32] E. Calabi, *Improper affine hypersoheres of convex type and a generalization of a theorem by K. Jörgens,* Michigan Math. J. **5** (1958), 105–126.

[33] E. Calabi, *Complete affine hyperspheres. I,* Symposia Mathematica, Vol. X (Convegno di Geometria Differenziale, INDAM, Rome, 1971), 19–38. Academic Press, London, 1972.

[34] E. Calabi, *Métriques kählériennes et fibrés holomorphes,* Ann. Sci. Ecole Norm. Sup. (4) **12** (1979), no. 2, 269–294.

[35] E. Calabi, *Extremal Kähler metrics*, "Seminar on Differential Geometry", 259–290, Ann. of Math. Stud., **102**, Princeton Univ. Press, Princeton, N.J., 1982.

[36] E. Calabi, *Extremal Kähler metrics. II*, "Differential geometry and complex analysis", 95–114, Springer, Berlin, 1985.

[37] F. Catanese and C. LeBrun, *On the scalar curvature of Einstein manifolds,* Math. Res. Lett. **4** (1997), no. 6, 843–854.

[38] A. Chau and L.-F. Tam, *Non-negatively curved Kähler manifolds with average quadratic curvature decay*, Comm. Anal. Geom. **15** (2007), no. 1, 121–146.

[39] J. Cheeger, *Some examples of manifolds of nonnegative curvature*, J. Differential Geometry **8** (1973), 623–628.

[40] J. Cheeger and D. Gromoll, *The structure of complete manifolds of nonnegative curvature*, Bull. Amer. Math. Soc. **74** (1968), 1147–1150.

[41] J. Cheeger and D. Gromoll, *On the structure of complete manifolds of nonnegative curvature,* Ann. of Math. (2) **96** (1972), 413–443.

[42] B.-L. Chen, S.-H. Tang and X.-P. Zhu, *A uniformization theorem for complete non-compact Kähler surfaces with positive bisectional curvature*, J. Differential Geom. **67** (2004), no. 3, 519–570.

[43] B.-L. Chen, S.-H. Tang and X.-P. Zhu, *Complete classification of compact four-manifolds with positive isotropic curvature*, arXiv:0810.1999, 2008.

[44] B.-L. Chen and X.-P. Zhu, *On complete noncompact Kähler manifolds with positive bisectional curvature*, Math. Ann. **327** (2003), no. 1, 1–23.

[45] B.-L. Chen and X.-P. Zhu, *Volume growth and curvature decay of positively curved Kähler manifolds*, Q. J. Pure Appl. Math. **1** (2005), no. 1, 68–108.

[46] X. Chen, C. LeBrun and B. Weber, *On conformally Kähler, Einstein manifolds*, J. Amer. Math. Soc. **21** (2008), no. 4, 1137–1168.

[47] H.-B. Cheng, L.-T. Cheng and S.-T. Yau, *Minimization with the affine normal direction,* Commun. Math. Sci. **3** (2005), no. 4, 561-574.

[48] S.-Y. Cheng and S.-T. Yau, *Differential equations on Riemannian manifolds and their geometric applications*, Comm. Pure Appl. Math. **28** (1975), no. 3, 333–354.

[49] S.-Y. Cheng and S.-T. Yau, *On the regularity of the Monge-Ampère equation* $\det(\partial^2 u/\partial x_i \partial x_j) = F(x,u)$, Comm. Pure Appl. Math. **30** (1977), 41–68.

[50] S.-Y. Cheng and S.-T. Yau, *On the existence of a complete Kähler metric on noncompact complex manifolds and the regularity of Fefferman's equation,* Comm. Pure Appl. Math. **33** (1980), no. 4, 507–544.

[51] S.-Y. Cheng and S.-T. Yau, *The real Monge-Ampère equation and affine flat structures*, "Proceedings of the 1980 Beijing Symposium on Differential Geometry and Differential Equations, Vol. 1." 339–370, Science Press, Beijing, 1982.

[52] S.-Y. Cheng and S.-T. Yau, *Complete affine hypersurfaces. I. The completeness of affine metrics,* Comm. Pure Appl. Math. **39** (1986), no. 6, 839–866.

[53] J. Cheeger and D. Gromoll, *On the structure of complete manifolds of nonnegative curvature.* Ann. of Math. (2) **96** (1972), 413–443.

[54] *Proceedings of the United States-Japan Seminar in Differential Geometry, Kyoto, Japan, 1965,* Nippon Hyoronsha Co., Ltd., Tokyo, 1966 ix+177 pp.

[55] S.S. Chern, *The mathematical works of Wilhelm Blaschke,* Abh. Math. Sem.

Hamburg **39** (1973), 1–9.

[56] S.S. Chern, *Affine minimal hypersurfaces,* Proc. US-Japan Seminar on minimal submanifolds and Geodesics, Kaigai, Tokyo (1978), 17–30.

[57] S. Choi, *Real projective surfaces,* Thesis (Ph. D.), Princeton University, 1988.

[58] S. Choi, *Convex decompositions of real projective surfaces. I. π-annuli and convexity,* J. Diff. Geom. **40** (1994), no. 1, 165–208.

[59] S. Choi, *Convex decompositions of real projective surfaces. II. Admissible decompositions,* J. Differential Geom. **40** (1994), no. 2, 239–283.

[60] K. Corlette, *Flat G-bundles with canonical metrics,* J. Diff. Geom. **28** (1988), no. 3, 361–382.

[61] P.C. Craighero and R. Gattazzo, *Quintic surfaces of* $\mathbf{P}^3$ *having a nonsingular model with* $q = p_g = 0$, $P_2 \neq 0$. Rend. Sem. Mat. Univ. Padova **91** (1994), 187–198.

[62] A. Derdziński, *Self-dual Käler manifolds and Einstein manifolds of dimension four*, Compositio Math. **49** (1983), no. 3, 405–433.

[63] M.P. do Carmo, *Riemannian geometry*, Birkhäuser, Boston, 1992.

[64] S.K. Donaldson, *Twisted harmonic maps and the self-duality equations,* Proc. London Math. Soc. (3) **55** (1987), no. 1, 127–131.

[65] S.K. Donaldson, *Anti self-dual Yang–Mills connections over complex algebraic surfaces and stable vector bundles*, Proc. London Math. Soc. (3) **50** (1985), no. 1, 1–26.

[66] S.K. Donaldson, *Infinite determinants, stable bundles and curvature*, Duke Math. J. **54** (1987), no. 1, 231–247.

[67] T. Eguchi and A.J. Hanson, *Asymptotically flat self-dual solutions to Euclidean quantum gravity,* Phys. Lett. **74** B (1978), 249–251.

[68] T. Eguchi and A.J. Hanson, *Self-dual solutions to Euclidean gravity,* Ann. Phys. **120** (1979), 82–106.

[69] D.A. Eisenman, *Intrinsic measures on complex manifolds and holomorphic mappings,* Memoirs of the American Mathematical Society. **96** (1970) i+80 pp.

[70] J.-H. Eschenburg, *New examples ofmanifolds with strictly positive curvature*, Invent. Math. **6** (1982), no. 3, 469–480.

[71] J.-H. Eschenburg, *Inhomogeneous spaces of positive curvature,* Diff. Geom. Appl. **2** (1992), no. 2, 123–132.

[72] J.-H. Eschenburg and M.Y. Wang, *The initial value problem for cohomogeneity one Einstein metrics,* J. Geom. Anal. **10** (2000), no. 1, 109–137.

[73] F.T. Farrell, *The Borel conjecture,* "Topology of high-dimensional manifolds, No. 1, 2." (Trieste, 2001), 225–298, ICTP Lect. Notes, 9, Abdus Salam Int. Cent. Theoret. Phys., Trieste, 2002.

[74] M.H. Freedman, *The topology of four-dimensional manifolds,* J. Diff. Geom. **17** (1982), 357–453.

[75] D. Fried, W. Goldman and M.W. Hirsch, *Affine manifolds with nilpotent holonomy,* Comment. Math. Helvetici **56**(1981) 487–523.

[76] R. Friedman and J.W. Morgan, *Complex versus differentiable classification of algebraic surfaces*, "Proceedings of the 1987 Georgia Topology Confer-

ence (Athens, GA, 1987)". Topology Appl. **32** (1989), no. 2, 135–139.

[77] R. Friedman and J.W. Morgan, *Smooth four-manifolds and complex surfaces,* Ergebnisse der Mathematik und ihrer Grenzgebiete (3), **27**, 1994. x+520 pp.

[78] P. Furness and E. Fedida, *Sur les structures pseudo-riemanniennes plates des variétés compactes*, C. R. Acad. Sci. Paris Sér. A-B **286** (1978), no. 3, A169–A171.

[79] A. Futaki, *An obstruction to the existence of Einstein Käler metrics*, Invent. Math. **73** (1983), no. 3, 437–443.

[80] A. Futaki and T. Mabuchi, *Uniqueness and periodicity of extremal Käler vector fields*, "Proceedings of GARC Workshop on Geometry and Topology '93", 217–239, Lecture Notes Ser., **18**, Seoul Nat. Univ., Seoul, 1993.

[81] A. Futaki and T. Mabuchi, *Bilinear forms and extremal Käler vector fields associated with Käler classes*, Math. Ann. **301** (1995), no. 2, 199–210.

[82] R. Geroch, *Positive sectional curvature does not imply positive Gauss-Bonnet integrand,* Proc. Amer. Math. Soc. **54** (1976), 267–270.

[83] G.W. Gibbons and S.W. Hawking, *Action integrals and partition functions in quantum gravity,* Phys. Rev. D **15** (1977), 2752–2756.

[84] G.W. Gibbons and S.W. Hawking, *Gravitional multi-Instantons,* Phys. Lett. **78** B (1978), 430–432.

[85] G.W. Gibbons, S.W. Hawking and M.J. Perry, *Path integrals and the indefiniteness of the gravitation action,* Nucl. Phys. B **138** (1978), 141–150.

[86] W. Goldman, *Convex real projective structures on compact surfaces,* J. Diff. Geom. **31** (1990), 791–845.

[87] A. Gray, *Invariants of curvature operators of four-dimensional Riemannian manifolds,* "Proceedings of the Thirteenth Biennial Seminar of the Canadian Mathematical Congress", Vol. 2, pp. 42–65, 1972.

[88] D. Gromoll and W. Meyer, *On complete open manifolds of positive curvature,* Ann. of Math. (2) **90** (1969), 75–90.

[89] D. Gromoll and W. Meyer, *An exotic sphere with nonnegative sectional curvature,* Ann. of Math. (2) **100** (1974), 401–406.

[90] M. Gromov, *Volume and bounded cohomology,* Inst. Hautes Études Sci. Publ. Math. No. 56 (1982), 5–99 (1983).

[91] M. Gromov, *Pseudoholomorphic curves in symplectic manifolds*, Invent. Math. **82** (1985), no. 2, 307–347.

[92] M. Gromov and H.B. Jr. Lawson, *Spin and scalar curvature in the presence of a fundamental group. I.* Ann. of Math. (2) **111** (1980), no. 2, 209–230.

[93] M. Gromov and H.B. Jr. Lawson, *The classification of simply connected manifolds of positive scalar curvature*, Ann. of Math. (2) **111** (1980), no. 3, 423–434.

[94] M. Gromov and W. Thurston, *Pinching constants for hyperbolic manifolds*, Invent. Math. **89** (1987), no. 1, 1–12.

[95] K. Grove and S. Halperin, *Contributions of rational homotopy theory to global problems in geometry,* Inst. Hautes Études Sci. Publ. Math. No. **56** (1982), 171–177 (1983).

[96] K. Grove, B. Wilking and W. Ziller, *Positively curved cohomogeneity one*

manifolds and 3-Sasakian geometry, J. Diff. Geom. **78** (2008), no.frm[o]–, 33–111.

[97] R.C. Gunning, *Affine and projective structures on Riemann surfaces,* "Riemann surfaces and related topics: Proceedings of the 1978 Stony Brook Conference.", 225–244, Ann. of Math. Stud., 97, Princeton Univ. Press, Princeton, N.J., 1981.

[98] S. Halperin, *Spaces whose rational homology and de Rham homotopy are both finite-dimensional,* Algebraic homotopy and local algebra (Luminy, 1982), 198–205, Astérisque, 113–114, Soc. Math. France, Paris, 1984.

[99] R.S. Hamilton, *Four-manifolds with positive curvature operator,* J. Diff. Geom. **24** (1986), no. 2, 153–179.

[100] Z.-C. Han, L.-F. Tam, A. Treibergs and T. Wan, *Harmonic maps from the complex plane into surfaces with nonpositive curvature,* Comm. Anal. Geom. **3** (1995), no. 1–2, 85–114.

[101] S.W. Hawking, *The path integral approach to quantum gravity, General relativity,* An Einstein century survey, 746–789, edited by S.W. Hawking and W. Israel, Cambridge Unversity Press, Cambridge-New York, 1979.

[102] S. Helgason, *Sophus Lie, the mathematician,* The Sophus Lie Memorial Conference (Oslo, 1992), 3-21, Scand. Univ. Press, Oslo, 1994.

[103] M.W. Hirsch and W.P. Thurston, *Foliated bundles, invariant measures and flat manifolds,* Ann. Math. (2) **101** (1975), 369–390.

[104] N. J. Hitchin, *Compact 4-dimensional Einstein manifolds,* J. Diff. Geom. **9** (1974), 435–441.

[105] N. J. Hitchin, *Polygons and gravitons,* Math. Proc. Cambridge Philos. Soc. **85** (1979), no. 3, 465–476.

[106] N. J. Hitchin, *Harmonic spinors,* Advances in Math. **14** (1974), 1–55.

[107] E. Horikawa, *Algebraic surfaces of general type with small C_1^2. I,* Ann. of Math. (2) **104** (1976), no. 2, 357–387.

[108] F. Hirzebruch and K. Kodaira, *On the complex projective spaces,* J. Math. Pures Appl. (9) **36** (1957), 201–216.

[109] M. Imada, *Sasaki-Einstein metrics on $S^2 \times S^3$*, Tokyo J. of Math. **35** (2012), no. 2, 367–373.

[110] M. Inoue, *On surfaces of Class* VIII_0, Invent. Math. **24** (1974), 269–310.

[111] M. Inoue, *New surfaces with no meromorphic functions*, "Proceedings of the International Congress of Mathematicians." Vol. 1, 423–426. Canad. Math. Congress, Montreal, Que., 1975.

[112] M. Inoue, *New surfaces with no meromorphic functions. II*, "Complex analysis and algebraic geometry," 91–106. Iwanami Shoten, Tokyo, 1977.

[113] M. Inoue, S. Kobayashi and T. Ochiai, *Holomorphic affine connections on compact complex surfaces*, J. Fac. Sci. Univ. Tokyo Sect. IA Math. **27** (1980), no. 2, 247–264.

[114] G.R. Jensen, *Einstein metrics on principal fibre bundles,* J. Diff. Geom. **8** (1973), 599–614.

[115] M. Kapovich, *On monodromy of complex projective structures,* Invent. Math. **119** (1995), no. 2, 243–265.

[116] M. Kapovich, *Convex projective structures on Gromov–Thurston manifolds,*

Geom. Topol. **11** (2007), 1777–1830.

[117] S. Kobayashi, *Hyperbolic manifolds and holomorphic mappings,* Pure and Applied Mathematics **2** (1970), ix+148 pp.

[118] K. Kodaira, *On homotopy $K3$ surfaces,* "1970 Essays on Topology and Related Topics (Mémoires dédiés à Georges de Rham)", 58–69.

[119] B. Kostant and D. Sullivan, *The Euler characteristic of an affine space form is zero,* Bull. Amer. Math. Soc. **81** (1975), no. 5, 937–938.

[120] D. Kotschick, *Einstein metrics and smooth structures,* Geom. Topol. **2** (1998), 1–10.

[121] M. Kontsevich and Y. Soibelman, *Homological mirror symmetry and torus fibrations,* "Symplectic geometry and mirror symmetry (Seoul, 2000)", 203–263.

[122] M. Kontsevich and Y. Soibelman, *Affine structures and non-Archimedean analytic spaces,* "The unity of mathematics", 2006, 321–385.

[123] P.B. Kronheimer, *The construction of ALE spaces as hyper-Kähler quotients,* J. Diff. Geom. **29** (1989), no. 3, 665–683.

[124] H.B. Lawson, Jr. and S.-T. Yau, *Scalar curvature, non-abelian group actions, and the degree of symmetry of exotic spheres,* Comment. Math. Helv. **49** (1974), 232–244.

[125] C. LeBrun, *Counter-examples to the generalized positive action conjecture,* Comm. Math. Phys. **118** (1988), 591–591.

[126] C. LeBrun, *Complete Ricci-flat Kähler metrics on $\mathbb{C}^n$ need not be flat,* Several complex variables and complex geomtry, Part 2 (Santa Cruz, CA, 1989), 297–304. Proc. Sympos. Pure Math., **52** Part 2, Amer. Math. Soc., Providence, RI, 1991.

[127] C. LeBrun, *On Einstein, Hermitian 4-Manifolds,* arXiv:1010.0238, 2010.

[128] A. Lichnerowicz, *Spineurs harmoniques,* C. R. Acad. Sci. Paris **257** (1963), 7–9.

[129] A.-M. Li, U. Simon and G.-S. Zhao, *Global affine differential geometry of hypersurfaces,* Berlin; New York: De Gruyter, 1993.

[130] J. Li and S.-T. Yau, *Hermitian–Yang–Mills connection on non-Kähler manifolds,* "Mathematical aspects of string theory." 560–573, Adv. Ser. Math. Phys., 1, World Sci. Publishing, Singapore, 1987.

[131] J. Li, S.-T. Yau and F. Zheng, *A simple proof of Bogomolov's theorem on class* VII_0 *surfaces with* $b_2 = 0$, Illinois J. Math. **34** (1990), no. 2, 217–220.

[132] J. Li, S.-T. Yau and F. Zheng, *On projectively flat Hermitian manifolds,* Comm. Anal. Geom. **2** (1994), no. 1, 103–109.

[133] C. Loewner and L. Nirenberg, *Partial differential equations invariant under conformal or projective transformations,* "Contributions to analysis (a collection of papers dedicated to Lipman Bers)." pp.245–272. Academic Press, New York, 1974.

[134] L. Markus, *Cosmological models in differential geometry,* U. Minnesota, 1962.

[135] M.J. Micallef, *Stable minimal surfaces in Euclidean space,* Ph. D theses, New York University, 1982.

[136] M.J. Micallef, *Stable minimal surfaces in Euclidean space,* J. Diff. Geom.

19 (1984), no. 1, 57–84.

[137] M.J. Micallef and J.D. Moore, *Minimal two-spheres and the topology of manifolds with positive curvature on totally isotropic two-planes,* Ann. of Math. (2) **127** (1988), no. 1, 199–227.

[138] M.J. Micallef and M.Y. Wang, *Metrics with nonnegative isotropic curvature,* Duke Math. J. **72** (1993), no. 3, 649–672.

[139] J. Milnor, *On the existence of a connection with curvature zero,* Comm. Math. Helv. **32** (1957), 215–223.

[140] J. Milnor, *On fundamental groups of complete affinely flat manifolds,* Adv. in Math. **25** (1977), 178–187.

[141] N. Mok, *The uniformization theorem for compact Kähler manifolds of nonnegative holomorphic bisectional curvature,* J. Differential Geom. **27** (1988), no. 2, 179–214.

[142] N. Mok and S.-T. Yau, *Completeness of the Kähler-Einstein metric on bounded domains and the characterization of domains of holomorphy by curvature conditions,* The mathematical heritage of Henri Poincaré, Part 1 (Bloomington, Ind., 1980), 41–59, Proc. Sympos. Pure Math., **39**, Amer. Math. Soc., Providence, RI, 1983.

[143] V. Moncrief, *Relativistic Teichmüller theory—a Hamilton–Jacobi approach to* $2+1$*-dimensional Einstein gravity,* "Surveys in differential geometry. Vol. XII. Geometric flows." 203–249. Edited by H.-D. Cao and S.-T. Yau. Surveys in Differential Geometry, 12. International Press, Somerville, MA, 2008. viii+347 pp.

[144] S. Mori, *Projective manifolds with ample tangent bundles,* Ann. of Math. (2) **110** (1979), 593–606.

[145] T. Nagano and K. Yagi, *The affine structures on the real two-torus. I,* Osaka J. Math. **11** (1974), 181–210.

[146] I. Nakamura, *On surfaces of class* VII_0 *with curves*, Invent. Math. **78** (1984), no. 3, 393–443.

[147] Lei Ni, *Ricci flow and nonnegativity of sectional curvature*, Math. Res. Lett., **11**(2004), 883–904.

[148] S.P. Novikov, *Topological invariance of rational classes of Pontrjagin*, Dokl. Akad. Nauk SSSR **163** (1965), 298–300.

[149] M. Özaydin and G. Walschap, *Vector bundles with no soul,* Proc. Amer. Math. Soc. **120** (1994), no. 2, 565–567.

[150] H. Ooguri and C. Vafa, *Summing up Dirichlet instantons,* Phys. Rev. Lett. **77** (1996), no. 16, 3296–3298.

[151] D. Page, *A compact rotating gravitational instanton,* Phys. Lett. B **79** (1979), 235–238.

[152] D. Page and C.N. Pope, *Einstein metrics on quaternionic line bundles,* Class. Quantum Grav. 3 (1986), 249–259.

[153] G. Perelman, *Proof of the soul conjecture of Cheeger and Gromoll*, J. Diff. Geom. **40** (1994), no. 1, 209–212.

[154] M.J. Perry, *Gravitational instantons*, "Seminar on Differential Geometry", 603–630, Ann. of Math. Stud., **102**, Princeton Univ. Press, Princeton, New Jersey, 1982.

[155] P. Petersen and T. Tao, *Classification of almost quarter-pinched manifolds*, Proc. Amer. Math. Soc. **137** (2009), no. 7, 2437–2440.

[156] P. Petersen and F. Wilhelm, *Examples of Riemannian manifolds with positive curvature almost everywhere*, Geom. Topol. **3** (1999), 331–367.

[157] P. Petersen and F. Wilhelm, *An exotic sphere with positive sectional curvature*, arXiv:0805.0812, 2008.

[158] F. Podestà and L. Verdiani, *Totally geodesic orbits of isometries*, Ann. Glob. Anal. Geom. **16** (1998) 399–412; Err. ibid. **19** (2001) 207–208.

[159] A.Z. Petrov, *Classification of spaces defining gravitational fields*, Kazan. Gos. Univ. Uč. Zap. **114** (1954), no. 8, 55–69.

[160] A. Preissman, *Quelques propriétés globales des espaces de Riemann*, Comment. Math. Helv. **15**, (1943), 175–216.

[161] M. Roček, *Modified Calabi-Yau manifolds with torsion,* Essays on mirror manifolds, 480–488. Int. Press, Hong Kong, 1992.

[162] J. Sacks and K. Uhlenbeck, *The existence of minimal immersions of* 2-*spheres,* Ann. of Math. (2) **113** (1981), no. 1, 1–24.

[163] R. Schoen and S.-T. Yau, *On the structure of manifolds with positive scalar curvature,* Manuscripta Math. **28** (1979), no. 1–3, 159–183.

[164] R. Schoen and S.-T. Yau, *Proof of the positive action conjecture in quantum relativity,* Phys. Rev. Lett. **42** (1979), 547–548.

[165] R. Schoen and S.-T. Yau, *Conformally flat manifolds, Kleinian groups and scalar curvature,* Invent. math. **92** (1988), 47–71.

[166] I.M. Singer and J.A. Thorpe, *The curvature of* 4-*dimensional Einstein spaces.* "1969 Global Analysis (Papers in Honor of K. Kodaira)", 355–365.

[167] N. Seiberg and E. Witten, *Electric-magnetic duality, monopole condensation, and confinement in* $N = 2$ *supersymmetric Yang–Mills theory,* Nuclear Phys. B **426** (1994), no. 1, 19–52. *Erratum.* ibid. **430** (1994), no. 2, 485–486.

[168] K. Shankar, *On the fundamental groups of positively curved manifolds,* J. Diff. Geom. **49** (1998), no. 1, 179–182.

[169] W.-X. Shi, *Deforming the metric on complete Riemannian manifolds*, J. Diff. Geom. **30** (1989), no. 1, 223–301.

[170] W. Simon, *Nuts have no hair,* Classical Quantum Gravity **12** (1995), no.12, L125–L130.

[171] Y.-T. Siu, and S.-T. Yau, *Compact Kähler manifolds of positive bisectional curvature,* Invent. Math. **59** (1980), no. 2, 189–204.

[172] S. Smale, *Generalized Poincaré's conjecture in dimensions greater than four,* Ann. of Math. **74** (1961), 391–406.

[173] J. Smillie, *Affinely flat manifolds*, Thesis (Ph.D.), The University of Chicago, 1977.

[174] J. Smillie, *Flat manifolds with non-zero characteristics,* Comment. Math. Helvetici **52** (1977), 453–455.

[175] J. Smillie, *An obstruction to the existence of affine structures,* Invent. Math. **64** (1981), 411–415.

[176] P. A. Smith, *The Smith conjecture,* Papers presented at the symposium held at Columbia University, New York, 1979. Edited by J.W. Morgan and H.

Bass. Pure and Applied Mathematics, **112**. Academic Press, Inc., Orlando, FL, 1984. xv+243 pp.

[177] J. Sparks, *New results in Sasaki-Einstein geometry,* "Group actions on manifolds (Boulder, Colo., 1983)", 243–367, Contemp. Math., **36**, Amer. Math. Soc., Providence, RI, 1985.

[178] M. Stern, *L^2-index theorem on warped products*, Thesis (Ph.D.), Princeton University, 1984, 40 pp.

[179] S. Stolz, *Simply connected manifolds of positive scalar curvature*, Ann. of Math. (2) **136** (1992), no. 3, 511–540.

[180] A. Strominger, S.T. Yau and E. Zaslow, *Mirror symmetry is T-duality,* Nuclear Physics B **479** (1996), 243–259.

[181] D. Sullivan, *Infinitesimal computations in topology,* Inst. Hautes Études Sci. Publ. Math. No.**47** (1977), 269–331 (1978).

[182] D. Sullivan and W. Thurston, *Manifolds with canonical coordinate charts: some examples*, Enseign. Math. (2) **29** (1983), no. 1–2, 15–25.

[183] D. Sullivan and N. Teleman, *An analytic proof of Novikov's theorem on rational Pontrjagin classes*, Inst. Hautes Études Sci. Publ. Math. No. **58** (1983), 79–81 (1984).

[184] C.H. Taubes, SW $\Rightarrow$ Gr*: from the Seiberg–Witten equations to pseudo-holomorphic curves*, J. Amer. Math. Soc. **9** (1996), no. 3, 845–918.

[185] L.-S. Tseng and S.-T. Yau, *Cohomology and Hodge Theory on Symplectic Manifolds: I* and *II*, arXiv:0909.5418, arXiv:1011.1250.

[186] J. Thorpe, *Some remarks on the Gauss-Bonnet integral*, J. Math. Mec. **18** (1969), 779–786.

[187] G. Tian, *On Kähler–Einstein metrics on certain Kählerler manifolds with* $c_1(M) > 0$, Invent. Math. **89** (1987), no. 2, 225–246.

[188] G. Tian and S.-T. Yau, *Kähler-Einstein metrics on complex surfaces with* $c_1 > 0$, Comm. Math. Phys. 112 (1987), no. 1, 175–203.

[189] V. Tosatti, *Geometry of complex Monge–Ampère equations*, Thesis (Ph.D.)-Harvard University. 2009. 156 pp.

[190] A.E. Treibergs, *Entire spacelike hypersurfaces of constant mean curvature in Minkowski space,* Invent. Math. **66**(1982), 39–56.

[191] N.S. Trudinger and X.-J. Wang, *Affine complet locally convex hypersurfaces,* Invent. Math. **150**(2002), 45–60.

[192] K. Uhlenbeck and S.-T. Yau, *On the existence of Hermitian–Yang–Mills connections in stable vector bundles*, Comm. Pure Appl. Math. **39** (1986), no. S, suppl., S257–S293.

[193] E.B. Vinberg and V.G. Kac, *Quasi-homogeneous cones,* Mat. Zametki **1** (1967), 347–354.

[194] È.B. Vinberg, *Discrete linear groups that are generated by reflections,* Izv. Akad. Nauk SSSR Ser. Mat. **35** (1971), 1072–1112.

[195] G. Vincent, *Les groupes linéaires finis sans points fixes*, Comment. Math. Helv. **20**, (1947), 117–171.

[196] L. Verdiani, *Cohomogeneity one Riemannian manifolds of even dimension with strictly positive curvature. I*, Math. Z. **241** (2002), no. 2, 329–339.

[197] L. Verdiani, *Cohomogeneity one manifolds of even dimension with strictly*

positive sectional curvature, J. Diff. Geom. **68** (2004), no. 1, 31–72.

[198] M.Y. Wang and W. Ziller, *Einstein metrics on principal torus bundles*, J. Diff. Geom. **31**(1990), no. 1, 215–248.

[199] N.R. Wallach, *Compact homogeneous Riemannian manifolds with strictly positive curvature*, Ann. of Math. (2) **96** (1972), 277–295.

[200] E. Witten, *Monopoles and four-manifolds*, Math. Res. Lett. **1** (1994), no. 6, 769–796.

[201] J.A. Wolf, *Spaces of constant curvature,* McGraw-Hill Book Co., New York-London-Sydney 1967, xv+408 pp.

[202] B. Wilking, *Torus actions on manifolds of positive sectional curvature*, Acta Math. **191** (2003), no. 2, 259–297.

[203] B. Wilking, *Positively curved manifolds with symmetry*, Ann. of Math. (2) **163** (2006), no. 2, 607–668.

[204] B. Wilking, *A duality theorem for Riemannian foliations in nonnegative sectional curvature,* Geom. Funct. Anal. **17** (2007), no. 4, 1297–1320.

[205] S.-T. Yau, *Calabi's conjecture and some new results in algebraic geometry,* Proc. Nat. Acad. Sci. U.S.A. **74** (1977), no. 5, 1798–1799.

[206] S.-T. Yau, *On the Ricci curvature of a compact Kähler manifold and the complex Monge-Ampère equation. I,* Comm. Pure Appl. Math. **31** (1978), no. 3, 339–411.

[207] W. Ziller, *Homogeneous Einstein metrics on spheres and projective spaces,* Math. Ann. **259** (1982), no. 3, 351–358.

Made in the USA
Monee, IL
05 November 2026

67987189R00267